# PEOPLE, PLANTS,
# AND JUSTICE

# PEOPLE, PLANTS, AND JUSTICE: THE POLITICS OF NATURE CONSERVATION

Charles Zerner

EDITOR

Columbia University Press
New York

Columbia University Press
Publishers Since 1893
New York   Chichester, West Sussex

Library of Congress Cataloging-in-Publication Data

People, plants, and justice : the politics of nature conservation /
Charles Zerner, editor.
    p.  cm.
  Includes bibliographical references.
  ISBN 0-231-10810-9 (alk. paper) — ISBN 0-231-10811-7 (pbk. : alk.
paper)
  1. Environmental justice.  2. Natural resources—Management.  3.
Environmental management.  I. Zerner, Charles.
GE170 .P45 1999
333.7—dc21                                99-053777

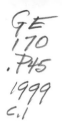

Cover: "The confluence of the Dirkhu and Tadi rivers at Kharanitar Village, Nuwakot
District, Nepal." © by Kevin Bubriski. All rights reserved.

All maps prepared by David Lindroth, with the exception of figure 12-1, page 310.

*In memory of my father, Charles S. Zerner (1897–1956)*

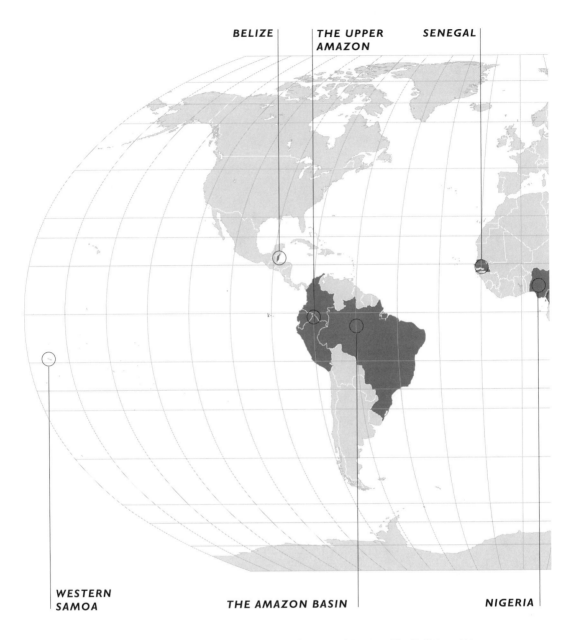

BELIZE

THE UPPER AMAZON

SENEGAL

WESTERN SAMOA

THE AMAZON BASIN

NIGERIA

Global Locator Map for Case Studies, *People, Plants, and Justice: The Politics of Nature Conservation.*

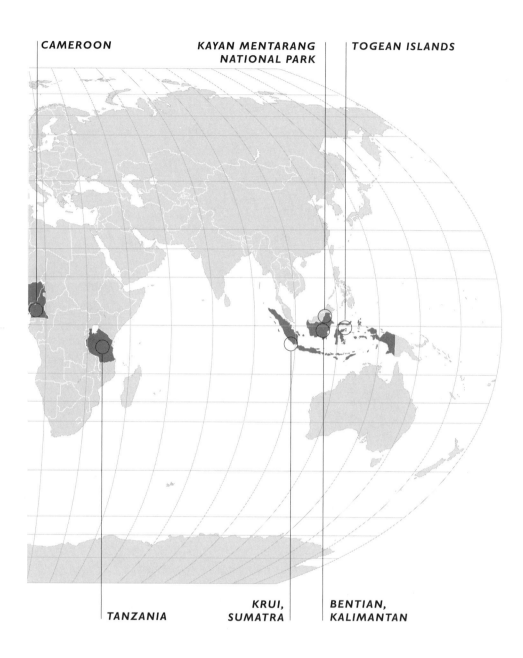

CAMEROON

KAYAN MENTARANG
NATIONAL PARK

TOGEAN ISLANDS

TANZANIA

KRUI,
SUMATRA

BENTIAN,
KALIMANTAN

# CONTENTS

Contents

# LIST OF ILLUSTRATIONS

# PREFACE AND ACKNOWLEDGMENTS

In 1993, when I became director of the Natural Resources and Rights Program of the Rainforest Alliance, varieties of market triumphalism permeated the world of conservation schemes. The dominant paradigm in international nature conservation stressed market links between sophisticated centers of global consumption, and remote, rural centers of undisturbed biodiversity and simpler social formations. In discussions of international trade policy, diplomacy, and development, global entrepreneurialism and newly greened forms of capitalism were believed to have salutary effects on the formation of democracies, perhaps even stimulating the growth of civil society, economic equity, and justice. These projects, and the visions that inform them, rarely attended to context-specific times, cultures, histories, economies, and distributions of power—and the consequences of specific market linkages.

*People, Plants, and Justice* was conceived as an exploration of the relationships between market-based schemes for nature conservation, or "sustainable" natural resource extraction, and issues of justice, rights, and power in specific contexts. The case studies were imagined as ways of asking questions about the larger world of international conservation policy presuppositions, projects, and planning, and of thinking based on market models. If these chapters provoke international conservation organizations, donors, nongovernmental groups, and environmental educators to examine the assumptions that animate their programs and projects, then the volume will have achieved its aims.

To ask these kinds of questions and to generate the kinds of context-specific insights I was searching for, I sought to establish links with scholars, advocates, and policy professionals engaged in research—in Africa, Asia, Latin America, and the South Pacific—at the edges of nature management, law, politics, geography, botany, and anthropology. The complicated back-and-forth was necessarily a kind of transnational collaboration. It has been a pleasure to engage vigorously with this group of scholars and practitioners. Our contact and colloquies have enriched me.

Anthony Cunningham drew on his vast knowledge of botany and botanical researchers to help me select potential case-study candidates—people and plants—for this book. Sarah Laird developed a casebook of potential chapters, geographic sites, themes, and plants. In the fall of 1994, six case studies were presented at the meetings of the Fourth International Congress of Ethnobiology, in Lucknow, India. Dr. S. K. Jain, host for the meetings, extended a gracious welcome.

I extend my thanks are extended to James Boyce, Janis Alcorn, Louise Fortmann, Toby Alice Volkman, R. Michael Wright, Mary Zurbuchen, Suzanne Siskel, and Sofia Perez, who provided comments on several drafts of the introduction. Many other colleagues have contributed to my understanding of the relationships among justice, markets, and nature management, directly and indirectly. I thank Michael Watts, Jim Scott, Christine Padoch, Richard Schroeder, Anna Tsing, J. Peter Brosius, Nancy Peluso, Peter Riggs, Jim Murtaugh, and Celia Lowe. Jim Nevin and Nancy Beth Rosenthal gave generously of their friendship, enthusiasm, and wonderful meals in Pine Hill. I alone am responsible for interpretations advanced here.

I am especially thankful to the Ford Foundation and to E. Walter Coward, Senior Director, Asset Building and Community Development Program, who graciously invited members of this group to meet at the Foundation in 1998, and to participants in that workshop: Janis Alcorn, Marshall Murphree, Richard Schroeder, Benedict Kingsbury, and Michael Watts (experts respectively in ethnobotany, community-based resource management, political ecology, law, and international studies), as well as Foundation staff, including Walter Coward, Larry Cox, Elizabeth Campbell, James Spencer, and Ellen Stein. Participants at this workshop engaged in a wide-ranging, imaginative, and intense discussion of the issues raised in this volume. Because initiatives articulated by key donors often shape the conceptual and pragmatic worlds of policy professionals, scholars, and advocates, we were pleased to locate this conversation, critically evaluating key concepts in social movements and nature management, at the Foundation.

The Ford Foundation generously supported the completion of this book, its publication, and dissemination to groups in developing countries. Frances Korten, then a program officer, and Walter Coward asked useful questions and provided support for a project that was unusual and provocative.

The support of several other foundations is also acknowledged with gratitude. The Rockefeller Brothers Fund, the Liz Claiborne–Art Ortenberg Foundation, and the Weeden Foundation provided core funding for the Natural Resources and Rights Program, while the Biodiversity Conservation Network (USAID/WWF/TNC/WRI), Earth Love Fund, and the Shaman Pharmaceutical Corporation provided funds for specific case studies. The Wenner-Gren Foundation for Anthropological Research provided funds for travel and presentation of several cases at the 1994 meetings of the International Congress of Ethnobiology.

I am grateful for the support of the members of the Rainforest Alliance, particularly Daniel Katz and Karin Kreider, who provided a home for an unusual program linking scholarly analysis, policy, and advocacy. Hilary Roberts, Eric Holst, Sofia Perez, Kristen Ohlson, and many other colleagues offered insights during many phases of program development. Ina Chaudhury assisted in coordinating early stages of this project. Over the past five years, Natural Resources and Rights Program interns Reed Smith, Kelly Kennedy, Kim Lopp, and Louis Putzel have assisted in the coordination of this project. I am grateful for their intelligence, generosity, and scrupulous attention to detail in coordinating the work of scholars and advocates working in Africa, Asia, Latin America, the South Pacific, and the United States.

I am grateful to Gloria Kirchheimer, my editor, who queried authors scattered across the globe. Her excellent eye and editorial hand helped clarify and ground these chapters. Marjorie Wexler completed the project of editing *People, Plants, and Justice* with skill and fastidious attention to detail. David Lindroth, cartographer par excellence, prepared all the maps.

To Toby Alice Volkman and Lucia Xingwen Zerner, wife and daughter, I am thankful that they are my fellow travelers, life companions. I hope Lucia may one day read and wonder at these stories and maps of faraway places, people, and nature.

# PART I

## Across the Terrain

# INTRODUCTION

## Toward a Broader Vision of Justice and Nature Conservation

*Charles Zerner*

Small-scale societies have long been engaged in the commodification of nature: extracting, producing, processing, and trading a diversity of products from a broad spectrum of natural environments. Scholars working in Southeast Asia, Africa, Latin America, and the South Pacific have amply documented the cultural modification of nature and the creation of complex commodity chains implicated in the global circulation of natural products. These chains have linked relatively remote communities with the material, cultural, economic, and ideological flows emanating from and flowing to global centers for centuries, and, in some cases, millennia.[1]

It is only recently, however, that global markets and the consumption patterns of northern elites have been programmatically linked to the welfare of remote communities in the South and improved environmental management and protection in the tropics. Over the last decade and a half, environmentalists in a variety of nongovernmental and governmental organizations, multilateral financial institutions, and corporations have sought to fashion and to implement a new family of environmentalisms based on markets, commodity flows, incentives, and the idea that people are fundamentally economic creatures.[2] This recent efflorescence of faith in *Homo economicus* is linked to two propositions that have grounded numerous programs and policies in tropical conservation. The way to move people is through their markets. The way to save wild "nature"[3] is through the eye of the market.

While the idea that strong links between global markets and remote communities would lead to increased environmental protection and community well-being is counterintuitive, we know that such links, in certain instances, have precipitated useful property rights, institutions, and collective environmental access systems including rules and sanctions.[4] Indeed, community-based natural resource management initiatives linking markets and income streams to collective property rights constitute a family of impressive experiments in economic justice, environmental governance, and democratization.[5]

But the environmental and social histories of markets in natural resources and commodities contain as many tales of destructive, uncontrolled resource extraction and environmental degradation, conflict, and social oppression as they do examples of social justice, good environmental management, and economic well-being. Hochschild's (1998) disturbing history of wild rubber extraction in the Belgian Congo, a regime linked to terror and social oppression in the late nineteenth and early twentieth centuries, stands as a cautionary tale.

Over the last two decades, the market has been positively valorized in numerous conservation projects, including debt swaps, ecotourism, ecocommodification and export, bioprospecting, video promotionals linking supermodels and rainforests,[6] purchases of acres of southern rainforest by northern schoolchildren, and other ecomarket hybrids that have proliferated across the globe.

The marketing of nature and nature protection, on the one hand, and a view of human nature and institutions as fundamentally economic, on the other hand, have permeated environmental theory, programs, and popular environmental imagery.[7] "Buy Your Conservation Here!" exclaims the EcoVitality web site (1998). Nature has become an emporium, a commercial warehouse awaiting its brokers. Conservation theory now analogizes nature to a stock market: we act to conserve nature because "wild nature" contains potentially useful "option values."[8]

The chapters in this volume exemplify approaches that attempt to integrate, to varying degrees, concerns for justice, power, markets, and the politics of resource and territorial control in the context of conservation projects. In their varied histories, they offer insights that should be useful not only to international conservation communities in the North and the South, governments, intergovernmental and nongovernmental advocacy organizations, as well as donors, but also to scholars and public intellectuals.

If there is a primary proposition that is to be distilled from *People, Plants, and Justice*, it is that there is no single market operating at a fixed scale and associated with particular social and environmental consequences. There are, rather, multiple, culturally shaped forms of markets that are inserted into, and articulated with, divergent economic, historical, and environmental contexts.[9] There is no entity, "the market," that exists outside of history, culture, and geographical context.

Consider these examples: rubber, damar resin, and gaharu or aloes wood, at three different sites: the Upper Amazon, Sumatra, and Kalimantan (Indonesia), and at two historical periods.

In chapter 4, "Outrage in Rubber and Oil: Extractivism, Indigenous Peoples, and Justice in the Upper Amazon," Søren Hvalkof documents a regime of social terror and bodily violence inflicted on communities of indigenous peoples in the highlands of Ecuador, Peru, and Columbia by European rubber companies, their Hispanic representatives in the Amazonian interior, and Caribbean mercenaries. Hvalkof documents a historic trauma, simultaneously material and psychological, visited on local peoples in an effort to force them into a system of debt peonage and rubber cultivation. This system was imposed to procure a docile and disciplined labor force in the service of European capitalist penetration of the Andean Amazon, directed at natural resource extraction.

However, Hvalkof's narrative of a particular instance of globally linked, capitalist extraction of natural latex in the upper Amazon does not and cannot represent the whole story of markets, extraction, and communities—social and environmental. The scale of each market, its political and cultural configuration and dynamics, the specific sites at which it articulates with local communities, economies, and the environment, and the historical moment must be distinguished from a variety of other market interactions and their consequences.[10]

Geneviéve Michon and Hubert de Foresta, for example, in chapter 7, "The Damar Agroforests in Krui, Indonesia: Justice for Forest Farmers," document the history of agroforestry production practices invented by local forest farmers living in the Krui coastal region of Sumatra in the late nineteenth and early twentieth centuries. The Krui system was, and continues to be, based on tapping damar resin from magnificent dipterocarps—among the giants of Sumatran coastal forests—and maintaining high levels of biological diversity. These practices continue to yield income for local communities on a sustainable basis, constituting a remarkable achievement in natural resource management, vigorous common property controls, and community well-being. The Krui system, moreover, articulates well with regional and world markets.

Stephanie Fried, in chapter 8, "Tropical Forests Forever? A Contextual Ecology of Bentian Rattan Agroforestry Systems," documents the ways in which Bentian Dyak communities of Southeast Kalimantan, responding to the external demands of coastal sultans for tribute during the eighteenth and nineteenth centuries, created a system of rotational agroforestry yielding commercially valuable rattan harvests while sustaining mature, well-managed tropical forests and rattan gardens.

The critical threats to community-based, market-linked natural resource management and extraction systems in both the Krui and the Bentian Dyak examples are found in the political-economic contexts in which these communities and environments are embedded. In the Krui case, regional political elites, allied with powerful timber interests, and supported by "purist" conservationist policies written into Indonesian forestry and conservation law, almost forced the exclusion of local communities from access to the forests they had created over the course of a century. In the Bentian Dyak case, government collusion with powerful timber concessionaires, as well as the ignorance of forestry management bureaucracies,[11] resulted in the destruction of gravesites and gravemarkers, as well as of thousands of hectares of productive rattan, fruit, and rubber gardens. Concession operations, halted only as a result of competition with other timber firms, almost led to the eviction of several communities from their historically managed, customarily claimed, and well-managed rattan forest gardens. Neither of these cases suggests that local communities' linkages with regional and international markets for forest products resulted in community disenfranchisement, pauperization, or forest degradation.

Neither demonization of markets nor blind acceptance of globalization and free market ideology will lead to just outcomes, useful analysis, or politically progressive collaborations in community-based natural resource management. Academic analysts, advocates, donors, and activists need to look at the kinds of markets that are being promoted, the power relationships and links between global, national, regional, and "local" levels of articulation, and the consequences of market insertion in specific contexts.[12]

Engagement with market economies may be inevitable. A host of specific questions, however, need to be asked. Engagement with whom? Engagement for the benefit of which sectors of society? Engagement with what forms of market economy and why? Engagement with what consequences? Conservationists and development planners need to assess the particular political, economic, and social consequences of generic global–local market interventions ("solutions") in particular contexts.

## THE CULTURAL NATURE OF MARKETS AND COMMODITIES

Ideas of the market, the commodity, and natural resources, moreover, are cultural conceptions invented, validated, and circulated at specific historical moments and within particular cultural-political contexts. Commodities and the practices associated with them need to be tracked historically and geographically as they pass through what Arjun Appadurai (1995a) has called "regimes of value." The same patch of trees may be valued by international conservationists and scientists as an embodiment of the world's precious biological diversity, mapped by an Indonesian commercial timber concession as merely another block containing so many cubic feet of exportable tropical hardwood, or seen and claimed by a local Dyak community as the site of a cultivated forest garden, inhabited by orchards of fragrant Durian trees and memories of family members gathering honey.[13]

Commodities change not only their nature but also their cultural meaning and value as they travel. There is no universal economic metric measuring these divergent forms of value, no calculus for conversion of that which is culturally incommensurable. Which institutions and individuals, moreover, are authorized to make judgements between competing claims based on different regimes of value? Until we understand the cultural nature and biography of commodities—their shape-shifting, value-changing nature as they pass through different cultural regimes of valuation and interpretation—attempts to talk about justice in the context of natural resource extraction will be limited to a narrow, albeit important, concern for economic valuation. In an era of market triumphalism, moreover, we must remind ourselves that the sphere of justice and rights—human rights, civil liberties, and property rights, among others—is a far more extensive terrain than that circumscribed by concerns for economic justice.

In an era in which thought and action are dominated by economic models, metaphors, and material flows,[14] conservationists must deploy the analytical methods of the social sciences and humanities, including revitalized area-studies and cultural analysis, to conceptualize, plan, and evaluate market-linked environmental management projects.[15] The multiple images and ideologies that ground conservation and environmental management programs, at global and national governmental levels (not only "local" levels), need to be seen through the lenses of culture and politics.[16] To ignore these perspectives is to invite undesirable outcomes, farce, and possibly tragedy in planning and implementing project interventions.[17]

## PROPERTY RIGHTS AND COMMODITY CHAINS

In chapter 6, "Rebellion, Representation, and Enfranchisement in the Forest Villages of Makacoulibantang, Eastern Senegal," Jesse Ribot provides an insight on markets, justice, and property rights: Property rights are perhaps a necessary condition for effective local community control. But property rights do not constitute the necessary and sufficient conditions for justice, community control of resources, capture of benefits, or effective environmental management. Ribot shows how most of the forest residents in the Makacoulibantang district of Eastern Senegal lack effective control over their forests, even though they have occasionally been effective in excluding outside laborers from operating in nearby forests. Control over spatial access, through clearly defined property rights and boundaries, has been emphasized as the key factor in effective community control over resources.[18] But Ribot shows how control over Senegalese forests resides less in the formal property rights to land, space, and territory, than in being able to control access to a variety of other determinants, including market outlets, the permitting process, transportation networks, and labor resources. Analysts and advocates need to frame the issue of access control in a broader, more multidimensional manner. We cannot assume that drawing perimeter lines around hypothetical community-based resource management areas and recognizing property rights within these zones, alone, will produce justice or effective resource management.[19]

To understand the economic justice issues implicated in market-linked conservation and resource management schemes, we must attend to the points at which leaps in value attach to the commodity as it moves through regional, national, and global channels of circulation. Ribot's chapter forces us to ask: At what point in these complex systems of natural resource extraction is value attached? How, by whom, and to whose benefit? Which individuals, people, households, communities, institutions, businesses, or government agencies, have control over the key sites where significant shifts in the value of "natural" commodities occur? Without some leverage over the jump in value at critical points in the commodity chain, projects supporting economic justice through incentives, and some degree of local control, will fail to achieve their aims.

## THE POLITICAL ECOLOGY OF CONSERVATION

Celia Lowe analyzes the dynamics of reef destruction, community impoverishment, and blame in chapter 9, "Global Markets, Local Injustice in Southeast Asian Seas:

The Live Fish Trade and Local Fishers in the Togean Islands of Sulawesi, Indonesia." Lowe underscores this insight: Look to the dynamics of the larger political economy to understand the forces driving local community practices. Lowe's chapter offers a strong corrective to a myopic focus on local communities in attributing responsibility for environmental degradation.[20] In the Togean Islands, at the behest of powerful, illegal, live-fish export cartels backed by the Indonesian military, poor people at the beginning of the commodity chain are inducted into debt, dependency, and destructive fishing practices. These communities are the objects of intimidation and extortion by government, military, and private sector officials. They are also targeted as the objects of intervention by international and national conservation organizations. The secret to controlling this destructive trade is not to lecture the desperately poor and politically weak fishers, but to direct policy and law enforcement at more powerful parties at regional and national levels, while making strategic alliances that include local communities.

What are the implications of a focus on larger scales of political economy and political ecology? Conservation organizations and intergovernmental bodies interested in making effective interventions to protect Napoleon wrasse and the reefs they inhabit in Indonesia, for example, could direct their attention to Indonesian governmental connections to fisheries cartels and traders, rather than targeting or victimizing the poorest, most politically vulnerable groups at the lowest point in the commodity chain. Lowe writes, "Further, by recognizing that the most substantive ecosystem abuses are not organized locally, but rather underwritten by a ramifying bureaucracy and business community, conservationists may find a basis for alliance with village people." This insight should be applied to a wide variety of resource management and conservation intervention cases. International conservation and environment/development interventions, moreover, can benefit from Lowe's analysis and policy recommendation:

> [Conservationists] might find they have success helping local people combat cyanide use by coming out fully in *support* of live fishing as the sustainable industry it potentially is. . . . The question would become not one of how to prevent local people from doing x or y . . . but rather how to help [them] respond to the power dynamics that reward gluttonous resource extractions in the name of development.

In the reefs of the Togean islands, in Senegalese forests, and elsewhere throughout the tropical developing world, the poor have few political or institutional controls over access to prized resources or ways of benefiting from fluctuations in the value of resources extracted from sites they manage and often claim. They are blamed, lectured to, and not infrequently threatened with criminal penalties for the environmental consequences of the acts of more powerful actors and institutions.

## DISPARITIES IN ECONOMIC, LEGAL, AND CULTURAL POWER

The fact is that the poor sell cheap. And because they sell cheap, the poor continue to suffer, their livelihood options are limited, and their lands are seized. Many market-linked schemes for conservation ignore vast disparities in political, legal, and economic power between macro-economic actors and local micro-scale communities, as if negotiations between large pharmaceutical firms and rural villagers were on an equal footing, or could be equalized through individual acts of good will.

"There is something about confronting the market and being poor," says Michael Watts, "which in some sense determines the fact that you sell poor. Poor farmers sell their grain at that time of the year when prices are lowest. And poor donors sell kidneys and other organs cheaply. No matter what the moral and ethical arguments about these markets is or whether they should exist, the point is that poor Indian women, through their fact of being impoverished, are selling their kidneys."[21] Issues of power and freedom—the capacity to choose whether or not to extract a local resource, be it a kidney from within one's own body, or a cluster of trees in a collectively claimed forest—are implicated in all these cases.[22]

## BIODIVERSITY PROSPECTING: EXTRACTION IN AN ERA OF INTENSIFIED GLOBAL CIRCULATION

These issues—the fact that the poor sell cheap, that they are blamed for the consequences of more powerful political and economic actors, and that they have little or no control over the resource at the critical junctures where the resource changes its cultural nature and jumps in economic value—are present as well in the cases in this volume that focus on biodiversity prospecting. The idea that equitable returns from extractions made by commercial bioprospectors should be returned to the local communities who are the "stewards" of biological diversity, and that these "equitable returns" will function as incentives to conserve local resources, has become an endlessly repeated mantra in conservation and environment/development circles.[23]

Bronwyn Parry's chapter, "The Fate of the Collections: Social Justice and the Annexation of Plant Genetic Resources" (chapter 15), provides a masterful analysis of the ways through which value is created in biological resources, and how values of genetic resources increase exponentially as they leave their site of origin, passing through "centers of calculation," and are materially, economically, and culturally transformed into new objects.[24] What was once a plant, or a plant extract—a bit of juice, leaf, or stem—is transformed into information that circulates endlessly in the channels of an increasingly baroque global information economy. Molecular fragments, the informational bits of former plants and other organisms, now travel in worlds economically, scientifically, and epistemically distinct and disjunct from their sites of origin and former nature(s).

But Parry's chapter is more than an analytical tour de force. Her account of the ways in which the "natural product" undergoes shifts, transmuting from plant stem to informational bit, circulating in a global economy, leaping exponentially in value, and changing its scientific and cultural nature as it travels away from its site, constitutes a significant critique of much conventional wisdom on bioprospecting and the idea of capturing value for conservation efforts. Parry articulates this insight:

By only partially acknowledging that biological material has been reconstituted as an informational resource, by continuing to pay out small sums of money for genetic samples as if they were stock resources, by continuing to rely on a "paper trail and a chunk of faith" to track successive uses of this material, the bioprospecting industry risks inviting accusation that these contemporary collecting projects not only mirror earlier colonialist phases of resource extraction but actually constitute a more sophisticated form of bioimperialism than any previously envisaged. (Parry, chapter 15)

If the poor sell their kidneys (as well as their Napoleon wrasse) cheaply, Bronwyn Parry demonstrates that natural resource extraction processes embodied in bioprospecting amplify the inequalities of trade by making it even more difficult to trace the path of profit. Parry's critique of the raison d'être of progressive bioprospecting schemes—the notion that "equitable" benefits shall be returned to host countries and local communities—suggests that energy directed at bioprospecting could be more usefully deployed elsewhere.

## THE POLITICS OF ENVIRONMENTAL IMAGERY

Candace Slater, in chapter 3, "Justice for Whom? Contemporary Images of Amazonia," focuses on the cultural force of environmental and social representations.[25] She explores the cultural and political history of images of the Amazon and its human communities. Her work forces the reader to reexamine the political and policy consequences of environmental representations. Slater demonstrates the need to take critical stock of the environmental images that are produced, how they are fashioned and for which consuming publics. What institutional interests are served by particular representations in specific cultural and political contexts? Her provocative contribution stands as a reminder that popular images of nature and natives that circulate in the press and in international conservation brochures—images of pristine forests and natural natives, poised in loincloths and described as stewards of nature—are often policy directives in disguise.

The production, proliferation, and circulation of "green" corporate environmental representations needs to be subjected to the same kind of critical scrutiny. This holds true, as well, for environmental representations produced and deployed by multilateral development banks and other financial institutions, intergovernmental agencies, and governments.

Although popular social imagery of the Amazon suggests a landscape of Indians and rubber tappers, Slater reminds the reader that these groups constitute only 5 percent of the entire Amazonian population, an estimated 23 million men, women, and children, including descendants of black slaves of African descent as well as Arabs, Japanese, and Sephardic Jews. Although popular environmentalist images of the Amazon suggest a region of vast green wilderness and natural wealth, Slater counterposes the reality of an increasingly urbanized Amazon landscape, including a diversity of settlements ranging from small towns to cities with over a million inhabitants. Given the striking social and environmental diversity of this region, Slater compels the reader to ask why so much attention, funding, and popular mobilization has been directed at natives and rainforests. At the center of her study is a question about justice and representation that might be asked of other regions: How do environmental representations potentially support a politics of exclusion?[26] How can more inclusive environmental representations be fashioned?

## *PEOPLE, PLANTS, AND JUSTICE:* A PROVOCATIVE COLLAGE OF CASES AND INSIGHTS

How should these varied chapters, recounting stories of natural resource extraction, injustice, markets, and communities throughout the tropical world, be interpreted? Are comparisons among such geographically, historically, and culturally diverse cases possible? These studies do not constitute a group of cases organized to provide systemic variance of one theme (for example, property rights and environmental management) while other factors are held constant. Rather, they constitute a juxtaposition, a provocative collage from a wide range of geographic regions, cultures, historical moments, environmental contexts, points of market insertion, and political organization. These cases do bear family resemblances to each other, however, and provide multiple points of comparison and insight.

The chapters published in *People, Plants, and Justice* are intended, in their singularity and in their juxtaposition, to provoke thought about already accepted, generic conservation and environmental "solutions" linked to markets and resource extraction. They challenge contemporary orthodoxies about communities (the "local" focal points of many conservation projects), markets (beneficial), minor forest product extraction (good), control over spatial access (critical/necessary and sufficient condition for "local" community control), global-local relationships and processes, and the links or "disconnects" between justice, conservation, and environment/ development.

All these studies constitute ways of asking questions about power, justice, markets, and nature management, through specific histories and in specific sites. They are interpretive screens that yield insights when calibrated to the specific contexts against which they are held.

These cases are illustrative rather than representative. Jill Belsky's chapter on ecotourism in Belize, "The Meaning of the Manatee: Community-Based Ecotourism Discourse and Practice in Gales Point, Belize" (chapter 11), for example, should be read as a history of ecotourism in a particular place at a particular historical moment: the Gales Point Belize ecotourism project in the 1990s. Belsky's study offers a specific instance of the effects of transnational ecotourism in a particular social and environmental site, an interpretive screen to hold up against other ecotourism projects. It suggests ways of interrogating ecotourism *in situ* and provides insights about some of its potentially disruptive or inequitable economic, political, and cultural effects.

How might we move beyond the insights offered here?

## KEYWORD: COMMUNITY

To unpack and to work with that social black box, the "local community," conservationists and environmental management professionals need to borrow a page from the analytical methods and insights of the social sciences and humanities.[27,28] What are the institutions and sources of authority over environmental access, rights, and management practices at a variety of levels, including, but not limited to, the microsocial "community"? What are the cultural, class, and ideological cleavages as well as points of articulation and alliance within and between communities at a variety of levels? Under what historical circumstances are community representations linked to environmental management outcomes, and why?[29] What forms of community representation emerge when international conservation or environment/development projects loom on the horizon?[30] How are these representations deployed, and with what consequences?

The question of local governance and community must also be addressed. Ribot's chapter (chapter 6) sharply poses the issue of justice and community in these terms:

> Justice—both distributive and procedural—in the context of participatory natural resource management is about *what* is devolved to *whom*. *"To whom"* is about the problem of representation. "Indigenous" and "local" do not necessarily mean representative or fair. Some process of inclusion or some form of accountable representation must be constructed if the notion of community—which is always a stratified ensemble of persons with different needs and powers—is to have a collective meaning. This story brings into question whether chiefs really do "represent" their villages in any accountable sense. It brings up the question of whether new natural resource policies should place powers in a chief's hands, strengthening this particular local—but not necessarily representative or just—institution.

Advocates, donors, and the academic community also need to be alert to the ways in which "community-based natural resource management" (CBNRM) is being socialized in the discourses and projects of bilateral aid agencies and international financial institutions, as well as in the politically creative projects of transnational and national nongovernmental groups. What are the purposes and consequences, social and environmental, of these deployments by powerful institutions?[31] In imagining strategies for local justice and empowerment, activists and analysts also need to take into account the ways in which communities seize and transform transnational movements into local opportunities, and, conversely, the ways in which transnational movements deploy representations of local realities on the national and international political stage.[32]

## CROSSING BORDERS: LOCAL SITES,
## TRANSNATIONAL CONNECTIONS

The roles of transnational environmental and human rights institutions in providing information, imagery, and strategic advice, as well as access to transnational channels of communication and political lobbying, need to be made visible in the conceptualization and implementation of "local" projects. What constitutes the "local," moreover, in an era of increased interconnections between global institutions, national governments, and communities?[33] What are "local" politics at a moment when international discourses and campaigns on human rights, indigenous and minority peoples, cultures, justice, and conservation proliferate across the globe, circulating and making connections in surprising ways?[34] What roles have transnational activists and engaged scholars, including many of the contributors to this volume, played in organizing, mobilizing and publicizing the dilemmas and struggles documented in these pages?

The potential efficacy of transnational advocacy networks is eloquently stated by Keck and Sikkink (1998:x):

> Where the powerful impose forgetfulness, networks can provide alternative channels of communication. . . . Transnational networks multiply the voices that are heard in international and domestic policies. These voices argue, persuade, strategize, document, pressure and complain. . . . By overcoming the deliberate suppression of information that sustains many abuses of power, networks can help reframe international and domestic debates, changing their terms, their sites, and the configuration of participants.

Under what conditions have transnational advocacy and issues networks been successful? What lessons can be learned from the histories of particular mobilizations? What kinds of linkages and collaborations—intellectual, institutional, and strategic—can productively unite social justice and environmental movements at the beginning of the twenty-first century?

## GOVERNANCE, REPRESENTATION, CITIZENSHIP

The question of local governance must also be addressed. In chapter 6, "Rebellion, Representation, and Enfranchisement in the Forest Villages of Makacoulibantang, Eastern Senegal," Ribot shows that devolution of authority for timber-cutting permits has not resulted in economic justice, democratic decision making, or sustainable resource management. Devolution, in the Senegalese forests, has created forms of decentralized despotism at the local level. Indeed, in numerous cases there appears to be little relationship between the substantive content of customary law and the structure of customary institutions on the one hand, and democratic process and institutions on the other.[35] What is the relationship between citizenship, representation, and customary law? What are the particulars of governance among Togean fishers,

Bentian forest farmers, and Cameroons communities, as well as the numerous communities mentioned in the other chapters? Until questions of representation, accountability, and transparency are dealt with on the local as well as the national institutional levels, the prospect of justice, in economic or other forms, is a moot question.

## ENVIRONMENTAL CONSEQUENCES OF COMMUNITY-BASED NATURAL RESOURCE MANAGEMENT

We also need more closely argued examples that test the assumption that community-based natural resources management results in unambiguously positive outcomes for both environments and communities. Michon et al. (chapter 7) present an excellent example, using data on biological diversity in managed and wild forests in Sumatra, Indonesia, that documents positive social and environmental consequences in that specific case of community management. In chapter 10 by Momberg et al., "Exploitation of Gaharu, and Forest Conservation Efforts in the Kayan Mentarang National Park, East Kalimantan, Indonesia," we are presented with an aspiration that might be paraphrased as follows: "We international conservationists think decentralization of power over access to community-held lands will increase local incentives for common property resource management enforcement, thus limiting outsider pressures on gaharu-bearing trees within the Kayan Mentarang National Park." But in this as in many other cases, we need stronger evidence on the issue of whether the desired outcomes have actually occurred.

## IMAGINING A BROADER LANDSCAPE OF ENVIRONMENTAL JUSTICE AND CONSERVATION

At the core of all the studies in *People, Plants, and Justice* is the animating assumption that democracy and justice go hand in hand with effective environmental management and conservation.[36] The association of social justice and conservation has, indeed, catalyzed a great deal of nongovernmental advocacy for recognition of local community rights to lands, resources, reef, wetlands, and seas in the tropical world as well as in northern temperate countries. During the past decade, the domestic environmental justice movement in the United States, impelled by concerns for both environmental management and social justice, has developed rapidly.[37]

The studies in *People, Plants, and Justice* from Africa, Latin America, the South Pacific, and Southeast Asia, all focus on justice and its multiple links to human communities, markets, and nature. But concerns for justice and environment, whether in the South or in the North, need to move beyond concerns for communities that are imagined to be rooted in space, indigenous, not fully connected to markets, and stewards of the environment.[38]

In a world in which capital, people, ideas, and media circulate across borders, we need to plan for a world of social and environmental landscapes in which the uprooted, the landless, and the migrant are as much the subjects of concern as is the indigene. This is an age of refugees and massive social dislocations. Received ideas of natives, endemicism, and fixed or "stabilized" social and environmental landscapes, need to be questioned and reconceptualized in ways that encourage politically plural, heterogeneous, and democratic outcomes.

The concept of biological diversity conservation needs to be repositioned in relation to fundamental human concerns. This means a shift from a somewhat unfocused enthusiasm for protection of all tropical biological diversity to a more nuanced and prioritized focus on the politics of food security and the political economy of poverty; crop germplasm conservation and the dynamics of global trade and trade agreements; livelihood and gender issues; the consequences of nuclear and biochemical weapons production, tests, and wastes; the production of toxic landscapes; and a focus on adequate and equitable access to uncontaminated water, air, food, housing, and public access to open space in cities as well as the countryside. We need a vision of environmental justice broad enough to encompass the migrant, the urban indigent, and the peasant agriculturalist, as well as the metropolitan, middle-class consumer and the indigene. The developing optic of environmental justice, moreover, needs to be articulated in ways that simultaneously support affirmations of regional particularity at the same time that they encourage emerging understandings of common ground for constructive struggle and collaboration across North-South borders.

In imagining this broader landscape of environmental and social justice, scholars and advocates need to think more inclusively and comprehensively about citizenship, civil society, and governance, and about their relationship to environmental entitlements.[39] What is necessary is a broad-gauged vision of justice and its links with the environment that explicitly engages inequitable concentrations of power, processes of democratization, and the formation of democratic institutions.

A broader intellectual and programmatic agenda for nature conservation in the late twentieth century will interrogate the implications of proliferating market triumphalism(s) for local livelihoods, rural-urban migration, citizenship, and immigration, as well as for economic justice and the environment.[40] This agenda would also raise questions about the unresolved relationships among national citizenship, ethnicity, and territory at a moment when indigenous territorial rights are increasingly and positively valorized as desirable in a variety of environmental and social contexts.

## A JUSTICE AGENDA FOR ENVIRONMENTALISM IN THE LATE TWENTIETH CENTURY

A justice agenda for environmentalism in the late twentieth century should examine the possible civil rights implications of increasingly intrusive environmental monitoring technologies on rights of privacy and freedom of movement.[41] It would probe the potentially problematic relationships between military and paramilitary power and "effective" conservation and environmental governance projects. This agenda would raise questions about how global environmental discourses and projects may result in coercion, political repression, or economic injustice and marginalization at the local sites where these agendas are operationalized and enforced. It would also ask whether the dissemination of "universal" property rights may result in unexpected and inequitable results at microsocial, regional, and national levels, as marginal peoples' social relationships to nature and natural resources are ignored or obliterated in the wake of the development juggernaut.[42]

A vision of justice and environment at this historical juncture, moreover, needs to encompass and to examine critically the cultural dimensions of competing visions and struggles over citizenship, rights, and cultural identity in concrete situations. A vision of environmental justice that is critical and reflexive needs to be articulated, and ultimately deployed, based on an understanding that conceptions of rights and rights holders—to property, territory, and ideas—are neither natural, universally accepted, nor self-evident. They are, rather, culturally constructed and historically shifting ideas, practices, and normative visions that are being inserted into highly variable contexts.

Images of nature and social relationships to it inform and animate the center of every environmental project and intervention.[43] But the politics of environmental imagery and its social justice implications are just beginning to be explored. The multiplicity of images of nature constructed and deployed by a bewildering diversity of environmental institutions, at a variety of scales of operation, is radically in need of critical scrutiny, not only by scholars but by advocates and donors. Every environmental campaign implies a specific politics of the environment—who is entitled to intervene and benefit, how, why, when, and where—and a culturally distinct body of images, and sometimes theory, of nature and normative social relationships to it.

The picture is further complicated by questions of scale. A broad-gauged vision of environmental justice will engage not only the specificities of "local practices" of "local communities." It will also engage and question the practices and discourses of environmental programs, campaigns, and interventions at a variety of levels, taking into account the complicated skein of linkages and consequences implicated in globalization processes.[44]

This list of themes is neither exhaustive nor definitive. It merely suggests the range of themes awaiting critical engagement.

All nature conservation and environmental management efforts are inevitably projects in politics. To ask questions about justice and environment in the late twentieth century necessarily means a vigorous engagement with politics, governance, and power. Certain species, landscapes, and environmental outcomes are privileged, while others are peripheralized or disenfranchised. Each park, reserve, and protected area is a project in governance: in drawing boundaries—conceptual, topographic, and normative; in implicating a regime of rules regulating permissible human conduct; in elaborating an institutional structure vested with power to enforce rules; and in articulating a project mission rendering the management regime reasonable, even natural. The task of critical scholarship and advocacy for international environmental justice is, in part, to examine regimes of nature management, to identify what kinds of politics and governance are imagined and implemented, and to assess the consequences. It is to place nature management projects on the scales of justice. The task of an engaged scholarship, moreover, is to formulate the theoretical groundwork and practical interventions that will integrate concerns for justice into the bewildering varieties of contemporary environmentalism.

## PLURALIST VISIONS OF CITIZENSHIP AND NATURE

The world of modernity constitutes a panoply of changing social and natural landscapes that cannot be confined within buffer zones—the stable, spatial fences, the *cordon sanitaire* of earlier "conservation" imaginings and policies that created hedges around nature and territorialized boundaries around culture. We are situated, inexorably, in a landscape of shifting configurations of nature/culture,[45] a landscape traveling within and across genomic, individual, communitarian, national, and international boundaries, confounding the social and environmental world watchers, monitors, and surveillance/information specialists attempting to keep people, things, and nature in their place. We need to imagine and to create more radically pluralist, democratic visions of nature and societal interconnections: visions in which the Creole and the hybrid, the mobile as well as the sedentary community, the provisional design as well as the ancient species are valued citizens of the changing state(s) of nature.[46] In this more expansive terrain, alternative visions of economy, markets, and social relations to nature should be given ample room to flourish.

## NOTES

1. Kopytoff (1986). On the flow of commodities on regional and global scales, see Appadurai (1995a,b); on the circulation of natural resource–based commodities in Southeast Asia, see Reid (1988), MacKnight (1969), Peluso (1983a), Wolters (1967), Andaya (1993), and Hall (1985).

2. See, for example, Clay (1990), Seibert (1989), and *The Economist* (1989). For critiques of green marketing and varieties of environmentalism linked to commodity flows and local communities, see Dove (1993b, 1994), Luke (1997), Salafsky et al. (1993), and McAfee (1999).

3. See Cronon (1995), Chaloupka and Bennett (1993), and Schama (1996) for a variety of contemporary analyses of cultural constructions of the idea of wild nature. See also Fairhead and Leach (1996) and Leach and Mearns (1997) for critical historical analyses of the cultural fashioning of African forests; on the cultural making of Latin American and Indonesian forests, see Padoch and Pinedo-Vasquez (1996) and Roosevelt (1980, 1997, and in press).

4. On the importance of property rights and local institutions in environmental management, see Ostrom (1990); see also Lynch and Talbott (1995). See Ribot (1995) for an analysis of the limitations of a property rights approach in Senegal.

5. On the history of community-based resource management in Zimbabwe, see Metcalfe (1994), Murphree (1993 and in press). For general overviews of some of the prospects and problems implicated in community-based resource management, see Brosius et al. (1998), Li (1996, 1997b), and Agarwal (1997).

6. See videotape "Supermodels in the Rainforest" (1994) for a remarkably lubricious linkage of commercially marketable bodies, enthusiastic testimonials, and rainforests.

7. Peet and Watts (1996).

8. See Reid and Miller (1989) for a formulation of natural resources as options.

9. See Hefner (1998a,b) for analysis and examples of the cultural and political constitution of markets as deployed in a variety of Southeast Asian settings.

10. See Peluso (1992b) for a historical account of the violence and power relationships implicated in the history of Javanese teak plantations.

11. See Dove (1983) on the relationship between economic and political interests of the Indonesian government forest bureaucracy and its "scientific" assessments of swidden agroforestry practices.

12. See Schroeder (1995 and in press) for particularly incisive analyses of the ways in which generic afforestation programs, coupled with structural adjustment initiatives, are reproducing locally supplied corvée labor for newly minted environmental programs.

13. Greenough and Tsing (1994).

14. See Appadurai (1996) for an attempt to theorize contemporary globalization processes and their manifestation in media, finance, ethnicity, and other domains of the social. See Peluso (1992a) for an analysis of the unintended consequences of several specific global conservation programs.

15. For an account of the "crisis in area studies" and an effort at revitalization, see Volkman (1998).

16. See, for example, Kuletz (1998) and Tsing (1993).

17. In the case of Hvalkof's horrific narrative, it is clear that the historical memory of the terror inflicted on local peoples inducted into the global rubber economy continues to play a central role in rejection of market-linked international conservation and development schemes.

18. See Ostrom (1990), Lynch and Talbott (1995), McCay and Acheson (1987), Berkes (1989), and Western and Wright (1994) for widely circulated formulations of the linkage between property rights and effective environmental management. On debates over property rights to germplasm, see Brush and Stabinsky (1996). See also Brush (1999).

19. The inscription of territorial boundaries or perimeters around ethnically identified communities presents possibilities for reactionary identity politics, as well as possibilities for local empowerment [see Watts, chapter 1; see also Zerner (1996), Malkki (1992), and Peters (in press)].

20. A process Dove (1983) characterizes as the "political economy of ignorance."

21. Personal communication, 13 March 1998, at "Workshop on People, Plants, and Justice: Lessons and Insights," New York: Ford Foundation. See also Rothman (1998) and Radin (1996).

22. Radin (cited in Rothman 1998:15) emphasizes poor South Asian women's lack of power in making the choice to have their kidneys extracted and sold to more affluent pur-

chasers: "Desperation is the social problem we should be looking at, rather than the market ban [on international sales of organs]. We must re-think the larger social context in which this dilemma is embedded" (Radin 1996:125). See Scheper-Hughes (1998) for an astute analysis of the international political economy of the global trade in organs.

23. For examples of neoliberal accounts of a biodiversity prospecting, see King et al. (1994) and Reid et al. (1993a,b). For critical accounts of bioprospecting, see Parry (chapter 15), McAfee (1999), Zerner (1996), and Zerner and Kennedy (1996). On regional alliances, shopping malls, and neoliberalism, see Marcos (1997).

24. See Appadurai (1995a,b) and Kopytoff (1986) on regimes of value and the cultural biographies of commodities.

25. For recent analyses of the cultural politics of environmental representations, see Lowe (chapter 9), Neumann (1995), and Zerner (1994b).

26. See Mitchell (1991) for an incisive analysis of the production of images of Egypt, by development agencies and technocratic experts, resulting in economic and political development interventions. See Escobar (1995) for a critical history of the concept of development and its deployment by multilateral financial institutions, and Northern governments and agencies.

27. See Agarwal (1997) for a thoughtful analysis of community representations in the community-based resource management movement.

28. See Li (1997b).

29. On the politics of community representations in environmental discourses, campaigns, and multilateral development bank projects, see Tsing et al. (1998). See also Tsing et al. (in progress) for a more comprehensive treatment of community representations and environmentalism in Africa, Asia, and Latin America.

30. On the cultural production of community, see Li (1996), Tsing (1999), and Zerner (1994).

31. For further consideration of these issues, see Tsing et al. (1998 and in progress).

32. On transnational issues, networks, and politics, see Keck and Sikkink (1998) and Brysk (1994a,b, and in press); see Tennant (1994) and Kingsbury (1998) on the history and proliferation of imagery and law regarding international human rights, indigenous and minority peoples, campaigns, and institutions; see Kingsbury (1998) for a constructivist account of indigenous and minority peoples' rights within the international human rights framework.

33. See Keck and Sikkink (1998); see also Brysk (1994a,b).

34. See Appadurai (1996) on the rhizomatic, fractal character of transnational cultural and economic flows.

35. On devolution of management authority to local communities and the danger of ethnically based political competition over resources and authority, see Peters (in press).

36. A broad environmental justice agenda, focusing on democracy and democratization, equitable dispersion of power, and a broad group of social and economic justice goals, needs to be distinguished from an environmental governance agenda. Environmental governance takes as its primary goal the effective management of environments and focuses on the institutional, political, and economic means and processes through which these management outcomes may be achieved and controlled. An environmental governance agenda is, from a social justice perspective, profoundly ambiguous. Taking environmental outcomes as its ultimate desiderata, environmental governance regimes may be undemocratic, socially repressive, or economically oppressive. Other environmental governance regimes may explicitly support social justice goals and outcomes.

37. On the environmental justice movement in the United States, see Szasz (1994), Pulido (1996), and Bullard (1994).

38. See Malkki (1992) and Peters (in press) for a critique of contemporary linkages of ethnicity, territory, and indigenous populations.

39. On environmental entitlements, see Leach et al. (1997).

40. On the threat that the North American Free Trade Agreement (NAFTA) poses to global food security, *in situ* genetic diversity, and the livelihoods of hundreds of thousands of Mexican peasant farmers, see James Boyce (1996, 1997). Boyce speculates that these outcomes may be linked to the possibility of massive peasant migration from rural areas to Mexican cities and across national borders.

41. On monitoring, surveillance, and the social implications of geographic information systems (GIS), see Pickles (1991, 1995) and Curry (1995). On nongovernmental use of GIS imagery as a tool to galvanize and focus national and global opinion on the causes and impacts of the Indonesian forest fires in 1997 and 1998, see Harwell (2000).

42. On struggles over nature, resources, and the cultural construction of property rights in a context of rapid "development" and resource extraction in Southeast Asia, see Zerner (in press).

43. On the new ecology generally, see Botkin (1990), Pickett et al. (1992), Worster (1990), and Zimmerer (1994). See Zimmerer (1998) for an insightful analysis of the lessons of the new ecology for conservation practice in developing countries.

44. It would include, for example, examining the assumptions, images, and environmental interventions sponsored by multilateral financial institutions, transnational conservation organizations, and bilateral aid agencies, as well as projects mobilized by nongovernmental environmental and social organizations, national environmental ministries, parastatal organizations, and private sector businesses.

45. For a masterful, provocative analysis of culture/nature conjunctions and fusions, see Haraway (1991).

46. See Williams (1980) for a critical examination of the history of the concept of nature in Western philosophy. See also Schama (1996) for a critical analysis of English, German, and American conceptions of nature, forests, and ideal landscapes, and Cronon (1995).

# CHAPTER 1

## Contested Communities, Malignant Markets, and Gilded Governance: Justice, Resource Extraction, and Conservation in the Tropics

*Michael J. Watts*

The poor and forsaken are still condemned to lie in a world of terrible injustices, crushed by unreachable and apparently unchangeable magnates on which the political authorities, even when formally democratic, nearly always depend. Do people really think that the end of historical communism . . . put an end to poverty and the thirst for justice? In our world the two-thirds society rules and prospers without having anything to fear from the third of poor devils. But it would be good to bear in mind that in the rest of the world, the two-thirds (or four-fifths, or nine-tenths) society is on the other side.

*Norberto Bobbio (1996:45)*

Mapping the contemporary landscapes of resource extraction and biodiversity prospecting in the Tropics—whether damar extraction in Sumatra, ecotourism in Belize, oil exploitation in Amazonia, or ecocolonialism in Samoa—reveals the fundamental fractures and fault lines, the political tectonics as it were, of sustainable development in the late twentieth century. On the one hand, these cases are irreducibly *local*. They speak to the efforts by individuals, households, communities, indigenous groups, nonstate groups of various stripe, all of whom articulate a historical and cultural set of claims over the access to and control over territorial resources, to secure livelihoods from their local environmental inheritance. On the other hand, resource extraction and biodiversity are profoundly *global*. The cases speak to actors, agents, processes, and organizations that, even if they have local and particular geographical nodes (the World Bank has its headquarters in Washington, D.C., Genentech in

San Francisco, the International Union for the Conservation of Nature and Natural Resources (IUCN) in Gland, Switzerland), are nevertheless transnational in reach and scope, serving constituencies that are decidedly not local in the same way. In this sense, many if not all of the issues raised in any book pertaining to extraction and conservation occur in *globalized local sites*. The manner in which local and global interests and forces intersect around questions of the resource and use and conservation in tropical environments provokes challenges both for social theory and for practical politics and public policy.

To say that resources (or the environment) in the Tropics are contested and negotiated by competing constituencies is hardly original. But environment politics has become a battleground of a quite particular, and complex, sort. On the one side is the nature of environmental problems themselves, often global or regional in character, which transgress national borders (they may not be, in any simple sense, place specific). In the same way, these environmental crises and problems are constituted as parts of global discourses or epistemic communities (Haas 1990), often typically in moral, technocratic/scientific, and managerial languages (Buttel and Taylor 1992). On the other side is an enlarged panoply of actors who engage in environmental management, regulation, and governance. The gradual opening and flourishing of civil society in parts of Africa, Latin America, and the former socialist bloc has contributed to the proliferation of all manner of local green movements and nongovernmental organizations (NGOs)—to a thickening of civil society, as Jonathan Fox (1994) puts it. However, these community-based and grassroots initiatives are not "environmental" in any simple way since they are often driven by poverty and justice concerns, human and political rights, customary culture and indigenous identity, and so on. Neither are they solely local environmentalisms since they typically invoke transnational environmentalism and its armory of institutions, whether it be the world of the international NGOs (World Wildlife Fund or Greenpeace), international green networks and advocacy groups [Rainforest Action Network (RAN)], or the efforts of newly "greened" multilateral legislative bodies and institutions [the Biodiversity Convention, the Global Environmental Facility (GEF) of the World Bank, or the World Trade Organization (WTO), for example]. Somewhere in the middle stands something called the state, or national, policy making.

A central, but by no means exclusive, concern of this book is the debate over the meanings and practices of community-based environmental management of tropical resources. It is typically presumed that the "community"—a contested and complex entity in itself, as we shall see—has historical attachments and entitlements over environmental resources, indigenous institutions, and customary rights and practices, which control access to and regulation of resources, and that it is the repository of alternative environmental knowledges (so-called *indigenous technical knowledge*, or ITK) that undergird sustainable community management. National policy making—and the armory of state apparatuses and ministries regulating forests, or petroleum, or germ plasm—stands as a sort of counterpoint resting on formal legal and political jurisdictions and systems, and increasingly global scientific discourses (i.e., sustainable development), that is to say expertise formed around *epistemic scientific communities*, as Peter Haas (1990) calls them. These highly stylized systems of governance, knowledge, and institutionalized interests and power come into contention in local, national, and multilateral environmental arenas.

Embedded within resource extraction and biodiversity are the knotty questions of property, human rights, institutions and capabilities, social capital and governance, international law, representation and accountability, and the role of the market. I cannot possibly do justice to all of them—which is in any case far beyond my competence—but I do wish to explore a number of concerns as they appear in the case studies before us. I wish to do so, however, by adding a (brief) case study of my own, one that is entirely consistent with the themes explored in this volume, although, unlike many of the studies, it has been something of a ground zero for international environmental activism. I refer to the struggle of the Ogoni people in the Niger Delta and the efforts by its murdered leader Ken Saro-Wiwa to confront both the international oil companies and a vicious military state in the name of ethnic autonomy, political decentralization, sustainable development, social justice, and indigenous resource extraction. I offer it here as a case to complement the others in this book and to throw into sharp relief a number of concerns that are perhaps less visible in the chapters that follow.

## THINKING OGONI: ENVIRONMENT, EXTRACTION, AND JUSTICE

Standing at the margin of the margin, Ogoniland (like Chiapas in Mexico) appears to be a socioeconomic paradox (figures 1-1, 1-2). Home to six oil fields, half of Nigeria's

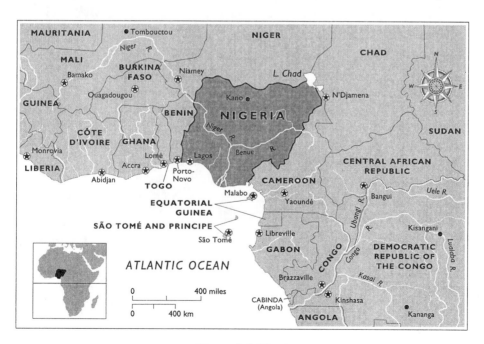

**Figure 1-1** Nigeria

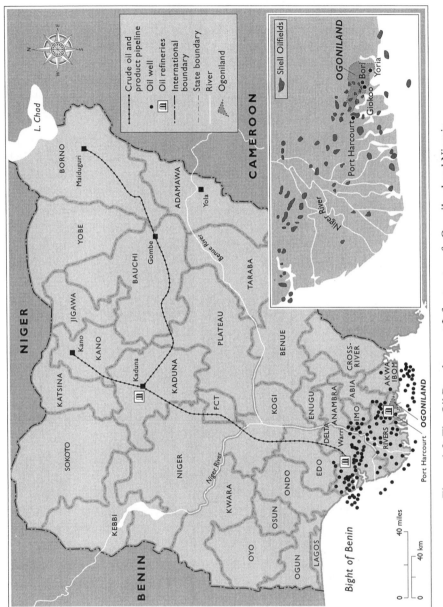

**Figure 1-2** The Oil Development Infrastructure fo Ogoniland and Nigeria.

oil refineries, the country's only fertilizer plant, a large petrochemical plant, Ogoni-land is wracked by unthinkable misery and deprivation. During the first oil boom, Ogoniland's fifty-six wells accounted for almost 15 percent of Nigerian oil produc-tion[1] and in the past three decades an estimated $30 billion in petroleum revenues have flowed from this Lilliputian territory; it was, as local opinion had it, Nigeria's Kuwait. Yet, according to a government commission, Oloibiri, where the first oil was pumped in 1958, has no single kilometer of all-season road and remains "one of the most backward areas in the country" (Furro 1992:282). Few Ogoni households have electricity, there is one doctor per 100,000 people, child mortality rates are the high-est in the nation, unemployment is 85 percent, 80 percent of the population is illit-erate, and close to half of Ogoni youth have left the region in search of work. Life expectancy is barely fifty years, substantially below the national average. In Furro's survey of two minority oil-producing communities, over 80 percent of respondents felt that economic conditions had deteriorated since the onset of oil production, and over two-thirds believed that there had been no progress in local development since 1960. No wonder that the systematic reduction of federal allocations and the lack of concern by the Rivers government was, for Ogoniland, part of a long history of "the politics of minority suffocation" (Ikporukpo 1996:171).

If Ogoniland failed to see the material benefits from oil, what it *did* experience was an ecological disaster—what the European Parliament has called "an environmental nightmare." The heart of the ecological harm stems from oil spills—either from the pipelines that crisscross Ogoniland (often passing directly through villages) or from blowouts at the wellheads—and gas flaring. As regards the latter, a staggering 76 percent of natural gas in the oil-producing areas is flared (compared to 0.6 percent in the United States). As a visiting environmentalist noted in 1993 in the delta, "some children have never known a dark night even though they have no electricity" (*Village Voice,* November 21, 1995:21). Burning twenty-four hours per day at temper-atures of 13–14,000 degrees Celsius, Nigerian natural gas produces 35 million tons of $CO_2$ and 12 million tons of methane, more than the rest of the world (and ren-dering Nigeria probably the biggest single cause of global warming). The oil spillage record is even worse. According to Claude Ake, there are roughly 300 spills per year in the delta, and in the 1970s alone the spillage was four times greater than the much-publicized Exxon Valdez spill in Alaska. In one year alone, almost 700,000 barrels were spilled, according to a government commission.[2] The Ogoni are typically seen as a distinct ethnic group, consisting of three subgroups and six clans.[3] Their popu-lation of roughly 500,000 people is distributed among 111 villages dotted over 404 square miles of creeks, waterways, and tropical forest in the northeast fringes of the Niger Delta. Located administratively in Rivers State, a Louisiana-like territory of some 50,000 square kilometers, Ogoniland is one the most heavily populated zones in all of Africa. Indeed the most densely settled areas of Ogoniland—over 1,500 per-sons per square kilometers—are the sites of the largest wells. Its customary produc-tive base was provided by fishing and agricultural pursuits until the discovery of pe-troleum, including the huge Bomu field, immediately prior to Independence. Part of an enormously complex regional ethnic mosaic, the Ogoni were drawn into internecine conflicts within the delta region—largely as a consequence of the slave trade and its aftermath—in the period prior to arrival of colonial forces at Kono in 1901. The Ogoni resisted the British until 1908 (Naanen 1995) but thereafter were left to stagnate as part of the Opopo Division within Calabar Province. As Ogoniland was gradually incorporated during the 1930s, the clamor for a separate political division grew at the hands of the first pan-Ogoni organization, the Ogoni Central Union, which bore fruit with the establishment of the Ogoni Native Authority in 1947. In 1951, however, the authority was forcibly integrated into the Eastern Region. Experiencing tremendous neglect and discrimination, integration raised longstanding fears among the Ogoni of Igbo domination.[4] Politically marginalized and economically neglected, the delta minorities feared the growing secessionist rhetoric of the Igbo and consequently led an ill-fated secession of their own in February 1966. Isaac Boro, Sam Owonaro, and Nottingham Dick declared an illegal Delta Peoples Republic but were crushed and were subsequently, in a trial that is only too reminiscent of the Ogoni tribunal in 1995, condemned to death for treason. Nonetheless, Ogoni antipathy to what they saw as a sort of internal colonialism at the hands of the Igbo, continued in their support of the federal forces during the civil war. While Yakubu Gowon did indeed finally establish a Rivers State in 1967—which compensated in some measure for enormous Ogoni losses during the war—the new state recapitulated in microcosm the larger "national question." The new Rivers State was multiethnic but presided over by the locally dominant Ijaw, for whom the minorities felt little but contempt.[5]

Traces of Ogoni "nationalism" long predate the oil boom of the 1970s, but they were deepened as a result of it. Ogoni fears of what Saro-Wiwa called "monstrous domestic colonialism" (1992) were exacerbated further by federal resistance to dealing with minority issues[6] in the wake of the civil war and by the new politics of post–oil boom revenue allocation. Rivers State saw its federal allocation fall dramatically in absolute and relative terms. At the height of the oil boom, 60 percent of oil production came from Rivers State but it received only 5 percent of the statutory allocation [roughly half of that received by Kano, Northeastern States, and the Igbo heartland, East Central State (see Khan 1994)]. Between 1970 and 1980 it received in revenues one fiftieth of the value of the oil it produced. In what was seen by the Rivers minorities as a particularly egregious case of ethnic treachery, the civilian Shagari regime reduced the derivation component to only 2 percent of revenues in 1982, after Rivers State had voted overwhelmingly for Shagari's northern-dominated National Party of Nigeria (Forrest 1995; Lewis 1996). The subsequent military government of General Buhari cut the derivation component even further at a time when the state accounted for 44.3 percent of Nigeria's oil production.

Ken Saro-Wiwa played a central role in the tactical and organizational transformations of the Movement for the Survival of the Ogoni People (MOSOP) during the 1990s. Born in Bori as part of a traditional ruling family, Saro-Wiwa was already, prior to 1990, an internationally recognized author, a successful writer of Nigerian soap operas, a well-connected former Rivers State commissioner, and a wealthy businessman. Saro-Wiwa was also president of the Ethnic Minorities Rights Organization of Africa (EMIROAF), which had called for a restructuring of the Nigerian federation into a confederation of autonomous ethnic states in which a federal center was radically decentralized and states were granted property rights over on-shore mineral resources (Osaghae 1995:327). Under Saro-Wiwa, MOSOP focused in 1991 on links to pro-democracy groups in Nigeria (the transition to civilian rule had begun under heavy-handed military direction) and on direct action around Shell and Chevron installations. It was precisely because of the absence of state commitment and the deterioration of the environment that local Ogoni communities, perhaps understandably, had great expectations of Shell (the largest producer in the region) and directed their activity against the oil companies after three decades of betrayal. There was a sense in which Shell *was* the local government[7] (*Guardian,* July 14, 1996:11), but the company's record had, in practice, been appalling.[8] A watershed moment in MOSOP's history was the drafting in 1990 of an Ogoni Bill of Rights (Saro-Wiwa 1992, 1989). Documenting a history of neglect and local misery, the Ogoni bill took head-on the question of Nigerian federalism and minority rights. Calling for participation in the affairs of the republic as "a distinct and separate entity," the bill outlined a plan for autonomy and self-determination in which there would be guaranteed "political control of Ogoni affairs by Ogoni people . . . the right to control and use a fair proportion of Ogoni economic resources . . . [and] adequate representation as of right in all Nigerian national institutions" (Saro-Wiwa 1989:11). In short, the Bill of Rights addressed the question of the *unit* to which revenues should be allocated—and derivatively the rights of minorities. Largely under Saro-Wiwa's direction, the bill was employed as part of an international mobilization campaign. Presented at the U.N. Subcommittee on Human Rights, at the Working Group on Indigenous Populations in Geneva in 1992, and at the meeting of the Unrepresented Nations and People Organization (UNPO) in The Hague in 1993, Ogoni became—with the help of RAN and Greenpeace—a cause célèbre (UNPO 1995). The hanging of Ken Saro-Wiwa and the Ogoni nine in November 1995—accused of murdering four prominent Ogoni leaders who professed opposition to MOSOP tactics—and the subsequent arrest of nineteen others on treason charges, represented the summit of a process of mass mobilization and radical militancy which had commenced in 1989.

What sort of articulation of Ogoni identity and political subjectivity did Saro-Wiwa pose? It was clearly one in which territory, community, and environment were the building blocks upon which ethnic difference and indigenous rights were constructed. And yet it was an unstable and contradictory sort of articulation. First, there was no simple Ogoni "we," no unproblematic unity, and no singular form of political subject (despite Saro-Wiwa's ridiculous claim that 98 percent of the Ogoni supported him). MOSOP itself had five independent units—an object of bitter dispute in itself—embracing youth, women, traditional rulers, teachers, and churches. In spite of the remarkable history of MOSOP between 1990 and 1996, its ability to represent itself as a unified pan-Ogoni organization remained an open question, particularly for

Saro-Wiwa. There is no pan-Ogoni myth of origin (characteristic of many delta mi-norities), and a number of the Ogoni subgroups engender stronger local loyalties than any affiliation to Ogoni nationalism. The Eleme subgroup has even argued, on occasion, that they are not Ogoni. Furthermore, the MOSOP leaders were actively opposed by elements of the traditional clan leadership, by prominent leaders and civil servants in state government, and by some critics who felt Saro-Wiwa was out to gain "cheap popularity" (Osgahae 1995:334). And, not least, the youth wing of MOSOP, which Saro-Wiwa had made use of, had a radical vigilante constituency that the leadership were incapable of controlling. What Saro-Wiwa did was to build upon over fifty years of Ogoni organizing and upon three decades of resentment against the oil companies, to provide a mass base and a youth-driven radicalism—and, it must be said, an international visibility—capable of challenging state power.[9] But it represented a fractious and increasingly divided "we," as the open splits and conflicts between Saro-Wiwa and other elite Ogoni confirm.[10]

Second, he constantly invoked Ogoni culture and tradition yet he also argued that war and internecine conflict had virtually destroyed the fabric of Ogoni society by 1900 (1992:14). His own utopia then rested on the re-creation of Ogoni culture—insofar as Africa's tribes are "ancient and enduring social organizations" (1995:191)—and suffered like all ur-histories from a mythic invocation of the past.

Third, ethnicity was the central problem of postcolonial Nigeria, i.e., the corrup-tion of ethnic majorities and its panacea (the multiplication of ethnic minority power). Invoking the history of exclusion and the need not simply for ethnic minority inclu-sion, but also that it serve as the basis for federalism, allowed Saro-Wiwa to totally ignore the histories and geographies of conflict and struggle among and between eth-nic minorities. The narrative of Ogoni exclusion and internal colonialism proved also to be partial (and exclusionary). Compared to many delta minorities, the Ogoni have fared well (with 12 percent of Rivers State population, the Ogoni accounted for one third of the state's commissioners). The Ogoni produce roughly 4 percent of Rivers State oil currently; two other small minorities with no political representation account for 68 percent. And finally, Ken Saro-Wiwa and the Ogoni hated modernity because *they could not get enough of it*: the Ogoni were angry because they could neither af-ford the cars nor use the roads, which were the icons of petrolic success. In this sense, the Ogoni story invokes Walter Benjamin's concerns with the fantasy and shock of modernity. In Ogoniland, it was the phantasmagoria of petrolic commodification (of wealth without effort), and the shock of modernity that frame the rise of MOSOP and what Benjamin himself, in his concern with utopias, called "the moment of awaken-ing." But it would be a mistake to simply assume that the Ogoni movement repre-sented any simple rejection of either. It can be read as an almost archetypical mod-ernist struggle over rights, and especially over state accountability, transparency, and local representation—much of which incidentally could be seen as a celebration of the sort of Green Conservatism of the New Right.

This is of course a potted history, but it speaks directly to *all* of the fundamen-tal concerns running through this book: globalization in the forms of transnational capital, transnational advocacy, and international legal and regulatory fora; the relations between resource and environment entitlements (narrowly understood) and other social, economic, and political concerns (land rights, political representation, state accountability and transparency, ethnic identity, social justice); the role and meaning of the community or the locality as a basis for political action and mobilization; the character and form of green struggles and movements (civil society and extracommunity linkages); institutions (local or otherwise) and the forms of governance around environmental change; and the questions of local initiatives and their relationship to development (and its alternatives).

## POSTCOLONIAL GREENS AND THE SEARCH FOR ALTERNATIVES

Against a backdrop of deepening global inequality and intractable southern poverty—according to the United Nations Development Program, the polarization of global wealth *doubled* between 1960 and 1989—it is perhaps inevitable that development as theory and practice is once more mired in debate and controversy. Many intellectuals and activists from the South have come to see development as a cruel hoax—a "blunder of planetary proportions" (Sachs 1992:3)—which is now coming to an unceremonious end. "You must be either very dumb or very rich if you fail to notice," notes Mexican activist Esteva (1992:7), "that `development' stinks." It is precisely the groundswell of *antidevelopment thinking*—oppositional discourses that have as their starting point the rejection of development, the end of Enlightenment, and the purported failure of development as a modernist project—which posits the likes of the Ogoni movement, and many of the movements and struggles documented in this volume, not simply as sites of "alterity" but of a radical reimagining of development itself.

Standing at the center of a broad and heterogeneous field of so-called alternatives to development (Pieterse 1996) is, broadly speaking, a philosophical sympathy toward poststructuralism, and correspondingly a strenuously enforced antipathy to the metanarratives of both neoliberal orthodoxy *and* Marxism. Antidevelopment efforts to imagine a "postdevelopment era" start from the capacity of the development imaginary to shape identity and produce particular sorts of "normalized subjects" in the South. Development here threatens diversity, homogenizing local traditions through the apparatuses of the state (investment, measurement, and planning). The local is eclipsed by the use of general conceptual categories and Western assumptions. A postdevelopment alternative, then, depends fundamentally on local spaces of self determination and autonomy (Sachs 1993; Escobar 1992a). Both articulations are draped in the presumptions of antifoundationalism and of a resistance to what are seen as universalizing discourses of the West. Escobar's work (1995) is central to imaginary postdevelopment because it links both of these lines of thinking and provides an explicitly theoretical framework for the study of development, environmental sustainability,

community movements, and indigeneity (or, in his terms, cultural movements turning on questions of identity). Specifically, he finds modern development discourse
to be an invention—more properly a "historically produced discourse" (p. 6)—of
the post-1945 era, the latest insidious chapter of the larger history of the expansion
of Western reason. This discourse is governed by the "same principles" as colonial
discourse but has its own "regimes of truth" and "forms of representation" (pp.
9–10). Development is about forms of knowledge, the power that regulates its practices, and the forms of subjectivity fostered by its impulses. Hegemonic development discourse appropriates societal practices and meanings into the modern realm
of explicit calculation, thereby subjecting them to Western forms of power-knowl-
edge. It ensures the conformity of peoples to First World economic and cultural practices. In short, development has penetrated, integrated, managed, and controlled
countries and populations in increasingly pernicious and intractable ways. It has
produced underdevelopment, a condition politically and economically manageable
through "normalization," the regulation of knowledges and the moralization and
technification of poverty and exploitation. What was and is missing from development, according to Escobar, is *people*.[11] People reenter the postdevelopment story
insofar as there has been "the resurrection, reemergence and rebirth of . . . civil
society" (Cohen and Arato 1992:29), and it is within these attempts to thicken civil
society that any alternative to development actually resides. Partly in response to the
collapse of state resources, and partly as an outcome of an uneven democratization
process, various forms of local and community movements have emerged in the interstices of the state-market nexus. These new social movements (NSM), many of
which are green or at least link resources, environment, and rights in complex ways,
are taken to be new insofar as they represent a sort of postmodern politics outside of
and in many respects antithetical to class or social-democratic party politics (Escobar 1992b). Many are also new in that they are an integral part of a widespread "environmentalization" of institutional practices (Buttel 1992). Enormously heterogeneous in character and scale (anti-dam movements, squatter initiatives, minority
cultural/indigenous movements), these grassroots movements often focus on efforts
to take resources out of the marketplace, to construct a sort of moral economy of the
environment (see Martinez-Alier 1990), often focusing more broadly on livelihoods
and justice. This multidimensionality is, according to Escobar (1992a), indicative of
a new mode of doing politics, so-called autopoietic (that is to say, self-producing and
self-organizing) movements that exercise power outside the state arena and that seek
to create "decentred autonomous spaces."

The *local community* and *grassroots initiatives* loom very large in current critical (poststructural) approaches to development. What they represent is certainly a
form of collective action, but more specifically and profoundly a "resistance to development" (Escobar 1995:216; Rahnema and Bawtree 1997; also Routledge 1994)
that attempts to build new identities. The implication is, of course, that these identities fall outside the panoptic gaze of the hegemonic development discourse as new
forms of subjectivity that stand opposed to modernity itself. As Escobar puts it

(1995:216), these movements are not cases of "essentialized identity construction" but are "flexible, modest, mobile, relying on tactical articulations arising out of the conditions and practices of daily life." To the extent that these movements are often "environmental" or "green," vast claims have been made on their behalf: they are a "revolt against development," "a new economics for a new civilization," "learning to be human in a posthuman landscape."

What then is the new content of such movements and what are their relations to development alternatives? There are, in my reading, at least five fundamental and overlapping aspects attributed to grassroots movements as vehicles of counter-modernity. First, they purportedly contain new sorts of politics and new sorts of political subjectivity. They are typically local, outside the organized state sphere, and "without one particular ideology or political party" (Escobar 1992b:422). They are "self-organizing and self-producing," exercising nonstate forms of power. Second, "cultural difference is at the root of postdevelopment" (Escobar 1995:225) and hence the movements are, above all, examples of popular cultural discourse. Minority cultural communities[12] figure centrally in both green and antidevelopment movements; indeed, the "indigenous" becomes the lodestar for the "unmaking of the Third World." Indian confederations in Latin America, or "ethnic" green movements in Africa often turn on the ways in which cultural identity is mobilized as "a transformative engagement with modernity" (Escobar 1995:219). Third, the movements employ, in creative ways, local or subaltern reservoirs of knowledge. Paul Richards's (1985) invocation of "inventive self-reliance" rooted in local African peasant knowledge is an influential case in point of this line of thinking. The proliferation of the field of "indigenous technical knowledge" (ITK) and the so-called actor-oriented interface analysis is another. In singing the praises of this subaltern science position, women's knowledge and nature are often central. In Shiva's words, "Women as victims of violence of patriarchal forms of development have risen against it to protect nature" (1989:xvii) and by virtue of their organic relationships to things natural have a "special relationship with nature" (1989:43). Indeed, for Shiva, feminine/ecological ways of knowing are "necessarily participatory." Fourth, local community and "tradition" are neither erased nor preserved as the basis for alternative development but are refashioned as a hybrid: hybridity entails "a cultural (re)creation that may or may not be (re)inscribed into hegemonic constellations" (Escobar 1995:220). This is the heart of the new political subjectivity that speaks to a "transcultural in-between world reality" (Escobar 1995:220). And finally, these movements produce a defense of the *local*: such a defense is a "prerequisite to engaging with the global . . . [and represents] the principal elements for the collective construction of alternatives" (Escobar 1995:226).

Much can be said about this body of work, which frames the sorts of issues displayed in this volume as alternatives to the market-driven, high-tide of neoliberalism (see Cowen and Shenton 1996). Identity politics is championed by Escobar, for example, because it represents part of an alternative reservoir of knowledge and because such ideas stand against the "axiomatics of capitalism." But there is surely nothing necessarily anticapitalist or particularly progressive about cultural identity: calls to localism can produce Hindu fascism as easily as Andean Indian cooperatives. Running through much of the literature that posits local environmental struggles as alternatives is an uncritical appeal to the "people"—that is to say, populist rhetoric—or "the community" or "tradition" without a sensitivity to the potentially deeply conservative, sometimes reactionary, aspects of such local particularisms. A striking feature of so much of the alternatives school is the constant uncritical appeal to the local, to place, and to the cultural (where cultural is synonymous with a self-consciously local sense of community). Yet as Pierre Bourdieu has noted, in discussion of "the people" and "popular" discourse, what is at stake is the struggle between intellectuals (1990:150). These debates among intellectuals celebrate, typically in a romantic and quasi-mystical way, the efficacy of all action/knowledge from below; moreover, they often forget that the "local" is never purely local but is created in part by extralocal influences and practices over time. There is precious little about the linkages and alliances between the local and the nonlocal, about the synergisms between public and private, or local and regional.

## KNOWLEDGE AND NARRATIVE

Knowledge figures quite centrally in a number of the chapters of this book, and, as I have shown, knowledge—or, more properly, institutionalized knowledges (discourses), which are the basis for particular truth claims or regimes of truth—is also key to current debates over, and poststructural critiques of, development and sustainability. But the field of knowledge, discourse, and narrative is a complex one, and I propose to briefly chart here five currents that run through *People, Plants, and Justice.*

### Indigenous (Subaltern) Knowledges

Here the knowledge issue turns especially on what individuals and groups (and de facto communities) know and practice with respect to resource extraction from, and conservation of, their local tropical environments. This indigenous technical knowledge is described in a number chapters, most compellingly in the accounts of Bentian rattan forests and damar agroforestry in Indonesia (but one might as easily point to work on artisanal marine fisheries in Brazil, or rangeland rehabilitation and use in the pastoral Sahel, or peasant understandings of soil and climate in south India). The body of knowledge is substantial (and there are a number of international organizations devoted to its generation, propagation, and use) and widely understood within academic and activist circles. It needs to be said, however, that its basic premise—that local actors know a great deal about local ecology and that this knowing is typically culturally "institutionalized" and "embedded" in a variety of persons, offices, rituals, and customary practices—is certainly not new. The early 1960s work on cognized models of the environment [Rappaport's *Pigs for the Ancestors* (1984)] was a sophisticated example of this, as was the early ethnobotanical research. The concern with "adaptation" in peasant societies in the 1970s is chock-full of such evidence (and one can push this historical frontier back *much* further: see Watts 1983). The questions become, then, (1) Why has this knowledge been so difficult to "legitimate"? (2) Under what circumstances can such knowledge/practice be institutionalized without co-optation or subversion? and (3) How might it be systematized in some way? In short, how might this approach get out of the cul-de-sac of "agroecological populism" (the peasants, the herders, the extractors know best and let's leave things as they are)?

Running through this literature, incidentally, is another set of core issues that needs to be highlighted: first, that this knowledge is unevenly distributed *within* local societies; second, that it is not necessarily right or best just because it exists (the problem of functionalism or reification; i.e., it can often be wrong or inappropriate); and third, traditional or indigenous knowledge may often be of relatively recent invention (which is to say, these knowledges are not static or stable but, as Paul Richards suggests, may be predicated on forms of experimentation). Indeed, it may not be indigenous as such but instead really *hybrid* [see Akhil Gupta's new book (1998) on Indian agroecology]. More problematically, this indigenous knowledge—typically counterposed to Western science in an inept and polarized fashion—sometimes carries its own freight. If "Western" soil science (a curiously inept and partial account of the genesis and construction of this knowledge) has its own defects, blind spots, and biases, it is not difficult to find within, say, Vandana Shiva's account (1993c, 1989) of Indian women as "natural" peasant scientists a massive dose of essentialism, mystification, and projection. Finally, let me say that indigenous knowledge has great weight—although this is less well often understood—because it speaks to the shared meanings by which social groups "integrate" [I am taking the term from Jürgen Habermas (1994)] as a condition for coordination, trust, cooperation, decision making, and therefore *governance*. However, it is not often well understood how this ITK is in fact related to governance as opposed to its ecological efficacy.

### Images, Narratives, Ecological Orientalism

Candace Slater's excellent chapter on Amazonia (chapter 3) reveals this aspect of knowledge (as image) quite clearly. She is able to show how there is a popular imagery of the region (perhaps transnational in appeal), and how this imagery has a history (albeit unstable, as she shows in the shift from Green Hell to Eden), and how literary, media, and other cultural machinery contributes to what I have elsewhere called a "discursive ecological formation" (Peet and Watts 1996; see also Guthman 1997). In a sense, the reference point here is the *Orientalism* of Said (1994), and specific environments (rainforest, Amazonia, the desert) are, as it were, "orientalized" in particular ways. In Slater's account, the Edenic or naturalized narrative always

silences (the Indians have no voice or no voice of their own) and these tropes exclude or "distort." But to invoke the principle of distortion raises as many questions
as it answers. Does not distortion imply a "truth"? In any case, one needs to know
a great deal about the details of who is constructing this image and why they are
more or less fragile (was the Indonesia rainforest subject to the same sort of construction as Amazonia, and if not, why not, and with what consequences?). Slater
ends with the provocation that there is an absence of competing images of Amazonia. True (perhaps), but under what conditions can competing images *really* compete? This is not unlike Emery Roe's (1991) complaint that development "stories"
have staying power (say, the tragedy of the commons) and we simply need new
stories (as if a new folk tale is ipso facto able to do the dirty work of jettisoning fifty
years of mismanagement and state policy). Which is to say, do these images really
have the power and effect implicit in these accounts of narratives? Are they "just"
images and irrelevant to the hard edges of political economy and environmental destruction? Melissa Leach et al. (1997) argue conversely that the images have simply
to be employed to combat other images, and that they should be judged less on their
"empirical truth" than on the policy purposes they serve. It seems to me that this sort
of strategic essentialism is very dangerous.

### Epistemic Communities

Here the knowledge is Western rationality and science, and more properly the communities of expertise (and typically transnational in scope) that revolve around environmental problems to be multilaterally regulated (say, the law of the sea). Peter
Haas (1990) has argued in the context of understanding regional (European Union)
and global (multilateral) conventions that the process of consensus building and collective action more generally is *knowledge based* and *interpretive*. That is to say, international regulatory cooperation is fueled by fundamental scientific uncertainty
about the environment, which ensures that governments seek out authoritative ad-
visors (experts) who, to the extent that they are part of epistemic communities, are
more important as regards political solutions than the content of the ideas per se.
Cross-national differences in state behavior are determined by the variation in the
penetration and institutionalization of experts (epistemic communities). Biodiversity and stratospheric ozone cooperation are seen in this way as instances of the cog-
nitive and bureaucratic power of scientific experts. This is an argument that has also
(curiously) been made for the North America Free Trade Agreement (NAFTA) by
Benton (1996), who argues that the trade and environmental constituencies brought
together around tariff reduction actually created a dialogue that had not hitherto existed and that turned on the making of an epistemic community of sorts. There is no
reason in principle why such arguments could not be made to apply to the international NGO fora in which the epistemic community of experts may be held together
by epistemological and scientific interpretations quite different from the "hegemonic" views held by Haas's communities. It remains an open question, however,

whether these epistemic communities in the haasian sense have the efficacy he imputes—as opposed to either the interstate power approach (U.S. dominant interests scuttle consensus) or indeed, as Raustialla (1997) has shown in examining the differences between Britain and the United States at the Biodiversity Convention, the key role of domestic regulatory and political structures and the differing influence of business.

## *Biopower and Ecological Revisionism*

The final knowledge question speaks less to what Foucault calls the "insurrection of subjugated knowledges" than to the ways in which particular models ("conventional models," as Leach calls them), rooted in particular institutions and practices, become hegemonic and are then subsequently contested. I am thinking here, for example, of how new revisionist ecology often meets on the same terrain as critical social science or political ecology studies. Leach and Mearns's (1997) reinterpretation of the forest–savanna mosaic is a case in point in which historical studies coupled with detailed local analysis of agroecology confirms what the new "nonequilibrium" ecology posits, namely that climax models of ecological stasis are unhelpful. These static models, however, do enter into administrative practice (colonial and post-colonial), which lends a particular weight to seeing Guinee's forest cover as "relic" (which Leach and Mearns see as basic for driving "repressive policies designed to reform local land use practice") rather than as the outcome of intentional management practice. A similar case could be made for the intersection of the new rangeland ecology with the critical studies of Sahelian agro-pastoralism. Swift (1997) has shown how the assumptions about desertification not only rest on remarkably spare evidence but on questionable models of the dynamics of semiarid rangelands—their resiliency and stability, in other words—which are (1) the expression of linear, cybernetic models of ecological structure and temporalities, and (2) attached to neo-Malthusian models of social change. The key here is that conventional wisdom is challenged as an embodied form of knowledge, and the challenge itself reflects a peculiar unity of local knowledge and practice with nonlinear models of new ecology. Out of this emerges a concern with pluralism (at the level of truth claims), with democracy (to open up the practices of policy making to other voices) and complexity/flexibility (of local conditions and historical dynamics). All of these right and proper concerns raise as many questions as they answer, of course: If policies simplify (as they surely must?), how are complexity and local variation to be built into policy, short of simply involving "uncertainty" (see Roe 1991)? Isn't this simply a populist myopia? Is calling for counternarratives really the answer in itself? I want to conclude with one implication that speaks to the crucial question of knowledges and their use:

> As far as knowledge is concerned, localness has paradoxical implications of systematicity. We cannot abandon the strength of standards, generalization, theories,

and other assemblages of practice with their capacity for making connections and at the same time providing for the possibility of systematic criticism. We need to recognize that "systemic discipline" and "local resistance" are two sides of the same coin; promoting systematicity is a local practice, and local resistance contains the impetus for systematization. If we do not recognize this joint dialectic of the local and the global, we will not be able to understand and establish conditions conducive to the possibility of directing the structure and circulation of power in knowledge systems, and conditions for promoting redistributions.

Watson-Verran and Turnbull (1994:136–137)

### *Multilateral Green*

A final zone of knowledge and discourse speaks to the multilateral (interstate) arena—to global environmentalism understood as management (Escobar 1995). On the one hand, as Kingsbury (1994) shows in his account of the incorporation of environmental issues into the WTO and trade debates, the process of greening these institutions has only just begun. Zerner's work (1994a) on the GEF is also relevant here, because it poses quite sharply the question of how environmental concerns are "socialized" and "institutionalized" into a complex organization like the World Bank. Robert Wade's (1997) book on the greening of the Bank, and McAfee's (1999) work on what she calls "green developmentalism" show precisely how discourses (such as gender and development, incidentally) are institutionalized in quite specific ways with quite specific institutional powers. Of course, much of this discourse turns on how the idea that nature has to be sold to be saved is constructed and legitimated (or justified).

## COMMUNITY AND COMMUNITY REGULATION, PARTICIPATION, AND RESISTANCE

> The complexity of Community this relates to . . . on the one hand the sense of direct common concern; on the other hand the materialization of various forms of common organization . . . Community can be the warmly persuasive word to describe an existing set of relationships . . . unlike all other terms of social organization . . . it seems never to be used unfavorably.
>
> Raymond Williams (1976:15)

The community looms large in the chapters in this volume, and indeed more generally in all manner of policy discussions over regulation, participation, and international law. But the community turns out to be—along with its lexical affines, namely, tradition, custom, and indigenous—a sort of keyword whose meanings (always unstable and contested) are wrapped up, in complex ways, with the problems it discusses (to paraphrase Raymond Williams). The community is important because it is typically seen as a locus of *knowledge*; a site of *regulation* and management; a source of *identity* and a repository of "tradition"; the embodiment of various *institutions* (say, property rights), which necessarily turn on questions of representation, power, authority, governance, and accountability; an object of *state control*; and a theater of *resistance* and struggle (of social movement, and potentially of *alternate visions of development*). It is, then, an extraordinarily dense social object and yet one that is rarely subject to critical scrutiny (not least by those, such as the poststructuralist critics, who sing its praises). It is often invoked as a unity, as an undifferentiated thing with intrinsic powers, that speaks with a single voice to the state, to transnational NGOs, or to the World Court. Communities are of course nothing of the sort.

One of the problems is that the community can express quite different sorts of social relations and forms, from a nomadic band to a sedentary village to a confederation of Indians to an entire ethnic group. It is usually assumed to be the natural embodiment of "the local"—configurations of households, lineages, longhouses—which has some territorial control over resources that are historically and culturally constructed in distinctively local ways. A community, then, typically involves a territorialization of history ("this is our land and resources which can be traced in relation to these founding events") and a naturalized history ("history becomes the history of my people and not of our relations to others"). Communities fabricate, and refabricate through their unique histories, the claims that they take to be naturally and self-evidently their own. In this sense, a parallel with nationalism might be in order (Berlant 1991), since they always involve forms of fantasy and invention (or imagination), and they are always shot through with power and authority—some do the inventing and imagining in the name of others who do not. This is why communities then have to be understood, as Tania Li (1996) and Florencia Mallon (1995), among others, have shown, in terms of hegemonies: not everyone participates equally in the construction and reproduction of communities, or benefits equally from the claims made in the name of community interest.

I began with my own account of one community—the Ogoni "ethnic minority" or "indigenous people"—precisely because it shows clearly something of the problematic status of a community identity. The Ogoni Bill of Rights as an expression of Ogoni interests had to be constructed out of a heterogeneity of interests and communities (the five Ogoni kingdoms). Differentiations of age, gender, and class—and what we might call internal cultural differentiation—made this construction of a community's ecological goal especially knotty. Indeed, Saro-Wiwa's death was in part a reflection of the fractured nature of "the community"—a fracture that was obviously played upon and deepened by both the complicity of the state and transnational capital. Saro-Wiwa (if I can borrow from Stuart Hall) was in the business of *articulating* an identity that was at once cultural and political—as a precondition for making claims about local environmental and resource management, compensation, and control. Contained within this articulation were, in practice, *several* Ogoni community movements with contrasting characters, trajectories, and political dynamics. And this, it seems to me, is exactly what is at stake in, say, the current revisionist work on Chipko by Guha and Gadgill (1995), by Priya Rangan (1995), by Linken-bach (1994), and by Sinha et al. (1997). Far from the mythic community of tree-hug-ging, unified, undifferentiated women articulating alternative subaltern knowledges for an alternative development [as Shiva (1993c) posits], we have three or four Chip-kos, each standing in a quite different relationship to development, modernity, sus-tainability, the state, and local management. It was a movement with a long history of market involvement, of links to other political organizing in Garawhal, and with aspirations for regional autonomy. Tradition or custom hardly captured what is at stake in the definition of the community.

There are a number of implications that stem from this insight. First, and most obviously, the forms of community regulations and access to resources are invari-ably wrapped up with questions of identity. Second, these forms of identity (articu-lated in the name of custom and tradition) are not stable (their histories are often shal-low) and may be put to use (they are interpreted and contested) by particular constituencies with particular interests. Third, images of the community, whether ar-ticulated locally or nationally, can be put into service as a way of talking about, de-bating, and contesting various forms of property (and therefore of claims over con-trol and access). Fourth, to some extent, communities can be understood as differing fields of power—communities are internally differentiated in complex political, so-cial, and economic ways—and to that same extent we need to be sensitive to the in-ternal political forms of resource use or conservation (there may be three or four dif-ferent Chipkos or rubber-tapper movements within this purportedly community struggle). Fifth, communities are rarely corporate or isolated, which means that the fields of power are typically nonlocal in some way (ecotourism working through lo-cal chiefs, local elites in the pay of the state or local logging companies, and so on). And not least, the community—as an object of social scientific analysis or of practi-cal politics—has to be rendered politically; it needs to be understood in ethnographic terms as consisting of multiple and contradictory constituencies and alliances. This can be referred to as identifying "stakeholders"—a curiously anemic term—but of-ten what is at stake is something that comes close to class analysis, or, at the least, the identification of wildly different forms of political power and authority.

If I have been suggesting that the community is often taken for granted—it is a sort of black box (and this may be as much the case for state agencies and transna-tional NGOs, as well as academics)—this is often the case for community move-ments and resistance. Resistance is "ethnographically thin." There is precious little in the way of analysis of the political rough-and-tumble of actual resource extrac-tion or conservation that can answer the previously posed questions (who is doing what and why, and in whose interest). Here I might refer to the excellent work of Brosius (1997) and Li (1996, 1997a) in Indonesia; they conducted *comparative com-munity* work in two seemingly similar local communities to show (1) how the type and fact of resistance varied dramatically between two communities that were in many respects identical "cultural' communities, and (2) how these differences turn on a combination of contingent but nonetheless important historical events. Brosius, working among the Penan, found that the radical differences in resistance to logging companies between the two communities turned on their histories with respect to colonial forces, their internal social structure, their autonomy and closed corporate structure, and the role of transnational forces (environmentalists in particular). As Neumann shows in chapter 5, questions of memory are key, but they play out in unexpected ways.

The point is that some communities do not resist (which disappoints the foreign or local academic) and may not have any local knowledge of or interest in doing so. By the same token, as Zerner shows (1994a), local "traditions" can be discovered (not necessarily by the community, being often driven by academic work of local traditions drawn from elsewhere) that can be put to the service of the new political circumstances in which villages and states find themselves. Indeed, we know that some groups within communities are happy to take on board essentialism and wrongheaded "local traditions" pedaled by foreign activists or investors, to further local struggles (as Roderick Neumann shows in his account of Tanzanian pastoralists).

If the meanings and images of the community are something that has to be understood as a contested construction—highlighting the identification of who does the construction, how, and with what consequences—this is no less the case at the national and international levels. States the world over invoke *customary law*—in Africa this is an integral part of what Ribot in chapter 6 calls the "decentralized despotism" that characterizes the colonial inheritance of Indirect Rule. In fact, the current market triumphalism spoken by multilateral institutions has furthered the praise-singing of the community on the grounds that it furthers decentralization of the state and market-driven individualism. But this is also deceptive. Community indigenous law is often promoted subject to its ability to conform with national law (this is the case in the Philippines and Indonesia). Custom in West Africa may simply reinforce the powers of community chiefs who de facto become state bureaucrats (Mamdani 1995). For the like of the World Bank, the traditional community necessitates the discovery of local corporate groups who can assist in the proliferation of property titles and individual property holding. The community can be, from the vantage point of the state, a way of investing [as Sara Berry (1994) suggested in a local setting] in meanings rather than in the means of production.

Kingsbury (1998, in press) has shown beautifully how the contested nature of the community has its counterpoint in international law over the cover term *indigenous* (and one might as well add *tribe* or *ethnicity*). The United Nations, the International Labor Organization (ILO), and the World Bank have, as he shows, differing approaches to the definition of indigenous peoples—which has implications for normative and justificatory issues (see Brush 1996). The complexity of legal debate raised around the category is reflected in the vast panoply of national, international, and interstate institutional mechanisms deployed and the ongoing debates over the three key criteria of nondominance, special connections with land/territory, and continuity based on historical priority. These criteria obviously strike to the heart of the community debate that I have just outlined and carry the additional problems of the normative claims that stem from them (rights of indigenous peoples, rights of individual members of such groups, and the duties and obligations of states). Whatever the current institutional problems of dealing with the claims of nonstate groups at the international level [and there are knotty legal problems as Kingsbury (1994) demonstrates], the very fact of the complexity of issues surrounding "the indigenous community" makes for, at the very least, what Kingsbury calls "a flexible approach to definition," and at worst a litigious nightmare.

## INSTITUTIONS AND GOVERNANCE, RIGHTS AND REPRESENTATION

Many of the chapters in this book show clearly that a focus on institutions is a necessary starting point as a way of analytically linking socially differentiated communities with biologically differentiated environments. Institutions—understood not simply as the "rules of the game" but as the habituated and regularized "rules-in-use" maintained by human practice and investment performed over time—are typically distinguished from organizations, understood as actors or players brought together for a particular purpose. However, while it is clear in a number of the case studies that institutions figure large in discussions of community-based management and conservation, the character of these institutional practices and their relationship to both the nature of the environmental and extractive problems they are designed to resolve and the exercise of power (and therefore questions of representation and accountability) is the area of greatest deficiency. Quite specifically, while we have a number of studies that demonstrate that informal forms of institutional access and control constitute flexible and ecologically adaptive forms of resource extraction—for example, chapter 7 by Michon, de Foresta, Kusworo, and Levang, and chapter 8 by Fried—it is not at all clear (1) what the forms of representation, authority, and accountability are in such local, or community, forms of management, (2) how such forms of management intersect in complex ways with all manner of other formal and informal (and especially nonlocal) institutions that have consequences for the environment, and (3) how the institutional matrix of conservation or extraction at a multiplicity of institutional levels can be grasped as *forms of governance*.

One way to approach institutions and their character is through the deployment of Amartya Sen's (1980) theory of entitlement. Leach et al. (1997) have suggested that entitlements provide a way of linking what Sen calls "capabilities" with institutional design and performance. *Entitlement* refers to effective command over alternative commodity bundles that derive from a person's endowments. Environmental entitlements can then be seen to be the "sets of benefits derived from environmental goods and services over which people have *legitimate effective command* and which are instrumental in achieving well-being" (Leach et al. 1997:9, emphasis added). Environmental entitlements are thus a subset of a larger group of entitlements that collectively provide the means by which basic human needs are met and by which people can experience well-being. To pursue preferred forms of social life—what Sen calls freedom or well-being—presumes that persons can participate in the forms of life in which they find themselves. Basic human needs are those universal preconditions for a successful and critical participation in a form of social life—what Sen calls a capability set. In Sen's language, the freedom to lead different types of life is reflected in the person's capability set, which contains a number of functionings representing the various alternative combinations of beings and doings, any one combination of which a person can choose. Capability is a space of functionings that defines a person's state of being. Functionings represent parts of a state of a person—in particular, the various things that he or she manages to do or be in leading a life. The different functionings of a person will be the constituent elements of the person's being, seen from the perspective of his or her welfare. Functionings entail *procedural* and *material* preconditions for enhancing need satisfaction. The former relate to the ability of the persons to identify and appropriate needs satisfiers in a rational way and the availability of the means for political participation and claims-making, whereby people express their felt needs and dissatisfactions—which itself implies forms of democratic resolution. The latter refers to the capacity of economic systems to produce and deliver necessary and appropriate capabilities and to transform them into final need satisfaction.

All of this sounds very abstract, but what it highlights are the means by which differentiated social actors gain access to, and control over, resources through institutionalized practices. Sen starts with an endowment and a set of entitlements and ends up, via capability, with a series of political and economic preconditions for well-being in general—and environmental health and conservation in particular—to be achieved. Institutions are the social forms through which endowments, entitlements, and capability are linked. Institutional facts are the mutual understandings about social reality that can be stated in the form "$x$ counts as $y$ in context $z$" (Searle 1995). Institutions (formal or informal) are necessarily about shared meanings (often contested, often taken for granted) through which persons habituate themselves to the natural world. They are, however, categories and practices—forms of knowledge and power—which imply forms of governance—that is to say, the means by which social interaction (around resource extraction or conservation) is structured and brought to closure. Governance highlights mechanisms by which participation—and the exercise of power and authority—is secured. It signals not the shared meanings per se so much as the way that power is exercised, how persons participate in that exercise, and how accountable and representative that process is.

Much of the work on indigenous communities has properly emphasized the forms of what Jürgen Habermas (1987, 1994) calls social integration—the shared meanings (which is typically the cultural stuff of custom or tradition) by which institutions can function. Many of these sorts of community institutions are of course relatively informal (and may operate outside the confines of the state, if not of the market). Much of the ethnographic work on community management is entirely consistent with the work on social capital, trust, and conventions, which emphasizes the embedded intracommunity ties that generate trust, cooperation, and social networks (Evans 1996; Boltanski and Thevenot 1991). Other institutional configurations, especially those that operate transnationally, or that link community (informal) institutions with extracommunity (formal) organizations, may operate on the basis of what Habermas calls "system integration," which is to say, through disembedded mechanisms such as money or authority.

All of this suggests that there are, as Habermas, Offe (1985), and others have shown, distinctive forms of governance by which institutions operate—and each mechanism has quite different implications for what we might call representation, democracy, and accountability. These mechanisms—rational communication, influence, prestige, authority, and money—are typically mixed up and rely on shared meanings in quite different ways. Money and authority (typically associated with markets and states as institutions) are created by anchoring institutions such as property rights and bureaucratic rules. They are dependent on shared meanings but stand at a remove from, say, influence or prestige, which, as modes of governance, are rooted in networks of limited rather than rational (open and free) communication. In all of these configurations, the institutions that contain various rights are always shaped and constituted by struggles; that is to say, the rules of the game are always contested and negotiated in various ways.

What might all this mean for environmental extraction or conservation? Chapter 6 by Ribot, it seems to me, opens up a number of important avenues for analysis. He is of course examining essentially state institutional arrangements that shape access to and control of fuelwood. In his view, the state deploys law as a form of rural control; local appointed authorities backed by the state create something like Potemkin villages, in which there is nothing like local representation. Community participation is in fact disabled by forms of state intervention—and in his view, by the continuance of the colonial model of rule through "decentralized despotism" [the term is from Mamood Mamdani (1995)]. Ribot argues that participation without locally accountable representation is no participation at all. As he has put it (1998a:4), "When local structures have an iota of representativity no powers are devolved to them, and when local structures have powers they are not representative but rather centrally controlled." What passes in Mali or Niger or Senegal as community participation is circumscribed by the continuing power of chiefs backed by state powers, by the lack of open and free elections, and by the decentralized despotism of postcolonial regimes. In the case of institutions that involve state–community linkages then, it is influence and prestige, coupled with authority and money, that fundamentally frame the forms of governance—and hence who participates and who benefits.

A number of the chapters in this book invoke community institutions in their various forms that are largely outside the state (here, lineage or longhouse or community common property institutions are key) or institutions that involve complex overlapping between village and extralocal institutions (for example, ecotourism or the national efforts to include communities, researchers, and capital in plant genetic collection). However, understandings of the mechanisms of governance are often weak. On the one hand, the customary forms of resource use, while strong on knowledge or their declaration of common forms of property, are almost totally opaque on the questions of governance and how this relates to some form of political representation and accountability (by gender, age, lineage). On the other, the fact that local community regulation consists of complex and overlapping institutions (some of which may not be environmental per se but which have environmental consequences)—and hence contain nested and multiple forms of access and control embracing formal and informal practices—makes the analysis of governance, and its relation to democracy, all the more important.

Chapter 6 (see also Ribot 1998b) throws the question of access—or entitlements—into stark relief. Who participates in the benefits generated by particular environmental entitlements, and who participates in what decisions concerning functionings? He has raised, in short, the question of capability (to return to Sen), which is necessarily about the hard questions of how institutions are governed. In the case of the environment, the questions of governance speak not only to whether entitlements confer an equitable benefit, or how entitlements can be challenged or changed, but also to how forms of governance are themselves shaped by environmental scale, environmental certainty, and environmental grain or variation. It is here that the agency of the environment is of consequence for institutional design and for the appropriate of forms of governance [this is indicated by Peluso (1996) in her account of how the life cycle of trees shapes property form]. On both these fronts—the relations between environment and governance, and the relations between political representation and accountability and mechanisms of governance—much of our work is relatively silent. Indeed, it is surprising that so much of the work that rightly sings the praises of local knowledge or community/customary control has been seemingly uninterested in the forms of social power, the forms of bargaining, and the mechanisms of political representation that they contain.

Finally, the work of Peter Evans (1996) on social capital is especially relevant because his concern with what he calls *public–private synergies* speaks to the ways in which multiple institutions of control and access associated with the state and with civil society, operating at different scales and levels, can operate synergistically. It sharply poses the question of how public institutions can be coherent, can be credible, and can have organizational integrity, and how the institutions of civil society can engage in accountable ways with the public sphere. In the case of environmentalism, however, these public–private synergies cross cut international boundaries and pose sharp questions for both multilateral regulation and for transnational activism. The case of NAFTA is especially interesting because it highlights a number of processes that have yet to be fully understood. One is political in regard to the NGO community, since it is clear that the green activists in the United States were split (in complex ways) around NAFTA and the environment (see Dreiling 1997), which raises questions of the internal politics of the green oppositional movements (nationally) in relation to multilateral green regulation. Second, one of the consequences of NAFTA, however, has been that it has deepened the process of cross-border NGO organizing and activity in unprecedented ways. And, not last, it has highlighted the problems of building accountable and transparent green multilateral regulation. As of 1997, three petitions have been filed to test the procedural dimensions of NAFTA's environmental institutions, but they have been shown to be deeply flawed. As chapter 14 by Laird, Cunningham, and Lisinge shows, there are often *no* institutionalized frameworks to cater to many of the processes and actors at work in the world of intellectual property and drug or seed collection.

Transnational advocacy groups (TNGOs)—and transnational environmental organizations in particular—appear in a number of the case studies, but there remains much to be learned from their experiences (see Keck 1995; Keck and Sikkink 1998). Roderick Neumann's account of how TNGOs in Serengeti played a catalytic role in conjunction with the emerging indigenous pastoralist NGOs is an instructive case. However, the landscape is much more diverse and differentiated than this case would suggest. On the one hand, a number of the large TNGOs have themselves been shaped by the changing political and market-driven winds in the West, producing a sort of in-house corporate environmentalism ("green corporatism") within the larger TNGO community. This itself raises the questions of (1) how large TNGOs as major donors change the domestic politics and structure of the local NGO communities in the South, (2) how foreign and local NGOs actually build political strategy and alliances (and here chapter 4 by Hvalkof is much more salutary, referring to the "ambiguous" role of foreign NGOs that produce "pragmatic and short-lived alliances"), and (3) how social capital is constructed in North–South inter-NGO collaborations. Bailey's (1998) work on the activities of the World Wildlife Fund in Ecuador highlights the tensions between transnational and local NGO green activism—and that there is a necessary unity of interest between them. The significance of territorial control and demarcation by local communities—the property issue once again—in contradistinction to the international biodiversity canon, which rests on protected area practices, is especially vivid (see chapter 4).

## MARKETS, ENCLOSURE, JUSTICE

One of the consequences of the neoliberal counterrevolution of the 1980s, and the extraordinary degree to which capital movements have been facilitated by the global deregulation, is that "market solutions" to environmental problems are now legion. Economists may, for example, account for the existence of environmental "externalities" in terms of the absence of property rights—which is to say that the global commons under threat (the seas, stratospheric ozone depletion, and so on) require an *enclosure*. Indeed, the parallel with the English enclosures by which the commons were subject to what has been called the long theft of commoner rights is instructive because a number of the chapters in this book document the ways in which contemporary forms of enclosure operate (Bronwyn Parry's excursus on the corporatization of collections, for example, in chapter 15). But the key point is that, even from within the professional ranks of conventional economics and from within the political ranks of the free marketeers, there is an engagement with the idea of a green economics or a green conservatism. John Gray's book *Beyond the New Right*, for example, is built around the idea of why there is a "natural" confluence between high Tory and green ideas (counterposed of course to the devastating consequences of actually existing socialisms, such as they are, on the environment). Here, the green conservatism argument turns on a recognition that unrestricted capitalist growth cannot be sustained ecologically and that some form of stationary-state capitalism does require state intervention. However, as Gray makes clear, green conservatism must (1) promote the radical *deepening* of the benefits of ownership (enclosure), and (2) recognize the role of political decentralization (local environmental regulation) and the thickening of civil society as a necessary complement to the nation-state. I raise these issues to drive home the point that, while the greening of economics recognizes the problem of market-driven externalities, it nonetheless still insists on market- and ownership-based forms of valuation and regulation. What is strikingly absent, of course, is any discussion of social justice (a liberal "digression," as John Gray sees it) as opposed to the market concerns with efficiency and Pareto-optimality.

One way to think about the question of justice and the environment is derived from Karl Polanyi's book *The Great Transformation* (1947). His point is that it is precisely the way in which the market is disembedded (and, conversely, how social relations are embedded in the market, i.e., the commodification of the life world) that threatens land, labor, and the environment—and hence capitalism itself. Using the case of England, he charts how violent and exploitative was the process by which land, labor, and environment became commodified (the production of their "fictitious" qualities). In a sense, this process of disembedding land, environment, and labor—of wrenching them out of social contexts typically characterized by all manner of moral, social/communitarian, and spiritual relations—returns us once more to enclosure, to the long theft, and to matters of justice. Polanyi refers to the debates around these issues in the late eighteenth century as the "discovery of society." Perhaps the debates around the global commons speaks to the discovery of a global society.

In many of the case studies here and elsewhere, the local link between environment and social justice is clear and unambiguous. Neumann shows how protected areas are compromised by the historic theft and appropriation of local land rights, and Hvalkof details the role of historical memory in the construction of a popular consciousness in which the violence of the past, and the need for retribution, is a foundational principle for struggles over resources in the Upper Amazon. Indeed, much of the new political ecology and environmental history work stresses the point that indigenous communities have histories in which nonlocal forces have had a transformative impact on resources and environment. This incorporation into the market arrives in diverse forms (foreign companies, colonial states, missionaries) and with varying degrees of violence and political oppression. All of which is to say that the desire to maintain forms of livelihood, resource extraction, and conservation by communities is invariably shaped by the history of past exploitation as much as by the ostensibly new forms of market involvement. Access and control—the politics of resource use and conservation—seem to stand at the heart of the case studies presented here. They can be understood, to return to Polanyi, as reembedding the market or preventing this disembedding from taking place.

One way to think about environmental justice is to see the conflicts and movements documented in this volume as expressions of resistance to the ways in which environmental externalities are being internalized (if at all). Whether there is toxic dumping or compensation for petroleum-induced damage of rainforests, communities are contesting the fact that externalities are not being internalized by the perpetrators (companies or governments). This is what Martinez-Alier (1997) calls ecological distributional conflicts: the politics of who picks up the costs (how costs are shifted, in other words) of ecological damage. What remains less clear, however, is how environmental justice operates at the global as opposed to the local level. Part of the problem pertains to the identification of a single polluter in the case of, say, enhanced greenhouse effects (and hence an international polluter-pays principle or an international law of tort would not apply in a simple way). The other part pertains to the questions of how the economic values of externalities are determined in the context of property rights and the maldistribution of power and income. In other words, the process of the "internationalization of the internalization of externalities" (the awful language is taken from Martinez-Alier 1997) proves to be complex for a number of reasons: the fact that "the poor sell cheap" (the Larry Summers principle); the fact that transnational capital has enormous power and influence (both to resist legal challenges and to lobby in multilateral fora); and the fact that multilateral institutions are resistant to the claims of the South.

However, the focus on environmental justice (the equitable access to resources, and the equitable distribution of the burdens of pollution and the costs of ecological rehabilitation and conservation) is now both a local and a global issue. Furthermore, the two realms are often linked; for example, local community control can make an appearance in international legal settings over the definition of indigeneity and the nature of their normative claims. However, environmental justice must still confront the knotty problems of valuation (how should costs be calculated?) and ownership (should local genetic resources be rendered as private property?). But bringing equity considerations into the analysis of value and property does remind us that any form of money valuation depends on distribution and power. In this way, ecological distributional conflicts question sharply the centrality not of price and valuation as much as the role of those who act against externalities, and the political alliances and mobilizations constructed around this opposition.

## COERCION, VIOLENCE, AND SURVEILLANCE: CITIZENSHIP AND CONSERVATION

There is one final current running through *People, Plants, and Justice* that, if somewhat unevenly represented, is of enormous importance for theory, policy, and activism. The issue of coercion—and its relation to violence—has of course been raised by Peluso (1992a) in her account of the ways in which conservation policy is often attached to (and can be seen as part of) a long history of authoritarian state interventions rooted in colonial policies. Conservation as a set of ideologies and practices that, in the late twentieth century, appear through the agency of multilateral institutions and transnational environmental NGOs and advocacy groups can then become forms of legitimation for coercive and antidemocratic politics [the idea, for example, that saving nature ("survival") requires fighting fire with fire]. Insofar as the state figures centrally in the ways in which resources are regulated and managed, it is inevitable that coercive action produces friction at the local level in a multiplicity of forms. A numbers of the chapters in this volume chart these over a gamut of local strategies, from the everyday resistance and poaching described by Neumann (chapter 5) in his account of how the historical appropriations associated with national parks have been resisted, to the rebellions of Senegalese charcoal producers against state-licensed merchants (Ribot's chapter 6), to the bloody and murderous histories of rubber tappers and other resource extractors in the upper Amazon charted by Hvalkof (chapter 4).

In practice, of course, coercion extends beyond the state, and violence cannot be understood solely in terms of the local "moral" community versus the "fascist" state apparatus. Transnational mining or timber companies, local merchants or contractors, local elites who straddle private-sector business and the local state, the panoply of quite differentiated agents within the various apparatuses of the state machinery, and the vast array of NGOs, civics, and transnational advocacy groups make for a heady mix. Indeed it is precisely the complex alliances of actors around conservation and resource extraction that make the fields of power in which coercion and violence are located so intractable. Indeed, one of the most pressing tasks is to understand the contours of what is coercively undertaken in the name of conservation and to determine both the complex alliances (and fissures) and the forms of power and hegemony associated with each. Peluso (1998) has suggested that one can identify coercive exclusionary resource exploitation (for example, the slick alliance of oil companies and the Nigerian state that Saro-Wiwa referred to), coercive exclusionary conservation (alliances between states and multilateral green institutions or international conservation NGOs), and coercive inclusionary conservation (alliances between states, NGOs, industry, and trade organizations that require participation in green production schemes in conservation and development projects). All of this is very preliminary, of course, but it highlights the ways in which the broader politics of resource extraction and conservation must be rooted in the alliances, networks, and pacts in which a multiplicity of transnational, national, and local actors participate—to map these different configurations—and as a consequence to come to an appreciation of their coercive character (and implicitly their capacity for violence).

There are three issues that seem to me to be important here. The first has to do with the politics of conservation and in particular the ways in which nondemocratic forms of governance are legitimated. Here there is an interesting parallel with the discourses on population control and especially the distinction made by Amartya Sen

(1990) between "override" and "collaborative" approaches to family planning. The former turns on the ways in which population is constructed as a threat—a mortal threat to growth, equity and so on—that demands radical action to overcome the inevitable and deleterious consequences of demographic momentum. This produces the single-child policy in China or the authoritarian disincentive systems of Singapore. On the other hand, collaboration turns on the ways in which state and civil society provides conditions in which individuals can make reasoned choices (as in Kerala, where the conditions of female autonomy and security are secured) and where the technologies for such choices are widely available. This parallel draws attention to the ways in which conservation is constructed and legitimated, and also to the ways in which market incentives and disincentives can indeed have coercive qualities (i.e., there are a number of often subtle ways in which coercion can be expressed and institutionalized).

Second, there is the extent to which conservation policies are now employed (in the name of override) by the armory of new geographic information systems (GIS) and the means of surveillance to ensure compliance. Here again, the relatively low cost and flexibility of such technologies, coupled with the fact that they inevitably take the form of some sort of collusion with local militaries and militias, suggest that the powers of coercive conservation are becoming more pernicious under the watchful eyes of the multilateral institutions. There is very little of this in the chapters in this volume, but it is a crucial area for investigation, not only because these technologies are means for "better" compliance and policy implementation (which threaten civil and political liberties) but also because they are the means by which property rights, boundaries, and territories are determined in contexts of local cultural flexibility and juridical ambiguity. Not only is surveillance coercive but it is also the vehicle by which the late twentieth century enclosure movement advances. It is in this context that the work on countermapping by local communities assumes great importance not only as a form of resistance but also as a way in which the crude forms of the conservation panopticon can be confronted.

Finally, there is the question of violence, and how state or private forms of coercion mark a slide into something beyond individualized footdragging and subversion, that is to say into violent collection action. Here, the case of the Ogoni with which I began this chapter is instructive, but there are other perhaps less dramatic examples of communities confronting logging companies or local comprador classes. Violence typically takes the form of state violence against communities— although we need to recognize, and here Ogoni is again exemplary, that intra- or intercommunity violence, sometimes precipitated by state collusion, goes with the territory of coercive conservation—but this raises the question of why local and relatively isolated community actions can elicit such ferocity on the part of the state. Ogoni, after all, produced an insignificant amount of Nigerian oil by 1990 and Saro-Wiwa was a well-connected apparatchnik with connections to the military! So why would such a "green" movement represent such a threat to the military government? Of course, a part of the answer turns on the strategic attack on the slick alliance that

MOSOP articulated, but I think that the real power of the local here resides on the challenges to sovereignty, to prevailing senses of citizenship, indeed to the very idea of Nigeria itself (to radically decentralize, as Saro-Wiwa suggested, on the basis of ethnicity would effectively shatter the Nigerian federation). All of which is to say that violence needs to be located on the larger canvas of the sorts of political claims made by communities against coercive policies.

Once again, then, I return to the question of governance, since this necessarily runs through any account of coercion and violence. Conservation and resource extraction become arenas in which some very big fish are fried: rights, participation, nationality, citizenship. It is a strength of this volume that these issues are laid out with such clarity and prescience. And yet it remains the case that these very issues represent the most profound challenges for both social science and the conservation movement as a whole.

## NOTES

A version of this paper was presented to a workshop on "People, Plants, and Justice," Ford Foundation, New York, March 13, 1998.

1. According to the Nigerian government, in 1995 Ogoniland produced about 2 percent of the Nigerian oil output and was the fifth largest oil-producing community in Rivers State. Shell maintains that total Ogoni oil output is valued at $5.2 billion before costs!

2. Ogoniland itself suffered 111 spills between 1985 and 1994 (Hammer 1996:61). Figures provided by the Nigerian National Petroleum Company (NNPC) document 2,676 spills between 1976 and 1990, 59 percent of which occurred in Rivers State (Ikein 1990:171), 38 percent of which were due to equipment malfunction. Between 1982 and 1992, Shell alone accounted for 1.6 million gallons of spilled oil, 37 percent of the company's spills worldwide. The consequences of flaring, spillage, and waste for Ogoni fisheries and farming have been devastating. Two independent studies completed in 1997 reveal total petroleum hydrocarbons in Ogoni streams at 360 and 680 times the European Community permissible levels (Rainforest Action Network 1997).

3. Ogoniland consists of three local government areas and six clans, whose members speak different dialects of the Ogoni language. MOSOP is in this sense a pan-Ogoni organization.

4. As constitutional preparations were made for the transition to home rule, non-Igbo minorities throughout the Eastern Region appealed to the colonial government for a separate rivers state. Ogoni representatives lobbied the Willink Commission in 1958 to avert the threat of exclusion within an Igbo-dominated regional government that had assumed self-governing status in 1957, but minority claims were ignored (Okpu 1977; Okilo 1980).

5. The Ogoni and other minorities petitioned in 1974 for the creation of a new Port Harcourt State within the Rivers State boundary (Naanen 1995:63).

6. What Rivers State felt in regard to federal neglect, the Ogoni experienced in regard to Ijaw domination. While several Ogoni were influential federal and state politicians, they were incapable politically of exacting resources for the Ogoni community. In the 1980s, only six out of forty-two representatives in the state assembly were Ogoni (Naanen 1995:77). It needs to be said, however—and it is relevant for an understanding of state violence against

the Ogoni—that the Ogoni have fared *better* than many other minorities in terms of political appointments: in 1993, 30 percent of the commissioners in the Rivers State cabinet were Ogoni (the Ogoni represent 12 percent of the state population) and every clan has produced at least one federal or state minister (Osaghae 1995:331) since the civil war. In this sense, it is precisely because the Ogoni *had* produced since 1967 a cadre of influential and well-placed politicians (including Saro-Wiwa himself) that their decision to move aggressively toward self-determination and minority rights was so threatening to the Abacha regime (Welch 1995).

7. Prior to the cessation of operations in 1993, Shell was the principal oil company operating in Ogoniland, pumping from five major oil fields at Bomu/Dere, Yorla, Bodo West, Korokoro, and Ebubu.

8. In 1970, Ogoni representatives had already approached the Rivers State government to approach Shell—what they then called "a Shylock of a company"—for compensation and direct assistance [a plea that elicited a shockingly irresponsible response documented by Saro-Wiwa (1992)]. Compensation by the companies for land appropriation and for spillage has been minimal and is a constant source of tension between company and community. Shell, which was deemed the world's most profitable corporation in 1996 by *Business Week* (July 8, 1996:46) and which nets roughly $200 million profit from Nigeria each year, by its own admission has provided only $2 million to Ogoniland in forty years of pumping. Ogoni historian Loolo (1981) points out that Shell has built one road and awarded ninety-six school scholarships in thirty years; according to the *Wall Street Journal*, Shell employs eighty-eight Ogonis (less than 2 percent) in a workforce of over 5,000 Nigerian employees. Furthermore, the oft-cited community development schemes of the oil companies began in earnest only in the 1980s and have met with minimal success (Ikporukpo 1993). In some communities, Shell began community efforts only in 1992, after twenty-five years of pumping, and then by providing a water project with a capacity of 5,000 gallons for a constituency of 100,000! (*Newswatch,* December 18, 1995:13).

9. Following the MOSOP precedent, a number of southeastern minorities pressured local and state authorities for expanded resources and political autonomy: the Movement for the Survival of Izon/Ijaw Ethnic Nationality was established in 1994, the Council for Ekwerre Nationality in 1993, and the Southern Minorities Movement (twenty-eight ethnic groups from five delta states) has been active since 1992. The Movement for Reparation to Ogbia (MORETO) produced a charter explicitly modeled on the Ogoni Bill of Rights in 1992. These groups directly confronted Shell and Chevron installations (Human Rights Watch 1995; Greenpeace 1994) and in turn have felt the press of military violence over the last four years. The point is simply that MOSOP was a flagship movement for a vast number of oil-producing communities and threatened to ignite a blaze throughout the oil-producing delta.

10. Saro-Wiwa was often chastised by Gokana (he himself was Bane), since most of the Ogoni oil was in fact located below Gokana soil. In other words, on occasion the key territorial unit became the clan rather than the pan-Ogoni territory.

11. Escobar (1995) sees poverty as invented and globalized with the creation of a battery of transnational "welfare" institutions at Bretton Woods (in New Hampshire) and in San Francisco following the signing of the United Nations charter. The discourse of national and international planning and development agencies was able to constitute a reality "by the way it was able to form systematically the objects of which it spoke, to group and arrange them in certain ways, and to give them a unity" (p. 40). Patriarchy, ethnocentrism, gender, race,

and nationality were embraced in this discourse (pp. 43–44) at the same time that economists were privileged within its ranks. This rule-governed system (p. 154) has remained unchanged at the level of practice, although the discursive formations have been unstable. In all of this, modernity's "objectifying regime of visuality" (p. 155) turned people of the South into "spectacles," and the "panoptic gaze" of development became an apparatus of social control (pp. 155–156). Development is constructed in large part through keywords—"toxic words"—which really mean something else: "planning" normalizes people; "resources" desacralize nature; "poverty" is an invention; "science" is violence; "basic needs" are cyborgs, and so forth (Sachs 1992).

12. "The greatest political promise for minority cultures is their potential for resisting and subverting the axiomatics of capitalism and modernity in their hegemonic forms" (Escobar 1995:224). Escobar, however, has little to say about what constitutes minority (Hindu nationalism? the Islamic Salvation Front in Algeria?) or what indeed is nonhegemonic modernity or capitalism.

# CHAPTER 2

## Beyond Distributive Justice: Resource Extraction and Environmental Justice in the Tropics

*Richard A. Schroeder*

For several years, Frank Momberg, Rajindra Puri, and Timothy Jessup, three self-described "conservationists" closely affiliated with the World Wildlife Fund (WWF), have been involved in research connected with the creation and maintenance of Kayan Mentarang National Park in East Kalimantan, Indonesia. It is thus with considerable alarm that they report in this volume (chapter 10) that a resource extraction rush is underway in Kalimantan. Large numbers of "exogenous" collectors have surreptitiously infiltrated remote sections of Kalimantan in the vicinity of Kayan Mentarang in search of gaharu, a resinous heartwood used in the manufacture of incense and medicinals throughout Southeast Asia.

Gaharu is contained only in certain species, and, indeed, within select individual trees of the genus *Aquilaria*. As Momberg et al. describe it, local residents, who rely on gaharu to meet a variety of personal needs, have tended until recently to be quite selective in their harvesting techniques, extracting gaharu from resin-bearing trees and leaving other Aquilaria specimens unmolested. By contrast, "outsiders" entering the area are seen as a dual threat. A sharp increase in gaharu prices (and the lack of economic alternatives) has led these groups to adopt indiscriminate collecting practices, felling all *Aquilaria* trees they encounter in a frenzied rush to capture rich bounties. At the same time, the outsiders represent competition for local collectors. Many residents of the area have accordingly abandoned comparatively benign extraction techniques, joining the rush to harvest dwindling supplies of gaharu before they are lost for good.

In response to what they see as a rapidly deteriorating situation, Momberg et al. offer what seems a reasonable suggestion. They propose strengthening tenure security for local groups, and providing carefully controlled access to gaharu sources on park lands. The ability to exclude outside groups would, in their view, restore the economic incentives for local residents to resume more sustainable harvesting techniques. While the restoration of tenure security would appear to offer a means of addressing a complex set of resource management problems, closer inspection reveals the environmental justice concepts underpinning the proposal of Momberg et al. to be quite narrowly construed. Momberg et al. are at pains to explain that they support greater local control over forest resources "not only because of the economic incentives for conservation it seems to offer, but also because . . . any attempt at conservation based on the unfair taking of people's property without due compensation is bound to fail" (chapter 10, page 000). They support local tenure claims less because they feel that local groups are *entitled* to extract a livelihood from their forests than because the alternative represented by the gaharu rush poses dramatically greater risks to conservationist interests.

Where Momberg et al. do engage connections between people, plants, and justice, their proposals stress the importance of economic incentives in helping to promote sound environmental management among Kalimantan residents, an approach widely shared by many in the conservationist community. Indeed, the gaharu rush has taken place against the backdrop of a dramatic proliferation of approaches to environmental management premised on intensified commodity production. Proponents of such disparate environmental policies and practices as debt swaps and environmental conditionalities, buffer zones, and extractive reserves, bioprospecting, and ecotourism ventures have all embraced the idea that the way to save nature is to buy and/or sell it (McAfee 1999). From this market-centered perspective, the demands of justice are most directly served by distributing the economic benefits of conservation more evenly.

Expressed in this way, the goals of distributive justice seem straightforward and unimpeachable. I argue in this chapter, however, that the simplicity of this formula is misleading, and that there is a need to examine the political dimensions of distributive justice remedies more carefully. Steps toward the redistribution of wealth may well be warranted in many (or even most) circumstances, but the evidence collected in this volume suggests that a purely distributive approach may actually be counterproductive. Rather than resolve justice issues, such an approach could work to *protect* powerful and wealthy interests, and to prevent more radical alternatives from being realized.

I begin by reviewing arguments set forth in several chapters here, concerning the difficulties of operationalizing distributive justice goals. My contention is that many of these difficulties stem from a theory of justice that is insufficiently robust to overcome countervailing political pressures. I suggest that a stronger sense of justice could be based on the notion of compensation for (1) the forfeiture of rights; (2) the loss of physical property (e.g., land); (3) the sharing of intellectual property (local knowledge); or (4) the provision of labor services; that is, compensation would be given on the basis of specific entitlement claims set forth by wronged parties. Significantly, each of these claims implies a dimension of economic justice that goes beyond the simple sharing of wealth and profits. In the second part of the chapter, I review contributions to the volume that seek to move beyond the distributive justice paradigm altogether, toward a consideration of alternative conceptions of justice more directly centered on questions of rights to livelihood, the abuse of state power vis-à-vis local property claims, reparations for environmental damages, the establishment of durable democratic political processes, and recognition of differences in cultural values. I argue, with Parry (chapter 15), that these issues cannot be resolved by simply "rejigging the distributive mechanism" but must instead be addressed on their own terms. In the concluding section, I argue that the widespread use of distributive economic incentives by conservationist and corporate interests threatens to obscure a range of critical justice considerations that do not lend themselves to easy reconciliation by distributive means. I acknowledge that there is considerable scope for strengthening existing approaches to distributive justice by linking them to clearly defined entitlement claims, but I maintain that core entitlements surrounding such issues as land, labor, and livelihood are not easily bought off and cannot be summarily dispensed with if the interests of justice are to be fully served.

## DILEMMAS OF DISTRIBUTIVE JUSTICE: A PHARMACOPOEIA OF PROBLEMS

The intensification of concern over the justice issues raised by resource extraction derives in part from the recent groundswell of interest in bioprospecting. The search for biological and genetic materials suitable for commercial development has been led by pharmaceutical companies seeking to locate and develop new drugs for use in the treatment of disease. Questions of distributive justice arise from the recognition that the firms responsible for these activities sometimes earn tremendous profits, and that most of their prospecting efforts are conducted in parts of the world where resident populations are desperately poor. In such a context, the idea that some sort of monetary or in-kind compensation is in order if the demands of distributive justice are to be served is most compelling, and resident groups and relevant branches of local government have sought to make this case wherever possible.

Typically, distributive mechanisms involve royalty payments or taxes of some sort, which are either funneled into community development programs or into state coffers, from which they are theoretically disbursed in efforts to offset conservation costs (see a more detailed discussion in chapter 14 by Laird, Cunningham, and Lisinge). Both potential outlets suit the interests of conservation proponents. Their argument is that wherever the value of forests, marine resources, and other "natural" landscapes can be demonstrated to resident populations, the prospect for saving them increases proportionately. Consequently, conservationists have become deeply involved in brokering agreements between the bioprospecting industries and the traditional occupants of given territories in the hopes that economic incentives will work to further conservation objectives (see chapter 13 by Cox).

Neither the justice implications of these agreements nor their environmental outcomes are self-evident, however. If the contributors to this book share any insight at all from their work in a wide variety of research contexts, it is that the apparently simple political demands of distributive justice are tremendously difficult to realize in practice. With this general caution in mind, it is worth reviewing the problems posed by demands for distributive justice more systematically.

## Plants and Borders

Successfully meeting the demands of distributive justice hinges first and foremost on determining the geographical distribution of the plant or resource in question. One problem would-be benefits administrators often face is that plants and other resources are rarely confined to a single geographical unit. This should not present a problem in and of itself, since, theoretically, benefits from commercial development of a given resource could be distributed to *all* areas of origin. It has, however, become a problem in the context of a market-oriented political economy premised on multisourcing of production sites. For under a global economic regime, there is no guarantee that the country that "originally supplied the material will be the one resupplying it in quantity for commercial production" (Parry, chapter 15). What is lacking in circumstances where the people in a given locality do not hold monopoly control over a resource is, thus, not the legitimacy to claim a share of benefits, but the economic *leverage* necessary to wrest favorable terms of extraction from recalcitrant industries. Under such conditions, the prospects for voluntary and equitable sharing of benefits are minimized. No leverage, no justice.

## Provenance

A closely related problem growing out of recent advances in genetic and cell culture technologies is that of actually determining site of origin. The production processes into which the extraction of plant specimens feeds no longer depend on the use of bulk quantities of material. Instead, as Parry demonstrates in her brilliant chapter, they may rely on the constituent parts of the original specimens, which are broken up and recombined over and over again until the essence of the source material is effectively lost. Alternatively, the key commodity is not the physical specimen at all, but the information contained in its genetic makeup, which scientists employ in efforts to synthesize useful compounds. Indeed, Parry argues that in some cases, "prospecting" for marketable biological and genetic compounds can now be done more profitably within existing collections than in the relatively inaccessible and uncontrolled natural settings where the specimens originated. These developments have made tracing biological and genetic materials up and down the production chain next to impossible, and frequently render the question of distributive justice moot. The practice of "re-mining" existing collections suggests that any leverage people in the areas of origin of key compounds might have had over decisions regarding the sharing of economic benefits may have been lost long ago. Furthermore, long delays in developing "commercial" uses for particular compounds break the connection many activists and developers have sought to establish between conservation and development, and undermine the efficacy of distributive benefits mechanisms in producing desired environmental outcomes altogether.

## The Proprietary State

Where the origins of resources can be reliably ascertained, advocates of distributive justice face the equally difficult problem of determining who is qualified to receive benefits. Many approaches rely on existing political or territorial authorities with jurisdictions covering various scales—states, regions, districts, communities/villages—to guide the distribution of benefits. Depending on how this determination is made, however, the assignation of benefits can result in different levels of political jurisdiction vying with one another for control over economic dividends. The case study by Laird et al. (chapter 14) of Cameroon, where a potential anti–human immunodeficiency virus (HIV) compound captured a great deal of international interest in the late 1980s, is a case in point. Rather than let the national university control funds generated by bioprospecting partners, the most centralized of state organs—the office of the prime minister—intervened to wrest control of negotiations away from university officials. Rationalized in the name of the "national good," such steps can often be shown to be motivated by simple greed on the part of corrupt officials. In practical terms, they may produce confusion, delays, and intragovernmental factional fighting, all of which can discourage further prospecting efforts and promote the development of alternative sources—in other locations, or via synthetic means—more firmly under corporate control (for a description of efforts along these lines, see chapter 12 by Cunningham and Cunningham).

## Elite Beneficiaries

The problem of powerful political and economic interests monopolizing distributive benefits also plays itself out at the community scale. Local elites, such as the managers of the Belizean ecotourism preserves described by Belsky (chapter 11), often end up controlling the "community" benefits generated by exploitation of local resources. Belsky describes the efforts of wildlife biologists Robert Horwich and Jon Lyon to promote "community conservation" in the Gales Point area of Belize, a coastal zone best known for its population of endangered manatees. While proponents of the proposed biosphere reserve represented the project as a "win-win situation"—they claimed the goals and objectives of conservationists would be met even as local residents reaped economic benefits—Belsky demonstrates that the actual implementation of the project had a dramatically different outcome. Rather than enhancing community cohesion, the project produced considerable intracommunity tension as the benefits from the creation of tourism infrastructure in Gales Point quickly became concentrated in the hands of local elite. Among many other ironic injustices, Belsky shows how those most dependent on the natural resources packaged and sold by ecoentrepreneurs—hunters, fisherfolk, and trappers—were in fact least likely to benefit from the new tour packages promoted in the area. In many respects, then, the felicitous phrase, "win-win," applied here to ecotourism, but frequently used to tout the multiple benefits of bioprospecting arrangements, debt swaps, and various other sorts of "conservation-with-development" programs, simply means that the dominant parties engaged in promoting new resource management schemes "win" repeatedly, both going into negotiations and coming out.

## Establishing Residence and Indigeneity

A further problem with the implementation of distributive justice involves determining who is a legitimate resident of a given area and thus entitled to share benefits. Questionable claims of residency may surface in circumstances where the benefits distributed are substantial, e.g., in the case of funds generated via hunting fees and tourism revenues under Zimbabwe's CAMPFIRE program (Metcalfe 1994). The notion of "indigeneity" is sometimes employed as a guideline to help prioritize recipients, but this concept often raises as many problems as it solves. As Laird et al. note, the concept of "indigenous rights" is especially fraught in many of the areas richest in biodiversity, notably forest areas with complex settlement histories and patterns of resource stewardship in force. The definition of residence and/or the claim of indigenous identity can also be difficult to establish wherever patterns of economic migration take certain community members out of the locality on a seasonal or interannual basis (compare Giles-Vernick 1999). One solution promoted by Laird et al. is to avoid defining the term *indigenous* too narrowly. For example, "The Amazonian concept of undisturbed autochthonous groups makes little sense in West Africa where most forest areas have a long and complex history of settlement and re-settlement" (Laird et al., p. 000; Slater's description of the 23 million inhabitants of Amazonia problematizes the use of the term *indigenous* in even that setting, however—see chapter 3). Thus, at a minimum, any claim to economic benefits based on residential or indigenous status must account for historical and geographical demographic changes if it is to be applied justly.

## Distribution Along the Commodity Chain

Several contributors to this volume structure their essays around the analysis of particular commodity chains. At the center of these chapters stand complex, vertically integrated production, processing, and distribution systems that connect remote localities to centers of economic and political power on regional and global scales. Ribot's case (chapter 6), for example, details how benefits derived from the charcoal production industry in Senegal inhere less in formal property rights over localized land resources than in the control over market outlets, transportation networks, labor resources, and, most particularly, the license and permitting processes organized through the state (see also Ribot 1998b). These are the pressure points at which profits are squeezed out of the forests of central Senegal, and around which, ultimately, the most salient questions of distributive justice frequently arise. Rarely, if ever, do these political economic dimensions of resource extraction figure in deliberations over policies related to the sharing of economic benefits.

## *Historical Geographies of Distributive Injustice*

Laird et al. note the way Cameroonian foresters once used the term *forest reserve* to designate sites "reserved" for logging concessionaires. Subsequent efforts to create reserves for conservation purposes have accordingly been firmly resisted by Cameroonian peasants, who continue to associate the term with unjust land allocation policies. The critical question in this case is, thus, whether past experiences have not already and ineluctably predisposed specific groups of resource users to resist contemporary extraction efforts. As Laird et al. put it, "Is there a good chance that royalties, and up-front benefits like schools and roads and health clinics, will be more forthcoming, or better applied to long-term conservation and development objectives, than those previously or currently supplied by timber, mining, and oil companies?" The historical record shows how groups of peasants, pastoralists, forest dwellers, and fisherfolk have been repeatedly stripped of their localized knowledge, conscripted into *corvée* labor gangs organized for resource extraction purposes, and duped by promises of compensation. People sharing such experiences are "unlikely to be fooled by old ideas dressed up" as responsible and equitable benefit-sharing measures, "even if outside policymakers will be" (Laird et al., chapter 14, p. 000).

   In retrospect, it seems apparent that the difficulties associated with attempts to distribute the economic benefits of resource extraction often derive from the fact that the underlying concepts of justice employed by relevant policymakers are insufficiently robust to withstand political pressure. Several of the distributive justice mechanisms listed previously, for example, are based on accidents of geography: specific groups were deemed qualified to receive benefits simply because they happened to reside somewhere in the vicinity of economically desirable resources. Several more fundamental questions need to be addressed if the interests of justice are to be served.

   The case of the rosy periwinkle plant (see discussion in Laird et al., chapter 14) illustrates my argument perfectly. Located in several tropical countries, including the Philippines, Jamaica, and Madagascar, the rosy periwinkle plant has been collected and processed for its medicinal properties in the treatment of childhood leukemia. Madagascar has taken the lead in providing periwinkle to the international market and claims the lion's share of the benefits being distributed following its commercialization, yet the plant's medicinal purposes were better known much earlier in the Philippines and Jamaica. The question thus arises: Should (the people of) Madagascar collect the majority of benefits derived from the development of periwinkle when its beneficial properties were first discovered elsewhere? Underlying this question is the recognition that different theories of justice apply to each set of national interests. The case for Madagascar seems to hinge on an accident of geography, as just described, whereas the case for the Philippines or Jamaica rests on the notion of compensation for intellectual labor performed in developing relevant medical treatments (compare the case of *Prunus africana* described by Cunningham and Cunningham, chapter 12). The latter is arguably a more robust theory of justice.

The case studies included in this book would suggest that we can extrapolate from this example to a variety of other situations. Some form of compensation, monetary or otherwise, would seem warranted in cases where specific groups of resource users have agreed to forfeit use rights, relocate out of extraction zones (or preservation areas), or provide labor in support of the extraction effort. In each of these cases, compensation would be based on a specific entitlement claim, rather than on the abstract argument that wealth should be distributed more evenly. The likelihood that those who bore the brunt of a resource extraction effort, or otherwise contributed to that effort's success, would actually receive a share of benefits would increase proportionately.

It is worth noting at this juncture, however, that the redistribution of economic benefits takes place only with the acquiescence of groups or individuals who control the wealth in question. Thus, restricting the demands of justice to distributive claims may in fact do little to address the fundamental disparities of power embedded in a given situation. Indeed, the limits to distributive justice are especially clear in cases where, instead of originating with the demands of aggrieved parties, proposals for benefits distribution emerge as part of self-serving public relations efforts orchestrated by corporate officers or conservation officials. In such circumstances, distributive action is not meant to serve the interests of justice at all but is directed instead at buying off opposition to corporate and/or conservation policies. These strategies may be highly effective—the recipients of corporate or conservation largesse may (temporarily) comply with policy prescriptions—but they should not be confused with efforts to serve the demands of social justice.

## BEYOND DISTRIBUTIVE JUSTICE

I turn now to a series of case studies that do not fall under the distributive justice rubric. Unlike the position taken by Momberg et al. (chapter 10), who approach social justice as a means of meeting conservation objectives, or that advanced by distributive justice proponents, who attempt to rectify problems involving the skewed distribution of wealth and resources, several contributors to this volume feature rights claims that are incommensurable with any form of cash or in-kind compensation. These studies involve claimants who cannot be "made whole," in legal terms, through compensatory remedies. They are, accordingly, groups that are not easily bought off.

## *Rights to Livelihoods*

Momberg et al. have analyzed the potential outcomes in the Kayan Mentarang case in terms of their significance for conservation goals. To get a sense of how differently the Kayan Mentarang case might be framed, it is instructive to compare Momberg et al. with two companion pieces, from Indonesia by Michon et al. (chapter 7) and Fried (chapter 8). These chapters, which deal with the cultivation of damar, another source of fragrant resins, and rattan forests, respectively, foreground political conflicts that pit rural livelihood seekers against the extractive propensities of the Indonesian state and private logging concessionaires. In each of these cases, forest-based production systems critical to the well-being of Indonesia's rural populations and essential to the political task of (re)appropriating forest spaces have been left unmapped by state officials. Invisible on the landscape to the (sometimes willfully) untrained eye, these anthropogenic forests are perpetually at risk of being logged over or cleared for more "rational" uses such as plantations.

In the interests of helping rattan and damar foresters bolster their tenure claims, both Fried and Michon et al. describe how peasant groups manage their forests over different temporal scales to meet interseasonal, interannual, and intergenerational economic needs. The authors also show how damar and rattan production efforts are part and parcel of much more elaborate agroforestry and forestry systems, production regimes that are as valuable for features that "will never be valued as commodities" (Michon et al., chapter 7, p. 000)—the creation of habitat for other forest plant and animal species, the stabilizing of soil resources—as for their degree of integration into commodity markets.

In providing their respective profiles of vibrant and vital forest management systems, Michon et al. and Fried attempt to place the Bentian Dayaks and the damar agroforesters back on the map. While both agree with Momberg et al. that justice concerns of damar and rattan cultivators, and others who derive economic benefit from the extraction of nontraditional forest products, include the need for state recognition of basic property and resource use rights, they justify their positions in starkly different terms. Fried and Michon et al. insist that the value in protecting such rights lies in helping to meet the multiple social needs of rural forest communities. Thus, their primary concern is not simply that serious damage might occur to the resource base if local land use practices are not protected, as Momberg et al. fear, but that the rights of access, histories of occupation, and social formations organized around the broad distribution of forest benefits might otherwise be erased from the political ecological landscape altogether. The differences in approach are fundamental.

## *Dispossession and Criminalization*

One of the issues Momberg et al. take for granted is the establishment of Kayan Mentarang National Park. All of their solutions to the gaharu rush are premised on the park's continued existence, and on controlling access to resources contained within it. Neumann's contribution (chapter 5) focuses our attention on the other side of the people and parks divide, introducing a set of disputes that pit rural community groups against the interests of the conservation elite, including actors within state governments and the major international conservation agencies. The rural groups in question are primarily Maasai pastoralists located on rangelands along Tanzania's northern border with Kenya, who have been dispossessed of their rights to land and other resources in the interests of establishing parks and protected areas. Neumann's approach moves directly to denaturalize both the laws governing park creation and regulations pertaining to resource use and access in surrounding areas. His historical argument explores the "conflicting and contradictory ideas about African relationships to nature" that helped shape colonial land law and eventually gave rise in 1959 to Tanzania's national park ordinance, which set a critical precedent in designating "areas where all human rights must be excluded" (chapter 5, p. 000).

Against this backdrop, Neumann concedes that Tanzanian pastoralists have long engaged in illegal hunting, grazing trespass, and fuelwood theft from national parks and game reserves but argues that these actions are often misrepresented and misunderstood: "Standard explanations of park law violations often focus on overpopulation and ignorance of conservation values, thereby emptying them of their political content. To the contrary, park and forest violations may often be highly politicized acts of resistance to and protest of protected area policies" (Neumann, chapter 5, p. 000). What is at issue, in other words, is the very act of criminalization itself, its scope and the selectivity of its targets. By definitional fiat, the Tanzanian state has negated the long occupational history of the pastoralist groups from their rangelands, a step various groups, including the newly formed Maasai and pastoralist nongovernmental organizations (NGOs) Neumann describes, have continued to resist with all the means available to them.

Lowe's description of the live fish trade in the Togean Islands (chapter 9) provides an interesting parallel to Neumann's analysis. The live fish trade is increasingly controlled by the wealthy owners of heavily capitalized fish camps who supply regional markets in Singapore and Hong Kong. Fish camp managers provide local subcontractors with diving equipment and cyanide, which fishermen (mostly younger men and boys) use to indiscriminately stun all fish in a given reef location and then selectively capture species prized by elite diners in wealthy Asian capitals. The effect has been to destroy many reefs, disrupt the supply of fish stocks of all sorts, and invalidate local knowledge concerning "species, currents, locations, equipments, and baits," i.e., the "reservoir of natural history and naturalist thinking" (Lowe, chapter 9, p. 000) that underpins artisanal fishing in the area. As Lowe demonstrates, however, the central role of fish camp managers in these destructive processes has been lost in the efforts to regulate live fish capture. Most legislation designed to control cyanide fishing applies only to the fishermen themselves, leaving the managers of the system to ply their trade without constraint. Consequently, whole groups of fishermen linked by ethnicity rather than by any similarity in fishing practices have been targeted for extortion by police engaged in enforcement efforts.

## Environmental Reparations

Hvalkof draws our attention to Ecuador, where the livelihoods of several native groups have been seriously affected by industrial waste generated by the oil industry (chapter 4). He documents a legal case mounted on behalf of these groups against the Texaco corporation with respect to that corporation's long history of "negligence, recklessness, and intentional misconduct" in connection with oil extraction in the area. A kind of industrial sludge known as "production water" generated in the process of oil pumping has had particularly damaging environmental consequences, destroying the fish populations in many of the area's rivers and contributing to high rates of morbidity among human residents (compare Watts's description of the Ogoni struggle in Nigeria, chapter 1). This attempt to establish corporate liability through the courts and to win damage payments for injured parties is one of a handful of cases worldwide that eschew distributive justice mechanisms in favor of some form of retribution for corporate crimes.

## Enfranchisement and Due Political Process

As part of his analysis of the charcoal commodity chain in Senegal, Ribot (chapter 6) provides extensive detail on the political structures controlling charcoal production. Most charcoal originating from central Senegal is produced by migrant laborers, and subsequent sales are controlled by urban-based woodfuel merchants. The implication of such extensive nonlocal controls is that the landholding rights of local residents have been reduced to a nominal status, i.e., locals maintain *access* to forests, but lack the effective *control* necessary to convert their landholding rights into profitable livelihoods. One of the key constraints on the establishment of greater local control is a permitting and quota-granting process organized by the Forest Service. This process is thoroughly dominated by charcoal merchants and village chiefs. While the latter would seem to be in a position to advocate on behalf of their constituents, the chiefs often act to undermine those they represent. As Ribot argues, this circumstance derives from the fact that the appointment of local-level chiefs often takes place outside legitimately democratic channels. Thus, residents of forest communities remain disenfranchised. Ribot argues that "local" political processes do not necessarily produce just economic outcomes if "effective enfranchisement" is lacking. In his words, "neither representation without power nor power without representation" can resolve these difficulties (Ribot, chapter 6, p. 000).

## Cultural Concerns

In his detailed two-part morality tale concerning attempts to establish bioprospecting covenants with Samoan villagers, Cox (chapter 13) argues that "Western" conservationist priorities focused on ecosystem protection, legal process, and intellectual property rights are unduly myopic. In his view, unless conservationists pay considerably greater attention to more culturally informed notions of justice ("respect for traditional leaders, human dignity, and observance of cultural forms"), the prospects for striking successful resource-use covenants are quite dim. While in general, Cox tends to romanticize "indigenous" groups and essentialize "Western" arguments (his chapter fails to acknowledge the influence of political ecologists and environmental justice advocates on debates concerning bioprospecting, for example), his core argument that religious concerns and questions of fundamental human dignity are all too often swept away in political dealings concerning access to valuable plant and animal resources applies to many resource extraction efforts.

The foregoing list covers a disparate set of issues. These concerns originate from different political perspectives and may differ in their prospects for bringing about social justice. What they have in common, however, is their irreducibility. These are claims that typically cannot be converted through any sort of economic calculus into some measure of distributive economic benefits. They are in some sense too essential to the livelihood practices or identity claims of groups residing in target areas to dispense with in such summary fashion. Indeed, the issues are so fundamental that the target groups often defend them fiercely.

Taken together, they remind us that the focus on distributive justice that features so prominently in the policies and practices of many environmental organizations and corporations engaged in resource extraction addresses but one dimension of justice concerns. Viewed individually, each issue claims priority over distributive concerns, either as the basis for an entitlement claim raised in determining distributive outcomes, or as a separate and at times countervailing issue entirely. The alternative justice concerns cannot, in any case, be ignored by advocates seeking to link the interests of people, plants, and justice.

## CONCLUSION

The justice considerations wrapped up in the distribution of economic benefits to local groups whose livelihoods are affected by resource extraction are complex. On one level, the distribution of cash or in-kind benefits to impoverished groups residing in the vicinity of resource extraction zones may issue a long overdue corrective to profligate and destructive attempts to extract wealth from these territories. And the redistribution of wealth under these circumstances holds out the prospect, at least, of helping resident groups meet a range of critical economic and social needs. In this light, distributive justice mechanisms are not only compatible with, but must be seen as essential to, the demands of social justice.

On another level, however, the distribution of economic benefits to localized groups appears as little more than a condescending gesture on the part of powerful political and economic interests designed to manipulate public sentiment and quell popular dissent. Benefits packages are wholly out of proportion with the wealth generated by the extraction efforts in question, and they bear little connection to the losses—displacement, dispossession, insult, injury, harassment, and outright terror in some instances—suffered by aggrieved parties. This brand of "trinkets and beads" justice (Hvalkof, chapter 4) primarily serves public relations purposes, or it is cynically employed to summarily dispense with justice obligations and facilitate other causes.

What is clear from the foregoing is that an *exclusive* emphasis on distributive justice mechanisms overlooks, and potentially undermines, a wide range of fundamental claims to resources, place-based identities, and livelihoods. There is also ample precedent for benefit programs to lose their compensatory character and thus fail to serve the demands of justice altogether. In such cases, when the distribution of benefits is used to gain leverage over local groups rather than to generate genuine political legitimacy for conservation activities, prospects for forging critical links between people, plants, and justice are forfeited. Conservation groups that mistake the gesture of benefit sharing for genuine engagement with the demands of social justice accordingly pay the price, directly or indirectly, of continued resistance from the resident groups whose resource management practices they seek to influence.

# PART II

## On Location: Case Studies

# CHAPTER 3

## Justice for Whom? Contemporary Images of Amazonia

*Candace Slater*

An immense, primordial forest teeming with birds and snakes and jaguars. A sky-blue river with a half-built dam ringed by a ragged circle of protesters. Indians in feather halos who lock arms with sun-burnt rubber tappers to protect a glistening wall of trees. These are the images that many middle-class Americans have today of the Amazon, but they were not always so. Representations of the sort that regularly appear today in U.S. newspapers, television documentaries, and movies, and on calendars, T-shirts, and cereal boxes have changed significantly over the past three decades, and even more over the past century. In the following pages, I outline some of these changes, while continuing to argue that a large number of contemporary portrayals remain rooted in far older notions of Amazonia—by which I mean all of that approximately 3.5 million square mile territory within which the Amazon River and its various tributaries lie (figure 3-1). The word "image" refers here not just to graphic descriptions but to figures of speech and, by extension, to recurring ideas.

Because issues of social and ecological justice are so bound up in, and so dependent on, particular notions of this region presently home to countless plants and animals, as well as to well over 20 million people, these images matter. So do both the real differences between past and present representations of Amazonia and the equally real persistence of much older ideas about the Amazon as a realm of nature whose precise definition varies over space and within time.[1] Chief among these recurring ideas are those of a vast store of natural wealth that remains inaccessible for the moment but which nonetheless promises an extraordinary transformative potential that demands action from without and, often, from above. I will return to this loose, and always shifting, constellation that I call "El Dorado" as this discussion progresses.

Although I focus on a number of those depictions of the Amazon most common today in the United States, these enjoy far wider circulation, often resurfacing in

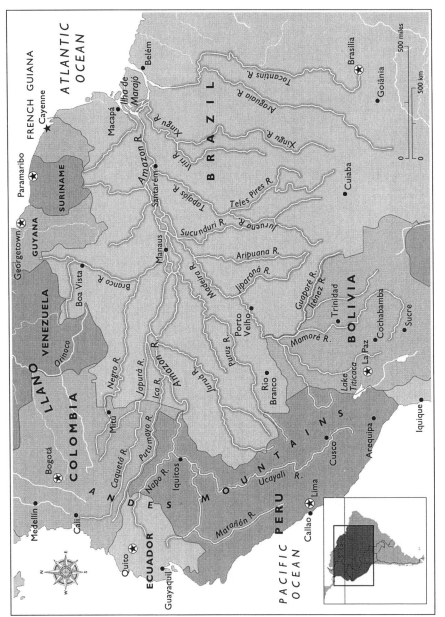

**Figure 3-1** The Amazon Region

official policy statements and academic writing. Then, too, while the concern for Amazonia may take different guises in the United States and in Peru or Brazil or Venezuela, similar images often crop up in representations of this region from the seven (or nine, depending on how one counts them) nations within which the Amazon River and its various tributaries lie.[2]

Many of the portrayals to which I refer have a readily documentable physical basis. There are very definitely Indians, rubber tappers, and any number of snakes, palms, vines, and fishes in Amazonia. My point here is not that many of the most compelling contemporary images of the Amazon are, in themselves, false, but rather that they remain limited and dangerously exclusive. Clearly, much depends on who is using a particular image toward what ends. Many choices reflect conscious political or commercial objectives. Not infrequently, however, the most apparently neutral descriptions (the term *Amazon rain forest*, for example) encode larger assumptions about the human relationship to a variously defined nature, and about the outside world to Amazonia, that their users not only do not necessarily intend, but which they might summarily reject were these presented in explicit terms.

## DEPICTIONS OF AMAZONIA IN THE 1970S AND 1980S

For much of the twentieth century, and, indeed, much of its history, the Amazon appears in popular portrayals as a nightmarish jungle of the sort that lives on in comic book adventures of voluptuous warrior women armed with bows and arrows (or, from time to time, machine guns) and mass market movies such as *Anaconda*, in which an enormous snake gulps down whole boatloads full of people. Home to jaw-snapping caimans and blood-thirsty piranhas, this jungle predates, and still excludes, humans, since the Indians who make their home there are usually as ferocious as the piranhas. Although its wild and unknown quality inspires fascination, it remains intensely dangerous—a vast, supremely alien space.

At the same time that this turn-of-the-century Amazon is a precise geographical location, it is also an unabashed province of the imagination. Arthur Conan Doyle's (1995) *The Lost World* and Jules Verne's (1952) *Eight Hundred Leagues on the Amazon*[3] are just two examples of adventure stories that describe a tangled realm of nature within which its authors feel no need to tread. Certainly, their readers make no demands for any sort of firsthand testimony. If the dinosaurs who cross paths with Conan Doyle's Professor Challenger do not actually roam about the Amazon, they are nonetheless entirely true to perceptions of the region's monstrous immensity and its propensity to stun, crush, and, above all, gobble up the unwary traveler. *Lost World* confirms that the jungle remains prehistoric in the sense of an obstinate refusal to be defined in human terms.

The late 1960s and the 1970s witness important changes in the sorts of representations of the Amazon that reach a general public.[4] A burgeoning environmental movement in the United States and other industrialized countries, together with new government development policies in Brazil and elsewhere that trigger a wave of burning and destruction, do much to convert the terrifyingly chaotic jungle into a fragile, marvelously complex rain forest.[5] No longer prehistoric, but breathtakingly primordial, this forest is not a "lost," so much as a virginal, world. Instead of fear or the determination to subdue its hostile disorder, it invites reverent wonder and assiduous preservation.

The threatened Eden that replaces Green Hell demands not just care or wonder, but also detailed and dispassionate study.[6] A botanical term that first begins appearing in dictionaries at the very end of the nineteenth century, *rain forest* has a distinctly quantitative, scientific side. Technically, the name refers to any woodland that receives a hundred or more inches of rain a year. The wooded area that gets 99.9 inches of precipitation is thus not a rain forest. Neither is the craggy boulder upon which endless storms descend.

The scientific quality of the Rain Forest (I use capitals to signal an imaginative conception that may have little to do with an actual place) only enhances its tremendous appeal. By no coincidence, *rain forest* enters popular usage in the early 1970s at precisely the same moment as both *environment* and *diversity*. Rain forests are nothing if not biologically diverse. So too, of course, are jungles, but their rife disorder yields here to a dazzling abundance that seems to belie grim accounts of species destruction and environmental degradation. Where unabashedly fictional professors once surveyed the tracks of equally fictional dinosaurs, real-life scientists now pour for days over a single ant heap or scrutinize the undersides of leaves with the assistance of a raft of finely calibrated instruments. Ostensibly focused on practical solutions to all-too-concrete problems, they do much to instill a sense of the Rain Forest as an answer to deep-seated fears about the planet's future.

The generalized switch from inward-spiraling jungle to cornucopia-like rain forest greatly broadens the range of plants and animals deemed worthy of attention.[7] The insects formerly dismissed as one more of the jungle's scourges, or the tree frogs once too tiny to prompt notice, now—at least in theory—become as worthy of notice as brilliant butterflies or howler monkeys. Not just too distant, but suddenly, too vulnerable to instill fear, jaguars and caimans find their way into the boxes of animal crackers that children cart off in school lunches. Although beetles and fungi remain conspicuously absent from endangered species T-shirts, even they acquire a new importance, if not as potential cancer cures, then as one more part of an intricately interdependent tissue of creation.

At the same time, the once savage human inhabitants of the forest begin emerging as the bearers of ancient cultures with a mystic, timeless aura. No longer primitives in desperate need of civilization, they become the keepers of unchanging secrets that hold out an antidote to a long list of present problems.[8]

Where they once inspired curiosity or horror, the Amazonian natives now prompt nostalgia for that golden age in which human beings are seen to live in harmony with nature. Not infrequently, the reports that describe the first encounter between previously uncontacted tribes and anthropologists or various government officials compare the Indians to the hippies then roaming the streets of San Francisco and Sao Paulo.[9] In their proclaimed desire to return to nature and the pleasures of a simpler era, the members of the so-called counterculture appear to share a series of common beliefs and values with these "primitive" peoples.

To the extent that all the components of this new and eminently attractive Rain Forest invite meticulous attention, it is not so different from the Amazonia of the nineteenth-century naturalists. Unlike its predecessors, however, this new incarnation is extremely fragile. Calls to "Save the Rain Forest" underscore the vulnerability of the dense jade curtain of trees that begins appearing in newspaper headlines and TV documentaries, on packages of bubble bath and candy, and on posters, buttons, and a variety of other promotional materials produced by environmental organizations.

Fragile in its own right, the finely balanced forest that emerges in the 1970s reveals profound connections to a larger world. The burning of trees in distant Amazonia suddenly becomes the cause of droughts that wither roses in U.S. backyards. Although scientists will quarrel about the extent, the effects, and even the existence of the so-called greenhouse effect, the theme of an interconnected planet finds strong popular acceptance. Not just a rain forest, Amazonia emerges as *the* rain forest and, as such, a symbol for all of endangered nature.

Differences between the turn of the century and the 1970s are particularly obvious in comparisons between an author such as Theodore Roosevelt in his *Travels Through the Brazilian Wilderness* (1914) and later writers such as Alex Shoumatoff, whose *The Rivers Amazon* first appears in 1978.[10] Although Shoumatoff looks directly to the nineteenth-century British naturalist Henry Walter Bates for inspiration (his title consciously echoes Bates's *A Naturalist on the River Amazons*), his book invites comparison and contrast with Roosevelt as much, or more, than it does with Bates.[11]

Essentially journalistic writers who aim at a more or less general U.S. readership, Roosevelt and Shoumatoff share a lively interest in Amazonian nature. Their travelogues, however, are less scientific reports than adventure stories, portions of which first appear in major literary magazines. (Roosevelt publishes excerpts in *Scribner's Magazine*; Shoumatoff, in *The New Yorker*.) Both men go to the Amazon on self-appointed fact-finding missions. And both are very definitely in search of something.

Despite these immediate resemblances, however, Roosevelt and Shoumatoff are very different people. One is a fifty-five-year-old U.S. president who lights out for what he calls "the last frontier" after loosing his bid for a third term in office; the other is a twenty-nine-year-old freelance writer when he sets out for Amazonia.[12] Yet more important, not only do the two men occupy distinct social positions but they live in very different times. For instance, in contrast to Roosevelt, who envisions an Amazonia crisscrossed by railroad tracks and telegraph lines, Shoumatoff expresses alarm at the rapid pace of change. If Roosevelt finds the natives to be savage, but fascinating, Shoumatoff finds them fascinating and life back in the United States savage. ("I would see if life as it is presented in the United States is the way it has to be, or if there are alternatives," the latter says hopefully.)[13] Roosevelt describes a threatening nature that demands the utmost of the "true wilderness explorer"; Shoumatoff, a threatened nature that demands protection.[14]

Perhaps most significant, Roosevelt appears to find what he is seeking. Not only does the expedition succeed in pinpointing the previously unknown source of the River of Doubt, but this river is rebaptized on the spot in his own name. Shoumatoff, for his part, goes home largely empty handed in his quest for an alternative to the very civilization whose advent Roosevelt so extols. While the one concludes his account with an affirmation of "astonishing progress," the other ends with irony and self-deprecation.[15] ("So how was Africa?" demands a newspaper vendor upon catching sight of the newly returned traveler.)[16]

## REPRESENTATIONS OF AMAZONIA IN THE 1990S

The differences between the late 1990s and the early 1970s are by no means as dramatic as the divergences between the end of the twentieth century and its beginning. And yet, while the central image of the Amazon remains a sea of green, this leafy realm undergoes a number of important transformations. Perhaps most obvious, the wondrous Rain Forest becomes an increasingly familiar marketing tool. At the same time, this once largely empty space suddenly becomes peopled. Not only does the cast of residents expand to include a number of non-Indians, but these forest dwellers begin speaking for themselves.

The marketing of Amazonian nature is immediately evident in the emergence of a series of rain forest products that cater to consumers who regard themselves as environmentally aware. Not by coincidence, these products are often sweet not just in terms of taste (ice cream, cereal, cookies, candy bars, and juice drinks), but also of sound ("Rainforest Melodies," "Sounds from the Rainforest") and smell (aromatic teas, Amazon potpourris, rain forest shampoos, body gels, perfumes, room fresheners, and, rather improbably, "sea salts.") Along with this new wave of actual commodities, a growing number of environmental organizations (Rainforest Action Network, Rainforest Alliance) adopt the rain forest name.

The shift in spelling from "rain forest" to "rainforest" that occurs in the late 1970s and 1980s almost certainly reflects its expanded currency as an idea. An excellent example of what Pierre Bourdieu has called "symbolic capital," the Amazon rain forest becomes newly valuable not simply for the goods that it produces but for the positive associations on which various enterprises attempt to capitalize.[17] As English adjectives tend to be a single word, the elision marks its growing use as an adjective. Names such as "Rainforest Crunch," "Rainforest Sounds," and terms like "our rainforest policy" confirm the passage from ideas of the forest as a place, to the forest as a state of mind. ("Raw Vanilla," reads the label on the perfume bottle, "it's like meeting in the rain forest. The fresh organic power of the rainforest captured in a fragrance for men.")[18]

We have seen that early Rain Forest representations often dwell exclusively on the flora and fauna, and the image of a wall of trees remains widespread in 1990. All half-dozen flavors of the Mistic Rain Forest juice drink, for instance, sport labels featuring various combinations of parrots, monkeys, butterflies, and great green leaves. These already familiar portraits of a primeval nature untouched by human beings, however, mingle with a growing spate of other representations of Amazonia and the Amazonian rain forest in which men, women, and children play a newly active role.

Archaeological research of the 1980s and 1990s that unearths evidence of advanced civilizations in an area once considered the exclusive province of simple nomads finds repercussions in a wider sphere.[19] Demographic studies suggesting the existence of once-elevated populations in what today looks like "virgin forest" have a similar effect.[20] In addition, and most important, the growth of grassroots movements in Amazonian countries creates a new focus on the human presence within nature.[21]

The "peopling" of the rain forest is obvious in posters and children's book illustrations, such as one in which an enormous tree full of parrots, monkeys, and snakes grows out of the left foot of a tiny Indian. It is equally clear in Rain Forest mobiles in which a copper-colored boy with bow and feathered arrow dangles beside jaguars, butterflies, and orchids. Although these sorts of images start cropping up in the 1970s, they become yet more prevalent as time goes on.[22]

This growing presence of people in popular depictions of the Amazon is yet more evident in newspaper reports that describe the impassioned protests of Indians before the World Bank and the U.S. or Brazilian Congress. Accompanying photographs of Amazonian natives arrayed in war paint, both for the real and more symbolic battle against intruders, leave no doubt that the forest is somebody's home.

Moreover, while Indians retain a particular fascination for outsiders, they are not the only people to appear in contemporary representations of Amazonia. Beginning in the late 1980s, a number of other inhabitants of the region start to make a more or less regular appearance. The initial image of a rubber tapper in the movie version of *At Play in the Fields of the Lord*, although no such image exists in the novel, bespeaks a newly close association between the Amazon and an expanded cast of forest dwellers.[23]

The rise to international prominence of Chico Mendes, the head of the Amazonian rubber tappers' union, and Mendes's subsequent murder by a local rancher a year later, definitively marks the beginning of a new era in a larger world's conception of the region.[24] The political alliance of a number of so-called *povos da floresta* or "forest peoples" finds echoes in images of Indians and rubber tappers standing shoulder to shoulder against a long line of bulldozers.

The Indians who appear on a world stage in the 1970s are largely voiceless beings. Sometimes, they speak another, more telluric and yet distinctly foreign language that boasts five different words for wind.[25] Often, however, these natives say absolutely nothing. The fact that others must do the talking for them reinforces their close relation to the likewise mute and needy plants and animals among whom they appear.[26]

Although the tradition of speaking for Indians has by no means entirely disappeared, the forest dwellers of contemporary representations are nonetheless considerably more likely than those of the past to express themselves directly.[27] The existence of a widely circulated Rubber Tappers' newsletter is just one example of Amazonians' growing insistence on speaking in their own voice.[28] The increasingly common presence of representatives of grassroots groups at meetings sponsored by various national and international environmentalist and human rights organizations signals a new political reality that cannot help but influence popular conceptions of Amazonia.

## THE LEGACY OF EL DORADO

The genuine changes associated with the last three decades ensure that the Rain Forest of the 1970s and its contemporary counterpart are by no means identical. We have seen that a yet greater gulf divides the deep, dark jungle of the beginning of this century from present-day images of an Amazonia that serves as a home to humans as well as other living creatures.

And yet, while these changes are real and unquestionably important, it would be dangerous to overlook ties to the past. For one thing, older ideas of both the chaotic jungle and the uninhabited Rain Forest live on in popular imaginings (and at least some news reports, policy statements, and academic studies). For another, even the peopled forest can be seen, in many ways, as one more variant on a far older and immensely powerful El Dorado theme.

By El Dorado I mean less the fabled city of the gilded king ardently sought by the first European explorers than a recurring, if decidedly fluid, nexus of ideas about natural wealth, momentary inaccessibility, and an intense transformative potential that find early and particularly vivid expression in accounts of this golden realm.[29] El Dorado, the place, thus implies, from the beginning, a broader vision of nonhuman nature that invites, if not demands, human domination. Frequently the site of the original Golden Kingdom, Amazonia—unlike most other parts of the Americas—successfully repels waves of intruders for many centuries. As a result, the region continues to embody a fabulous, if resistant, nature perceived as inviting or demanding action from without and above.

The ideas associated with the golden city of El Dorado change over time and take on distinct forms in different epochs, with particular elements of this open constellation assuming varying importance at particular historical junctures. (The idea of transformative potential, for instance, is particularly obvious in ideas of Amazonia as a "land without men for men without land,"[30] while notions of inaccessibility weigh heavily in portraits of an enormous swamp.) In the case of the first explorers, "natural wealth" takes on the guise of the gleaming treasures presided over by a ruler so rich that he himself becomes a precious object, thanks to the golden talc with which he coats his skin. (*El Dorado* actually means "The Gilded One," thereby implying an uneasy fusion of a rich land and its people in the figure of the golden king.) This gold's transformative potential resides in its ability to be recast into an infinity of other forms. Unfortunately for those often penniless adventurers who dream of how the precious metal will convert their hard existence into a life of leisure, a variety of obstacles block the road to El Dorado. Chief among these is that group of fierce women warriors believed to be the Amazons who first appear within Greek myth.[31]

Later, in the nineteenth century, Amazonia's wealth ceases to take the form of gold, and becomes instead a far more diffuse, sublimely abundant nature whose secrets the European naturalists toil to catalogue and thereby comprehend. The results of their painstaking labors will one day allow them to wed European civilization to tropical nature, thereby giving birth to "the perfect race of the future."[32] For the moment, however, their quest remains frustrated, in good part by present-day Amazonians' lack of "higher principles" and their accompanying refusal to utilize their all-important capacity for reason.[33]

At the beginning of the twentieth century, this same, sublime nature becomes the jungle I have already described. At the same time that this distinctly un-Yosemite-like wilderness invites development, it serves as a last bastion of prehistoric life forms. In line with this vision of the jungle as at once off-putting and alluring in its cache of ancient secrets, Roosevelt presents the Nambikwara Indians both as untrustworthy children in desperate need of guidance (he describes with obvious disapproval how an old woman sits upon a spoon in an attempt to steal it) and as nothing less than "Adam and Eve before the Fall."[34] For him and his contemporaries, the region's transformative potential lies in its ability to confirm the march of human progress. While awaiting the advent of Western civilization, this last frontier transforms the man brave enough to face its wild beasts, poisonous insects, and torrential waterfalls from a mere adventurer into a true "wilderness-winner."[35]

Genuinely distinct from past representations in the various ways I have already mentioned, the Amazonia of the 1990s nonetheless suggests its own sort of El Dorado. The Rain Forest whose value lies not just in plants, animals, and minerals of immediate economic utility, but also, and above all, in its marvelous biodiversity, is nothing if not a collection of natural wealth. A last refuge for life forms that have become extinct elsewhere, this forest shores up an older world's abundance. By reaffirming through its mere existence the original human bond with nature, it offers a way back into a richer, more harmonious past.

The El Doradian Rain Forest's transformative potential assumes many guises. Among these are gold to pay the interest on a burgeoning national debt, and minerals that can fuel worldwide production, and tiny insects in whose bodies reside cures for ills not yet discovered. Probably the most important of these present dreams of what Amazonian nature could be, however, resides in that bevy of competing ideas and objectives subsumed under the term *sustainable development*.[36] In most cases an uneasy fusion of economic and environmental interests, the term most commonly denotes a series of pragmatic measures intended to utilize natural resources without destroying them (Lélé 1991). At the same time, invocations of "sustainable development" frequently signal a much broader, almost religious vision in which Amazonia appears not just as a place or even a string of problems, but rather as a model for a different sort of world (McCormick 1989). Now an ideal laboratory, now a model for socialist ecology, it may as easily be a "green cathedral" as a "splendid blueprint for the planet's future."[37]

In accord with these varying definitions of the Rain Forest's transformative potential, the inaccessibility of its riches may lie not just in ill-conceived colonization policies, too many or too few tax incentives, or the lack of scientific research, but in less tangible reasons. Some of the most commonly cited of these are the selfishness of the industrialized nations toward their less developed counterparts (a frequent theme in countries with Amazonian territories) and the presence within the region of ill-intentioned and destructive people.

## THE IMPLICATIONS OF EL DORADO

My claim for the existence of an ongoing set of narrative expectations that resurface in different, often partial ways in superficially very different stories about the Amazon raises an obvious question. If that geographic place called Amazonia admittedly possesses a vast array of hard-to-get-at natural riches that suggest a variety of uses, why should the recurring presence of related images constitute any sort of problem? After all, the Amazon, with its plethora of species and of human cultures, *is* fabulously rich.

I began this discussion by suggesting that the trouble with a number of current images of Amazonia lies not so much in what they include, but rather in what they exclude or distort. The underlying assumption of a sharp division between the human and the nonhuman fuels the former's domination of the latter.[38] By the same token, the idea of Amazonia as a realm of nature that demands action from the outside reinscribes much older patterns of domination and subordination that create particular difficulties in regard to issues of social and ecological justice.

The stress on natural *wealth* that lies at the heart of many different sorts of El Dorados, for instance, makes for the exclusion of other, less spectacular land formations in favor of the Rain Forest. Although Amazonia is composed of flood plains, brush, and savannah as well as woodlands, the richly diverse forest tends to crowd out these other sorts of spaces. While specialists are eminently aware of Amazonia's complex geography, the popular (and, often, scholarly) fixation on the Rain Forest cannot help but encourage a devaluation of that larger whole within which Amazonia's forests play just one, albeit important, part.

The same insistence on nature's treasures inspires portrayals of a not just rich, but radiant woods that offers a striking contrast to the jungle's murky gloom. Although the thick canopy of actual rain forests make them dark and—to the untutored eye—often monotonous places, the Rain Forest of calendars and posters is more often than not a light-struck wonderland. An outright fabrication, this dazzling forest's luminosity becomes a promise of redemption from environmental ills.[39]

The concern for *natural* wealth explains the still very strong equation of the Amazon, not just with the region's forests but with all of tropical nature.[40] The stress on biological, geographic, and economic factors greatly overshadows any sense of the region as a historical, political, and cultural entity. As a result, even with the growing presence of people within portrayals of the forest, the human presence remains distinctly circumscribed. Most non-Amazonians are surprised—and rarely pleased—to learn that the region is currently home to many millions of men, women, and children.

Moreover, although the range of human groups who appear in depictions of the region has definitely increased, there remains a heavy emphasis on "natural people" who continue to form part of the flora and the fauna. Chief among these are the Indians and rubber tappers who, together, account for no more than about 5 percent of the present Amazonian population.[41] Although the political and cultural importance of these groups extends far beyond their actual numbers, to concentrate solely on them means to exclude the overwhelming majority of the people who presently live in Amazonia, as well as to ignore the larger social universe within which these smaller groups exist. Moreover, much like "Rain Forest," the apparently unitary "Indian" is actually an umbrella term that includes many different sorts of people with distinct languages, separate cultures, and often varied aspirations.

Over half the Amazonian population now makes its home in cities, ranging from small towns of 3,000 to metropolises of 1.5 million persons.[42] Fixed divisions between "rural" and "urban" do violence, however, both to cities ringed by forest and to the large number of "commuters" who move regularly between the city and outlying countryside. Because, for instance, the factories of Manaus's free trade zone prefer to hire women, who are more apt to work for lower wages, the men often remain in the interior, regularly bringing fish and produce to their families. It is common for men and women to live for several months or years in a city before returning to the often tiny communities where they were born.

If the El Doradian insistence on nature and natural peoples leaves little place for cities except as sources of problems, it also tends to exclude various groups who have played an important role in Amazonia's history. These include Arabs, Japanese, Sephardic Jews, and, above all, persons of African descent.[43] Although the descendants of black slaves comprise a very large percentage of the population in some parts of the Amazon (the Brazilian state of Pará, the Guianas, and the northern coast of Colombia, for instance), their imported status makes them seem less a part of nature, and thus less truly Amazonian. When blacks appear in representations of the region, they tend to be described as "traditional" communities who have long made their homes in the rain forest.

We have seen that some Amazonians are far more wont to speak for themselves today than they were in the past. And yet, if they occupy a new space in the pages of U.S. newspapers, the language they speak is almost always Environmentalese. Although groups such as the rubber tappers and various Indian communities have proved adept at using this vocabulary for their own ends, they nonetheless must invoke nature to be heard.[44] When they cease to appear as guardians of nature or endangered species, they are rejected as unauthentic, or pointedly ignored. Once media darlings, the Kayapó, for instance, cease to be "real Indians" as soon as they begin to sell the rights to trees and minerals on their lands. While one may not like the tribe's present actions, the idea that they are somehow less genuinely Indian when they start displaying what appears to others as environmentally suspect behavior reaffirms the still less-than-human status of native peoples.[45]

The El Doradian need for explanations of the continuing inaccessibility of Amazonia's riches sometimes identifies authentic problems such as the existence of short-sighted policies and inadequate research on a host of issues. It also, however, encourages scapegoating. Time and again, representations of the region pit natural against unnatural peoples, defenders against destroyers, and the entirely good against the very bad. There is little room for mixed motives—and, thus, the great majority of human actions—in pictures of saintly forest guardians and pyromaniacal intruders hell-bent on torching every tree in sight.

A particularly clear example of this sort of scapegoating can be seen in various presentations of Amazonian gold miners, or *garimpeiros*, in the U.S. and international press. Although miners do pollute streams and rivers with mercury, spread malaria and other communicable diseases, and, sometimes, attack Indians, headlines such as "Deep in the Heart of Amazonia, Ragtag Miners Rape the Virgin Forest" obscure the fact that most are very ordinary people who are often victims of a lack of other opportunities.[46] Carpenters and fishermen, taxi drivers and jobless migrants, the great majority of *garimpeiros* try their luck from time to time in the gold camps before returning home. Although some men (the so-called *peones rodados*) move from site to site over a period of years, they comprise a relatively small percentage of the approximately one million miners in Amazonia today. Moreover, while a good number of *garimpeiros* are from outside the region, many others are longtime residents who may include Indians and other "forest peoples."

The El Doradian insistence on transformative potential means that Amazonia is never sufficient unto itself. Thus, although the ideas and motives surrounding programs of sustainable development are often complex and varied, they often reveal a familiar insistence on the region as a source of answers for the outside world. As the "noble cause that will unite all humanity," Amazonia becomes a panhuman project that must be entrusted to those most able in every sense to see it through.[47]

Even impassioned calls to preserve the Rain Forest—at first glance, overt rejections of El Doradian exploitation—may impose protection as a form of control. No longer a hostile, harpy-like forest from which wealth is to be systematically extracted, Amazonia becomes a vulnerable, virgin (and thus, passive, if still female) territory to be jealously preserved from those who might have other visions and other uses. "Save Our Rainforests" say the bumper stickers on a crowded highway. "Your nickel buys 500 acres of Amazonian rainforest," proclaims the label of a brightly colored Rain Forest meter, implying that this ownerless expanse can and should be "sold" to distant guardians. Conveniently far from home and the complexities that inevitably cloud domestic issues, this Amazonia becomes an answer to needs that may—or may not—coincide with the desires of those who actually live within its borders.

What then should be done about El Dorado if its presence is so pernicious? Get rid of any hint of anything remotely resembling a rain forest? Stop showing footage of Indians on television? Ban glistening trees from calendars and keep parrots off cereal boxes?

These suggestions are intentionally absurd. The problem is emphatically not Indians, rain forests, or parrots, but rather the unexamined ideas that underlie particular depictions of them. The problem with existing images resides within the persistent notion of Amazonia—and Amazonian peoples—as an object for outsiders' use. And the problem, despite real progress on some fronts, is a continuing absence of competing images, other ideas, and genuinely different voices that would force prevailing notions into the open and make them more subject to discussion and debate.

Sometimes, to be sure, simplification is necessary. And yet, if pared-down and largely predictable versions of complex realities may be desirable and even essential in certain cases, there also has to be a time and place for fuller, more multifaceted accounts. That time is now. The alternative to the radiant Rain Forest is not mud, misery, and mosquitoes but a yet more fascinating and compelling drama that involves forests and cities, plants and animals and human beings. While some may argue that the general public has no interest in anything beyond virgin nature, it is hard to imagine safeguarding the future of the great trees and jaguars without a clear sense not just of the world within which these presently exist, but also of the all-too-human reason that their fate seems so important.

What is needed? The answer, at least on a very general level, is obvious. We—by which I mean both Amazonians and non-Amazonians—need to render visible the recurring tale or tales of El Dorado, both by actively identifying their varied forms in different times and different places, and by consciously seeking out these other images and voices. We need a more vivid awareness of the varied interactions between people and plants (be these acres of wild orchids or strings of automobile factories), as well as of the ways in which particular facets of the region play off against each other and the world beyond them. Above all, we need justice, but before we ask what is right and fair in the case of Amazonia, we need to ask, "Justice for whom and in whose terms?"

## ACKNOWLEDGMENT

I thank the Rockefeller Foundation for supporting the preparation of this chapter at its Bellagio Study Center. The chapter is part of a larger study, *Entangled Edens: Images of Amazonia* (in press).

## NOTES

1. Population figures vary widely, depending in part on what percentage of Brazil is included in the Amazon. The U.N.-commissioned report *Amazonia Without Myths* (United Nations 1992) gives a figure of 21,411,000 for the seven watershed countries, with an additional 442,000 for the "Greater Amazon." The total figure is almost certainly greater today.

2. The actual watershed countries are Bolivia, Brazil, Colombia, Ecuador, Guyana, Peru, and Venezuela, but Suriname and French Guiana are also often included among the Amazonian nations. Brazil is by far the largest Amazonian nation in terms of both territory (just over two-thirds of the total territory) and population (17 of the region's total of at least 22 million people).

3. The 1952 reedition of the Verne book contains a preface by Theodore Roosevelt.

4. For a discussion of the different connotations of the key words *wilderness*, *jungle*, and *rain forest*, in relation to Amazonia, see Slater (1995:114–131).

5. For a description of these policies, see Mahar (1979, 1988).

6. The term *Green Hell* finds early expression in Rangel (1914).

7. Colombian novelist José Eustacio Rivera's classic description of the Amazon as jungle bears the telling title *La Voragine*, or "*The Vortex.*"

8. One of the earliest expressions of the Amazon as an antidote to civilization is Claude Lévi-Strauss's *Tristes Tropiques* (1992). First published in 1955, the book recounts the author's field experiences of the 1930s.

9. A spate of articles from 1972 and 1973 that appear in *Veja*, the Brazilian equivalent of *Time* or *Newsweek*, include frequent comparisons of the Amazonian Indians to the hippies. See, for instance, the interview with Cláudio Villas Boas (1973).

10. See Roosevelt (1994b) and Shoumatoff (1986).

11. See Bates (1988). The book first appeared in 1863; the most recent edition includes an introduction by Alex Shoumatoff.

12. Roosevelt (1994b:333).

13. Shoumatoff (1986:vii).

14. "Such a man, the real pioneer, must have no strong desire for social life and no need, probably no knowledge, of any luxury, or of any comfort save of the most elementary kind" (Roosevelt 1994b:333).

15. "Everywhere, there was growth and development. The change since the days when Bates and Wallace came to this then poor and utterly primitive region is marvelous" (Roosevelt 1994b:345).

16. Shoumatoff (1986:216).

17. For a fuller explanation of the concept of symbolic capital, see Bourdieu (1993).

18. Copyright Coty Fragrances, 1996. Ads for this men's cologne have appeared in various general circulation magazines including *Sports Illustrated*.

19. For an overview of some of these developments, see Anna Roosevelt (1994a).

20. Present estimates of the pre-Contact Amazonian population range from 6 to 15 million persons.

21. For a summary of the emergence of these movements in Brazil, see Hecht and Cockburn (1990:161–191).

22. All these items were available at The Nature Company, a California-based nature boutique, in January 1996.

23. See Matthiessen (1965). The 1992 movie version is by the Saul Zaentz Company (Universal City, Universal Release).

24. For a review of a number of the publications about Mendes that appeared after his death, see Maxwell (1991). One of this multitude of books is Shoumatoff (1990).

25. Shoumatoff (1986:161) notes that the Aika people possess five different words for *wind*.

26. Donna Haraway (1992) addresses the question of "who speaks for the jaguar."

27. See Ramos (1988).

28. The *Boletim do Conselho Nacional dos Seringueiros* is published on a regular basis.

29. For a discussion of the original El Dorado, see Hemming (1978). See also de Holanda (1985).

30. "A land without men for men without land" was a slogan of the Brazilian government's colonization drive in the early 1970s. The idea was to relieve demographic pressure, and potential social upheaval, in the heavily populated and increasingly mechanized south by giving Amazonian lands to colonists from other parts of the country. For one account of this plan's rousing failure, see Smith (1982).

31. The Amazons are regularly associated with great stores of material wealth. In fact, they are said to weep tears of pure silver. (See Weckmann (1996). Explorers' contracts often contained clauses enjoining them to seek out these female warriors.

32. "The superiority of the bleak north to tropical regions, however, is only in their social aspect; for I hold to the opinion that, although humanity can reach an advanced state of culture only by battling with the inclemencies of nature in high latitudes, it is under the equator alone that the perfect race of the future will attain to complete fruition of man's beautiful heritage, the earth." Bates (1988:377, see note 11).

33. See Bates (1988:38).

34. This caption accompanies a picture of a naked Nambikwara couple in the original edition of Roosevelt's *Through the Brazilian Wilderness* (1994b).

35. Roosevelt (1994b:332).

36. The term *sustainable development* became popularized in 1987 with the publication of a report called *Our Common Future* in World Congress on Environment and Development (1987).

37. The "green cathedral" reference is from de Onis (1992). The "splendid blueprint" is from a United Nations report (Commission on Development and Environment for Amazonia 1992:xiv).

38. A very different vision of the relationship between the human and nonhuman nature in Amazonia is evident in folk stories of Enchanted Beings, or *Encantados*. See Slater (1994b:233–256).

39. See the analysis of the McDonald's rain forest in Slater (1995:126–129).

40. For a consideration of the specifically "tropical" quality of this nature, see Arnold's discussion of the invention of tropicality (1996).

41. Statistics vary greatly, depending in good part on definitions of the terms *indigenous* and *Indian. Amazonia Without Myths* (United Nations, Commission on Development and Environment for Amazonia 1992:28) gives a total figure of 935,949 for the total Amazonian indigenous population, but a number of the census data are quite old. (The single highest component figure—that of almost 400,000 for Venezuela—dates all the way back to 1972.)

42. For an introduction to urban Amazonia, see Browder and Godfrey (1997). In 1991,

almost 60 percent of residents of the Brazilian Amazon lived in cities. Belém, Manaus, and Porto Velho can all be considered regional metropolises.

43. For a discussion of the importance of blacks in the history of Brazil's most populous Amazonian state, Pará, see Salles (1971).

44. See Keck (1995).

45. See Ramos (1994:153–171) for a discussion of this phenomenon.

46. See Slater (1994a:720–742).

47. The quote is from *Amazonia Without Myths* (United Nations, Commission on Development and Environment for Amazonia 1992:91).

# CHAPTER 4

## Outrage in Rubber and Oil: Extractivism, Indigenous Peoples, and Justice in the Upper Amazon

*Søren Hvalkof*

This chapter examines the cyclic boom-and-bust history of extractivism in the Upper Amazon of Ecuador, Peru, and Colombia (figure 4-1). It focuses on the social relations of production created by the extractive industries, characterized by debt bondage, serfdom, and slavery—a system that reached its zenith during the rubber boom in the beginning of this century and that has continued in certain regions until the present day. The specific horrors of the atrocities committed against the indigenous population in the Putumayo in this period by a British-owned rubber company is spelled out. Although this case is well documented and has been analyzed repeatedly (Casement 1913; Collier 1968; Gray 1990; Hardenburg 1913; Luxemburg 1968; Taussig 1985; US Department of State 1913), in the debate over rainforest conservation it seems to be forgotten that a regime of terror with the character of a holocaust was taking place all over the Upper Amazon of Ecuador and Peru only two generations ago. This was the direct result of the extractive boom economy, where terror against the debt-bonded indigenous population was the only means to dramatically increase short-term productivity, as the bust of such economies is always imminent. Reproduction of the labor is of no concern in cycles of fifteen to twenty years.

Rather than a tasteless attempt at selling the aesthetics of violence, the detailing of some of the monstrosities committed is intended to contextualize the conservationist discourse in the light of justice. Although those responsible for the cruelties were identified and internationally denounced, none were ever made accountable. Instead, they were honored as active entrepreneurs, thus legitimizing such conduct, creating a formula for general rapport with the native population that has continued up till the current oil bonanza in the Amazon, epitomized in the Texaco case of Ecuador. The outrageous crimes of the extractive industry have left an enormous moral debt in the Western world that weighs particularly heavily on the shoulders of environmentalist

83

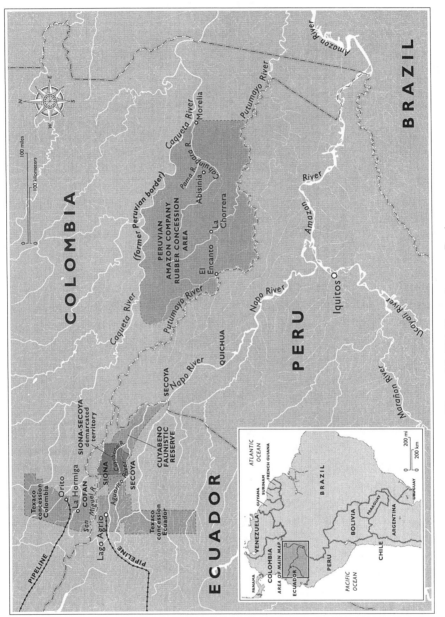

**Figure 4-1** The Upper Amazon Region

organizations imposing conservation measures in what rightfully is the territory and future of the indigenous peoples. We cannot afford to forget holocausts.

In light of this history, the appropriateness of governing strategies for biodiversity conservation, and especially the pertinence of "green marketing," i.e., commodifying tropical forests by attributing a commercial value to their resources under the assumption that this will motivate conservation and sustainable use, is questioned. It is argued that such strategies reinforce external territorial and social control, as well as the reproduction of a nature/culture dichotomy maintaining a hegemonic status quo. However, the same process of globalization has also nurtured alternative strategies by the affected indigenous groups, who have succeeded in gaining international support for large-scale indigenous land-titling and organizing, creating a new political space, and generating democratic momentum tending toward indigenous self-development and relative autonomy. Such postdevelopment initiatives are overruling local power structures and recapture social control, giving local populations a bargaining position that puts them at odds with the canon of conservation. If this postmodern reality is not reflected in the strategies for bioenvironmental protection, the choice will be between biodiversity and democracy.

## FLASHBACK AND CIGARS

One late Sunday afternoon in July 1995, I was sharing a couple of cigars with some of the locals in a small palm-thatched house in the Siona Indian settlement of Puerto Bolivar on the Cuyabeno River in the Ecuadorian Amazon. It had been a rather dramatic day. The Indians had succeeded in curbing yet another extractivist enterprise exploiting the resources in their territory—one of the very few incidents of provisional victory in the sad history of the Siona. They had rejected a Chinese oil company that was to begin its exploration operations. The oil company representatives had arrived smiling at the community meeting, expecting an obliging attitude as they unveiled their plans for seismic surveys, the establishment of numerous helicopter platforms, and other infrastructure. They also intended to present the usual treats and trinkets for the Indians which, in the contemporary model, correspond to primitive school buildings and a stack of corrugated iron sheets. But they were never allowed to finish their presentation or to enter into any negotiations or agreements. Angry community members, old and young, women and men, stood up one by one and talked vehemently against allowing the company to operate at all. They forcefully rejected the legality of the permission for exploration given by the Ecuadorian central authorities and proclaimed the oil company and its representatives "not wanted." This was an unusually strong reaction from this small Siona group.

The Siona community of Puerto Bolivar is enclosed inside a large nature reserve, the Faunistic Reserve of Cuyabeno, created in 1979 without regard for the territorial interests and rights of the local indigenous populations.[1] A few months prior to the oil meeting, the Siona, together with their Secoya neighbors, had signed a treaty with the reserve's Ecuadorian administrators acknowledging their territorial rights inside the protected area as well as stipulating their rights and responsibilities in the management of the reserve. The agreement included the framework for a detailed management plan for the protected area/indigenous territory. They had also agreed on creating a zone of maximal protection, an area that even the Indians themselves would leave undisturbed.

The treaty was the result of the territorial demarcation and legalization project the Siona and Secoya organizations, ONISE and OISE, together with the indigenous umbrella organization for the Ecuadorian Amazon, CONFENIAE, had developed and implemented since 1991. The project had been haunted with all manner of problems,[2] but despite all the odds they had carried it through to final success after five years of struggle. On April 30, 1995, the last signature on the agreement was added by the director of Instituto Ecuatorian Forestal y de Areas Naturales y Vida Silvestre (INEFAN), and the treaty was assured.

The men had worked for weeks and months in the forest and swamps, in cold rain and burning sun, cutting the border trail and doing the topographic surveying. Their wives had been alone with all the daily work in the houses and the fields while the teams were working on the demarcation. They had all contributed and taken part in this tedious operation. The process of demarcation, legalization, and obtaining an agreement had raised their political consciousness and had eventually elevated their organizational skills. It was their territory and their effort. And now, for the first time, they had a legal basis for defending it. After thirty years of devastating experiences with oil companies and exploration activities in their area, they knew what to expect.

It had therefore outraged them to find out that the ink on the treaty had barely dried before it was broken. As one person expressed it, "They sign with the hand and erase with the elbow." Just a month after the final signature, the very same institution that ratified the treaty, INEFAN, issued an oil exploration permit (dated June 8) to the Chinese oil company and the Petroecuador consortium, allowing them to carry out large-scale oil exploration and prospecting surveys in the zone of maximal protection of the Cuyabeno Reserve/Siona-Secoya territory. The act violated several articles in the treaty, of which the most important was the guarantee that no contracts with third parties, or other activities, could take place inside the affected sector without the consent of the Siona and Secoya indigenous organizations; furthermore, it had been agreed that no activities related to petroleum exploitation could take place, unless it was stipulated in the management plans of the reserve; and, last, no activities at all, apart from scientific investigation, could take place inside the zone of maximal protection.

The remarkable consensus on the refusal to allow the oil exploration, by both community members and representatives from the indigenous organizations, was primarily based on the powerful argument that the violations imperiled the value of the entire treaty—and thus the legalization of their hard-won territory.

The oil company delegation left immediately after the short but agitated meeting, visibly shocked by the rejection and massive confrontation.[3] There was relief and a feeling of empowerment among the Siona of Puerto Bolivar after the meeting. Their territorial project had been proven valid.

The cozy chat following the meeting soon drifted off the subject of the day and turned to a more meaningful discussion of the spiritual quality of my cigars—and the dizziness they caused; a state of mind that was interpreted as favorable for hunting, as the game would be less inclined to escape a human being affected by strong cigar smoke. A young man whom I had never met before was especially interested. He wanted to try out the cigars the same night. The others commented that he was very knowledgeable about hunting magic and added that he was half Huitoto Indian from the Putumayo River. Without really realizing the gravity of my question, I injudiciously asked him if he had ever heard any of the stories about the atrocities committed against the Huitoto during the rubber boom at the turn of the century. He had indeed, and began to relate what he had heard from his grandfather; this was soon followed by similar accounts from the Siona. In impassioned voices, they told revolting stories about how the local rubber patrons and their contractors had murdered, tortured, and abused their folks. The accounts were not limited to the Huitoto tragedy but embraced the Siona, the Secoya, and the Cofán, all present in and around the Cuyabeno Reserve.[4] The coziness of the evening vanished. The discussion soon turned back to the oil companies and it became clear that for these humble and unpretentious people there was no major difference between the conduct of the rubber patrons in the first decades of this century and of the oil companies and their representatives now. It represented a continuum of nonindigenous expansion motivated by extractive interests—rubber plants and oil plants were both part of the idiosyncrasies of white and mestizo culture. But the modest triumph earlier this day did denote a faint possibility of some kind of justice. As night fell, a few of the men went hunting with a handful of my cigars.

## Ethnographic and Historical Fragments

The Siona and Secoya are western Tucano-speaking groups, descendants of the once numerous "Encabellado" nation[5] that, at the time of the conquest in the eighteenth century, inhabited the tributaries of the vast area between the Napo River and the upper Putumayo and Caquetá Rivers, an area that today is divided between Ecuador, Colombia, and Peru. According to the *Handbook of South American Indians* (Steward 1949), they numbered approximately 16,000 at the time of the first contact. Today the total number of Siona and Secoya in the three countries is some 1,000 to 1,200, of which some 700 live in six communities in Ecuador along the Aguarico River and tributaries such as the Cuyabeno (Zamarenda 1996). The rest are equally distributed with a group of Siona in Colombia and a group of Secoya in Peru (for an analysis of the territorial development of the western Tucanoan in Ecuador, see Vickers 1983).[6]

Their neighbors, the Cofán, have lived west and north of them in the headwater region of the Aguarico River and to the north of the San Miguel and Putumayo Rivers, today divided by the Colombian–Ecuadorian border.[7] At the time of contact, there were several thousand, as the Jesuits have a record of some 6,500 baptized by 1604 (Barriga López 1988:58). It has not been possible to classify their language unequivocally.[8] The total Cofán population today in Ecuador and Colombia is approximately 700.

The causes of this drastic population decline are indisputably to be found in the expansion of the nonindigenous society. The incitement and economic mover for colonizing these areas has always been the extractivist exploitation of renewable as well as nonrenewable resources. The history of the Siona, the Secoya, and the Cofán is not unique in this sense. The majority of the indigenous groups in this region have been deeply affected by the structural characteristics of exploitation linked to the extractive economies that were established throughout the Upper Amazon and the so-called Montaña[9] soon after the first European contacts.

The first colonizing attempts were carried out by the Jesuit mission during the early seventeenth century, when the Cofán population was concentrated in so-called Reducciones. The Jesuits had been sent by the Spanish viceroyalty to pacify the Cofán who had been attacking colonist settlements and towns (Barriga López 1988:25–74). The Franciscans, backed by the military force, continued the conquest and pacification of the Cofán through the next century, now and then spurring indigenous resistance and local revolts.

For the Siona and others of the Encabellado nation, the Jesuits established as many as seventeen missions between 1709 and 1769 (Vickers 1981:53). By then, the process of trying to relocate and concentrate the population had caused many conflicts. The Encabellado, like the Cofán, rebelled in 1744, and for the same reason: the priests had intervened intolerably in indigenous daily life. There was also widespread suspicion that the mission had slavery on their hidden agenda, and that the practice of teaching the children Quechua in the mission schools was designed to make them more marketable since it was the lingua franca of the region.[10] Many similar indigenous revolts and rebellions of varying scale and magnitude took place all over the Upper Amazon of the Spanish viceroyalty until the epoch of the national independence and the creation of the modern states.[11]

Despite the failure of the different missions to establish a Christian utopia, two centuries of insistent intervention had a profound effect on indigenous societies. Epidemics of European diseases had taken a heavy toll, and the peoples had been pushed around, relocated, and resettled, causing interethnic conflicts over land and resources. Territories had been opened for further exploration, resources (including labor) mapped, languages studied, and spontaneous colonization and resource exploitation begun.

## EXTRACTIVIST REGIMES

### Systems of Slavery

Slavery has always been the preferred system of production by the colonizers of the Amazon. In a recent report from Anti-Slavery International (Gray 1997:183–186), three main types are identified: (1) chattel slavery, or the partial or total ownership of one person by another; (2) debt bondage, or labor in payment of a debt that is never liquidated; and (3) serfdom, or a status of tenant bound to live and work on another man's land but not free to change. In the Upper Amazon, we find variations and combinations of these forms, modified according to local conditions.

From the very first Spanish expeditions in the Upper Amazon, the identification and exploration for extractive products was given high priority. The search for valuable minerals such as gold is well known (in fact the Cofán were known for their gold panning). Also, botanical products, especially those for medical purposes, played an important role in the colonial economies of the region. With the industrial and scientific revolution in Europe, the demand for such products grew, and new social relations of production were created in which the indigenous population was given a central role in the extraction of products. The Indians were the only ones with the detailed local knowledge that enabled them to identify the desired habitats and species. They were also the only labor force available to the extractive industry.

The procurement of indigenous labor came to be guaranteed through debt bondage, the *enganche*,[12] which means to hitch a horse to a carriage or to hook up something. The basic procedure was that the patron who wanted the labor made an advance payment, most often in kind, to the peon who later would be required to pay off his debt with work. The amount of work required to pay back a debt was determined by the patron and varied according to his estimate at the time of reimbursement, and failure to comply with the work "contract" was met with severe (physical) punishment and retribution. Debt bondage was used on the Spanish haciendas in the Andean highlands to secure agricultural labor. In the Upper Amazon, it developed into a particular system of accumulation, the system of *habilitación*, which was intimately connected to the extractive economy and to the national and international markets. The *habilitación* system functions as a hierarchy of interconnected debt relations in an exploitation chain. From the top of the system, the exporter or commercial house prepays a contractor to deliver the product at a given time. He habilitates several subcontractors, who habilitate local patrons, and so on, until we reach the ultimate producer at the bottom of the chain, most often an Indian and his family.

The system is set up in such a way that there is an accumulation of debt at the production level. The lower the position in the hierarchy, the less the accumulation and surplus from the transaction. In fact, it is only the top strata of the hierarchy that can make capital gains in this system (cf. San Roman 1975:129; Chibnik 1994), thus avoiding competition from would-be barons and upstarts.[13]

A related system of chattel slavery developed parallel with the debt bondage system, with which it overlapped. Since debt was inheritable, patrons benefited from accumulated debts through generations. Indebted indigenous families constituted real capital and could be traded in very much the same way as bonds on the market. Thus patrons could redeem their own debts to other patrons by handing over their bonded families or serfs. There are several documented cases of gambling debt where Indians form part of the stake. This commodification of the indigenous population rapidly evolved into a slave trade run by the contractors involved in extractive enterprises. With growing colonization and with changes in production and labor requirements, slavery and slave trafficking became rampant in the Upper Amazon and developed into an independent economic activity, an "extractive" industry of its own. Armed slave raiding, called *correrías*, became widespread in the nineteenth and twentieth centuries, and a market for Indian slaves, mostly children and young women, was established in the town of Iquitos, the fastest-growing commercial center of the entire Upper Amazon region of today's Peru, Ecuador, and southern Colombia (Bellier 1994; Lucena Salmoral 1977; San Roman 1975).

The slaves were mostly exported to Brazil or sold to other patrons who were establishing themselves with estates along the major waterways in neighboring areas. The systems of chattel slavery and debt bondage reinforced each other. If Indians were free of bonds to patrons, it also meant that they and their families were free prey for slave-raiders, whereas if they had been "commodified" through debt or serfdom, they were the property of a patron and under his protection. Most of the extractive industry was and is based on bonded labor.

The conceptualization of the indigenous peoples of the Amazon by the nonindigenous society in South America, as well as the relationship and rapport established during the extractivist booms, still dominates as the legacy of an extractivist economy. The systems of *enganche* and *habilitación* continues in the Peruvian Amazon, and slavery was common practice in certain areas until a short time ago (cf. Hvalkof 1986a, 1986b, 1987, 1989, 1990, 1994a, 1994b; Gray and Hvalkof 1990; Gray 1997).

## *The Products*

### Zarzaparrilla

Many different extractive products have seen the light of day and disappeared again overnight, following the whims of the international and national markets creating the boom-and-bust cycles in the Amazon. One of the first successful export products was the zarzaparrilla, a spined shrublike vine (*Smilax* species) growing along river banks, the roots of which are dried and used to prepare extracts. Zarzaparrilla is thought to have blood purifying and antirheumatic qualities and is a renowned remedy for syphilis, as is reflected in the Latin names of the two types exploited in the Amazon, *S. officinalis* and *S. syphilitica*. The medical qualities of the zarzaparrilla were known as early as in the sixteenth century and were incorporated into the popular pharmacopoeia of Europe and the colonist society of South America (Dominguez and Gómez 1990). It was also brought to North America, and because of its sweet and spicy taste it was ultimately used in the production of soft drinks and soda pop, by, among others, the Shakers of the northeastern United States, where it still is sold under the name Sarsaparilla. Also, it has recently had an almost explosive revival within alternative medicine.

Zarzaparrilla became a major export product, tons being shipped yearly to Brazil and Europe from Iquitos. It was a typical indigenous extractive product (it was never grown commercially), and because the demand was rapidly increasing and the

growths were being depleted so quickly, the Prefect of Iquitos in 1859 found it necessary to take the conservation measure of ordering a total closure of its extraction
in certain river areas (San Roman 1975:101).

The extraction and commercialization of zarzaparrilla and the slave trade soon
became an integrated enterprise:

> Some white residents of the Marañon banks are accustomed to go up part of the
> Napo River with some peons to extract zarzaparrilla; and while the Indian peons
> they brought with them are occupied with the extraction of this root, the patrons go
> inland to hunt and capture by surprise the tribes . . . in their settlements. They used
> to carry out incursions called *correrías* just when the shipment of zarzaparrilla was
> ready to embark.
>
> Villavicencio 1984:368; author's translation

### Quinine

Another extractive product that developed early into a profitable venture was
the bark of the cinchona tree (*Cinchona officinalis*), also known as quina or cascar-
illa. Its active ingredient, quinine, was known to be most effective as an antipyretic
and was used all over Europe and its colonies and in the United States as a remedy
for malaria.

The social relations of quina extraction were the same as for the zarzaparrilla,
but the ecological degradation and its economic consequences worried the Spanish
viceroyalty. The rapid destruction of the quina stands in what is now Ecuador led a
superior colonial administrator, counseled by a French botanist, to propose regulations aimed at counteracting the destruction of the quina sources and to establish a
controlled reforestation program as early as 1735 (Dominguez and Gómez 1990:30).
In 1820 two French chemists succeeded in isolating the active alkaloid, quinine, and
with the introduction of steamboats in the mid-nineteenth century on the Amazonian river systems, the gate was opened for the first bioprospecting projects sponsored
by the major colonial powers of Europe. Both the Dutch and the British entered the
race to establish quina plantations in their colonies. By the late 1880s, European
plantation production had ousted the less productive extractivism of South America,
and the Amazon cinchona bark economy collapsed. Meanwhile, a war of independence had been fought in the old Spanish colonies, and the new states of Colombia,
Ecuador, Peru, and Bolivia were founded, and the Upper Amazon was divided by
new but poorly defined national frontiers. The new nation-states never succeeded in
catching up with the European quina plantations despite several attempts.[14]

### Other Plant Products

Other commercial products that have been extracted from the Upper Amazon
by bonded indigenous labor include bombonaje palm fibers for "panama hats"
(which are not produced in Panama but in Peru and Ecuador);[15] chambira (*Astro-
caryum chambira*) fiber cords used for making hammocks, nets, and bags; and the

yarina palm (*Phytelephas microcarpa*) fruits (drupes) called *tagua* and used as a substitute for ivory in the production of buttons, necklaces, and other minor items—so-called vegetable ivory, a major export item in the last century now undergoing a renaissance in Ecuador for the past ten years![16] There were also aromatic plants and woods including vanilla (*Vanilla aromatica*), Peru-balm (*Miroxylon peruiferum*), and Copaiba-balm (*Copahifera officinalis*), to mention a few of the many extractive plant products, all dependent on indigenous knowledge and labor. I should also mention the extraction of very valuable tropical hardwoods such as mahogany, Spanish cedar or deodar, and Brazilian rosewood, which accelerated with the spread of steamboats and which was—and still is—completely dependent on cheap Indian labor.

### Gold

Of the nonvegetable extractive products, placer gold was common and was panned by the Cofán Indians in the upper Aguarico (hence the name, meaning rich water) and San Miguel Rivers at the time of the conquest. The availability of gold led to an invasion by Spanish adventurers, gold panners, and colonists in the sixteenth and seventeenth centuries, establishing a gold extraction economy based on forced Siona labor (cf. Dominguez and Gómez 1990:243–245).[17]

### Rubber

The most disastrous of all the extractivist activities was the production of elastic rubber around the turn of the century, accentuating the perverted cruelty of primitive accumulation disguised as modern noncolonial internationalized capitalism.

There are a number of different species of plants and trees that produce the saps that, when dried, have the characteristics of elastic rubber. The two most common species in the Amazon are the *Castilla* species,[18] of which we have several varieties, and the *Hevea* species, of which nine varieties have been identified (cf. Dominguez and Gómez 1990:93). There is much confusion over the names and identities of the many different species, but basically the most commonly extracted in the Upper Amazon are the *C. ulei*, usually referred to as *caucho*, which produces a high-quality latex, and parts of the *H. brasiliensis*, in the lower Amazon, which also produces a high-quality latex referred to as *jébe* or *shiringa*.[19]

In the Upper Amazon of Peru, Ecuador, and Colombia, rubber was mostly tapped by cutting and/or draining the tree completely to produce as much as possible in the shortest possible time. This caused the rapid depletion of these highly valued species, which led to tapping varieties of lesser quality, the so-called jébe débil, or weak rubber. In contrast to this destructive practice, in Brazil the rubber was tapped by incisions and the rubber sap collected in small containers, allowing the tree to survive and keep producing. This was a result not so much of higher ethical standards but of the fact that labor was scarcer in the lower Amazon basin, requiring a more rational and sustainable system of extraction. In the Upper Amazon, the norm was to produce as much as possible as quickly as possible, exploiting the abundant native labor to the limit of its destruction.

The new boom in rubber production continued the extractivist relations of production established in the preceding booms of zarzaparrilla and especially quina extraction, boosted by an exploding international demand for, and exponential growth
in the market of, raw rubber. The coincidence of the introduction of the steamboat
in the Amazon, the collapse of the quina economy in 1885, and the invention of the
vulcanization technique and of the pneumatic tire in the industrialized countries (and
of the bicycle and the automobile) led to this boom. Its zenith was reached between
1900 and 1914, after which the whole Amazonian rubber economy again collapsed
as a result of the market takeover by newly established rubber plantations in the
British, Dutch, and French colonies in Asia and the Pacific.

Indisputably, the rubber boom is the single period in the history of the Amazon
that has had the most devastating and disastrous effects on the indigenous population. In this short time span, possibly hundreds of thousands of indigenous inhabitants of the Upper Amazon became involved in rubber extraction as debt peons and
slaves. The rubber gatherers penetrated even the remotest area of the Amazon
forests, contacting, enslaving, or killing whomever crossed their tracks in their fre-
netic quest for the "black gold,"[20] leaving behind a trail of mutilated bodies and dis-
located families. This short historical period produced the most radical changes in
the demographic and settlement patterns in the Amazon. Forced relocation of local
groups and rubber serfs by patrons left a hitherto unseen ethnic patchwork in the rub-
ber-producing zones, and violent epidemics, especially measles, had detrimental effects. Tens of thousands of Indians fleeing enslavement, genocide, and diseases created a completely new ethnic mapping in the montaña of the Upper Amazon. The
demographic patterns and locations of indigenous settlements and populations we
find today are a direct result of these social disarticulations, dislocations, and forced
migrations.

## *The Atrocities*

The most well known and best-documented case of the atrocities committed against
the indigenous population during the rubber boom comes from the upper Putumayo
River on the border between Peru, Colombia, and Ecuador. It is a very complex and
dramatic story involving high-level British and American external and internal politics; European and American corporate interests in extractive industries in Latin
America; South American geopolitics; an acute border dispute between Colombia
and Peru and eventually Ecuador; Peruvian national politics; and early international
human rights activism. It is not possible within the limits of this chapter to reveal all
the intricacies of the Putumayo scandal, but I can summarize the main events.

Had it not been for the fortuitous presence of two young American travelers[21]
with sufficient courage, and the moral and political acumen to react, the world outside Peru would probably have never known of the outrageous details of rubber gathering activities in the Upper Amazon, nor would they have aroused the attention of
the international media. Walter E. Hardenburg and W. B. Perkins had both been

working as railway engineers in the Cauca valley in southern Colombia. In October 1907, they quit their jobs to journey from the Pacific coast of Colombia across the Andes and the entire Amazon basin to the Atlantic by river. It became an eventful excursion. Traveling down the Putumayo River, they unwittingly canoed right into the lion's den, the rubber-tapping concession area of the Peruvian company Casa Arana, owned by the London-based Peruvian Amazon Company, locally referred to as "the civilizing company." The unfortunate American adventurers arrived just as the company was engaged in an attack on the last Colombian settlement in the concession area, backed by sixty Peruvian soldiers and a gunboat, an attack that finally consolidated Peruvian control over a large area belonging to Colombia between the rivers Putumayo and Caquetá, forcing out Colombian settlers and rubber gatherers alike.[22] Hardenburg and Perkins, who were nearly killed in this incident, were violently detained and held as captives on one of the company's estates. They both succeeded in getting away, months apart, and Hardenburg pursued the case by bringing to light what he had experienced. The disclosure caused an international outcry and scandalized the Peruvian and British governments.

The rubber extraction enterprise J. C. Arana y Hermanos was incorporated in 1903 by the Peruvian merchant Julio Cesar Arana with his brother Lizardo Arana and his brothers-in-law Pablo Zumaeta and Abel Alarco (cf. Hardenburg 1913; Gómez et al. 1995; Gray 1990). The extraction of the crude rubber was organized according to the systems of *habilitación* and *enganche*. Each level in the chain of exploitation was organized on a commission basis through independent subcontractors (Hardenburg 1913:203). This system was very productive and highly successful, and in 1907 Arana secured new capital from British investors, transforming his company into a British corporation.

The company had two main delivering and depositing stations, the estates of La Chorrera and El Encanto on the Putumayo River. From there, the loads were shipped to Iquitos for export to Europe. The rubber was collected at some twenty company posts, so-called sections, scattered in the vast extraction area. The heavy bundles of rubber (weighing well over 100 pounds each!) were carried by enslaved Indians overland from the outposts to the main stations once every three months. To oversee the rubber extraction activities and guarantee the delivery of the maximal quantity, the company recruited 196 Barbadians with experience from the Caribbean sugar plantations, with 300 mestizos and whites manning the stations and posts (Gray 1990:1). The labor force in the rubber-gathering activities consisted mainly of the Huitoto, Bora, and Andoke Indians, but other indigenous groups were also exploited by the Casa Arana, alias Peruvian Amazon Company. The company rapidly cultivated a reign of terror against the indigenous population to secure a stable labor supply at the lowest expense possible.

In an affidavit, Hardenburg stated that the abuses committed were based on his own observations during his stay in the region and as a hostage at El Encanto, as well as on interviews he conducted with the various parties and personnel involved. When he at last succeeded in getting to Iquitos, he also collected a number of sworn testimonies by former employees of the company and others documenting the atrocities taking place in the Putumayo (cf. Hardenburg 1913).[23] Following is his summary of the situation in the extraction area controlled by the company:

1. The pacific Indians of the Putumayo are forced to work day and night at the extraction of rubber, without the slightest remuneration except the food necessary to keep them alive.
2. They are kept in the most complete nakedness many of them not even possessing the biblical fig leaf.
3. They are robbed of their crops, their women and their children to satisfy the voracity, lasciviousness, and avarice of this company and its employees, who live on their food and violate their women.
4. They are sold wholesale and retail in Iquitos, at prices that range from £20 to £40 each.
5. They are flogged inhumanly until their bones are laid bare, and great raw sores cover them.
6. They are given no medical treatment, but are left to die, eaten by maggots, when they serve as food for the chiefs' dogs.
7. They are castrated and mutilated, and their ears, fingers arms and legs are cut off.
8. They are tortured by means of fire and water, and by tying them up, crucified head down.
9. Their houses and crops are burned and destroyed wantonly and for amusement.
10. They are cut to pieces and dismembered with knives, axes and machetes.
11. Their children are grasped by the feet and their heads are dashed against the trees and walls until their brains fly out.
12. Their old folk are killed when they are no longer able to work for the company.
13. Men, women, and children are shot to provide amusement for the employees or to celebrate the *sábado de gloria* [Easter Saturday], or, in preference to this, they are burned with kerosene so that the employees may enjoy their desperate agony.

In addition to this, during my subsequent investigations in Iquitos I obtained from a number of eye-witnesses accounts of many of the abominable outrages that take place here hourly, and these, with my own observation are the basis of the indictment.

Hardenburg 1913:185

Hardenburg's account contains appalling details qualifying the above summary, for example,

By way of amusement these employees of the company often enjoy a little *tiro al blanco*, or target-shooting, the target being little Indian children whose parents have been murdered. The little innocents have been tied up to trees, the murderers take their positions, and the slaughter begins. First they shoot off an ear or hand, then another, and so on until an unlucky bullet strikes a vital part and puts an end to their sport.

Ibid. 206

Add to this the constant sexual abuses against women and children by the company's contracted administrators and field personnel.

## *The Case*

Hardenburg presented the case to the American consul in Iquitos, Guy T. King, who did not take any interest in his report, simply stating that "he could do absolutely nothing" for them (ibid. p. 195).[24] Highly motivated to seek justice, he continued his journey to London in September 1909, where, after some futile attempts to gain interest in the case, he succeeded in presenting the material to the chair of the Anti-Slavery and Aborigenes Protection Society, who immediately took action.[25] The Anti-Slavery Society started lobbying and campaigning about the case in Britain, while the humanistic magazine *The Truth* published Hardenburg's documentation in a series of articles (Olarte 1911). Because of the usual perplexities in British domestic policy as well as the insistent push by the Anti-Slavery Society, the case was propelled into the public, and soon the Foreign Office was forced to take interest in the case. The Foreign Secretary consulted (retired) Captain Thomas Whiffen, who had spent several months on the Arana estate in the Putumayo as a convalescent, to hear his opinion on the articles in *The Truth*, and they were fully corroborated.[26]

The Peruvians, including the Peruvian Consul, vehemently denied all charges with allegations that the whole case was a setup to blackmail the company. The British (and one French) shareholders and board members of the company[27] disclaimed any responsibility or knowledge and tried to wash their hands of the matter (cf. Gray 1990:11). Public indignation, however, did call for action. Both the company directors and the Foreign Office set up committees to investigate the allegations, the two eventually working together. The British government appointed Sir Roger Casement, their British Consul to Brazil, as chief investigator. Casement had investigated similar scandals in the Belgian Congo. As an Irishman with strong sympathies to the nationalist cause in Ireland, and as a homosexual, he had strong feelings for stigmatized social groups and colonized peoples (Gray 1990:6).[28]

The commission began its inquiries in Putumayo in September 1910, and the first investigative report was sent to the Foreign Office February 1911; it was published as an official Blue Book in 1912. The Casement Report supported the veracity of Hardenburg's accusations and added thirty new testimonies and interviews substantiating the allegations. It is a vile account of the numerous atrocities committed, the character of which is so disgusting as to exceed the imagination. One testimony will give the reader an idea of the character of these crimes. The following excerpt is from a statement given by James Chase, a native Barbadian, to the joint commission constituted by Consul Casement, Chief Manager Tizon, and Chief of the Commission, tropical agriculturalist Barnes, at the rubber station Entre Ríos, October 13, 1910. The incident was corroborated by another Barbadian, Stanley S. Lewis, who saw the company chief overseer at one of the stations "commit the atrocious crime . . . the details of which are unprintable" (Casement 1913:308). We will however print an excerpt of it:

He states that about four months ago whilst engaged at Abisinia he was sent on a commission toward the Caqueta River. The Expedition set out from Abisinia, and was commanded by a man he calls Jermin Vazquez, whose right name is found to be Fernand Vasquez, but usually called Filomene Vasquez. There were also two other Peruvians, one named Armando Blondel, and the other Esias Ocampo, he himself, the deponent James Chase, and eight Indian "muchachos" [the trained murderers] all from the station of Abisinia. All were armed with Winchester rifles, he himself having 36 cartridges. They were sent by Agüero [chief of section] to go toward Gavilanes, an Indian "house" on the River Pama, a tributary to the Cahuinari, which itself is a tributary of the Caqueta. They were sent to look for fugitive Indians who had run away from the district of Morelia, of which Armando Blondel was subchief. The date would be about May, 1910. They were also to hunt for a Bora Indian named Katenere, a former rubber worker of the district of Abisinia, who had escaped, and, having captured some rifles, had raised a band of fellow Indians, and had successfully resisted all attempts at his recapture. Katenere had shot Bartolomé Zumaeta, the brother-in-law of Julio C. Arana, and was counted a brave man and a terror to the Peruvian rubber workers. The expedition set out from Morelia, and at the first Indian "house" they reached in the forest they caught eight Indians, five men and three women. They were all tied up with ropes, their hands tied behind their backs, and marched on farther. At the next house they reached they caught four Indians, one woman and three men. Vasquez, who was in charge, ordered one of the "muchachos" to cut this woman's head off. He ordered this for no apparent reason that James Chase knows of, simply because "he was in command, and could do what he liked." The "muchacho" cut the woman's head off; he held her by the hair of her head, and, flinging her down, hacked her head off with a machete. It took more than one blow to sever the head—three or four blows. The remains were left there on the path, and the expedition went on with the three fresh male prisoners tied up with the others.

Casement 1913:317–318

They split up into smaller units, continued their raid, and succeeded in taking Katenere's wife as captive. After a shoot-out with some Bora men, three "muchachos" and their leader "Henrique" were killed, as well as some Bora men. The units rejoined and the punitive expedition continued:

They had then, Chase states, 12 Indians as prisoners, who included Katenere's wife, and also of the original party that left Abisinia, two Indians, who were in chains, who had been brought as guides to point out where Katenere and his fugitive people were living. These were some of Katenere's men who had not succeeded in escaping when he got away. The whole party set out to return to Morelia through the forest, having lost "Henrique" and his rifle. Soon after they began their march in the morning they met in the path a child—a little girl—who was said to be a daughter of Katenere by another wife he had once had, not the woman they now held as a prisoner. This child, Chase states, was quite a young girl, some 6 or 8 years of age. She was frightened at the sight of the armed men, the Indians in chains and tied up, and began to cry as they approached. Vasquez at once ordered her head to be cut off. He knew it was Katenere's child because Katenere's wife, in their hands, told them so. There was no reason that Chase knew for their crime save that the child was crying. Her head was cut off by a "muchacho" named Cherey, a Recigiro Indian boy. He was a quite young boy. They came on about half an hour's march past that, leaving the decapitated body in the path; and as one of the women prisoners they had was not walking as fast as the rest, Vasquez ordered a "muchacho" to cut her head off. This was done by the same boy Cherey in the same way, he flinging the woman on the ground and chopping her head off with several blows of his machete. They left this body and severed head right in the path and went on again toward Morelia.

Ibid. 318–319

The expedition continued, cutting off heads and executing slow-walking prisoners. By the time they arrive at Morelia, they had killed thirteen Indians on this single trip alone. The rest of this man's testimony is a narrative of a series of deadly floggings, killings, and other perversities, such as cutting off a man's ears for fun and burning his wife alive before his eyes, killing an Indian man by crushing his testicles while in the stocks, hanging Indians in chains by their throats while torturing them, and using a woman for target practice. And this is only a few pages out of hundreds containing testimonies of such atrocities. The report estimates that between 10,000 and 40,000 local Indians in the Putumayo rubber fields lost their lives in a time span of around ten years, during which there was an extraction of some 4,000 tons of crude rubber.[29]

Before it was published in England, the report was made known to the Peruvian government, which began its own investigation. This was carried out by Judge Carlos Valcárcel (1915) and Judge Rómulo Paredes, accompanied by a Spanish physician, resulting in a 3,000-page report adding even more aggravating evidence of larger massacres and mass graves. Based on this evidence, Judge Paredes ordered the arrest of 237 named persons involved with the company, including Julio Arana himself and his Iquitos manager and brother-in-law Pablo Zumaeta, one of the principals charged in the indictment. No arrests were ever made.

The United States government had also shown an interest in the case. As early as 1907, the U.S. Consul at Iquitos, Charles C. Eberhardt, filed a report on the general conditions in the Putumayo, substantiating the conditions of slavery but showing that neither were the cruelties the work of American citizens nor did they affect American interests, much to the relief of the State Department. But an article in the *London Times*, July 1912, summarizing the Blue Book, caused the U.S. House of Representatives to adopt a resolution calling upon the Secretary of State to initiate a U.S. inquiry of the alleged slavery and atrocities "tending to show the truth or falsity" . . . "that . . . the worst evils of the plantation slavery which our forefathers labored to suppress are at this moment equaled or surpassed. They are so horrible that they might seem incredible were their existence supported by less trustworthy evidence" (U.S. Department of State 1913:3). In coordination with the British Foreign Office, the new U.S. Consul to Iquitos, Stuart J. Fuller, was sent to Putumayo together with the British consul G. B. Michell in April 1912 to report on the situation. Despite efforts by the company's directors and supporters to impede and distract the investigation, Fuller filed a report shortly after the visit, which was published in February 1913. The British Consul also filed an update report. Both ascertained that the situation was still the same and that the allegations seemed true.

Julia Arana, Pablo Zumaeta, and friends such as the Peruvian Consul to Brazil, Dr. C. Rey de Castro, who was indebted to Arana, initiated a counterattack by publishing a series of smaller books with documentation striking a clear nationalist and anti-imperialist tone (Arana 1913; Rey de Castro 1913, 1914; Zumaeta 1913), and in general by trying to discredit all the authors of the different reports. According to them, it was a dirty attempt by foreign nations with imperialist ambitions to smear the image of Peru and take sides in favor of Colombia and Ecuador in the border dispute.

But the company was soon forced into liquidation, initially with a liquidator and chief creditor named Julio C. Arana! (Gray 1990). The Amazonian rubber bonanza was over as soon as the new European plantations in Asia began to yield cheaper crude rubber. Despite all the diplomatic activity, high-level intervention, piles of documentation and evidence, and hundreds of affidavits, international publicity, and campaigning, not one offender or accessory to the genocide in Putumayo was ever tried or convicted, and all charges were dropped. World War I had broken out and North America and Europe had other problems to contend with. In the Upper Amazon, the local patrons were forced to diversify after the collapse of the rubber economy. Exploitation took on the more normal forms of debt peonage and trade in children and women on the Iquitos slave market. The Casa Arana was an integral part of the regional power elite with direct links to the upper levels of Peruvian society and the government in Lima. The notorious company manager Pablo Zumaeta was elected mayor of Iquitos in 1914. Julio Cesar Arana became senator of Loreto in 1922, and the Peruvian government recognized Arana's landholdings of some 6 million hectares in the Putumayo, disregarding Colombian claims, not to mention those of the indigenous population (Gray 1990:32). In a report published by the U.S. Department of Commerce in 1925, of a crude rubber survey in the Amazon basin, exploring the potentials for the establishment of a U.S. rubber plantation industry, the name Pablo Zumaeta appears in the acknowledgements (Schurz et al. 1925).

The border dispute with Colombia (and Ecuador) continued, and in 1930 Colombian troops occupied part of the Putumayo. Nine years later, the Arana family was awarded US$200,000 in compensation for lost land. Andrew Gray mentions that reports of exploitation from the area continued into the 1960s and 1970s, and Survival International campaigned in 1974 to release a group of Andoke from debt bondage from a certain Sr. Zumaeta (Gray 1990:33). The descendants of these and other rubber barons are still free-roaming patrons in the Upper Amazon, living off the indigenous labor engaged in the extraction of tropical hardwood.[30]

## The Rationale

The regime of terror exercised in the Putumayo was not accidental, nor was it a single incident of madmen's work, nor a satanic cocktail of unfortunate, chance circumstances.[31] It was a deliberate and carefully executed strategy. Sixty "specialists" from a completely different cultural and social background were brought in from Barbados to oversee the ruthless exploitation. They had no possibility of withdrawing and were completely dependent on the company.[32] Orphan Huitoto boys were brutalized and trained to be the most obedient and cruel instruments in the terror against their own peoples.[33] This was the result of a well-established system of political and economic control, adapted to the specific conditions of labor, the type and quality of rubber trees, logistics, and regional geopolitics: the political ecology of rubber tapping.

The only way colonists and other patrons could "legally" procure indigenous labor in the Amazon was through debt peonage. No indigenous family depended on wage labor or other forms of employment for their social reproduction. The crucial key to the indigenous economy was and still is subsistence production. All permanent colonists in the Amazon know that they have to allow the indigenous society to reproduce itself by its own means in order to maintain minimum costs of labor, which is necessary for their own survival as colonists. But this also implies that the indigenous people have a high degree of economic independence and flexibility in relation to the patrons. They may work temporarily to obtain certain nonsubsistence goods, but there are no economic constraints preventing them from leaving if they are dissatisfied with the conditions. The only way to shackle indigenous labor is by force, and the only way to rationalize and legitimize this is through debt relationships. Thus, it has always been intrinsic to the colonist economies of the Amazon that the indigenous worker receive loans or advance payment, mostly in kind.

Thus indebted, the "legality" of deploying violence to exploit labor is in place. This is why the henchmen of the Casa Arana threatened to shoot anyone who did not want to receive an advance "gift" (Casement 1913; Taussig 1984). Using force against debtors to compel repayment one way or another has until recently been legal practice in most of the countries in the Upper Amazon, backed by the public authorities, and it has been common knowledge that an indebted person is not free but subject to the will and whims of the creditor (cf. Hvalkof 1986a, 1986b, 1994a; Gray and Hvalkof 1990; García 1996). In fact, the trafficking in indebted individuals and families was until very recently common practice in the Amazon of Peru, and the core pretext for actual slave trade, as has recently been documented (Gray and Hvalkof 1990; Hvalkof 1994a; Gray 1997).

The commodification of the indigenous people implied objectifying and dehumanizing them (and vice versa). This process is well known and has been reproduced in different forms since the conquest.[34] But the Indians were not just passive objects in this process. Several smaller or larger uprisings against the rubber extractors took place from the very beginning, with a major incident in 1917 when the Peruvian military was called in. Many Indians chose to escape to areas free from rubber gatherers, if possible, or to seek refuge with more acceptable patrons.

Although the extraordinary economic gains in rubber extraction motivating hyperexploitation were no longer possible after 1915, rubber gathering continued on a minor scale until the 1940s. Many of the colonists and patrons continued with other activities, for instance, extraction of tropical hardwoods (which in certain areas became a major industry), cultivation of "barbasco" (*Lonchocarpus nicou*) for production of the insecticide rotenone,[35] agricultural production, and cattle raising. Although no new "boom" occurred, the social relations of production based on bonded labor remained the same.

### The Secoya, Siona, and Cofán

The Secoya, Siona, and Cofán were living close to the concession area of Casa Arana, Limited, in the upper Putumayo and upper Napo River regions. We know that they suffered a steep population decline and were split into several subgroups and pushed out of their earlier territories.[36] A Catholic missionary notes "that from the Napo to the Marañon not one single family is free [from patrons]" (Cabodevilla 1989:17). Because of constant harassment and mistreatment, combined with an intolerable border conflict between Peru and Ecuador in 1941, most of the Secoya living in Peru decided to escape and move to their relatives, the Siona, in the Aguarico and Cuyabeno areas of Ecuador. Another group stayed with the patron in Peru as his peons until his death in 1949.

The Cofán suffered a similar fate. From a population of some 15,000 in the eighteenth century, they have been reduced to some 300 to 400 today, split into four different local groupings. An Italian–Ecuadorian missionary described the Cofán situation in the 1920s:

> The Peruvian rubber patrons, coming up the tributaries from the Amazon, penetrated the easterly regions of our Republic; not only did they take with them the rubber that grows there, without paying the minimum of tax, but they also seized in large scale and by force, the savage Indians who inhabit these regions, and they brought them out to be sold in the Amazonian towns. There was a rubber worker (péon) who paid off all of his debt of 300, 400 or even more soles [Peruvian currency] with male and female children hunted in the jungles and delivered afterwards to his creditors. . . . The miserable Ecuadorian Indians are being persecuted, hunted, some dead and some captive, to be sold afterwards. What ignominy! For 20, 30, 40, 50, 60 soles, according their sturdiness and physical development.
>
> As quoted in Barriga López 1988:73–74; my translation

The ebbing away of rubber extraction did not secure the peace for these Indians. Colonization and the agricultural frontier were rapidly moving eastward from the Andes, both in Colombia and Ecuador, and the split and fractioned indigenous groups had only a few years of relative truce before a new extractive venture once more would cause radical changes in their lives: the search for oil.

### The Oil Boom

With the collapse of the rubber economy, the Ecuadorian state began to develop a colonization policy and oil exploration of the Orient. Merging with new military/

oil/settler interests, the government built the first major road into the Amazon from the Andes in 1931, and in 1939 it constructed the first company airport and town named Shell, after the owner. The Ecuadorian state secured a monopoly over all non-cultivated land areas, backed up by laws, regulations, government departments, and programs for the colonization of the Ecuadorian Amazon. Oil exploration became synonymous with *civilization* and *progress*, and a developmentalist frenzy broke loose in the region.

The Texaco Petroleum Company had started operating in the Colombian section of the upper Putumayo and San Miguel region in 1963 when "Los Texas," the local nickname given the oil company workers, invaded the Cofán and Siona territories. They came out of nowhere, literally dropping from the sky. Helicopters flew in all the building materials, heavy machinery, personnel, and whatever else was needed, and in no time Texaco had constructed an amazing infrastructure, including a pipeline across the Andes to the Pacific port of Tumaco on the Colombian coast and a road that connected to the rest of the Colombian road network. The small and peaceful Siona village Orito, accessible only via the San Miguel River, was transformed into a booming oil town, the oil capital of the Putumayo, with "an enormous street, nearly two kilometers long, flanked by long rows of `cantinas' and `prostibulas' [brothels]. At one time Orito was the most expensive town in Colombia" (Lucena Salmoral 1977:30). The entire area was invaded by fortune hunters, workers, and colonists from other areas of Colombia, converting the upper Putumayo into a typical boom town hell, fueled by fast dollars, fast booze, fast women, and not so fast chicken stew.

By 1971, the production peaked at 120,000 barrels a day. A few years later, it had declined drastically, and by the end of the decade production fell to 15,000 barrels a day. The Colombian boom was over, leaving behind a social, economic, and ecological disaster. The Spanish anthropologist Lucena Salmoral recounts a couple of anecdotes from his fieldwork in the area, illustrating some of the impact on indigenous families:

> During the golden epoch of petroleum some helicopter pilots landed close to the indigenous houses in some clearings, inviting the "Indians" to enter their contraptions, where they had a good time in a mix of astonishment and fear.
>
> Later the pilots invited the "Indian girls" to do the same and ended up transporting them to the Bodega, then situated along the road under construction between Orito and La Hormiga, where they were forced into prostitution. The daughters of Rogelio Criollo, for example, ended up at the Bodega every Saturday, brought in by helicopters that landed in a clearing in front of their father's house. Alaila, Trinidad, and Victoria (their names) were so "sick" in 1970 that, although they were dressed up in gowns from Medellin and with make-up in their faces, no Cofán wanted them as their spouses. More menacing was that they had succeeded in infecting the whole indigenous community.
>
> Lucena Salmoral 1977:34; my translation

The oil boom had forced the Indians to give up their subsistence activities for a dependency on cash income from oil workers. A self-sufficient people were turned into puppets in the American oil company's sideshow (cf. Lucena Salmoral 1977:33–34). When Texaco had extracted what they wanted, they moved on to the Ecuadorian side, leaving a trail of destruction.

Texaco began its Ecuadorian operations in 1967, transferring its expertise and equipment from Colombia. History repeated itself. In a very short time, Texaco and its associates and subcontractors had established an enormous infrastructure of access roads, company towns, camps, airports, heliports, drilling sites, pipelines, and heavy equipment. By 1972, another 300 miles of pipeline crossed the Andes to the Ecuadorian Pacific coast city of Esmeraldas. Along the pipeline, a road was constructed right into the center of the Cofán, Siona, and Secoya territory, causing landless colonists from the Andes to overrun the entire area. An estimated 300,000 colonists have settled in the area from the 1970s to today, generating immense social, economic, and ecological problems. Boom towns such as Lago Agrio—with the same characteristics as described for Orito—appeared overnight and grew out of control. Close to thirty different oil companies have been or are still working in Ecuador since the boom of the 1970s. The oil-colonization nexus is an integral part of the official national Ecuadorian development policy, according to which assimilation is seen as the most obvious way of getting rid of the indigenous population that stands in the way of "progress."

This "progress" is today expressed in a $1 billion per year oil business accounting for half of the country's export earnings and 62 percent of its fiscal budget. Most of the revenue goes directly to interest payments on Ecuador's $12.6 billion foreign debt. From 1972 to 1990, 1.4 billion barrels of crude oil were pumped through the Trans-Ecuadorian Pipeline to the benefit of Texaco and its shareholders. In 1990, Texaco's installations in Ecuador were handed over to the Ecuadorian state-owned Petroecuador, which continued the operations.

The adverse social effects of the oil extraction economy are not the only problems faced by the indigenous communities. The environment, the resource base of the local population, has been seriously damaged as well, resulting in a string of health-related problems and affecting the biological diversity of the Napo River watershed.

The most direct result of the contamination with crude oil and the "invisible" poison, the highly toxic "production water," is a decreasing food production. The ichthyological resources, which were the main source of high-quality protein and a major economic asset, have decreased almost to the point of extinction as a result of Texaco's practice of dumping the volatile production water into the rivers and wetlands. These effects are especially evident in the Aguarico and San Miguel Rivers as well as in the major Napo and Putumayo Rivers.

The production water is a major issue in this case, as it is in all crude oil extraction. As oil is extracted from the underground, water begins to fill the porous strata from which the oil was extracted. Thus, as oil is pumped up, increasing amounts of this highly toxic water rise to the surface with the oil. It is supersaturated and contaminated with all kinds of metal salts, including high concentrations of heavy metals. Although this production water is separated from the oil at the well head, it may still retain from 100 to 5,000 parts per million (ppm) of oil. Most oil-producing countries in the world have legislation that treats production water as toxic waste and requires that it be treated as such, even if it needs to be pumped back into the well—the so-called reinjection, which is the most commonly used method to get rid of it. Ecuador and Peru have no such legislation. Texaco and, later, Petroecuador have been simply dumping the water directly into the environment.[37]

Texaco's reasons for not controlling production water are obvious. It would cost approximately $1 million per well to pump it back, and there are some 300 wells to attend to. The cost of production of one barrel of crude in Ecuador is presently around $1. If the production water had to be pumped back, the cost would rise to about $6 per barrel, increasing the production costs to the same level as the average for the United States and Europe (Attorney C. Bonifaz, personal communication, February 1997). Texaco lawyers have resorted to a grotesque social-Darwinian reasoning by arguing that since laws controlling production water were not institutionalized in the United States until the 1960s, there should be no reason to comply with similar standards in a country such as Ecuador, which is to be regarded at the developmental stage of the United States in the last century, long before the mentioned legislation existed.

To the dumping of this toxic waste, add the indiscriminate burning of natural gas in enormous flares producing toxic smoke and soot fallout, the existence of multiple open disposal pits full of crude waste, and the oil company's bizarre practice of disposing part of their crude waste by pouring it directly onto the dirt roads, and we have both polluted air and water, with high precipitation washing the crude waste off the roads and directly into the water systems (Bonifaz 1996).

### The Lawsuit

The Cofán, Siona, and Secoya have tried to counteract this latest attempt to steal their livelihood and health. In 1992, they received unexpected backing from an Ecuadorian-born attorney and former chemical engineer, Cristobal Bonifaz, practicing in the United States. Mr. Bonifaz initiated research on the environmental and social consequences of Texaco's work in Ecuador. The alarming findings led to the development of a team of specialists composed of doctors, scientists, journalists, and attorneys willing to donate their time and effort to document the case[38] (see Bonifaz 1996). After field missions and additional studies, the following initial conclusions were drawn (ibid. p. 3):

- The local population faces serious illnesses directly attributable to the oil company's contamination of the region.

- The drinking water is highly contaminated with toxins, posing an increased high risk of cancer for the local population in the range of 1/1,000 exposures to 1/10,000 exposures. The acceptable risk of cancer allowed by the U.S. Environmental Protection Agency in drinking water in the United States is 1/1,000,000 exposures.
- Twenty years of oil drilling practices have caused widespread destruction of the Amazon rainforest and have endangered the lives of tens of thousands of people.

After establishing the liability of Texaco, Bonifaz, backed by a larger U.S. law firm, filed a class action lawsuit in November 1993 against Texaco, Inc., with 46 plaintiffs from the Ecuadorian Amazon on behalf of 30,000 indigenous people and other residents of the region. But the suit was not filed in Ecuador. It was filed in the federal court in White Plains, New York, home of Texaco's corporate headquarters. It demanded that Texaco be held accountable "for its negligence, recklessness, and intentional misconduct in its twenty years of oil drilling operations in the Ecuadorian Amazon" (ibid.).[39]

Since then, the case has been thrown back and forth between the U.S. juridical system and the involved parties. Texaco has tried to get the case dismissed on the grounds that it does not belong in the U.S. but in Ecuadorian courts. The plaintiffs have argued that they cannot get any justice in an Ecuadorian court and insist that the litigation belongs in the United States. In April 1997, the government of Ecuador communicated to the U.S. judge that Ecuador supported the position of the plaintiffs in the litigation. Nonetheless, Judge Rakoff of the District Court of White Plains, New York, dismissed the case in August, stating that Ecuador had entered the litigation too late. A parallel litigation in Peru was simultaneously dismissed on identical grounds. The case was immediately appealed to a higher court, where hearings were held in spring 1998. After five years of hearings, no definite position has yet been taken by the judges on whether to allow the litigation in a U.S. court.

Texaco has tried various strategies from buying off indigenous leaders and organizations with million-dollar deals and making an economic settlement with the Ecuadorian government, to putting pressure on Ecuador through the U.S. State Department not to support the case of the plaintiffs and threatening to sue the Ecuadorian state as co-responsible for the damage (which Texaco admits!). The successive Ecuadorian governments have gradually changed position from one indistinguishable from that of Texaco, to one supporting the plaintiffs' claim of litigation in the United States.

A most interesting aspect of this case, however, is the mobilizing effect it has had among the Indians and poor settlers in Ecuador, creating a democratizing momentum in the Ecuadorian public. The American and international media have taken a great deal of interest in this case, and the injured peoples have discovered that they are able to get both national and international support for their case, and that actions and protest can have an impact on policy and decision making in our postmodern world.

Texaco has argued that the case is a Pandora's box, threatening to open the door on liability not only for Texaco's operations throughout the world, but also for the operations of all other U.S. multinational corporations. It is manifest that the implications of this case are not limited to just the Amazon but have profound global effects, setting normative precedents for oil exploitation and other extractive enterprises as well as for the accountability of multinational corporations in general. There are great expectations among the indigenous people that the U.S. legal system at last will do them some justice.

## ALTERNATIVES

### Territorial Rights: The Key to Control

Although increasing globalization is often associated with neoliberal economic policies, and in our case with the expansion of multinational extractivism, there is another side to globalization: accessibility to alternative global resources for the weakest populations. Indigenous groups and other people formerly in a "no-bargain position" in local and national power hierarchies have successfully exploited the new international resource base in the form of nongovernmental organizations (NGOs) and other organizations favorable to their causes. Through such international financial and political support, local power structures and exploitative social relations have been profoundly changed or entirely eliminated through specific actions. In particular, collective land titles and the securing of indigenous territories has been a means of major importance, creating new political spaces for democratic processes.

One example: In October 1996, the venerable Anti-Slavery International in London[40] honored an indigenous organization of the Peruvian Amazon with their 1996 Anti-Slavery Award. The award was given to the Regional Indigenous Organization of Atalaya (OIRA) for "its work in freeing thousands of Asháninka, an indigenous people from the Peruvian Amazon, from debt bondage." The venture that generated this emancipation was a major indigenous land-titling project, identifying, demarcating, and legalizing Indian communities and territories.[41] The project was a response to abuses by local patrons and contractors, mostly descendants of rubber barons, who gained control of the area during the rubber boom, and who now were involved in extensive lumber extraction, cattle raising, and coca production for the Colombian drug mafia—and, as usual, trade in indigenous labor.[42] The project established more than 120 independent native communities in larger continuous indigenous territorial blocks, with collective private ownership and legal titles (deeds) to each community. A highly participatory process throughout the project generated profound structural changes in the regional hierarchy of power, setting thousands of debt peons free and giving them land necessary for their socioeconomic and cultural reproduction (cf. Garcia et al. 1998). This success has generated a great deal of optimism among other indigenous groups regarding territorial legalization as the key to self-government, self-development, and control of their own resources.[43] But the optimism may be short-lived. Scores of American and European oil companies, such as Oxy, Shell, Mobil, Elf, and dozens more, are presently streaming to Peru, which is busy parceling out its vast Amazon region into concession areas for the exploration and extraction of crude oil.[44] The very fragile democracy developing in the newly titled indigenous areas is severely endangered by massive pressure from the oil companies, which actively tries to undermine the new indigenous organizations by making partial "trinkets-and-beads settlements" with individual communities, setting aside their representative organizations, or simply buying off indigenous leaders.[45]

No matter how disturbing this new extractive bonanza may seem, it is an unmistakable sign of altered social and political relations that these companies bother at all to negotiate anything with the indigenous inhabitants who, a few years ago, were not even registered as citizens in their respective countries. As collective landowners, they have now become a political factor that cannot be bypassed, even though subsoil rights exclusively belong to the state. Indigenous opposition to any extractive activity is a growing problem for the companies.

## Green Capitalism

During the last decade's discussion of sustainable development in the Amazon and other tropical forest areas, a peculiar developmentalist strategy has emerged: the "strategy of commodification." It is based on the idea that if a sufficiently high commercial value is ascribed to the tropical forest areas and its biological resources, the states covering these areas will be encouraged to protect the forest and develop sustainable uses. It is also presumed that the local inhabitants can benefit by selling local (sustainable) forest produce (cf. Reid and Miller 1989:89; Gray 1991).

Given the history of extraction, it seems highly unlikely that increasing the market value of any extractive product facilitates wise use—on the contrary, such conjunctures have accelerated ruthless exploitation, questions of democracy aside. The problem with the "green capitalist strategy" is not the trading relationship itself. The problem is one of control.

Local populations will become increasingly dependent on the changing demands and the whims of an external market, a market they do not understand and over which they have absolutely no control. Local people, indigenous or not, living inside the would-be protected areas, are not the focus of these strategies. They are not a part of the strategy-formulating body, and they do not have any social control. They are hardly thought of as societies.[46]

The role of the different environmentalist NGOs in this latest phase of extractive expansion is rather ambiguous and varies greatly. Important alliances between indigenous organizations and environmentalists have been established in specific situations, such as in the Cuyabeno case, but they have always been strategic alliances of a very pragmatic and short-lived character.[47] Most often, there are profound conflicts of interest between the environmentalist and the indigenous movement. The key issue is social control: Who is going to control the resources, the areas, the production, and the sociocultural reproduction?

## *The Nature of Indians*

In general, the place of the indigenous people and peoples in the discourse of sustainability and biodiversity conservation is defined by classical dualism. Dichotomies of mind/matter, subject/object, or nature/culture, for example, organize the universe. *Nature* always falls on the *object* side of the divide, irrespective of its actual construction and character.[48] The Amazonian Indians have always been classified and defined as an integral part of Amazonian nature by outsiders. Even when the semantics indicate the opposite, such as in "indigenous cultures," it is understood that these are "nature-grown," unlike nonindigenous, or "Western," culture. The resulting objectification of the indigenous peoples has had profound and far-reaching effects, and it is probably one of the constituting prerequisites for the permanence of the outrageous situations I have depicted here.[49]

Indigenous cultures are looked on as particular "species," and cultural diversity (i.e., the "primitive cultures") is treated as an integral part of the biodiversity. This implies a view of culture and indigenous society as a collection of static cultural traits. The assumption is that indigenous populations are sufficiently "primitive" and "intimately adapted" to the environment, thus playing no depredatory role. In case these populations do not appear to conform to these "primitivity" standards as part of "nature," such populations are not regarded as "belonging" in "natural" areas potentially to be protected.[50] As long as the indigenous populations can be defined as "nature," they are accepted there, but the moment they begin to make political demands, wanting to decide on their own future, their presence is questioned. Nature does not act politically.

The extractive industries have analogously regarded the native Amazonian population as a natural resource on a par with other plant and animal species that can periodically be exploited. In the case of the oil industry's activities, the indigenous population is not seen as a resource since the industry is not dependent on its labor. But, as "nature," Indians are seen as a type of pretty weed, which can and will be removed if in the way. The consensus on this conception of indigenous people as objectified nature is widespread and bridges the big conservative environmental organizations to the Shining Path revolutionary movement.[51]

If sustainability is to be judged by a criterion of democratic development and if conservation of biodiversity should be other than a pretext for conserving a brutal status quo, it is in the Amazonian context crucial that priority be given to the actual incorporation of the indigenous input, developing a political platform that will favor and endorse an active political role for the indigenous peoples as autonomous participants. Apart from the very important work of international legislation,[52] immediate actions can be taken to initiate a pragmatic process of democratization and empowerment of indigenous populations in the Amazon through the following:

Demarcation, legalization, and titling of indigenous territories as collective private ownership; and

Facilitating legal proceedings of multinational corporations' liability to local populations.

## THE COLOR OF THE FUTURE

The process of demarcating and titling indigenous lands and territories may drastically alter the bargaining position of indigenous peoples, putting a stop to atrocities and exploitation and initiating a process of democratization in the regional context. Territorial control may be a powerful tool for managing extractive economies, and it is imperative for any measure of environmental protection. However, many environmental organizations are reluctant to support indigenous territorial rights, especially if they coincide with protected areas promoted under the canon of biodiversity.[53] It is obvious that productive activities of indigenous peoples do not per se guarantee an unchanged biological environment. But to use this as a pretext for the reproduction of a power hierarchy, which reduces such populations to mere objects in claimed biological processes controlled by self-appointed ecological guardians, is to consolidate the systemic structures that enable such genocidal and ethnocidal practices as we have described for the extractive industries of rubber and oil.[54]

The only democratically acceptable and viable solution to the problem of bioenvironmental conservation in areas such as the Amazon is to support indigenous land titling, and the organizing and development of sustainable economies tending toward maximizing the local populations' control of their social reproduction. Thus, by equalizing the indigenous group's bargaining position with that of the conservationist or other counterpart, protective measures may be negotiated as an interactionist endeavor. But the weakest part must have the option of veto.

The color of the future is green—as green as dollar bills. But a change in the shade of green is on the agenda, giving the hitherto incompatible concepts of peoples, plants, and justice in the history of Amazonian extractivism a chance to merge.

## ACKNOWLEDGMENTS

Thanks to my colleagues Dr. Charles W. Brown, University of Massachusetts, and Dr. Hanne M. Veber (also my wife), University of Copenhagen, and my research assistant Edith Kramer, University of Massachusetts, for inestimable help in editing and reducing the manuscript to a reasonable size. I also owe a debt of gratitude to my colleague Dr. Andrew Gray in Oxford, who kindly supplied me with his manuscript for an article on the Putumayo Atrocities, which he presented in 1990 at a seminar on State, Boundaries, and Indians at Oxford University. Many of the details and intricacies of the case come from this excellent study (Gray 1990).

This article is dedicated to the British anthropologist and indigenous rights activist Andrew Gray, who disappeared after a plane crash new Vanuato in the Pacific Ocean, May 9, 1999. He was on a networking trip to indigenous organizations for IWGIA (International Work Group for Indigenous Affairs). Andrew Gray has been a leading figure in supporting indigenous rights internationally and a board member of Anti-Slavery International, Forest People Program, and IWGIA. Closest to his heart was the Harakmbut peoples of Madre de Dios, in the Peruvian Amazon, where he spent years of his life with his family. Andrew Gray has been a key source for much of the information on the Putumayo case in this article and planned to do new research in the Putumayo area to follow up on the rubber boom atrocities. His death is an irreparable loss to all human rights work.

## NOTES

1. La Reserva de Producción Faunistico del Rio Cuyabeno was established on July 26, 1979 (Acuerdo Interministerial No. 322, Registro Oficial No. 69 of November 20, 1979). It initially covered 254,760 hectares but was extended to 648,760 hectares in 1991 (Acuerdo Ministerial 0328, Registro Oficial No. 725 of July 12, 1991). The territorial claims of the indigenous populations living inside or bordering the protected area were still unresolved.

2. Problems such as lack of experience and management capabilities; weak organizational structures; changing leadership in the different organizations involved; internal ethnic conflicts and power struggles; an anti-indigenous, hostile, and scheming public and nongovernmental organization (NGO) sphere, and a corresponding lack of legislation; technical difficulties; and political problems. Bad weather caused innumerable delays, along with changes of schedules and topographic crews, and so forth.

3. A few days later, ONISE, OISE, and CONFENIAE issued a press release denouncing the treaty, which intimidated the issuing authority and for now has kept Petro Chino out of the way.

4. Quichua and Shuar people have also moved into the area during the last decade and are now immediate neighbors of the Siona and Secoya in the Aguarico River and north of the reserve towards the Miguel River.

5. *Encabellado* was a Spanish denomination meaning the hairy ones, because of their custom of taking up their long hair "in elaborate braided coiffures" (Vickers 1981:51). Later in the nineteenth century they were referred to as Piojé (ibid.). The eighteenth-century Spanish colonial administrators and Jesuit missions in the area referred to Encabellado as a nation, basically referring to a conglomerate of linguistically and culturally related groups. It is unclear exactly which groups were considered part of the Encabellado nation, as there are many other groups, factions, subgroups, or clans all speaking some related language or dialect of the western Tucanoan bloc. The French ethnologist Irene Bellier has identified some fifty different indigenous group names in the early historical documents covering the area between the Napo River and the Caquetá (Bellier 1994).

6. A third closely related group, the Tetetes, who also lived in the Cuyabeno zone, was exterminated through a conflict with the Siona by the 1960s. Other related ethnolinguistic groups of the western Tucanoans, such as the Coreguaje, Macaguaje, Tama, Corijona, and

Mai-Huna, are still living in Colombia and Peru (Bellier 1994). The total number of descendants of the Encabellado nation today is estimated to be some 2,500 in the three countries (Payaguaje 1990:6).

7. Historically, the Cofán have been referred to as *Cushmas* by travelers, missionaries, and administrators. The term refers to their traditional dress, the cotton tunica called *cushma* in Spanish. This dress has been and is still of widespread use among many groups in the Upper Amazon of Colombia, Ecuador, and in particular Peru (Veber 1992, 1996). The term *Cushmas* as a name for the Cofán is misleading, especially where the Siona also wear cushmas and adornments that to the inexperienced outsider look identical.

8. Estimates of 60,000 to 70,000 have been given (Friede 1952:203; Costales 1983:84), which seem unfounded and highly exaggerated. Regarding the language, it was for some time primarily considered a branch of the Chibchan but is today classified as an independent language (Lucena Salmoral 1977:10–12).

9. The Montaña region is the vast U-shaped mountain forest area, forming the transition from the eastern Andes to the lowland Amazon basin, covering Colombia, Ecuador, Peru, and parts of Bolivia.

10. For an excellent summary of the history of the mission period, see Vickers (1981).

11. Although no coherent study of the development of slavery and slave trade in this region has ever been carried out, there are countless references in the historical documents indicating that slavery and the slave trade were an increasingly important factor in the development and consolidation of the regional economy and extractive activities in the Upper Amazon and Montaña region (e.g., Hvalkof and Veber 1997; Izaguirre 1922–1927; Larrabure y Correa 1905–1909; San Roman 1975).

12. The Spanish colonists had originally imported the debt peonage system from feudal Europe where it was the dominating system for exploiting peasant labor and production.

13. This *habilitación* system was also suited to the kind of extractive production where the patron is not present to organize and control the work and labor. The chattel slavery system requires that the slaves be under surveillance, which means establishing them physically in the extractive areas. With the growth in the extractive economies from the eighteenth century on, this followed gradually.

14. A French geographical expedition to Ecuador (of which the previously mentioned French botanist was a part) collected a large number of quina seeds and small plants and shipped them to the botanical gardens in Paris. Most of the plants died, and quina turned out to be difficult to grow. The different samples were classified, however, and the information systematized in a report issued in 1738. Another French-Spanish botanical expedition specifically studying the quina traveled around Peru, Bolivia, and Chile from 1777 to 1788. The growing interest in Europe in the economic potential of quina spurred several succeeding expeditions.

15. Hats woven from Bombonaje (*Carludovica palmata*) fiber were one of the few products that could be manufactured in the region, with decentralized production, since any skilled indigenous weaver could make them at home. "Panama" hats were the largest export item in the mid-1850s, and traders made huge profits on this high-priced product. As with all other export items, hats were mainly sold to Brazil, from which they were shipped to Europe and the United States. We have only scattered statistics, since only loads on larger vessels are registered. For example, the total registered export to Brazil of bombonaje hats from the largest Amazon province of Peru (Loreto) from January to November 1855 numbered 23,708 hats,

for a total value of 71,124 pesos. In the first 6 months of 1858, a total of 32,088 hats were exported from the town of Nauta (in the same province) for a total value of 98,467 pesos, a stable price of some three Brazilian pesos apiece.

16. Tagua was not a commercial item until a German major, lacking ballast, shipped some tons of tagua nuts with him back to Germany. In Hamburg, a combination of ingenuity and business acumen turned the tagua nut into buttons and small ornamental objects, and by 1865 tagua had become an inexpensive substitute for ivory (Dominguez and Gïmez 1990:255). A new industry had seen the light of day. Ecuador and, later, Colombia became major exporters of buttons, figurines, and smaller decorative objects, mainly sold to Europe and the United States. The button industry eventually died out with the introduction of modern synthetics, but tagua production and export was revived some 10 years ago in Ecuador and Brazil in the production of "green jewelry" as an alternative to ivory banned by the Convention on International Trade and Endangered Species (CITES). It is being sold with small certificates guaranteeing its ecological soundness. I have no information regarding the social relations of its extraction and production.

17. Colonized gold extraction penetrated into the Caquetá, Putumayo, Napo, and Zamora-Chinchipe Rivers in the nineteenth century and is to a certain degree still taking place in these areas, mostly as an indigenous subsistence activity. For a detailed analysis of the effect of similar gold extraction on the indigenous communities in Madre de Dios in southern Peru, see Gray (1986).

18. Synonymous with *Castilloa* sp. In the Upper Amazon, we have only one variety, *Castilla ulei.* But two other varieties are found in the wet forests of the Pacific coast of Ecuador and Colombia: *C. elastica* and *C. tunu.*

19. Other species tapped in Latin America are in the genera *Spium* and *Ficus.*

20. *Oro negro* was synonymous with crude rubber, referring to the color of smoked rubber balls. Later, *black gold* referred to crude oil.

21. It was in fact only one of them, W. E. Hardenburg, who undertook the arduous effort of revealing the atrocities to the British and American public.

22. The annexed territory was later given back to Colombia under international pressure.

23. Two Iquitos-based weekly newspapers, *La Felpa* and *La Sancción*, had published a series of shorter articles, testimonies, and columns illustrated with caricatures, and with political, satirical content about the criminal deeds in the Putumayo shortly before the arrival of Hardenburg. The publisher and editor Benjamin Saldaña Roca had been forced by Arana and his government allies to close the newspapers and flee from Iquitos. Hardenburg succeeded in getting copies of these newspapers through Saldaña Roca's son, who had hidden them inside musical scores secreted in a grocery store in Iquitos (Collier 1968; Gray 1990). Hardenburg was thus was able to follow up on the names of the sources of the information, managing to collect sixteen transcribed testimonies sworn under oath (Gray 1990:4).

24. Hardenburg remarks, "It is to be noted that, although a year and a half has elapsed since these outrages were committed, the American Government, in accordance with its immemorial custom and in spite of our appeals, has so far done absolutely nothing on our behalf" (Hardenburg 1913:195).

25. For an excellent and detailed description of the unfolding of the case in London, see Gray (1990).

26. Andrew Gray notes the eccentric personality of Whiffen, "who spent his year in the

Amazon wearing pajamas and bedroom slippers" looking for the lost French explorer Robuchon, another eccentric "with the habit of sending his horrendous `flashing toothed' and `blood-shot eyed' Great Dane into Huitoto malocas before entering them" (Gray 1990:6).

27. The directors of the Peruvian Amazon Company were Henry M. Read, London, W.; Sir John Lister-Kaye, Bart., London, W.; John Russel Gubbins, Esq., London, N.W.; Baron de Sousa Deiro, chairman, Manchester; M. Henri Bonduel, banker, Paris; Sr. Julio César Arana, major shareholder, Iquitos; Sr. Abel Alarco, managing director, London, E.C.

28. Sir Roger Casement was later hanged in the Tower of London for high treason because of his attempt during World War I to negotiate support from the German government for the armed liberation of Ireland from British rule. He was secretly set ashore in Ireland by a German submarine but was arrested a few hours later. The story of Casement's life is a very dramatic and amazing account of an emotionally very complicated personality with a strong feeling for social justice.

29. We are talking more precisely about a *depopulation*, which does not necessarily mean that they all have been killed directly by the company's employees. Rather, they could be victims of the combined effect of killings, epidemics, collapse of the indigenous social and productive system, and migration.

30. Although systematic terror in the Putumayo went berserk, the case is not unique. Terror, slave raids, and cruel exploitation were the general situation all over the Upper Amazon, especially along the navigable rivers. The scale, scope, and perversity varied, but we have abundant references to similar abuses in travel accounts, missionaries' reports, and even official documents of various sorts. However, systematic investigation has never taken place except in the present case, which still is the only detailed documentation of a total system of rubber extraction from that period.

Another well known regime is that of the fabled rubber baron Carlos Fermin Fitzcarrald, who controlled an enormous area of the upper Ucayali and Urubamba drainage areas in Peru. He was heavily engaged in slave raiding and waged an extermination war against the Harakmbut-speaking groups in the area for opposing rubber gathering in their territory. This charismatic figure later became immortalized and romanticized world-wide in the German film maker Werner Herzog's 1980 film *Fitzcarraldo*. I have recorded first-hand descriptions of similar atrocities from other areas of the Ucayali basin.

31. The Putumayo case is well known and has been the object of several interpretations since its disclosure. More recently, the American anthropologist Michael Taussig wrote a book and some articles about the case and its "semiotic" context, looking at how the narrative was constituted and which roles the different actors ascribed to each other in the Putumayo drama. The cultural construction of terror and fear is obviously an important and legitimate issue to address, and Taussig does it very skillfully. But as a poststructural deconstructionist, Taussig gets caught in the subjective mirror-cabinet of relative truth, of discourse and counterdiscourse, turning the question of cruel genocide into cool aesthetics—"torture and terror as ritualized art forms"—and cynical exploitation into a philological venture into the "culture of terror" and the "space of death" in Casement's report. Justice is a footnote in an academic discourse (Taussig 1984, 1985) (see also Gray 1990 for a comparative analysis of the different interpretations). Genocide becomes a comfortable unreality, a problem of narrative. The problem becomes ontologized, de-politicized, and reduced to a matter of communication and intercultural understanding.

32. Casement wrote, "The Barbadians were no savages. With few exceptions, they could read and write, some of them well. They were much more civilized than the majority of their supervisors; they were certainly more humane."

33. A well-known strategy used by repressive regimes all over the world to produce willing executioners.

34. Taussig focuses on the myth of indigenous cannibalism as being promoted by the whites to create frenzied antagonism as well as allegedly forced cannibalistic rituals among the cruel young henchmen of the company as a way of creating a perfect space of fear (Taussig 1984). I see it rather as the perfect way of dehumanizing the subject/object, reducing it to flesh or a natural resource—part of wild Nature. Stories of body-snatching patrons called pistacos, pelacaras, or sacaojos are very common all over the Upper Amazon and Montaña. The most recent version came from some Asháninka who had been captured by the Shining Path guerrillas and worked as their slaves. They claimed their captors had eaten slain Asháninka, partly as a symbolic act and partly because they had no food.

35. Many of the former rubber estates reorganized their economy toward rotenone production. Large areas were planted with this native plant, used for fish-poisoning by indigenous groups all over the Amazon. These "barbasquerías" depended on indigenous labor as in rubber extraction. Production stopped with the introduction of DDT on the world market.

36. One of the most affected of the Eastern Tucanoan group were the Macaguajes living in the lower Putumayo, who were practically exterminated, with only some fifty families left around 1985 (Cabodevilla 1989:17; see also Bellier 1994).

37. Of the daily production of 170,000 barrels of production water, only 30,000 are being reinjected, leaving 140,000 barrels to be dumped daily. In addition to the metal salts and heavy metals, the residual oil amounts to approximately 600 to 30,000 gallons being discarded daily into the water systems of the upper Ecuadorian Amazon (cf. Koon 1995).

38. Most of the scientific team members came from Harvard University's Medical School and Law School.

39. The lawsuit also demands that Texaco be held accountable under the Alien Tort Claims Act for violating international human rights and environmental laws.

40. This human rights organization is the modern continuation of the Anti-Slavery and Aborigines Protection Society, which actively lobbied for British intervention in the Putumayo scandal.

41. The project was carried out from 1990 to 1993, with a follow-up project from 1993 to 1995, both financed by the Danish foreign aid agency DANIDA. It was implemented by the regional indigenous organization of the Atalaya province in the upper Ucayali River basin and the national indigenous organizations in Peru (OIRA and AIDESEP). The aid was channeled and supervised by the International Work Group of Indigenous Affairs (IWGIA), Copenhagen, an international NGO supporting indigenous peoples.

42. In 1986, the indigenous umbrella organization for the Peruvian Amazon, AIDESEP, began collecting evidence of the abuses committed by the patrons and giving legal aid to the victims. Based on the evidence compiled by AIDESEP and others, an official multisector commission was formed to investigate the denunciations. It issued a report officially acknowledged by the Ministry of Justice in August 1989.

43. One should bear in mind that in the immediately neighboring areas in Peru, thousands of indigenous peoples are still living in debt bondage to local patrons and contractors involved with lumber and coca. Hundreds of families have also been abducted by the Shin-

ing Path guerrillas, serving as serfs and cannon fodder. Thousands of Asháninka are still missing as a result of the conflict between the Shining Path and the Peruvian military, and there is a general refugee problem, which can be solved only through titling of community land on which to reconstruct an indigenous life. Unfortunately, the Peruvian government has been reluctant to back this, opting instead to exploit the temporary dislocation to promote colonization.

44. Mineral prospectors and bioprospectors have followed in the wake of this new bonanza. Presently, the extraction of *uña de gato* or cat's claw (*Uncaria tomentosa*), a spined vine used to boost the human immunological defense system, is a fast-growing industry in the sphere of alternative medicine and a major export item in Peru. It is all extracted from naturally occurring plants, and large areas are being completely stripped. This new boom has created many completely unrealistic economic expectations among local indigenous organizations, the result of propaganda from the pharmaceutical industry. A state agency for buying and export has been set up in an attempt to control this market.

45. Although the Peruvian public as well as environmentalist interests are divided on the issue, a surprising number of individuals and NGOs have been co-opted by the oil companies. Researchers from reputable European and American institutions, academics, old radicals, and activists with years of experience in indigenous and environmental rights, and large international environmental NGOs are now working as consultants for the oil industry, "mediating" with the indigenous society. The argument is well known: it is better that they do it with their expert knowledge than if the companies did it on a less well informed basis. There is, however, the snag that they all are accountable to the oil companies who pay their fees and remunerations, leaving the question of responsibility to the realm of personal ethics in a moralist universe that the history of the extractive industries in the Amazon clearly has proven highly dubious.

46. The final version of the "Global Biodiversity Strategy" presented at the United Nations Conference on Environment and Development (UNCED) in Rio in 1992 by the World Resources Institute seems to reflect this. Although the report explicitly recognizes the special rights of indigenous peoples and their importance for the implementation of the strategy, and in general acknowledges the value of cultural diversity, it has mostly turned out to be rhetoric, characterized by a conventional top-down approach to conservation. This may not be so strange considering that of the nearly fifty organizations concerned with the initiative of establishing the strategy of biodiversity, not one is accountable to local people (Gray 1991:64).

47. For an excellent analysis of the relationship between indigenous organizations and environmentalists in the Brazilian Amazon, see Conklin and Graham (1995).

48. Nature is what you can do something to: You can exploit it, harvest it, hate it, love it, save it, protect it—or you can subdue it, domesticate it, control its powers, or utilize it in a rational way.

49. The reproduction of this "naturalization" and "objectification" of the indigenous people is constantly taking place, a process in which the environmentalist agencies play a central part.

50. Indigenous Amazonians have several times expressed their concerns to me about the tendency of foreigners to increase demands for the protection of the rain forest. As one Quichua woman in Ecuador once expressed it, "The day is near when the ecological Rambo comes in and dictates to us how to live and behave in our own forests." Under cover of the global principle of sustainability, hegemonic world order may be reinforced.

51. The attitudes and behavior of the Shining Path (SL) guerrillas in the indigenous parts

of the central Peruvian Amazon are conspicuously similar to those of the rubber patrons. SL is still organizing *correrías*, raiding Asháninka settlement to procure Indian labor, and exercising a regime of terror comparable to that of Casa Arana, torturing and executing Indians.

52. An important contribution toward this goal is presently being developed in the United Nations in the form of an Indigenous Rights Declaration.

53. The idea of biodiversity and the ideological and epistemological construction of the concepts of *nature* and *rain forest* are relevant in this context but are unfortunately not within the scope of this article. For a discussion of this topic see Escobar (1996, 1997), Hvalkof and Escobar (1997), and Benavides (1992, 1993).

54. To impose protectionist policies and ecologically "sound" production techniques disregarding the rights and interests of local population inevitably leads to "green" totalitarianism. "Green" capitalist extractivism does not differ from any other form of capitalist production. Its goal is to accumulate, not to secure democracy and self-determination. The vision of the ecological Rambo conserving the political status quo may not be far-fetched.

# CHAPTER 5

## Land, Justice, and the Politics of Conservation in Tanzania

*Roderick P. Neumann*

This chapter examines the ways in which questions of customary rights of access, social justice, and protected area conservation have been entangled throughout Tanzania's modern history (figure 5-1).[1] It is guided by the assertion that environmental conservation in Africa generates its own unique politics that are deeply rooted in the history of colonial occupation. The politics of conservation revolve around contested notions of land and resource rights among different segments of African society and between African society and the state in concert with Western conservationists. Using Tanzania as an exemplar, the chapter traces the origins of modern resource conservation to the initial imposition of European colonial power and authority. While developing the conservation history of Tanzania into the contemporary period, it simultaneously traces the parallel development of peasant and pastoralist resistance to the loss of ownership of and access rights to land reserved for nature protection. The intermittent skirmishes and petty thefts of the colonial era have, today, given rise to new forms of local political movements, organized around the defense of locally defined and locally remembered customary land rights. Simultaneous with and related to these developments, the programs and policies of international conservation nongovernmental organizations (NGOs) and state agencies have begun to move toward a greater recognition of local concerns about the distribution of the costs and benefits of protected area conservation.

In colonial Tanzania (then Tanganyika), wildlife conservation was characterized by often coercive state policies that altered African settlement and land and resource use patterns (Neumann 1998). These policies were strongly influenced by the intervention of international conservation NGOs (Neumann 1996). International conservationists concentrated their efforts on encouraging colonial governments to create a system of national parks in Africa that would provide sanctuary from African and white hunting. By the end of the colonial period, they sought to prohibit all forms of human use and occupation. Thus the most significant effect of the colonial state's

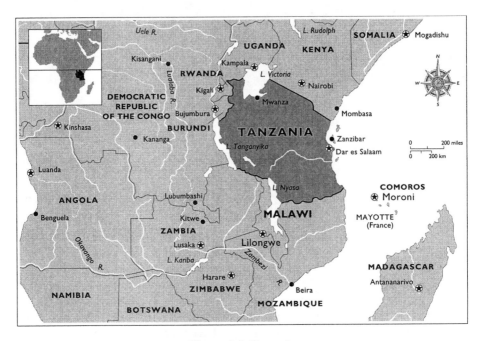

**Figure 5-1** Tanzania

protected area conservation policies on African populations was the extinguishment of rights of access. Mass relocations were often carried out in the process of protected area establishment, typically without feasible relocation plans or recourse through appeals. As one colonial officer remarked, the Sukuma peoples "were arbitrarily deprived of [their lands in Serengeti National Park] years ago, without any prior consultation."[2] Pastoralists have been the most severely affected by protected area establishment, as grazing lands that once overlapped with prime wildlife habitat are now off limits (Kjeckshus 1977; Arhem 1984, 1985; Diehl 1985).

The conditions of colonial rule limited the potential for organized political action by peasant farmers and pastoralists who objected to losing ancestral lands to national parks and reserves. Thus, "everyday forms of resistance" (Scott 1985) such as illegal hunting, grazing trespass, and fuelwood theft prevailed in colonial protected areas (Neumann 1998). Scott argues that these types of actions are aimed not at reforming the legal order but at "undoing its application in practice" (1987:447). While this conceptualization of resistance tends "to lump together many actions with different intentions and outcomes" (Isaacman 1990:32), it nevertheless advances our understanding of local responses by recognizing explicitly that wildlife conservation is political. Standard explanations of park law violations often focus on overpopulation and ignorance of conservation values, thereby emptying such violations of their political content. To the contrary, park and forest violations may often be highly politicized acts of resistance to and protest of protected area policies.

The disruption of African land use practices resulting from nature preservation programs became a point of political friction not only between Africans and Europeans, but also within the colonial government. British colonial ideology was far from monolithic and there was vocal, if not powerful, opposition to preservationists' ideas. From territorial governors down to the district officers, many administrators criticized forest and wildlife conservation proposals for disregarding African claims of customary rights. Underlying many of the debates conducted within the colonial government were conflicting and contradictory ideas about African relationships to nature. At the risk of oversimplifying British colonial-era views of African culture, we can clearly identify two ideological currents, sometimes merging but more often flowing in counter-fashion. On the one hand, there was a romanticization of pre-European African society, which included ideas of moral innocence,[3] a respect for African bush-skills, and a generalized notion of the noble savage, a mythological construction that Europeans evoked repeatedly in colonialist encounters with India, Africa, and the Americas (Grove 1995). On the other hand, there was a modernizing mission, whereby Africans would be freed from their backwardness and become efficient producers within the sphere of the British colonial economy. The former ideological current became important in the debates over whether or not to recognize customary rights of occupation and use in national parks.

Historically, struggles for justice and customary rights of access in national parks and protected areas have influenced and been influenced by shifts in the larger political terrain. The chapter follows these interactions through time by focusing specifically on the case of Serengeti National Park in northern Tanzania. Serengeti was the first national park in colonial Tanzania and so was the site of the initial debates and conflicts over customary rights and conservation goals. It has continued to be the locus of new initiatives and strategies on both sides of the conflict. From the imposition of colonial rule, through the post-World War II nationalist movement, to the most recent efforts to "democratize" national politics, Serengeti has been at the center of questions addressing the interrelationships among land rights, justice, and state conservation policies. The chapter therefore begins with a synopsis of the colonial origins of state conservation policies and their effects on African customary rights of access. This is followed by a closer examination of colonial conservation through the case history of Serengeti. The last section brings the Serengeti case into the contemporary period, reviewing the recent mobilizations for land rights and new developments in conservation approaches. A brief conclusion speculates on the potential of these developments to simultaneously improve protected area conservation and social justice in the region.

## HISTORICAL PATTERNS OF RESOURCE CONTROL AND RESISTANCE

### Colonial Land and Resource Conservation

Following Germany's defeat in World War I, the League of Nations "mandated" most of German East Africa to Britain, which renamed it Tanganyika Territory. The 1922 League of Nations Mandate agreement specified the following:

> In framing the laws relating to the holding or transfer of land and natural resources, the administrating authority should take into consideration native laws and customs, and respect the rights . . . of the native population.[4]

The Mandate subsequently provided an important influence on the system of land tenure that the Colonial Office chose for Tanganyika, outlined in the 1923 Land Ordinance. Under the Ordinance—modeled on the land law developed earlier in the Nigeria colony—all land was declared "public" (i.e., Crown) and controlled by the governor. Freehold land, less than 1 percent of the territory throughout colonial rule, was exempted from the category of public lands (James 1971). Public lands were also alienated as leaseholds, which, in many respects, functioned like freeholds. Africans were granted "rights of occupancy" on public lands, which would remain under the control of the "Native Authorities" through customary tenure systems.

Although the Land Ordinance affirmed the importance of customary claims, it nevertheless contradicted the Mandate by declaring the entire territory to be "public lands" (James 1971). This declaration was the basis for claiming all of the territories' natural resources—the wildlife, forests, and minerals—to be the property of the Crown, administered by the territorial governor (Neumann 1998). This was true even for resources on lands recognized as being under African "rights of occupancy," particularly if they had commercial value (Neumann 1997). It also meant that the government had the power to designate game and forest reserves out of African-occupied lands. The 1921 Game Preservation Ordinance more or less regazetted the Germans' game reserves and created new reserve categories. In "complete game reserves," no hunting was allowed and the governor had the power of "prohibiting, restricting, or regulating" entry, settlement, cultivation, and the cutting of vegetation. The legal framework for administering the territory's forests was established by the 1921 Forest Ordinance, which also incorporated all the previously designated German Forest Reserves. It instituted a series of prohibitions in the forest reserves, including cutting or removing trees or forest produce, firing, squatting, grazing, and cultivating.[5] One significant (and ultimately contentious) concession was the free use by Africans "of any forest produce taken by them for their own use only."[6] The type of forest produce, however, was restricted to trees without recognized commercial value.

Proposals for the creation of national parks in East Africa also began in the early years of the twentieth century. In Tanganyika Territory, the possibility of establishing a system of national parks was first brought forward by local game officials in the 1920s, although the strongest push for the idea originated outside the colony. The first indication of outside interest in national park establishment came in the form of a confidential letter sent in 1928 from the secretary of state for the colonies to Governor Cameron concerning a proposal to create an international agreement to protect African wildlife.[7] The plan for an international agreement was the product of an influential conservation group in London, the Society for the Preservation of the Fauna of the Empire (SPFE). In 1932, the SPFE took on the task of chairing a "Preparatory Committee" to draft a set of proposals for an international agreement.

The Convention for the Protection of the Flora and Fauna of Africa was held in London in 1933 and resulted in an international agreement that closely followed the SPFE's Preparatory Committee proposals. It is clear from the Convention agreement that the primary interest of the SPFE and other European preservationists was to create in Africa a system of national parks. A section in the London Convention obligated signatories—which included all of the European colonial powers—to investigate the possibilities for national parks in their respective colonies.

The SPFE believed that human populations were undesirable in parks, but it allowed that preservation could not "be pushed to a point at which it seriously conflicts with the material happiness and well-being of the native population."[8] A "native" presence in the parks might be tolerated, if these indigenous communities were found to be living in what European preservationists deemed a state of harmony with nature. The organization stressed, however, that any human activities would be controlled by park authorities.[9] If they deviated from the ideal of the natural savage, their presence became problematical with respect to the preservationists' vision. European preservationists' ideas of "traditional culture" were formed from longstanding stereotypes about African race and culture, particularly the notion that some Africans were considered to be living in a natural state. Moreover, it was the Europeans' prerogative to determine the character of primitive culture. That is, just as there was a particular European conception of unspoiled nature, which Africa represented, there was an interrelated concept of primitive human society. As will be seen, those Africans whose behavior did not fit with preconceptions of primitive man could not be allowed to remain in the national parks (the symbol of primeval Africa) regardless of their claims to customary land rights.

### Resistance to Parks from Within and Without

Throughout the period of the British Mandate, administrative officers continually criticized the national park proposals much as they did the game and forest laws. From their perspective, for example, hunting by African residents inside the proposed parks was acceptable, since "when such natives have enjoyed customary rights of hunting there is no reason or justification for depriving them of these rights."[10] Far and away the most prevalent theme of the comments of Tanganyikan officials was a concern that the parks and other protected areas would interfere with local customary rights.[11] Specifically mentioned were rights to grazing, hunting, and minor forest products. One district officer wrote that the SPFE's recommendations could not be seriously considered since they "pay no regard to native interests."[12] Consequently, Acting Governor Jardine wrote to the secretary of state for the colonies in 1933 asking that a clause be added "to the effect that the protection of vegetation in national parks does not interfere with the rights at present enjoyed by the native inhabitants to pasture or to forest produce."[13] The secretary's eventual reply points out that nothing in the national park definition makes "native" habitation inconsistent, "provided that they are controlled by Park authorities."[14]

While these debates continued within the government, Africans were defending the customary rights listed here, through actions that the state read as boundary encroachment, livestock trespass, and poaching. As might be imagined, some of the tensest confrontations involved Africans exercising their customary hunting rights and game officers trying to stop them. Field officers frequently complained of having hunting weapons turned against them. Hunters who were arrested and turned over to the native courts were rarely prosecuted. For example, a game patrol in the Sukuma section of Serengeti was castigated before the chief's *baraza* (native court) for arresting locals. The chief was quoted as telling the patrol, "We are not prevented to kill game in what you call the game reserve."[15] Cases such as this, where the entire community, including the government-salaried native authorities, refused to recognize the validity of national park and conservation laws were not uncommon (Neumann 1998).

Within the colonial government, debates over the level of recognition of African customary rights influenced the enforcement and prosecution of the unpopular conservation laws. Cognizant of the prevailing community acquiescence regarding the open violation of conservation laws, forest and wildlife officials felt communal fines were the solution. The 1921 Collective Punishment Ordinance allowed the governor to impose fines on an entire village or community if any of the members were involved in harboring criminals. The Conservator of Forests and others repeatedly requested the government to invoke the ordinance for cases of forest burning.[16] Weighing the political risks involved in applying the ordinance, the government repeatedly refused. Another proposed solution was to leverage compliance by holding community leaders responsible for violations in their areas of jurisdiction. Game Warden Philip Teare, for example, urged "headmen of a village to be held responsible for fire."[17] In one case of "disorder among the Ikoma" wherein communities were openly challenging wildlife conservation laws, the provincial commissioner punished the chief by withholding half his salary.[18] Many of these actions by the state simply raised the level of animosity between conservation officials and African populations. The history of Serengeti National Park offers an ideal case study to examine how these types of conflicts and local and national politics mutually influenced one another.

## THE COLONIAL ORIGINS OF SERENGETI NATIONAL PARK

Throughout the 1930s, the Colonial Office in England pressed the Tanganyikan government to comply with the terms of the London Convention. Specifically, it pressured Tanganyika toward establishing a system of national parks. Colonial game officials focused their attention on the Serengeti region near the border shared with Kenya. Most of the Serengeti had already been placed under special protection when it was designated a complete game reserve under the 1921 Game Ordinance, but the SPFE and others wanted a national park. As the pressure for park establishment increased, members of the colonial administration in Tanganyika voiced their concerns over the displacement of populations and the curtailment of customary access rights.

Some officials argued that African land and resource rights were protected under the Mandate and that any restrictions on access to park lands would adversely affect their livelihoods.

The government in Tanganyika eventually drafted a new game ordinance that included a clause declaring Serengeti a national park. A special committee was established by the governor to review the bill and it reiterated government concerns over indigenous rights in the proposed park. The committee recommended "that the requirements of the National Park not be allowed to interfere with existing grazing or water rights."[19] Thus, the residents of the area were allowed to remain and Serengeti became a national park in the revised 1940 Game Ordinance. The level of legal protection and administrative control was still unsatisfactory to the SPFE and associated conservationists, however, and they lobbied hard for an entirely new and distinct national park ordinance.

The final result of the conservationists' efforts was the passage of the 1948 National Park Ordinance, which created strict legal protection for Serengeti National Park and established an autonomous governing body—the Serengeti National Park Board of Trustees—to oversee its administration (figure 5-2). Significantly, the

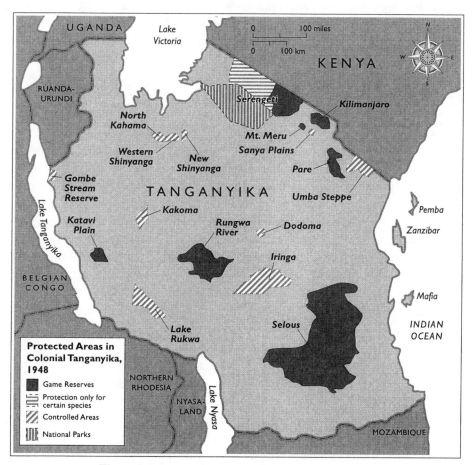

**Figure 5-2** Protected Areas in Colonial Tanzania, 1948.

ordinance explicitly permitted the unhindered movement of people "whose place of birth or ordinary residence is within the park."[20] The proposed park boundaries were immediately disputed by Africans living nearby and they were not finalized until 1951.[21] Although the secretary of the new Serengeti National Park Board of Trustees reassured the government that "the rights of the Masai . . . to occupy and graze stock in the Park are unaffected by the Ordinance,"[22] less than a week later the new park warden wrote that the trading post and cattle market in Ngorongoro must be removed.[23]

There soon followed "constant and vexatious clashes of interest"[24] between the Africans who used or lived in the proposed park—the Maasai, Ndorobo, Ikoma, and Sukuma peoples—and the park officials. One of the more intriguing documents from this period is a detailed statement prepared in the name of the "Masai of the national park," outlining their historic land claims, grazing and water tenure system, and subsequent disputes with the park administration. The document states that the Maasai have claimed the territory surrounding the Ngorongoro portion of the park at least since the beginning of the nineteenth century.[25] These ancestral claims were summarily dismissed by international conservationists.[26] Confrontations between African hunting parties and game rangers—who warned that "the situation will develop into guerrilla warfare"[27]—were not uncommon. The Ikoma, who had a strong tradition of hunting, were particularly defiant and willing to attack rangers attempting to enforce game laws.[28] They openly boasted to the rangers that they would continue to hunt as they pleased and threatened to use poison arrows against anyone trying to stop them.

By the early 1950s, a special administrative post had to be set up in Ngorongoro Crater because the "Masai were openly defying the Park laws, and the political situation had consequently become explosive and a magnet for agitators."[29] In 1951, the same year that Serengeti's boundaries were finally set, the Mau Mau Emergency was declared in Kenya, just a few hours' drive from the park headquarters. The fighting there between the African Land and Freedom Armies and the colonial government continued throughout the period of greatest unrest in Serengeti. Tanganyika authorities feared that their subjects, including Maasai, would be "infected" with Mau Mau through contact with migrant Kikuyu laborers from Kenya. In the "general state of unrest" resulting from Mau Mau and the troubles at Serengeti, the Maasai, it was observed, had "gone in for widespread cattle raids and the *moran* [warriors] have got out of hand."[30] The disputes that arose between park residents and the park administration thus evolved within the context of a rising tide of African nationalism fueled by the struggle for land and resource rights.

The government amended the Park Ordinance in 1954 to expressly deny any right of occupants to cultivate, and to give the governor extraordinary powers to prohibit any other activities deemed undesirable. Those residents who were allowed to stay were placed under strict control to ensure that they remained "primitive." The chairman of the National Park Board of Trustees wrote in 1955 that Serengeti is to be "*reserved as a natural habitat both for game and human beings in their primitive state*" (emphasis added).[31] In the mind of the preservationist, the Maasai in the park were a colonial possession and could be preserved "as part of our fauna."[32] The Maasai were imagined to be living more or less harmoniously with nature because they were nomadic, did not hunt, and generally did not cultivate. When Africans did not live up to European stereotypes, attempts were made to make them conform and, in the context of Serengeti, these attempts generated more conflict. For instance, some Masai did in fact cultivate,[33] although the National Parks Director attributed this to the Maasai having become "much adulterated with extra-tribal blood."[34] Other land use practices that have been historically important to pastoralists were unacceptable as well, particularly the use of fire to manipulate vegetative growth.[35] This practice was outlawed within Serengeti National Park by the 1940 Game Ordinance, although the park administration was unsuccessful in its efforts to stop it. Fire thus became a point of struggle between park officers and resident Maasai, who protested their predicament by starting fires "with malicious intent."[36]

As unrest among park residents grew, the preservationists' position hardened and became less ambiguous over the issue of human occupation in a national park. "The interests of fauna and flora must come first," a park manager wrote, "those of man and belongings being of secondary importance. Humans and a National Park can not exist together."[37] Subsequently, the government appointed a committee of inquiry to review the issues, examine the various proposals for reconstituting the park, and make recommendations.[38] At the heart of the committee's final recommendations was an endorsement of the principle that human rights should be excluded in any national park. The committee recommended that the national park should be reconstituted in the Western Serengeti, and that the Ngorongoro Crater sector be excised from the park and managed as a special conservation unit, the Ngorongoro Conservation Area, where Maasai pastoralists would be allowed to stay. In 1959, the National Park Ordinance (Amended) was passed. Summing up the legislation, the chairman of the Board of Trustees wrote:

> Under this ordinance the Tanganyika National Parks become for the first time areas where all human rights must be excluded thus eliminating the biggest problem of the Trustees and the Parks in the past.[39]

## Postcolonial Conservation

During the colonial period, much effort went into preventing African involvement in protected area management and conservation. As independence approached, the international conservation NGOs recognized that new tactics were needed. The International Union for the Conservation of Nature and Natural Resources (IUCN, now the World Conservation Union), African Wildlife Foundation (AWF), and others made efforts to assist the decolonized nations to plan and manage their own national parks and conservation programs. Julius Nyerere, Tanzania's first president, embraced their ideas of wildlife conservation in a 1961 speech known as the Arusha

Manifesto. International organizations immediately focused their attention on Tanzania, funneling money and technical support for the establishment and management of protected areas. One result of the international involvement has been to create a class of conservation bureaucrats, trained in Western ideologies and practices of natural resource conservation. International conservation NGOs, in sum, aimed most of their efforts for "African involvement" at government officials and party leaders. There was literally no popular conservation movement within African society. Quite the contrary, since peasant support for the nationalist party in the 1950s was partly based on discontent with colonial natural resource conservation policies, there was widespread popular opposition.

Dislocations and the tightening of restrictions on resource access continued through the 1970s and 1980s. In a 1990 paper, Henry Fosbrooke—a former conservator of Ngorongoro Conservation Area, who has spoken critically of coercive conservation practices—recounted a 1974 eviction in Ngorongoro Conservation Area:

> Without explanation and without notice they ordered the immediate eviction of the inhabitants and their cattle. Their possessions were carried out by transport of the Conservation Authority and dumped on the roadside at Lairobi. (Fosbrooke 1990)

The most recent incidence of mass relocation took place in 1988 in the Umba-Mkomazi Game Reserve complex, when over 5,000 people were evicted, many by force after refusing to leave voluntarily (Mustafa 1993; Fosbrooke 1990). As with the case of Serengeti and Ngorongoro, the initial 1951 legislation that established the reserve had granted the continued occupation and grazing rights of resident pastoralists.

The colonial and postcolonial strategies of promoting nature preservation through coercion have not helped the conservationists' efforts to "educate" the masses. Local communities have historically resisted these dislocations. Government park reports from the 1960s contain numerous references to confrontations between park officials and local residents (URT 1964, 1965, 1966, 1967). Illegal settlement, grazing trespass, and cultivation encroachment remain a perennial problem for park administrators. From the colonial period up to the present, community leaders have made repeated requests for justice and the recognition of customary rights, in written and verbal appeals through official channels (Neumann 1998). We can thus trace a distinct continuity between the colonial and postcolonial situations. Laws still vary little from the colonial period, many of the boundaries are unchanged, forced relocations have continued, African personnel are trained in practices developed in the West, and international conservation NGOs continue to play a critical role. As a consequence of the coercive nature of wildlife conservation, everyday forms of resistance to national park establishment and management policies have persisted throughout the period of independence.

Local resistance, protest, and petitions have obliged international conservationists and state authorities to reassess coercive park and wildlife protection policies. In roughly the early 1980s, international conservation NGOs began to reorient some of their energies toward encouraging "local participation" and the redistribution of the benefits of national park tourism (e.g., McNeely and Miller 1984; Miller 1984). By the first half of the 1990s, these ideas had been institutionalized within African countries' natural resource agencies and within international conservation NGOs. Around the same time in Tanzania, there was increasing domestic and international pressure for "democratizing" the electoral process. Both of these processes, the search for a "new approach" to conservation (e.g., Baskin 1994; Fletcher 1990; Ramberg 1992) and the push toward "democratization," have altered the political context of the historic conflict between neighboring communities and conservation authorities at Serengeti.

## NEW VOICES, NEW APPROACHES AT SERENGETI

One consequence of the recent political changes in Tanzania has been the unmuzzling and amplification of historically silenced voices in the struggle for customary rights of access. To understand why this is so, we first need to briefly examine the national political context in which conservation policies were developed and practiced since independence. During most of the postcolonial period, all mass political organizations (e.g., women, workers, youth) were under the leadership of the sole legal party, *Chama Cha Mapinduzi* (CCM, Party of the Revolution). Tanzania was governed through a particular form of "statism" that was characterized by the expansion of the state sector (including nationalizations of banking, agriculture, and industry) combined with the state control of mass organizations through CCM (Kiondo 1992:34). Essentially, the mass of peasants and workers were denied the right to organize outside the party and thus could not counter the formation of a bureaucratic ruling class (Shivji 1992).

Tanzania's political landscape was recently reconfigured when in May 1992 the Tanzanian parliament approved a multiparty electoral system, accompanied by a loosening of the party's grip over mass political organizations. This has allowed for a mushrooming of new political parties and new activist groups organized around issues of justice, land rights, and the environment. There has been a dramatic increase in the number of domestic, environmentally oriented NGOs in Tanzania. Organization and coordination are increasing as evidenced by the founding in December 1992 of the Tanzania Environment NGOs Networking (TANEN), an assemblage of twenty individual NGOs.

Pastoralists whose property rights have been adversely affected by national park establishment, particularly at Ngorongoro and Serengeti, have also organized. Groups have been registering with the government as NGOs and establishing contacts with other NGOs and institutions, domestically and internationally. In 1992, the Pastoral Network of Tanzania was established by an alliance of pastoralists, NGO researchers, and donor representatives. The first and most visible example is the Korongoro Integrated Peoples Oriented to Conservation (KIPOC). The organization is concerned primarily with the defense of the culture and rights of "indigenous minority peoples"

(KIPOC 1992:1) in the Ngorongoro District (on Serengeti's eastern boundary) and their self-directed economic development. Specific attention is given to initiatives to "restore legal and political respect to community ancestral lands" (KIPOC 1992:7) and for "integrating community development with nature conservation" (KIPOC, n.d.). A second pastoralist organization, Inyuat e-Maa, was established in 1991 to promote self-directed "economic and cultural development" of Maa peoples in Tanzania (Inyuat e-Maa, n.d.). Like KIPOC, Inyuat e-Maa emphasizes "integrating community development with nature conservation" (Inyuat e-Maa, n.d.). And, like KIPOC, the organization recognizes the past and present threat of protected areas and agricultural expansion to their economy and culture (Oitesoi ole-Ngulay 1993:5). A third group, calling itself the Ngorongoro Conservation Peoples Saving Trust, has organized to rectify perceived injustices by the Ngorongoro Conservation Area Authority (ole-Saitoti 1994). The common interest binding all these groups is a concern with land rights and justice, often related directly to national parks and game reserves.

The newly organized pastoral groups emphasize, in varying degrees, the interrelationships between the recognition of customary land and resource rights, economic development, and the environment. The following, from KIPOC, is an example:

> The required focus of action is authentic measures geared to speed up the restoration social justice and environmental harmony. (KIPOC 1992:22)

Inyuat e-Maa similarly relates Maasai cultural survival to environmentally conscious development:

> We shall try to diversify our economy by protecting the abundant wildlife within our range resources and making use of it. We could create multiple land use units (livestock, wildlife and tourism) in such areas and claim hunting and camping fees. (Oitesoi ole-Ngulay 1993:5)

Most significantly, activists for pastoralist rights are using their newly empowered voices to highlight past injustices and displacements resulting from protected area conservation. Examples abound, but the essential representation is found in a KIPOC position paper:

> Meantime, under pressure from the powerful preservation lobbies in the North, extensive tracks of quality rangelands have been carved into wildlife preserves for exclusive use by wild animals and tourists, the latter from the affluent society. In that pursuit African regimes have carried on eviction of indigenous peoples and denied them access to resources vital to the viability of their flexible transhumance system of utilization of their land. These losses of land are accompanied by denial of access to critical sources of water and salt licks as well as sacred sites of worship and burial. This process of displacement, launched in the colonial era, continues to date with ever increasing momentum. (KIPOC 1992:14)

Clearly, activists for pastoralist rights are articulating a pointed critique of wildlife conservation practices. They offer a perspective on the history of nature protection that sharply contests the standard conservationist narrative of a morally directed mission that transcends politics (Anderson and Grove 1987; Caruthers 1989). Protected area conservation, as historically practiced in Tanzania, represents for these activists a threat to livelihood and so serves as a rallying point in grassroots political organizing. In this discourse, pastoralists are portrayed as victims of human rights abuses generated by conservation practices, rather than as encroachers on protected areas:

> The victims are shown relocation areas [and] the authorities do not bother to provide even the very basic humanitarian resettlement services. (KIPOC 1992:15)

Activists have learned that without legal security of tenure, agreements are meaningless:

> We have suffered a lot with our land being alienated, first, to protected areas (National Parks). . . . In much of our land or areas, the whole [land tenure] situation is chaotic and without rules. (Oitesoi ole-Ngulay 1993:5)

In response to a history of displacement, rural activists' rhetoric at times becomes incendiary:

> I am now blowing an alarm that unless something is done as soon as possible this land alienation will lead to social disruption. How long can one tolerate being treated as a non-citizen of the area in the land of their birth? (ole-Saitoti 1994:1)

International conservation NGOs are now concentrating their efforts on trying to cool down these conflicts with local communities and open new possibilities for dialogue and community participation. Almost simultaneously with the rise in pastoralist activism and political organizing, there has been a shift in conservation thinking toward integrating rural development issues with conservation goals in the region surrounding Serengeti National Park. Driven by concerns over "increases in poaching, unplanned fires, and illegal tree cutting" (Mbano et al. 1995:605), a collection of NGOs and government agencies convened a workshop in 1985 to launch the Serengeti Regional Conservation Strategy (SRCS). As it was seventy years ago, the question of "the compatibility of pastoralism and conservation" continues to be a pressing issue in the greater Serengeti region. The basic goal of the SRCS is to find "long-term solutions to the resource use conflicts threatening conservation" in the area (Mbano et al. 1995:605). The "new approach" to conservation represented by the SRCS is in part based on the assumptions that

> local communities are committed to conservation . . . through being directly involved in [wildlife] management and utilization and through receiving direct benefits [and that] local communities achieve sustainable use of other resource in the region through ownership of land and village-generated land use plans, thereby reducing pressures on the resources of the protected area. (Mbano et al. 1995:606)

SRCS has been funded and supported by a variety of sources including the Norwegian Agency for International Development (NORAD), the Frankfurt Zoological Society (FZS), and IUCN.

Also coming out of the SRCS workshop was an effort by AWF to increase the cooperation between the park and neighboring communities (Bergin 1995). From 1985 to 1990, AWF conducted a pilot program in three villages in the Ngorongoro District near the park's northeastern boundary that has since become an integral part of Tanzania National Park's (TANAPA's) protected area management strategy and is now called the Community Conservation Service (CCS). The basic idea is to employ a full-time TANAPA "community conservation warden" for activities directed toward listening to community concerns and finding and initiating possibilities for common-interest planning. CCS has important implications for the historic conflicts between state resource management agencies and the communities surrounding Serengeti National Park. In coordination with SRCS initiatives, it has created a new institution for channeling community concerns to protected area management. By doing so, it has the potential to at least reduce the level of animosity and mistrust that many peasants and pastoralists have held toward the national park.

Most important, SRCS has incorporated pastoralist concerns for customary land and resource rights into its activities. Communities on the northeastern boundary of Serengeti were interested in securing group title to village lands and reducing cattle raiding on their herds. As part of the SRCS, these local concerns were addressed in a village land titling initiative. The logic of this action is that "land tenure is an essential prerequisite to promoting effective land husbandry" (Mbano et al. 1995:609–610). The goals of SRCS were thus joined with the goals of the local land rights activists in a program of demarcating villages and securing title deeds. Securing title deeds for pastoralist communities on the boundary of Serengeti National Park was KIPOC's first major effort at community organizing and helped launch its campaign for land rights (KIPOC 1992:17).

## CONCLUSION

Analyzing the politics of conservation from a historical perspective elucidates an important force behind the emerging land rights and justice movement in rural Tanzania. Past protected area conservation policies have been complicit in creating the climate of land tenure insecurity and conditions of underdevelopment within which many rural African communities exist. The establishment of virtually every national park in Tanzania required either the outright removal of rural communities or, at the very least, the curtailment of access to lands and resources. The historical processes of colonialism and postcolonial nation-building thus shaped the basic relationship between peasant farmers and pastoralists and the conservation regime. From the perspective of pastoralist political activists, numerous injustices have been carried out by the state in the name of wildlife conservation. The fact that pastoralist voices speaking out against conservation as usual are now heard loudly at international conferences and workshops is in itself a remarkable historical shift in Tanzania's conservation politics. Rural activists have incorporated the potent rhetoric of sustainable development and human rights into their struggle, an action that heralds a new assertiveness.

It is not, however, simply that wildlife conservation policies and practices are subject to the political forces in which they have developed and evolved. This has long been recognized by conservationists. The point is that these policies and practices create their own politics that extend well beyond issues of wildlife conservation. It is not clear that the new approach fully addresses this fact. Rather, the main interest lies in finding innovative ways to secure existing protected areas against historically hostile neighboring communities. In the SRCS, for example, the "immediate responsibilities are to ensure the integrity of the boundaries against human encroachment" (Mbano et al. 1995:608). TANAPA's CCS, although it turns the focus of protected management toward rural community development, has fundamentally different aims from those of many of the land rights activists in that it does not address past dislocations. Neither CCS nor the other new initiatives in the SRCS include the restoration of lost access or utilization of park lands and natural resources.

Local resistance to the loss of access rights to land and resources has motivated new efforts by international conservation NGOs to redistribute tourism benefits and promote social welfare in communities adjoining protected areas. Continued pressure from "below" will necessitate further attention to questions of land rights and justice. Increasingly in contemporary cases, local groups, often through the formation of indigenous NGOs, are demanding autonomous control of land and resources, which they view as customary property rights that have been usurped by the state. In this context, "it is often sociopolitical claims, not land pressure per se, which motivate encroachments" into protected areas (Fairhead and Leach 1994:507). Local demands can be politically radical, and most international conservation NGOs and state authorities are reluctant to go so far as to grant sole control of forests and wildlife habitat to villages or other local political entities. Local participation and local benefit-sharing, however, are not the same as local power to control use and access. Yet, in the end, this is what many communities seek.

So far, pastoralists are the main social group organizing to redress the perceived injustices of wildlife conservation in Tanzania. Other affected groups, such as peasant farmers on other park boundaries, have not yet organized around similar issues. The potential exists, however, for a much more widespread and comprehensive political struggle over land and resource rights in protected areas, such as developed as part of the nationalist movement in the colonial period. Provided with new democratic openings, pastoralists are moving away from "everyday forms of resistance" and protest toward more organized and formalized forms of political action. It is difficult to predict what new structures and policies for wildlife conservation will emerge as a result of their activism. Land rights activists have, however, made it clear that wildlife conservation issues cannot be addressed without considering broader struggles for human rights and social justice.

## NOTES

1. This chapter is a revision and synthesis of two articles previously published in 1995, "Local Challenges to Global Agendas: Conservation, Economic Liberalization, and the Pastoralists' Rights Movement in Tanzania," in *Antipode* 27(4):363–382; and "Ways of Seeing Africa: Colonial Recasting of African Society and Landscape in Serengeti National Park," in *Ecumene* 2(2):149–169.

2. Coordinating Officer, Sukumaland Development to Provincial Commissioner, Lake Province 5/5/48, TNA Secretariat File 34819.

3. Philip Curtin (1964:225) explains that the "Christianized noble African" was a common representation of African people among antislavery writers of the nineteenth century.

4. Article 8 of the Trusteeship agreement quoted in James (1971:17).

5. Tanganyika Territory Forest Department, Forest Rules of 1921.

6. 1921 Forest Ordinance cited in "Extract from the Report by Sir Sydney Armitage-Smith on a Financial Mission to Tanganyika Territory," 26 September 1932, TNA Secretariat File 21559.

7. Confidential letter from Secretary of State for the Colonies Amery to Governor Cameron 24/2/28, TNA Secretariat File 12005.

8. Report of the delegates of the International Congress for the Protection of Nature, Paris, June 1931, to His Majesty's Government, TNA Secretariat File 12005.

9. "Note on the Convention," anon., n.d., TNA Secretariat File 12005.

10. Under-secretary of state to SPFE, 2 October 1939, TNA Secretariat File 12005.

11. Summary of "Observations on the Report of the Preparatory Committee," n.d., TNA Secretariat File 12005.

12. Comments on Major Hingston's "Report on a Mission to East Africa" 9/4/31, TNA Secretariat File 12005.

13. Acting Governor Jardine to Secretary of State Cunliffe-Lister, 1/8/33, TNA Secretariat File 12005.

14. Secretary of state to Governor MacMichael, 17/3/34, TNA Secretariat File 12005.

15. Quoted in "Report from the Game Department on Game Scouts' Patrol in the National Park," 24 October 1946, TNA Secretariat File 35773.

16. CF to CS, 7 April 1930, TNA Secretariat File 10948. CF to CS, 6 May 1937, TNA Secretariat File 24595.

17. Swynnerton, director of tsetse fly research, to CS, 25 April 1930, TNA Secretariat File 10948.

18. PC, Lake Province, to CS, 18 August 1938, TNA Secretariat File 13371.

19. Report of the "Special Committee Appointed to Examine the Game Bill, 1940," 16/4/40, TNA Secretariat File 27273.

20. Tanganyika Territory National Parks Ordinance, 1948.

21. Memorandum no. 82 for Executive Council, 22/8/50, TNA Secretariat File 34819.

22. P. Bleackley, secretary, Serengeti National Park Board of Trustees to Member for Local Government, Dar Es Salaam, 18/10/51, TNA Secretariat File 10496.

23. Minutes of the second meeting of the Serengeti National Park Board of Trustees, 23/10/51, TNA Secretariat File 10496.

24. Tanganyika Territory, "Proposals for Reconstituting the Serengeti National Park," Government paper no. 5 of 1958, p. 3, University of Dar Es Salaam Library, East Africana Collection.

25. Masai of the national park (signed in behalf of Masai with thumbprint of Oltimbau ole Masiaya), "Memorandum on the Serengeti National Park," 1957, (5 pp.) RH, MSS. Afr. s. 1237b.

26. Russel Arundel, "Petition to Alan Tindal Lennox-Boyd, Secretary of State for the Colonies," presented on behalf of the American Nature Conservancy, National Parks Association, and Wilderness Society, among others. FPS Box Af/X1/NP.

27. Game Ranger, Bangai Hill to PC, Lake Province, 25 July 1938, TNA Secretariat File 13371.

28. A game patrol arresting poachers in Serengeti near Musoma had to retreat in the face of Ikoma and Sukuma hunting parties who wanted to attack them. Report on Game Scout's Patrol in the Serengeti National Park from Game Department, Bangai Hill, Musoma, 24 October 1946, TNA Secretariat File 35773.

29. District Commissioner Masai/Monduli to District Officer Ngorongoro, 5/3/55, TNA Arusha Regional File G1/6, Accession No. 69.

30. Governor Twining, Tanganyika, to Governor Baring, Kenya, 15 December 1953, PRO CO 822/502.

31. Notes on a meeting between the chairman of the National Park Board of Trustees, the director of National Parks, and the provincial commissioner, Northern Province 28/3/55, TNA Arusha Regional File G1/6, Accession No. 69.

32. Barclay Leechman, chairman of the Serengeti National Park Board of Management, in the minutes of the Serengeti National Park Board of Management meeting, 23/7/53, TNA Secretariat File 40851.

33. Acting PC, Northern Province, to director of National Parks, 8/6/55, TNA Arusha Regional File G1/6, Accession No. 69. The PC pointed out that a recent census had determined that 82 out of 216 families cultivating in the Crater are Maasai.

34. Tanzania National Parks Director Molloy to the provincial commissioner of Northern Province, Report on Human Inhabitants, Serengeti National Park, 8/6/55, TNA Arusha Regional File G1/6, Accession No. 69.

35. Government professionals knew that fire was a critical element in Maasai pasture management, eradicating disease-bearing ticks and maintaining grasslands. Infested pastures would be temporarily abandoned and burned until the threat of disease had been removed. Fires in the highlands surrounding Ngorongoro Crater and elsewhere would open up forest glades of high-quality forage. Tanganyika Territory, Annual Report of the Department of Veterinary Science and Animal Husbandry, 1929, p. 8, and Tanganyika Territory, Annual Report of the Department of Veterinary Science and Animal Husbandry, 1933, p. 81.

36. National Park Director Molloy to District Commissioner Masai/Monduli, 8/12/55, TNA Arusha Regional File G1/6, Accession No. 69.

37. Wilkins, SNP Board of Management, to SNP Board of Trustees, 16/2/54, TNA Secretariat File 10496.

38. *Report of the Serengeti Committee of Enquiry, 1957* (Tanganyika Territory, 1957).

39. Tanganyika National Parks, Reports and Accounts of the Board of Trustees, July 1959 to June 1960.

# CHAPTER 6

## Rebellion, Representation, and Enfranchisement in the Forest Villages of Makacoulibantang, Eastern Senegal

*Jesse C. Ribot*

In the district of Makacoulibantang in Eastern Senegal (figure 6-1), scores of villages are actively blocking urban-based woodfuel merchants and their migrant woodcutters from working in surrounding forests. Their rebellion is partly to stop the destruction of a resource on which they depend for daily needs, and partly to reap some of the benefits from woodfuel production and commerce. Local villagers cannot enter the woodfuel (firewood and charcoal) trade, since, as it now stands, urban-based merchants employ migrant woodcutters and use state-allocated licenses and permits to control access to urban markets where the woodfuels are sold and consumed. Forest villagers have resorted to blocking direct access to forests, since this is about the only influence rural populations have on the woodfuel sector. However, while villagers can control forest access, without access to markets and forest labor opportunities, they themselves can reap few benefits from forest exploitation. They can keep others out of surrounding forests, but they cannot enter forest commerce themselves. In short, the fact that forest villages can control direct access to forests does not give them access to the benefits that flow from forest commerce.

Most villagers in the area do not want woodfuel being cut from surrounding forests. But despite their objections, some village chiefs, who are usually hereditary powers appointed for life, allow production in village forests. Their decision is based partly on payoffs from merchants, and partly on the social status of merchants, which makes it difficult for chiefs to turn them down. Villagers complain but do not challenge their chiefs' decisions. This schism between chiefs and villagers raises the issue of enfranchisement and accountability of representation in local resource control. Makacoulibantang is an example where control is indeed local, but the results are still negative for the majority of the population since the chiefs make the decisions and reap most of the benefits, while the social and ecological costs of forest

134

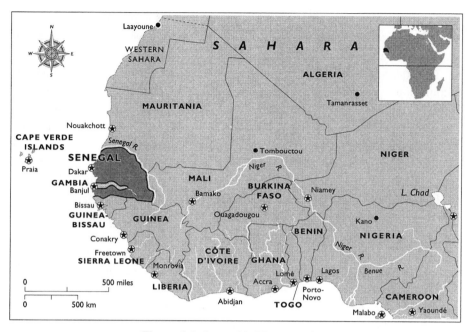

**Figure 6-1**  Senegal in Western Africa.

clearing are spread over the village as a whole—disproportionately affecting women and poorer households.

The story of resistance in Makacoulibantang illustrates two points concerning justice in local community natural resource use and management. First, without locally accountable representation, *local* control of forests (or of anything else) is not necessarily *community* control (see Ribot 1995, 1996). Villages are highly stratified, and elites—such as village chiefs—often make decisions that are not in the interest of the community as a whole. This is why locally accountable community representation—in the political sense—is an essential part of community participation. Second, community participation involves influence over the disposition of resources (natural, financial, etc.); access to values derived from their use, production, and exchange; and a voice in shaping the larger political economy in which practices and policies are made and contested. Effective enfranchisement requires locally accountable representation with real powers over resources, benefits, and policies. Neither representation without power nor power without representation can be considered community participation.

The story of Makacoulibantang illustrates a third, closely related point: control over forest resources—via property rights, threats of violence, or any other means—does not in itself confer benefits on local populations. Access to markets and labor opportunities are also necessary. Instituting local (or community) property rights, a common prescription in participatory development projects, is not in itself a sufficient policy tool for establishing local or community "participation" in forest benefits. Property rights are often presented as a means of giving communities or individuals access to the benefits from the resources around them. But property is fetishized to the exclusion of inspecting the many other dynamics involved in the devolution of benefits to local communities (see Ribot 1998b; Ribot and Peluso, in press).

Over the past decade, the idea of local participation in resource management has become popular in development circles (Cohen and Uphoff 1977; Satish and Poffenberger 1989; Peluso 1992b; Banerjee et al. 1994).[1] Participation in forestry often aims to devolve forestry decision making and benefits to rural populations, along with responsibilities for forest management. Such devolution is predicated on a number of assumptions about higher efficiency of local resource management as a result of greater local knowledge, lower transaction costs as a result of proximity to forests, and better decision making as a result of the integration into commercial decisions of opportunity costs (such as the loss of subsistence food, fodder, and game). Equity arguments are also often made. Devolving control of or benefits from forests to local populations, or even just incorporating their labor into forest management, is a complex matter. As this study indicates, transferring control or benefits to local populations must first address to whom that transfer will be made. Second, influencing such transfers requires identifying what exists to be transferred, and from whom and by what means, if meaningful decision-making powers and benefits are to be devolved.

This chapter points to enfranchisement as a core element of community participation in environmental management. It also puts property in place as only one mechanism at work in determining who can benefit from or protect natural resources.

## THE CASE OF MAKACOULIBANTANG

### Méréto—A New Spirit of Rural Resistance

I first stayed in the forest village of Méréto in 1986 to study the rural social and ecological consequences of the charcoal trade (see figure 6-2). In this forested area of Eastern Senegal, migrant Fulbe woodcutters from neighboring Guinea have been plying their trade since the mid sixties. They come to cut trees, partially burning them to produce charcoal for sale in the nation's capital, Dakar.[2] When I arrived in the mid eighties, many villagers complained about the woodcutting in surrounding forests, but few did anything about it. The forest villagers were upset, but passive, about having no rights to protect surrounding forests from woodcutters. When asked who owns the forests, they said with resignation, "the Forest Service." Officially, this was true. The forests are "national domain" under forest service management (RdS 1964, 1994). It was the Forest Service that could give licenses and cutting permits, and they gave them to urban merchants and their migrant workers (Ribot 1990, 1995.)

When I returned in July 1994, I found transformed attitudes about who should control and have access to forests. Villagers were complaining about woodcutters, while the woodcutters and their patron-merchants were complaining about village resistance. Villagers, however, were not just complaining about the woodcutters. They described chasing them from the forests and even trying to change the laws to ban charcoal production—the woodcutting being for charcoal. In short, formerly passive villagers seemed to be taking control of the issues of forest access: they were beginning to take matters into their own hands.[3]

In Méréto, villagers spoke in excited whispers, expressing amazement about how the neighboring rural community of Makacoulibantang had organized and kicked out the woodcutters. They explained that people there had initially succeeded in stopping charcoal production in their area. A series of meetings with the foresters, merchants, village chiefs, and local elected authorities ensued, resulting in the return of the woodcutters, but also in some areas being protected from charcoal production. Villagers were taking control of the forests. These stories put me on the road to Makacoulibantang, some thirty kilometers southeast of Méréto. I spent a week conducting interviews and returned in November 1994 for three weeks after having gathered available documents in Tambacounda and Dakar.

## Setting—Administrative and "Customary" Authorities

Makacoulibantang is a rural community (a territorial-administrative unit discussed later) at the edge of the largest charcoal production zone in Senegal. Maka is the village that is the capital of the rural community of Makacoulibantang. Figure 6-2, a map of the zone, shows the main charcoal production areas, the Makacoulibantang study area, and the cities where the charcoal produced in these forests is ultimately consumed. The villagers in this area are primarily millet and peanut growers. Makacoulibantang has

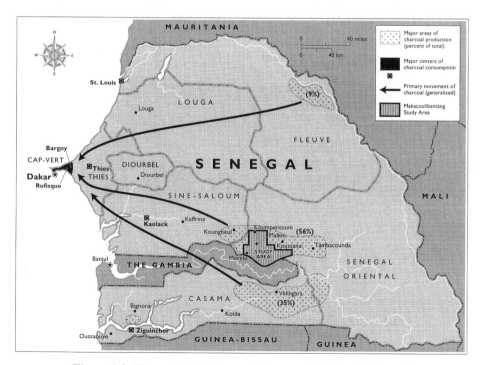

**Figure 6-2** Charcoal Production and Consumption Areas in Senegal.

no electricity, running water, or paved roads. The regional capital, Tambacounda, is eighty kilometers away, just over half on paved roads—a formidable distance for villagers to travel.

In official language, the rural community is considered the "base community" of rural political and administrative functions of the Senegalese government. Each of Senegal's 317 rural communities contains between ten and fifteen villages, and between 2,000 and 15,000 people. Two to seven rural communities are regrouped into arrondissements. Several arrondissements make up a department, and departments are the largest subdistricts of Senegal's ten regions.

The rural community, Senegal's smallest unit of local government, is governed by the rural council. Figure 6-2 shows where the rural councils are located in the larger political-administrative scheme. The important official actors at the local level are the rural councils, the subprefect and prefect, the Forest Service, and the Interior Ministry's Local Development Offices [multipurpose rural extension centers, centres d'expansion rurale polivalent (CERP); see figure 6-2]. The Forest Service, in theory, relates to the rural community through CERPs, which are responsible for rural development,[4] but CERPs play virtually no role in this relationship. At the most local level, the rural councils are the only elected bodies. The prefect and subprefect are administrative appointees of the interior minister, and the Forest Service staff are a professional technical service of the Ministry for the Protection of Nature. The CERPs are constituted of appointees of the various rural services and representatives from the rural councils.

Three-quarters of the representation on the councils of Senegal's rural communities is made up of members elected by universal suffrage and one-quarter of representatives chosen by a general council of cooperatives (commercial economic interest groups, mostly agricultural) operating within the rural community. Slates of candidates for the council of the rural community are presented by nationally registered political parties for election. The slate with the majority of votes is elected.[5] Once elected, the council chooses a president of the rural community from among its members. Independent candidates *cannot* present themselves for election. The president of the rural council officially represents (and is accountable to) the prefect, who is appointed by the minister of the interior (RdS 1992:4–13). The prefect must approve all decisions of the rural council, hence the prefect officially retains ultimate decision-making powers within the rural communities.

There are also "customary" rural authorities. Each village has a chief. This chief is proposed by the subprefect after consultation with the "heads" of the village's households, who elect a chief from among themselves. The chief is then named, for life, by a decree from the prefect. The decree must also be approved by the minister of the interior (RdS 1972:968). Officially, the chief works under the authority of the subprefect and the president of the rural council, and is charged with the application of laws, administrative decisions, and decisions of the rural council. The chief also collects a head tax of FCFA 1,000 (US$2) per year from each adult. The village chief keeps 4 percent and the remainder goes into the state coffers. In practice, most chiefs

gain their position through inheritance via a male lineage traced back to the founding family of the village, to warriors, or to families chosen by colonial powers to replace antagonistic local leaders.

Village chiefs and the rural councils of local state governance structures are frequently assumed to represent rural populations in participatory development and
natural resource management projects. Across the West African Sahel, *villages* are
the most common unit of social aggregation around which local use and management of woodlands is organized both by local populations and by outside agents.
Each village, ranging from less than 100 people to well over 1,000, typically has a
chief, and some have specialized chiefs overseeing forest use. There are also other
authorities within villages, such as imams, marabouts, sorcerers, non-village-based
pastoral chiefs, griots, and heads of certain castes. Colonial rulers, however, relied
on village chiefs, disproportionately shifting power to them. While other figures are
involved in resource management, most state and outside organizations still privilege chiefs as their primary village interface.

Given the manner in which chiefs are currently chosen, they neither represent
nor are they systematically accountable to the village as a whole. Most heads of
households are men, and most household compounds have about ten or twelve persons, which means that the chiefs are chosen by a small male minority. They are selected through a process that empowers less than 10 percent of the population and
systematically underrepresents women. These facts, plus appointment for life, compromise accountability. Although there can be various local mechanisms of accountability (see Spierenburg 1995; Fisiy 1992), the accountability of chiefs is by
no means assured—some are despots, others responsive leaders, depending on the
chief's personality, the specific history of the village in question, and its location in
a larger political economy.

### Building Tension in Makacoulibantang, 1990–1993

In 1990, villagers in Makacoulibantang and two neighboring rural communities
asked their rural council to protect the surrounding forests from migrant woodcutters (called *surga*).[6] The region around Makacoulibantang had been an official charcoal production zone in the late seventies and early eighties. After production in the
area was officially closed by the Forest Service in the early eighties, many woodcutters stayed on, working the forests illegally. Villagers always had had a tense relation with the migrant charcoal producers, but they had managed to coexist with
only scattered conflicts over the years. In 1990, tensions began to rise. Nothing special had changed within charcoal production itself, but the forest villagers were no
longer willing to put up with existing woodcutting arrangements.

The villagers had two sorts of complaints. First were the inconveniences of having migrant woodcutters in their forests. There is a history in this region of woodcutters destroying plants used for food, fodder, medicines, and dyes, as well as wood
for house construction and cooking fuel (Bergeret and Ribot 1990; Dia 1985; Niang

1985; Ribot 1990:82–85; Tall 1974). A rural councilor feared that deforestation would inhibit the rains (personal communication, Maka, June 1994; Ribot 1990).[7] Villagers also complained, "Surga chase women who are collecting forest products, they cut areas we use for grazing, and they start bush fires" (villagers of Maka, June 1994). Their complaints were not new,[8] but they were mounting. Second, a new set of complaints emerged. Villagers were asking for a share of the benefit from the lucrative woodfuel trade. They wanted labor opportunities in woodcutting—they wanted permits and quotas that would allow them to cut wood, carbonize, and sell the charcoal. They wanted to touch some of the revenues flowing from the commercialization of their forests.

While the desires of the rural population to halt outside commercial producers, to have forests for local use, and to be able to commercialize the forests themselves may appear contradictory, they are not. There is no reason that villagers—who use these forests for many different purposes—cannot include commercial exploitation as one of their uses. They also still value the forests for the other uses and may be able to conserve and exploit them at the same time—something like managed or "sustainable" (whatever that may mean) use. An advantage of the villagers' carrying out this exploitation, rather than allowing outsiders to do it, is that the contradictions between and within villages concerning the different uses of the forests must be fought and sorted out internally and locally. The local retention of the added income from commercial activities is, of course, another benefit. The situation is very similar to that of India's Chipko movement, which was highly misinterpreted (see Rangan 1996). Many outsiders saw the women of Chipko as wanting to protect the forests strictly for subsistence uses, but the movement also involved people who wanted to protect local forests from outside producers so that they could cut and sell some of the trees themselves. Just because Chipko was a local and indigenous movement did not mean that it was rooted in some kind of primordial conservationist ethic.

In 1990, the rural councilors joined the rural population against charcoal production in the zone (villagers, rural councilors, assistant to the subprefect, Maka, June 1994). The councilors told the prefect they wanted forests protected for use by the rural population. They also wished to see revenues retained for local development, pointing out that they "couldn't tax woodcutters and it was only foreigners [i.e., migrants] that were working." The rural councilors asked to charge a fee on woodfuel production to create a development fund under their jurisdiction.

Makacoulibantang's deputy in Senegal's National Assembly explained: "Villagers wanted to stop charcoal production in the area. They decided not to give the woodcutters a place to stay in the villages. Villagers wanted to fight woodcutters in the forests. The population wanted to force them out" (Deputy Kabina Kaba Jakhaté, July 1994). Villagers began chasing woodcutters out of the forests and complaining to their representatives. They also began to actively resist charcoal production.

Villagers discussed the issue with their councilors; the president of the rural community went to the prefect in Tambacounda, and in 1990 the prefect signed a decree closing the area to charcoal production (assistant to prefect, Tambacounda, June

1994). Some of the woodcutters stayed in the area in villages more friendly to them, and many left. The tensions subsided. But in 1993 the Forest Service scheduled new production "rotations" in the area.[9] This was normal forest management procedure and it was time again for production in the zone. The Forest Service scheduled production in Makacoulibantang for 1 January through 31 August 1993. The regional and departmental Forest Service agents told me that the forests of this region are managed forests in which it was time for the return of production rotations. They explained,

> If we did not go to Maka to produce, first the rotation would not be respected; and second, the other regions would react if the production was moved to their forests. Other rural communities would not accept what Maka rejected. We, the Forest Service, are the managers of the forest domain. (Koumpentoum, July 1994)

In early 1993, woodcutters started showing up with permits in hand. Villagers were angry. They thought charcoal production had been banned. The rural council again represented the villagers and demanded that the Forest Service not reopen the charcoal season. Some villagers threatened violence if more woodcutters showed up. But more woodcutters did show up and were installed by their merchant patrons in villages throughout the area. The merchants negotiated with village chiefs, and some chiefs allowed them in. In those villages that allowed the woodcutters to stay, most villagers did not concur with the chief's decision, but they did not challenge it.

Other villages did not admit the woodcutters. Many villagers cited the prefectorial decree as proof that they did not have to put up with the woodcutters any more. Those that resisted threatened woodcutters verbally, chasing them from the forests. Rural councilors recounted that "one Marabout [Islamic religious leader] refused to give access to the woodcutters. He took out his musket and threatened to shoot anyone cutting within three kilometers of his village." (No incidences of violence were reported.) Some chiefs and villagers also excluded the woodcutters by not allowing them to stay in their villages. Without village access, the woodcutters had no water, food, or lodging. They could not live and work in the forests on their own.

The result was a scattered pattern of production in the majority of the area's villages. Nonetheless, as villagers talked among themselves and the movement against the woodcutters heated up, there were more conflicts and more complaints coming into the rural councilors and being passed up the hierarchy through the subprefect and the foresters.

In March 1993, two months after the Forest Service agents' opening of the season, the Forest Service had the prefect sign a decree reopening the area to charcoal production. This move was designed to help avert the emerging conflict between villagers and the charcoal interests by withdrawing official backing from the movement.[10] Foresters explained (forestry officials, Tambacounda and Koumpentoum, July 1994) that the prefect had no authority to close or open production in the forests that were under the management of the Forest Service. Further, his decree was redundant, since before the Forest Service opened the season, production was officially illegal in the area anyhow. The foresters went on to explain, "Production rotations have to proceed as scheduled. . . . . The forests serve the whole nation. They are needed to provide fuel to the cities. The forests don't belong to villagers."

The prefect did not inform the local population of his second decree. The villagers and rural councilors were upset and continued to fight the decision. The resistance went on and many of the villages whose surrounding forests would have been brought under production were quiet. The situation remained tense.

### The First Meeting—March 1993

On 31 March 1993, the subprefect called, and presided over, a meeting in Maka to resolve the tensions. The following discussion draws from the minutes of this meeting, recorded by the director of the regional Forest Service office at Tambacounda. The cast of characters at this first meeting was remarkable. The fifty-four participants included four Forest Service agents and officials, eight forestry merchants (two of whom are powerful officers in the Merchants' National Union), eleven village chiefs, the president of the Arrondissement Council, twelve of Makacoulibantang's rural councilors (including the president), three other civil servants, and thirteen other villagers.[11] The Forest Service members and the forestry merchant delegation were listed at the front of the minutes as the official parties. All others were named in an appended list.

The meeting was opened by the subprefect at 11:00 A.M. After a brief introduction by the subprefect, the head of the Tambacounda Forest Service office, Mamadou Fall, opened by expressing the wish that this meeting would result in "objective" recommendations for a good organization of forest exploitation in the arrondissement. He emphasized "the fact that *forest exploitation is a governmental decision* of which the organization is *entrusted to the Forest Service.*" Next he described "the new Forest Code [being written at the time], which anticipates a remittance to Rural Communities that manage exploitation areas," adding "the ministerial decree organizing the exploitation season defines the open regions, and in each of these regions *the Inspector has the right to define the cutting plots, not the administrative or local authorities* [e.g., the prefect, councilors, or chiefs]." The minutes continue, explaining that Mamadou Fall then "insisted on the notion of *national solidarity,* which would permit the exchange of products among the different regions of the country, favoring *socioeconomic stability*"[12] (all italics added).

The foresters' introduction was completed by the regional hunting inspector, Ansoumana Diolla,[13] who said, "The reform [the proposed Forestry Code] has clearly defined the prerogatives of local collectives and [has shown] that the *management of forest formations belongs with the Forest Service, just as the revenues generated belong to the nation as a whole.* This justifies the importance of *national solidarity*" (italics added).

At this point, Yulbe Diallo, one of the most powerful charcoal merchants in Senegal and the president of the Regional Union of Forestry Merchants, stepped in. "He reassured the local representatives that the woodcutters will install themselves in the area to satisfy the *needs of a whole people* [i.e., the nation of Senegal]. In this manner he invited the participants to express their proposals toward good management of the *common good*, the forest" (italics added).

What followed were proposals and concerns from the rural population. The minutes read:

- Oussouby Laye, President of the Local Union of Cooperatives [mostly agricultural coops], feared that reforestation actions would not come to compensate the losses engendered by forest exploitation. He also mentioned risks of worsening rain shortfalls. . . . Then he expressed his wish for the creation of an intermediate treasury in Maka for revenues [presumably for reforestation efforts].
- Ousemane Ba [rural councilor] found a contradiction in the fact that forest management would be entrusted to the Forest Service and that of lands to the rural council.
- Yancouba Laye [rural councilor] hoped that the populations would take a substantial profit from the exploitation of their forests.
- El Hadji Saré Dia [chief of the a village of Boulimbou] affirmed that the populations are obligated to bow before the decision of the government, but asked that the agricultural zones be respected by the [forest] exploitation.
- Fodé Ba [rural councilor] of the village of Manigui Colibasso, was only distressed about the border zone which must not be opened for exploitation due to the risk of creating problems between Senegal and The Gambia.
- Mamadou Sow, chief of the village of Yérodoundé, declared that his village would refuse to lodge the woodcutters.
- Kael Diallo, president of the Rural Council of the Arrondissement, argued that no person present at the meeting is responsible for the decision to open the zone of Maka to exploitation; rather this decision emanates from the peak of the administrative pyramid. From this position, he wished for good harmony in view of a good organization of exploitation by channeling the woodcutters into precise zones, while reserving for the populations the forests from which they satisfy their daily needs for forest products, and for the animals the pasture areas for the rainy season.
- Soppata Keita [villager from Coulibantang] then declared that the decision by the authorities to open the zone of Maka to exploitation will triumph despite the disapproval of the population.
- Finally, El Hadji Bourang Diallo, president of the Rural Community of Maka, related the strategic importance of the Forest of Bokko for the population and animals and wished that it would be respected so that the populations could continue to exercise their usage rights on which their survival rests. He also announced that he was the author of the petition addressed to the authorities.

These proposals by the rural population certainly look like (1) integrated ("sylvo-agro-pastoral") environmental management, (2) rural benefit retention, and (3) disapproval that the authorities were not taking the people's wishes into account. They seemed to be fighting for a bit of "participation" in decision making over their forests and the benefits derived from them. What is remarkable is how close to the cutting edge of thinking on environmental management and integrated rural development the articulated needs and desires of the rural actors are. In sum, they requested (1) protection of the Forest of Bokko for village usufructuary and pastoral uses, (2) protection of a band along the Gambian frontier, (3) channeling of woodcutters into specified areas, (4) more reforestation, and (5) benefits for the rural population.

Mixed in with these management proposals, some strong disapproval of production in the region was voiced. The merchants' and foresters' responses followed the proposals and apparently threatening concerns. The minutes go on:

- Burum Gorkati and Soppowo Yimbe, both forestry merchants, then took the floor to *calm the doubting spirits of the people* (italics added).
- To all the questions posed, the director of the Forestry Sector of Tambacounda responded very clearly, which permitted the tension to drop.

Then the president of the Regional Union of Forestry Merchants asked the representatives [rural councilors] to provide a list of villages that do not wish to have woodcutters. [Thirty-three villages were then cited in the minutes.]

- Regarding all these villages, Yulbe Diallo [the Regional Union leader] agreed to move out all woodcutters so that they can be relocated in other areas ready to welcome them.
- The last word came from the subprefect who thanked the participants for the success of the meeting before closing the meeting at 3:00 P.M.

After this first meeting, there were few changes. Tensions remained high. Some villages kicked the woodcutters out, others let them stay. Discontent among peasants was widespread. Even in those villages where woodcutters were staying, everyone I spoke with did not want them. A year later, at the opening of the next season, tensions were still high. The merchants had not yet moved their woodcutters, and the Forest Service had not stepped in to resolve the problem.

## Second Season, Second Meeting—March 1994

On the fourth of February 1994, the prefect in Tambacounda established a commission composed of four Forestry agents and officials, a rural extension agent, the president of the arrondissement council, the president of Makacoulibantang's rural council, the local cooperative union president, and two high officials from the National Union of Forestry Merchants,[14] to resolve the charcoal production problems that had developed during the 1993 season. At the first meeting of the commission, on 5 March 1994 in Maka, the subprefect explained, "The prefect asked for the creation of a commission to resolve all the problems linked to charcoal production. Concerning Maka, the commission is to determine the zones to exploit, decide the production locations of quota holders, make an exploitation plan, and assure the control of woodcutters through the establishment of files and I.D. cards." (All of these were already the normal functions of the Forest Service.)

The subprefect opened the meeting, handing the floor to the Forest Service inspector from Tambacounda. After introductory remarks about the importance of the commission, the inspector announced several concrete actions he had decided to take:

1. The Forest of Bokke would be closed to production, "in taking account of the wishes of the people";
2. Lists of the woodcutters, along with their photos, would be dropped off at the offices of the subprefect, the chiefs of the concerned villages, and the Forest Service, to help control woodcutters;
3. A four- to five-kilometer band would be maintained between the Gambian border and production zones, to avoid problems; and
4. On 17 through 19 March 1994, the commission would tour the area to implement the commission's decisions and would diffuse information about them. Any woodcutters not following the deadlines established by the commission on this tour would later be fined.

The inspector ended by saying that the commission would rapidly evict the woodcutters still working in areas restricted from the previous season, reinstalling them in authorized zones. The merchants then expressed their pleasure with the plan, agreeing to move their woodcutters.

At the end of the meeting, the president of the rural council spoke. He inquired about "financial resources that the community could derive from woodcutting." The subprefect flatly responded that "forest exploitation escapes the jurisdiction of the rural council. Like mines, it's a domain reserved for the state to whom the taxes accrue."

The subprefect then closed the meeting, thanking "the inspector, his collaborators, and other participants." He said, "We have done useful work since we have deliberated in the interest of the country and her children. The prefect places much hope in the commission—we must not disappoint him." (This was the end of second meeting.)

### Outcomes and Implementation

These official pronouncements made the Forest Service, the merchants, and some village chiefs happy. Rural councilors and other villagers, however, insisted that they were still not in agreement. Villagers throughout the area insisted that they still did not want charcoal produced in their forests. One villager pointed out that nothing was done to enable villagers to work in charcoal production themselves. (When asked why labor opportunities and quotas were not discussed in the meetings, the villagers insisted that they had been discussed in the meetings and with the Forest Service agents. These discussions were not recorded.) Rural councilors were unhappy that none of the revenues would fund local development. The rural councilors said the "solution" was imposed: it was simply not an agreement. "The big patrons (Soppowo Ba and Yulbe Diallo) and the Forest Service (Inspector Fall from Tambacounda) made the population understand that it is the right of the State to carbonize and dispose of the trees as they like" (rural councilor, Makacoulibantang, June 1994).

In sum, two of the requests made by rural populations in the first meeting were granted at the outset of the second meeting: (1) the closing of Bokko, and (2) the protection of the border area. In addition, foresters promised that woodcutters would be tracked more closely. However, the rural council and villagers were not granted two other important requests: (1) access to revenues from charcoal production, and (2) access to labor opportunities. They were given neither permits nor quotas nor the right to charge a fee for the Council. The official outcomes looked on paper like a sound, although imposed, management plan for the area. However, in practice, even these plans and promises were never implemented.[15]

As late as November 1994, eighteen of the thirty-three villages that had originally requested the expulsion of woodcutters still had charcoal producers working in surrounding forests (rural council president and subprefect, Makacoulibantang, November 1994). Production continued in the "closed" areas. The rural councils continued to complain about this illegal production, claiming that no one was enforcing the agreement. The few concessions that the Forest Service had made were not upheld. Little seemed to have changed, except that many of the villages that originally complained were no longer saying anything.

The resulting production pattern after all these meetings was a mix of villages that resisted the charcoal producers and villages that hosted charcoal producers both inside and outside the protected areas. This outcome followed much more from what individual villages decided to do than from any rules laid down by the Forest Service or pleas made by rural councils and the populations for whom they spoke. However, the decisions of individual villages were not decisions of the villagers themselves, who, on the whole, continued to oppose charcoal producers. Rather, the chiefs were making decisions in a space between villagers and charcoal merchants.

### Pulleys in the Forest Well

The councilors explained that after the last meetings they advised chiefs not to accept the agreement—or, for that matter, the charcoal makers. However, according to the councilors, the merchants went to the village chiefs and paid them off with sacks of rice. As the rural council president said, "The merchants bought the villages."

Similarly, foresters explained, "When the [Forest Service] decree [opening the production season] was made, certain village chiefs, pastoralists, and rural community representatives did not want the charcoal makers," but "village chiefs accepted kickbacks and the villagers accepted afterwards."

The Regional Forestry Merchant Union president, one of the most powerful merchants in the market, recounted:

Patrons [merchants] give advances to the village chiefs [these are subsistence advances for the woodcutters]. The chiefs work with the means provided by the patrons with an agreement that, if they are caught [by the Forest Service] they are not to say who they work for. . . . There was a problem in Maka in that there were fraudulent charcoal makers. Village chiefs had surga [woodcutters] carbonizing for them. Village chiefs were already involved. They take a percentage of the profits. The villagers asked the subprefect what percentage they could have. The subprefect said they don't have any right to any percentage. . . . The villagers asked for the right to tax the surga. Patrons agreed that they should be able to assess them for the 1,000 FCFA (US$2.00) per year head-tax that village chiefs collect on villagers.

Both villagers and the rural councilors acknowledged that now they were able to levy a head tax on the migrant woodcutters. This agreement did not come up in the meetings but was mentioned by other villagers.[16]

Deputy Kabina Kaba Jakhaté painted a picture in which the foresters in addition to the merchants come into the area and install their own woodcutters in the villages. "Villagers wanted to fight them [the woodcutters] in the forests. The population wanted to force them out. . . . But the Forest Service comes with the woodcutters and pays FCFA 5,000 each [to the Chiefs]. Tambacounda's Forest Service has its own woodcutters." He said, "The chief is only a *pulley*—he facilitates but has no authority" (Dakar, July 1994).[17]

## For the National Good

In the Makacoulibantang story, state representatives—foresters and prefects—often evoke the role of forests in the "national good" in order to override local needs. This is a strategy as old as Jeremy Bentham's notion of "the greatest good for the greatest number for the longest time"—something that sounds suspiciously similar to notions of "sustainable development." "The forests serve the whole nation," cutting the forests is to "satisfy the needs of a whole people," and for these reasons "the management of forest formations belongs with the Forest Service, just as the revenues generated belong to the nation as a whole." The officials also evoke the common good, national solidarity, socioeconomic stability, and the interest of the country and her children, as reasons why rural populations should allow their forests to be exploited by outside commercial interests.

In Dakar, the minister for the protection of nature explained to me that if they had given a choice to the villagers in Makacoulibantang, the villagers would not have allowed any charcoal production. This would have caused problems for urban supply. He explained that villagers do not want to produce charcoal. I then related to him that villagers have repeatedly expressed their desire to make charcoal—"if they could get quotas and permits." Indeed, there were whole villages asking for the right to make charcoal. I added that these villagers were asking to take control of their forests so that they could profit from the charcoal trade. He then mused, "Senegal is a large and diverse country and what villagers may want in Maka does not reflect what they want elsewhere." He spoke of the Makacoulibantang story as a dangerous set of events that could spread—they had to be kept in place. "If villagers were given control of the forests," he said, "there would be fuel shortages in Dakar."[18]

This is the response I got at the end of the line of authority. For the time being, it appears, urban merchants will be allowed to exploit the forests in the name of rational management and the national good, while forest villagers will have to suffer the temporary, but socially and economically problematic, ecological consequences. In the end, the discourse of national good and environmental protection—used to justify the very existence of the Forest Service—serves neither. The forests are cut by merchants and protected by foresters to maintain a continuous supply of fuel to the cities. Villagers are left with a small amount of resources to protect their subsistence needs derived from forests, and with few resources to invest back into the local rural economy. Ironically, the only shortages that Dakar has ever seen were organized by merchants to protest infringements by policy makers on their privileged state-supported control over forestry markets. Forests, and villagers happy to cut them, are not in short supply.

## Chiefs and Merchants

Ultimately, it is the chiefs who mediate direct access to forests. While they cannot sell without the merchants, they can withhold the forests from the merchants and woodcutters by threatening the woodcutters with violence and by keeping them out of the villages. Woodcutters need village access in order to work. The merchants also need village roads to get into the forests to evacuate the charcoal to market. It is through the threat of violence and control of village infrastructure that chiefs control the forests. It is also through the loyalty of villagers that the chiefs can prevent villagers from independently giving woodcutters access to the forest or acting against them. But what chiefs do with their control of forest access depends on their relations with merchants and with forest villagers.

While chiefs control forest access, merchants control access to markets and labor opportunities. Merchants, who are usually urban based, are licensed by the Forest Service, which allocates charcoal production quotas to licensed woodfuel merchants. With a license and a quota in hand, the merchants can hire laborers and obtain production permits for them to work under their license and within their quota. They prefer to hire migrant Guinean Fulbe woodcutters rather than local villagers.[19] So, while chiefs control direct access to forests through control of village infrastructure and the threat of violence, merchants control access to markets and to labor opportunities through licenses, quotas, and permits. Merchants and chiefs must negotiate with each other if either is to benefit from commercial forestry.

The merchant-chief relationship is critical in explaining where and when production takes place. Historically, in Senegal's charcoal trade, merchants would come to the village chief and ask permission to install their producers in the area. This is an old practice to avoid conflicts between villagers and woodcutters. In the early

1970s, the Forest Service informally required charcoal merchants to strike an agreement with chiefs before beginning work in village forests. Today, an agreement is often worked out by a trusted charcoal producer working for the merchant. The usual arrangement involves the payment of FCFA 5,000 to 10,000 (US$10 to $20) per truckload of charcoal taken from the forests around a given village. Making these sorts of "arrangements" with chiefs is normal practice (Ribot 1990, 1995).

The relationship between merchants and chiefs is more complex than can be developed here. The village chiefs are embedded in various relationships with the charcoal merchants. Whether they work for the merchants or are just paid off by them, their decision to take on charcoal producers is a matter of both the social influence of merchants, and the revenues charcoal production can bring a village chief and other villagers. Some chiefs are bought off by the merchants. Others capitulate, given the high social status of many of the merchants, who are often political or religious leaders, or who are closely associated with important social figures such as powerful politicians, Forest Service officials and agents, the prefect, or Islamic religious leaders. Village chiefs find it difficult to turn down the request of these powerful personalities, particularly since they may depend on them in hard times or for labor opportunities that are secured through social relationships stretching from the village to the cities where villagers often migrate for work.

The chiefs' relationships to other villagers are also important. Most villagers were clearly against the woodcutters' presence. Yet, without the compliance of the villagers, chiefs would have trouble making decisions with which villagers tend to disagree. In this case, however, by accepting charcoal producers in their villages the chiefs were going against the tide of popular opinion. Why did villagers allow village chiefs to compromise their wishes? Villagers probably did not challenge their chiefs' decision, although they complained about it, because of the legitimacy and powers village chiefs still enjoy. Further, the alternative, the rural councils, are an administrative creation of the central government. Because the councils are seen as representing party politics and other notables, and because they have very limited decision-making powers, they are yet to have much local acceptance. Indeed, the attempt by councilors to establish a forest fee and a development fund is part of their struggle to gain some basis for local legitimacy.

The authority of chiefs, it is often argued, rests on their control over access to land (Downs and Reyna 1988; Watts 1993; Fisiy 1992). Here, it also rests on their control over access to forests. The attempt by the rural councilors to intervene in forest access may not be welcome by village chiefs. Indeed, anything that legitimizes the existence of the rural council may be a threat to the village chiefs, since the councils are an alternative form of governance to the chieftain system. The council system derives some of its administrative powers from the domain of chiefs. For the rural councilors, this charcoal conflict, where the rural population was united and the chiefs were split, presented an opportunity to speak for the rural populations. In so doing, the council could booster its legitimacy in the rural arena. However, as the results of this conflict show, the villagers and the rural councilors were pushed aside.

They were marginalized in a process that primarily involved the Forest Service and merchants (who act in each meeting as a unified voice) and the village chiefs. While rural councils are the official structure of enfranchisement of the rural populations, despite the undemocratic structure of electoral codes, they have no voice.

Villagers often do not feel that the rural councils represent them. The lists of candidates presented for elections are composed by parties based in urban areas and out of the range of the villagers' influence. In addition, villagers do not feel that the cooperatives, which appoint one-fourth of the council's members, represent them. [Lack of representation in cooperatives is not new. For an example in agricultural cooperatives, see O'Brien (1975).] The councilors interviewed feel that, because of the way they were chosen, they are not viewed as legitimate, which undermines the little power they have. Another issue is that of jurisdiction and powers, called competence. The councilors feel that part of their lack of legitimacy stems from the fact that they have so little power and so few resources that they are unable to perform even the smallest of duties. One pined, "We have nothing to offer the people we represent." Another said, "All we do is marriage counseling."

In the case of Makacoulibantang, the councilors may have good intentions or just strategic ones. It is difficult to separate out the structural position they occupy—which is disempowered, and selected by political parties outside of the villages—from their personal attributes or their social and political ties and aspirations. The councilors I spoke with were frustrated by having no powers and little legitimacy. While the parties they work for might favor merchant activities (some of the merchants being powerful party members), they themselves live within the communities they ostensibly represent. They would like to do something for their fellow villagers but have no powers and no voice. So, while the councilors take up the local cause, they make little headway. Whether they do it out of an identification with the local population, to try to garner some legitimacy for themselves and their parties, or both, is a difficult question. In the end, their expressed goals are not achieved.

There are many ties between the Forest Service and merchants—aside from the links formed through policies that regulate merchant activity, such as licenses, quotas, and permits. While the prefect and subprefect played a facilitating role in the meetings, both of the meetings took a form in which the Forest Service and the merchants came in as a dominant and unified voice. They began by telling the participants what their rights were and what they were not, and by moralizing about the "national good." They then listened to the concerns of the "other participants." Even the minutes reflect this hierarchy, since the two official delegations listed up front are the foresters and merchants, while all "other participants" are named in an addendum. In the second meeting, the Forest Service agents dictated the agenda. The merchants then gave their approval. When concerns were raised by a rural councilor, they were dismissed. The relationship between foresters and merchants is not simply collusion against the rural populations. Rather, it derives from the role of the Forest Service in maintaining the cheap supply of woodfuels to the urban areas. For the foresters, facilitating the merchant's access to the forests is an essential part of meeting their mandate. In addition, foresters have their own (formal and informal, legal and illegal) interests in forest commerce. For rural populations and rural councilors, this alliance is like a brick wall.

## Shifting Control and Power in Makacoulibantang

The rural councilors attempted to support a movement in which the rural population was turning against the charcoal merchants and their woodcutters, and the village chiefs were split. In appealing to higher authorities—the subprefect, the prefect, the Forest Service, their deputy, and the minister for the protection of nature—the rural councilors were trying to break the merchant-chief relationships that were frustrating the local population's desire to stop charcoal production in their area. In response, the Forest Service returned and asserted the rights of merchants to work charcoal in the zone, backing the merchants' claims. The merchants deflected the rural council with the Forest Service and went on doing business with village chiefs willing to work with them. Forest Service agents and officials represented themselves as upholding the law and serving the national good. On the surface, a compromise management plan was established. In practice, the old patterns of merchant-chief relationships continued to serve as the basis of forest access in Makacoulibantang.

In this struggle, the rural council was trying to change the structure of forest access control. They were trying to (1) tap into revenues from commercial forestry, (2) retain some of these benefits for the rural population as a whole (and certainly as a resource for the council and councilors—whether for public and political or for personal consumption), (3) gain access to labor opportunities for the local community, and (4) gain more control of decision-making balance of power and legitimacy between themselves and the village chiefs via forest control. In the process, they did support changes in the patterns of forest access, by speaking for what they perceived to be the needs of the rural population as a whole. But they did not change the broader set of political-economic structures and relationships now shaping the overall patterns of access to the flow of income and profits from Senegal's forests. They did not produce a devolution of benefits now captured by the merchants and by the migrant labor they hire.

What looked like a movement for local participation in integrated resource management policy wound up with business as usual. However, some gains were made. Under ordinary conditions, almost all villages would have hosted charcoal producers. In Makacoulibantang, it was only about half. So far, the movement has been partly successful. With all the agreements made, however, the villagers and rural councilors still feel that little has changed.

Ironically, with one hand in the forest, the Forest Service and the minister for the protection of nature are smothering what looked like a great example of a spontaneous participatory integrated resource management movement. With the other in Dakar, they are waving their new "participatory" forestry code, to be implemented over the next few years.[20]

The Makacoulibantang rebellion can also be read as a struggle over the dwindling basis of local rural power. Chiefs gain their power from land allocation. This power, however, has been eroded over the years by the shift in the rural economic base from agriculture toward more diverse sources of income, including urban migration and labor opportunities in nonagricultural production (e.g., charcoal making). These latter sources of power are in the hands of urban merchants. While the councils officially arbitrate over land disputes—part of an attempt in the 1970s to transfer some basis of local political power to the councils—few disputes are brought to them. Most are resolved within the villages [see Hesseling (n.d.)]. Now the councils are trying to gain a bit of leverage through forestry.

Forests take on a new importance as control over agricultural lands becomes less important. While the councils are attempting to enter the realm of forest control—for profit and power—village chiefs are taking it back, slowing the erosion of their power by widening their base beyond the fields into the forests. Chiefs and councils struggle with each other over this control as the balance of real economic power becomes more and more concentrated in the hands of urban elites. Devolving powers—or even small fragments of control and benefit—to local populations necessarily plays into the struggle between chiefs and councils. It also, of course, plays into the relationship between merchants and the local political economy, a relationship within which merchants can use the local splits to their advantage by choosing the local authorities with whom to conduct their business.

## FOREST COMMERCE IN SENEGAL: PUTTING MAKACOULIBANTANG IN CONTEXT

The story of Makacoulibantang is only a small part of a very complex set of relationships in which the local dynamics described here are embedded. Enormous profits are generated in the production and exchange of woodfuels from Senegal's forests. The profits that chiefs are reaping in exchange for village forests represent only a small percentage of the total profits derived from commercial forestry. Villages (villagers and chiefs combined) reap only 2 to 3 percent of the total profits from the woodfuel trade. The rest is concentrated among migrant laborers and urban venders who make a living wage, and merchants and urban wholesalers who together derive approximately 70 percent of all profits (gross income minus all costs and living expenses) reaped in the charcoal sector. Some of these traders make over US$100,000 per year. Each of these actors depends on different means, structures, and tactics for maintaining access to their share of the income and profit stream (see Ribot 1998b).

As is seen in this case, villagers use threats of violence and control over village infrastructure to leverage their tiny share of the commercial profits from the forests: villagers can exclude merchants, hence the merchants are obligated to pay them off. This ability to exclude is analogous to holding title or owning the forests, but it brings them only a small portion of the benefits that forests produce. Migrants and venders maintain access to labor opportunities through the cultivation of relationships with their patron merchants. Merchants maintain access to benefits from the forests by having access to labor opportunities and by controlling access to exchange. They do so through a mix of policy supports, credit arrangements, collective collusive action, knowledge, and social ties. Wholesalers maintain access to urban markets through credit arrangements, the cultivation and maintenance of a clientele, control of vending outlets, and knowledge of the urban markets.

After examining who benefits from the forests and how, it becomes clear that there are multiple layers of social aggregation and nodes of authority at which control of access to forest benefits is concentrated. The users of the forest are not only the forest villagers; they include various and well-organized user groups such as the merchants and wholesalers, and their dependent venders and migrant laborers. These user groups are related to each other in hierarchical and interdependent ways. In West Africa, this multilayered, multi-user-group forest commons appears to be very widespread. The single-user-group commons that the common property literature has so carefully searched out (e.g., Ostrom 1990) may be a delightful and easily studied find, but it is rare. The most evident form of collective action governing the forest commons of the West African Sahel is the collusion of layered and interlinked merchant organizations and networks, as they are shaped by selectively implemented state policies. Villages—as stratified as they are—also often act collectively to exclude others from the forests in their area (cf. Rangan 1995, 1996; Peluso 1992b), but other, nonvillage groups act collectively to exclude villagers from markets in which they could reap the commercial benefits of the resources under their direct control. In the Makacoulibantang case, direct control of forests renders little profit. It is through control over markets that profits accrue.

It is in this larger context that any attempt to increase local participation in the benefits from forests must be evaluated.

## CONCLUSION: PARTICIPATION, ENFRANCHISEMENT, AND LOCAL CONTROL

Participatory forestry is often presented as a solution to social and ecological problems. But what are rural populations to participate in, and who will participate? What will be devolved to whom in order to constitute community participation in natural resource management?

Through the issue of representation, there is an explicit link between environmental change[21] and enfranchisement. If there is no locally accountable representation, then *who* participates in community resource management? To whom can control of forests or of the benefits that flow from them be transferred? Devolution of control into the hands of locals who are not locally accountable will not achieve the efficiency or equity goals that participatory policies espouse—that is, they will not necessarily internalize social or environmental costs, nor will they improve distribution. If there is no system of locally accountable representation in place, then participatory approaches have no appropriate object of local concern with which to interact. They may transfer access control to some local person—a village chief or some other elite—but not to the community as a whole.

Means by which devolution is affected are equally important. For example, this story shows that direct control of access (whether through threats of violence, legal title of ownership, or other means) does not necessarily confer benefits. So, transferring property rights, a frequently proposed measure, may not resolve this problem. Other means for establishing local access to forest benefits must be explored— such as supports for market entry, the local ability (of some representative body) to tax, labor opportunities, and so on. Indeed, if there is really an interest in transferring benefits and decision-making powers to forest villages, policy analysts should be inquiring into who controls the markets and labor opportunities (and how they control them), rather than focusing solely on who controls forests.

Control of access to things does not confer control of benefits from them. The focus of many participatory approaches to property rights as the primary tool for devolving control to local communities is therefore not sufficient, particularly when there is no locally accountable community representative on whom such rights can be conferred. But it is also not sufficient because property rights—or any other forms of direct resource control—do not confer benefits from forests on local populations unless the populations also have access to markets and labor opportunities.

Justice—both distributive and procedural—in the context of participatory natural resource management is concerned with *what* is devolved to *whom*. The *what* must include the devolution of powers over the disposition of forests, of means to garner financial resources to operate with, of access to labor opportunities and markets, and of access to political processes. All are essential parts of what communities must participate in if they are to have effective control over the disposition of and benefits from resources around them. The *whom* involves the problem of representation. *Indigenous* or *local* does not necessarily mean representative or fair. Some process of inclusion or some form of accountable representation must be constructed if the notion of community—which is always a stratified ensemble of persons with different needs and powers—is to have a collective meaning. This story brings into question whether chiefs really do "represent" their villages in any accountable sense. It brings up the question of whether new natural resource policies should place powers in chiefs' hands, strengthening this particular local—but not necessarily representative or just—institution.

Representation must also be scrutinized in the rural councils. If the electoral codes are not reworked to make the rural councils locally accountable, then devolving natural resource management to the rural councils will result in the strengthening of another nonrepresentative institution. Chiefs and councils are both already deeply shaped by state law. Chiefs are officially elected and then appointed by the minister of the interior, just as the rural councilors are. The parties control the selection of candidates for the rural council. Reshaping the laws that govern their appointment affects only one entry point into national/local politics. Here it is the politics of potentially more equitable, more representative local empowerment. It is the politics of enfranchisement. Delving into electoral and administrative law and practice—that is, the politics of representation and control—is one important way to bring politics explicitly into the emerging field of political ecology.

## ACKNOWLEDGMENTS

Many thanks to Susanne Freidberg, Matthew Turner, and Charles Zerner for their constructive, insightful comments on this chapter.

## NOTES

1. Notions of participatory management have recently gained tremendous popularity in the Sahel (Freudenberger 1993:68,75; CILSS and LTC 1993; Ribot 1996; cf. Vedeld 1992 for pastoral systems; cf. Cleaver and Schreiber 1992 for agriculture).

2. Ninety percent of the charcoal is consumed in Dakar and 10 percent in other smaller urban centers. The woodcutters are also called lumberjacks and charcoal makers in other articles.

3. What caused these dramatic changes in local attitudes toward the forests is difficult to say. I suspect it was a combination of effects, including the fall of the Berlin Wall (which ushered in changes in international attitudes and funding, weakening African governments; it was also a symbol of the triumph of an important democratic movement; and it inspired such events as the 1991 Malian revolution); the revolution in neighboring Mali where foresters were violently chased from the rural areas; and the excitement around the elections in South Africa. Villagers were following these events through radio and word of mouth. In addition, national discussions of democratization were reaching the rural areas. This phenomenon is not specific to eastern Senegal. In Mali's 1991 revolution, foresters were the target of rural protest—some being killed in the process. After Mali's revolts, in a National conference to reconstitute the government, peasant representatives' first request was the elimination of the Forest Service (along with the *Commandant de Cercle*). In Niger, merchants are upset because they can no longer get access to certain forests that local populations are claiming for themselves (interviews, July 1994). Changes such as this seem to be taking place across the West African Sahel. There are contagious hopes and expectations popping up in the forests.

4. Senegal's Forest Service was set up by the French under colonial rule in the 1930s when the predecessor of today's forestry code (law) was written. At independence in 1960, the rules, regulations, and institutions were maintained much as they had been under colonial rule. Most of the same laws and institutional arrangements are still in place today (Ribot and Cline-Cole, in press).

5. As one villager put it, "The councilors are chosen by deputies in the National Assembly. Deputies choose people based on those who support them in their elections. The list is made by the deputy. The councils are chosen by the parties" (Koumpentoum, June 1994).

6. The three rural communities, Makacoulibantang, Endoga Babakar, and Kahäne (all in the arrondissement of Makacoulibantang) were against exploitation in their region. Maka was simply the capital where the meetings and my interviews occurred.

7. Note that whether deforestation is temporary or not is still an open question. While some of the studies cited indicate changes in species composition, all show that forest regeneration occurs. Woodcutters and villagers whom I interviewed in Senegal, Mali, Burkina Faso, and Niger all claimed that regeneration occurred after woodcutting.

8. In surveys and interviews in Daru Kimbu and four other nearby villages in 1986 and 1987, women recounted that before the arrival of charcoal producers, firewood had been available just outside the compounds, whereas after the first two years, firewood had to be gathered at distances of several kilometers, requiring anywhere from a couple of hours to half a day to collect. More than half the women saw the distancing of the firewood resources as a direct result of charcoal production; many expressed resentment. They also explained that charcoal production led to the disappearance of game birds and animals, and to the destruction of fodder (cf. Niang 1985:83; Tall 1974:68; Ribot 1990, 1995).

Village women complained that the presence of migrant charcoal producers drew down the wells, creating water shortages and water quality problems, causing tensions in the village. Villagers recounted how heavy truckloads of charcoal evacuated from the villages during the rainy season—outside the legal production season—tear up and rut the dirt roads so badly that villagers cannot negotiate them by horse cart, car, or minibus. It is thus difficult for them to take their products to market and to bring back the products they need. They also spoke of charcoal producers leaving with debts unpaid, fighting with women gathering firewood, and "chasing women" in the forests. Other researchers report similar problems around Senegal—including scarcity of useful species and commodities—associated with charcoal production (Dia 1985:43; Bergeret and Ribot 1990). Well over half the women interviewed wanted the charcoal makers out of their village and out of their forests, so that their forests could grow back. As one woman commented, "They will leave when there is no more wood. That will be soon."

Villagers, charcoal makers, and foresters also recount numerous conflicts between villagers and charcoal producers. Some have been violent (also see PARCE 1983:17; cf. Freudenberger 1993:63 on conflicts over gum arabic collection). In most cases, the villagers wanted the charcoal makers to leave or to pay the debts they had incurred, and fights broke out. In instances I witnessed, underlying tensions emerged over issues such as theft and adultery. In the early seventies, according to forestry officials, frequent conflicts between charcoal makers and villagers led the Forest Service to adopt an informal policy obligating charcoal *patrons* to gain the consent of the village chief before carbonizing in the area surrounding a village.

9. Arrêté ministériel no. 15411/MDRH/DEFCCS of 21 December 1992, organizing the 1993 season.

10. Arrêté no. 010/D.TC of 1 March 1993, reopening the area.

11. In addition to the subprefect, the minutes listed the following as "registered" participants: (1) from the Forest Service (Eaux et Foràts et Chasse), the director of the Forest Section Office of Tambacounda, the assistant to the Forest Service inspector of Tambacounda, the head of the Regional Hunting Division, the head of the Forestry Brigade of Koussanar (the neighboring arrondissement), and the head of the Forestry Brigade of Maka; (2) a delegation of forestry merchants (in this paper, *exploitant forestiers* is translated "forestry merchants"), composed of the president of the Regional Union of Forestry Merchants of Tam-

bacounda, the president of the Regional Union of Forestry Merchants of Kaolack, and six forestry merchants. A list of participants was appended at the end of the minutes. These included forty people from the arrondissement's villages: the president of the council of the arrondissement, the president of the rural council of Maka, eleven rural council members, eleven village chiefs, three civil servants (health clinic staff, school director, and agricultural extension agent), and thirteen persons named without titles.

12. The minutes state here, "He ended by evoking the deplorable image of trucks parked on the national highway near Koussanar and Koumpentoum, which, in a certain measure, had pushed technicians responsible for the organization and control of exploitation to propose to the prefect of Tambacounda that he sign a decree permitting exploitation in Makacoulibantang." It is not clear whether he is trying to evoke the national good that is suffering when the villagers make all these merchants sit and wait, if he is saying that he was forced to do this by the commercial interests, or both.

13. I have changed the names of all but high-level officials discussed in this story.

14. More precisely, the commission included the local Forest Service inspector, the Forest Service sector director, the Forest Service Koussanar Brigade director, the assistant to the director of the Maka Forest Service Brigade, the director of the Rural Extension Committee of Maka (CERP), the president of the council of the arrondissement, the president of the rural council of Maka, the president of the Local Cooperatives Union, the president of the Regional Union of Forestry Merchants, and the vice-president of the National Union of Forestry Merchants.

15. The one forester in the area was stationed in Maka. When I woke him at 9:30 one weekday morning to ask about the situation, he referred me to his superiors in Tambacounda.

16. The deputy in the National Assembly for this region indicated that today only between 10 and 20 percent of the tax ever gets collected (Jakhaté, Dakar, July 1994). There is a general tax strike across rural Senegal. In any event, this would not be a great benefit since only 4 percent (US$0.08) remains in the village as a salary for the chief.

17. At the level of the National Assembly, there are different factions with different interests in the charcoal market. The president of the National Union of Forestry Merchants is a deputy in the National Assembly. He is a member of the ruling Socialist party. The deputies from the Casamance have been against woodcutting there for a variety of political reasons, not the least of which is a secessionist movement in their region. Charcoal production has indeed been banned in some parts of the Casamance as a result of local protests combined with political fears that conflicts would flare up and spread.

18. Urban woodfuel shortages would be a political disaster for the minister and for the Forest Service. However, in fact, the only threat of shortages comes from merchants who withhold charcoal as leverage for policy changes they would like. The threat of woodfuel shortages in Senegal has nothing to do with the limits of the forests (e.g., deforestation leading to a woodfuel crisis). Indeed, forests are not even close to exhaustion, and regeneration after woodcutting is quite robust. It also has nothing to do with rural populations stopping the flow. Merchants have threatened in the past, and continue to threaten, the ministers and the Forest Service with shortages to eke out quotas and to keep the policy environment-friendly to their interests. The threat of shortages is a political tool of the merchants, not a matter of supply and demand. Urban woodfuel shortages are purposely created by merchants to get what they want.

19. The earliest merchants in the market were noble Fulbe from upland Guinea. They worked charcoal with their former serfs, also migrants from upland Guinea. This was a convenient arrangement since charcoal was looked on as a dirty and lowly activity, usually done

by a caste of blacksmiths. Further, when conflicts emerged between charcoal producers and villagers, the migrant charcoal producers were more tied to their merchants than to the village chief. In addition, migrants do not have competing labor demands on them and can be moved from place to place as the forests become exhausted.

20. See Ribot (1995) for an analysis of the new forestry code.

21. In eastern Senegal, the most important environmental problem caused by woodcutting appears to be temporary. While the forests appear to regenerate after woodcutting, the forest villages are faced with difficulties in gathering woodfuel, food, and fodder during the period between cutting and regeneration (see Ribot 1995, 1996; Bergeret and Ribot 1990). There are also many social problems that arise from hosting scores of migrant woodcutters in a village. These problems concern harassment of women in the forests, the exhaustion of the village well from the extra water demand, the failure of migrant workers to pay their debts, and the destruction of roads from the wood trucks.

# CHAPTER 7

## The Damar Agroforests of Krui, Indonesia: Justice for Forest Farmers

*Geneviève Michon, Hubert de Foresta, Kusworo,*
*and Patrice Levang*

The modern history of forests in Indonesia merges with a history of a continuous process of land and resource appropriation by the state at the expense of indigenous forest people, through a fair amount of ideological imperialism and a convenient use of legal and technical instruments as well as a touch of power abuse. From the very beginning of the Indonesian archipelago's history, the forest has represented the only large field for demographic, agricultural, economic, and geopolitical expansion, as well as the major instrument by which to attain wealth and power. While conflicts for forest appropriation or control in the past have occurred predominantly between equivalent groups of forest users and warriors, they presently involve structural opposition between the state and its political or economic elite on the one hand, and local communities on the other (figures 7-1, 7-2, 7-3).

In spite of the important contribution of wood industries to national development, the ecological, economic, and social damages related to forest management in Indonesia can no longer be concealed. Cases of resource exhaustion, violations of local populations' basic rights by forestry projects, and reports of ongoing local resistance are being publicized more and more through the local and national press. However, even though the discourses of policymakers at the highest levels integrate new objectives of social and environmental justice in forest management, forest development and conservation projects are still constrained by laws and regulations that still prevent the recognition of local people's practices and rights.[1] Actual benefits of local utilization and management of forest resources are seldom encompassed in a critical and nonpartisan way. The value of customary systems for controlling local forest management practices is either underestimated or misunderstood. The legal mechanisms for acknowledging local people's rights over forest lands and resources remain dramatically underdeveloped.

**Figure 7-1** Indonesia

In this current context, alternative successful forest management strategies developed by local communities may represent an important support for the development of formal recognition of local people's rights over forest resources. Among these strategies, the agroforestry strategy exemplified in this chapter could well be promising. In adapting the traditional modes of forest extraction through a logic of agricultural production, farmers have invented new and original agroforestry systems that reshape forest resources and structures. These complex agroforestry structures that combine forest species and tree crops have evolved from shifting cultivation practices and can be encountered in many forest farming systems all over the archipelago. As they are not, in spite of their appearance, natural forests but manmade agroforests—"forest gardens"—they could allow farmers to affirm, maintain, or regain control over forest lands and resources.

Starting from the broad context of local communities and national circles with competing ideologies, regulations, and practices regarding forest management, this chapter will focus on the history of an agroforestry landscape in the south of Sumatra, elaborating on the originality and efficiency of this particular strategy for forest resource control through agricultural development. We will then analyze the chronology of external pressures that threaten the agroforest, and farmers' reactions to repeated violations of their basic rights: from avoidance action to active resistance. As a conclusion, we will try to analyze how, in the present ideological and legal context that clearly does not favor local people's appropriation of natural resources, an integrated management of conflicts in forest resource use could evolve in a totally original way through the acknowledgment of the benefits derived from the integration of local forest management into agriculture.

**Figure 7-2** Sumatra

## CONFLICTS IN FOREST AREAS: FRAMING THE ISSUES

The analysis of local conflicts pitting indigenous communities against the state or its economic elite cannot be conducted without mentioning the ideological and political, as well as the practical, dimensions of the global differences in the way each protagonist conceives, appropriates, and manages forests. These two main categories of actors are differentiated by every aspect of forest development, from cosmogonies and representations of the world, political ideologies, and legal and institutional systems,

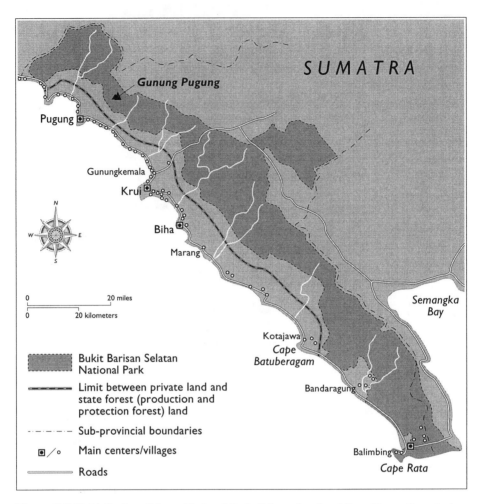

**Figure 7-3** Bukit Barisan Selatan National Park, Private Land and State Forest Boundaries, Sumatra.

to technical, labor, and financial capabilities for resource exploitation, and to negotiation and implementation power.

## *Ideology and Practice of Forest Management: Two Diverging Visions*

As the dominant land cover of the archipelago, forest has been a major element in cosmogonies and representations of the world for the various ethnic groups that presently constitute the Indonesian nation, as well as for their ruling authorities. Based on these representations, various management systems have evolved.

### Forest: In the Center of the World

For forest communities,[2] most of the society origin myths involve marriages between humans and forest spirits. Forest is the central place for both the spiritual and the economic life. Forest is considered a multiple-use resource and defined as a combination of facies, each being defined by a combination of potentially extractable resources and possible activities.

Forest utilization and management by local communities developed around two poles. The first includes subsistence resources: game and fish, plant foods, and material, and the forest dynamic itself as an essential resource for shifting cultivation. The second includes extractive resources, harvested for trade: incense, spices, and animal products of the precolonial trade; resins and latexes in the eighteenth and nineteenth centuries; rattan and timber since the beginning of the twentieth century. As a result of increasing space constraints during the last fifty years,[3] land also emerged as an essential forest resource for local communities. Interest in land and commercial resources represents the only common point between indigenous communities, the state, and private companies. But it is also the major cause of their conflicts.

Systems devised by indigenous communities for the management of forest lands and resources usually combine "production"—usually through forest clearing—and "harvesting"—in situ management of economic resources. But, contrary to what is commonly acknowledged by either policymakers or researchers, production is not limited to agricultural products, and forest resources are not managed only through harvesting: indigenous forest management often involves a considerable amount of production through active plantation of forest crops. Many of these forest production systems deserve more attention than they usually receive, as they do represent outstanding examples of sophisticated, multipurpose forest resource development.

### Forests on the Border of Civilization: A World to Control and Exploit

From the Javanese kingdoms of the tenth century to the Indonesian Republic, through three centuries of colonial administration, state authorities have tried, through their political and coercive power, to impose their representation and modes of forest management over indigenous communities. The perceived need for absolute state control over the forest is dictated by two imperatives: the economic importance of appropriating natural resources and the political imperative of assimilating alien cultures.

The official perception of the forest in modern Indonesia has been shaped both by the Javanese civilization, which valued clearings more than wilderness and permanent ricefields more than shifting agriculture (Dove 1985; Peluso 1992b), and by the occidental conception of Dutch colonial forestry. The underlying philosophy that seems to condition all forest policies and management regimes in modern Indonesia—as in any centralized political organization—is that forest is basically a domain that could easily escape state authority and control. Because of its very nature—a world based on principles alien to an organized state—and because of its geographic position in the periphery of the "civilized" world, forest, including forest lands, resources, and inhabitants, is perceived by the state as a fundamentally dissident area that must be strictly controlled. While forests contribute to the process of national development through their space and resources, the forest as an entity is not embedded as such in the philosophy of national development. State-sponsored agents of forest utilization are exclusively state or private companies. Forest people as a whole are globally denied any right to participate in forest management. They are considered squatters on state lands and plunderers of state riches.[4]

The economic utilization of forests by the state is characterized by an intense reductionism. For the last forty years, management has exclusively focused on mining the timber resource; most of the forest lands logged during the last two decades are no longer covered by forest. This chronic unsustainability of state-sponsored harvesting practices is another factor that differentiates national and customary systems of forest management.

## Policy, Legal, and Institutional Bases of Forest Management: Two Parallel Systems

Policy, juridic, and institutional levels also exhibit a juxtaposition of two different definitions and practices of forest appropriation that are not fundamentally incompatible but are in actuality mutually exclusive.

### National Modes of Control

Forest policies in Indonesia are burdened by several ambiguities: they must simultaneously ensure that the utilization of forest lands and resources serves as a major instrument in the building of national development, and they also need to enforce the protection of these resources for the present and future. This double task is assigned to a single state body, the ministry of forestry, which also has to harmonize profit building for the nation and social justice for the nationals. The strategies chosen to reach these political objectives have evolved along three main streams: separation of forest and agriculture; segregation of forest domains into production, protection, and conversion; and delegation of forest utilization and management to concessionaires, which releases the state from practical aspects of forest development and allows it to concentrate on the acquisition of revenue. As a consequence of these strategic choices, the first objective—development and profit—has been much emphasized, whereas the second—conservation and justice—has been considered an ideal target for the future.

These policies are implemented through the Basic Forestry Law issued in 1967. The law defines the extent of state forest lands, their functions, and their modes of utilization in a list that comprises many qualities and land-use categories, some unexpected,[5] but denies any right of existence to shifting cultivation. The law attributes the authority and jurisdiction over forest lands to the ministry that delineates the forest domains, defines their functions, and allocates—or denies—rights over them to (un)privileged users. Acquisition of revenue occurs through a system of regulations and legal taxation.

Like any national legal system, the forestry legislation constitutes a rigid body that is not easily susceptible to change. Accommodating local people's needs, or developing community forestry agreements, although theoretically desirable and possible, faces many legal impediments. Another chronic disease in the forestry system is the dramatic weakness in law enforcement: institutions in charge of the implementation of the forestry regulations are prone not only to political influence but also to collusion of any kind with those wealthy partners who do not really respect the laws.[6]

### Community Modes of Control: "Tradition" as an Evolving Body

Control of and access to forest resources by local communities usually involve a system of rights and regulations, ranging from pure forms of common property to exclusive private rights over lands or resources. The definition of rights commonly varies according to the nature of the involved resource(s) or the type of ecosystem. It also varies from one community to another, and for a given community it may vary over time. Indeed, unlike the national legal system, flexibility, mobility, and adaptability are the main characteristics of customary systems. Contrary to the common view, "tradition" in appropriation systems is not a rigid concept. Regulations and rights for resource control and access constantly evolve to accommodate external or internal changes—resource availability, destination, value, extraction or production techniques, and so on. Another important principle that differentiates customary and national legal systems is that, in most customary systems, use or property rights are usually accessed through labor investment, not through lobbying for or purchasing of concession rights. The main weakness of customary systems is the current lack of efficiency in controlling outsiders' abuses.

## Management in Practice: Implementation of Forestry Regulations and the Confrontation Between National and Customary Systems

All over its territory, including the portion designated as state forest land, the Indonesian constitution acknowledges customary rights over land and resources. However, this is most often ignored in the practice of forest management.

### Forest Land Designation, Delineation, and Mapping

The delineation of state forest lands has set aside 144 million hectares (ha), representing 74 percent of the whole nation's lands, for "forest" production and protection. Lands outside the forest domain are designated as "appropriated lands." This distinction between forest and appropriated domains implicitly designates forest lands as lands that can in no way be legally appropriated by either individuals or groups. As more than 95 percent of the land under customary control is—unfortunately!—included in state forest lands (Gillis 1988), indigenous farmers cannot expect to ever receive any legal land title.

The acknowledgment of customary rights stands "as long as these rights do not interfere with national interests." The formulation is ambiguous enough to allow the widest, as well as the most narrow, interpretations. In the real Indonesian world, customary rights unavoidably recede when government-sponsored projects or activities are carried out.[7] The practical interpretation of forest policies not only denies rights to forest lands and resources to local communities, it also denies the very existence of their forest production systems. State forest lands and their limits have been mapped for the whole archipelago, and these maps serve as basic documents for development planning. In harmony with state dogma, forest lands are intrinsically uninhabited: indigenous land-users do not appear on official documents. Rattan gardens (Weinstock 1983; Fried 1995), fruit forests (Michon and Bompard 1987a; Sardjono 1992; Momberg 1993; Padoch and Peters 1993; de Jong 1994), damar gardens (Michon and Bompard 1987b), rubber agroforests (Dove 1993; Gouyon et al. 1993), all the swidden and fallow systems of the outer islands, and sometimes even forest villages, simply do not exist. It is therefore easier for projects to ignore, or even erase, preexisting forest management systems.[8]

## Concession Rights to Resources and the Criminalization of Indigenous Forest Practices

The practical impact of the concession policy, aimed at controlling resource extraction and production through supervision and taxation, has many perverse effects for local communities. First, it clearly favors private companies at the expense of local communities. It has also facilitated the emergence of conglomerates led by timber tycoons closely connected to political elites. The technical, economic, financial, and political power of these tycoons gives them an unrivaled advantage in negotiations for land and rights,[9] as well as in the implementation of these rights in cases of conflicts with local communities. They easily obtain the support of forceful negotiators to solve these conflicts quickly and surely. The second consequence of the concession policies is that it clearly criminalizes unlicensed harvesting practices, not only for timber but also for the most profitable nontimber resources: rattan, birds' nests, sandalwood, eaglewood, etc. Communities that lived on free extraction of commercial forest products for centuries are presently outlaws: they would have to purchase temporary rights to harvest these products. But purchasing rights is, again, the privilege of elites who, again, by their political or economic influence, dispossess local communities of their most valuable resources.

## From Forests to Estates: Implications of the Latest Trends in Forest Land Development

The switch from logging to estate development started in the late 1980s, and the rate of change began to increase dramatically in the early 1990s. This chiefly involves timber estates, developed on "production forest" lands, and oil palm estates on "conversion forest" lands or empty "appropriated lands."

This new trend in development policies bears important implications for local communities. The first is the increased threat of displacement to give way to projects. After being leased to logging companies, indigenous lands are presently given to estate firms that not only have full legal rights over the land but also drastically transform them through plantation; the local people are considered merely a cheap labor force. The second consequence for local people is an unexpected "diversification" of the agents whose job it is to resolve conflicts, as well as increased opportunities for making new allies. After dealing exclusively with forest authorities under the forest law, local farmers must now also deal with regional administrators—either at the provincial, district, or subdistrict levels—who have full authority in the granting of conversion forest lands to private companies, and who, in addition, have the right to ask for a legal revision of state forest land boundaries.[10] However, these regional bodies also have the authority to defend and support local management systems and to acknowledge their customary foundations.

In this new phase of the land development game, the conflicts are not only between forest authorities and local farmers but also between "forest" and "agriculture."[11] Local communities could benefit from this rebalancing of forces, as it might appear to be an opportunity for them to assert their right of access and control over the lands they have developed. But they are more likely to be, once again, the main losers in the power game that builds up around forest lands. Caught between the various facets of state authority and the concupiscence and prerogatives of the estate plantation lobbies, local communities might well lose their last opportunity to have their basic rights acknowledged.

Tracing these conflicts takes us to the densely populated province of Lampung in the southern tip of Sumatra island. Westward, beyond the steep slopes of the Barisan range, the Pesisir subdistricts appear as an estranged appendix in the province. Apparently left behind in the intensive agricultural development that occurred in the central and eastern districts, Pesisir retains large tracts of forest. Pesisir farmers still rely on forest agriculture, but more than shifting cultivators, they are now true tree farmers who have developed highly original systems of forest cultivation over the years (figure 7-4).

The situation of the Pesisir farmers in the Krui area perfectly epitomizes that of forest farmers in Indonesia. Although they have occupied the land for more than five centuries, they still have no legal title to it: most of the lands they developed and manage are located on state forest lands. In spite of—or could it be precisely because of?—the ecological, economic, and sociocultural value of the forest management system they developed—a perfectly balanced forest plantation based on the most valuable species for foresters (figure 7-5), dipterocarps—they have encountered more trouble than effective support from the forest administration at the local, district, provincial, and national levels. Only recently did they begin to receive some kind of official recognition of their practices, which has yet to be translated into formal terms. Although covering more than 50,000 ha, their forest gardens are not yet on the maps. After being subjected to repeated (although relatively light) power abuse from the forest authorities, they are presently confronting the concupiscence of the private sector that tries to impose oil palm plantation over lands that are still

**Figure 7-4** Common Landscape in the Krui Area. Mature damar agroforests, initially developed by shifting cultivators, now cover about 50,000 ha, or 50% of the three Pesisir subdistricts total area.

under forest status. Conflicts of interest involve various government and non-government agents who all impact on local farmers, either through active support, purposeful ignorance, or direct confrontation.

## KEBUN DAMAR: BETWEEN FOREST AND GARDEN

Driving westward from the peneplain, a mosaic of dry fields and pepper plantations along the Sumatra highway, through the Barisan range, a succession of reddish hills extensively degraded by pioneer coffee growing, one suddenly enters another country: a land of trees that stretches all along the quiet descent to the Indian Ocean. The human mark on this forest landscape is not immediately obvious: some clearings bearing hill paddy, a few patches of fallow vegetation. Elsewhere, the land is a venerable jungle dominated by large trees. The area covers some 100,000 ha divided between a long coastal plain—130 km from the provincial border in the north to the southern Cape Cina in the Sunda Straits, which widens from north to south—and a steep hilly and mountainous area rising to a height of over 2,000 meters. It stretches over three administrative subdistricts.[12]

Wherever possible, irrigated ricefields, and associated permanent villages, have

**Figure 7-5** The Damar Tree. *Shorea javanica,* a big tree native to local forests in Sumatra, easily reaches 40–45 m high.

been established along the coastal plain, but the rude topography and the relatively low quality of the inland soils have limited the possibilities of further permanent agricultural food production. The hills have long been the domain of a classic agroforestry rotation: mosaics of temporary ricefields and coffee plantations with secondary, fallow vegetation. But for about a century or so, this traditional pattern of forest conversion to agriculture has evolved into a complex system of forest redevelopment. Planting valuable fruit and resin-producing trees in their swiddens, Pesisir farmers have managed to create a new forest landscape entirely tailored to their needs. This man-made forest, although forming an almost continuous massif, is made up of a succession of individually evolved gardens that the farmers have named after the dominant tree species, the damar[13] (Torquebiau 1984; Michon and Bompard 1987b; Michon and Jafarsidik 1989).

Damar gardens in the Pesisir represent totally original examples of sustainable and profitable management of forest resources, entirely conceived and managed by local populations. Originality lies in the ecological mastery of the main economic resource, the forest tree, not through conventional domestication, which usually involves modification of plant characteristics to achieve adaptation to a cultivated ecosystem, but through an almost total reconstruction of the original forest ecosystem in agricultural lands. Success is due to the proven reproducibility of the system over the long term as well as to its economic results and to its social bases. Today, more than 80 percent of the damar resin produced in Indonesia is provided not by natural forests but by the Pesisir damar gardens. Among the seventy villages scattered along the coast, only thirteen do not own damar gardens.[14]

Damar gardens can be analyzed as a forest, and, indeed, biologically they constitute a forest in their own right, a complex community of plants and animals and a balanced ensemble of biological processes reproducible in the long term through its own dynamics. The gardens can easily be mistaken—as they often have been by common observers—for a natural forest. But they definitely have been established not as a forest but as an agricultural production unit on an agricultural territory. They are part of lands that are agriculturally claimed by local people and are managed mainly as an agricultural enterprise. Occupying this vague interface between agriculture and forest—at least by the conventional perceptions that modern science has promoted—they fully deserve the name of agroforests (Michon 1985; de Foresta and Michon 1993).

## A Tree Plantation Modeled as a Forest-Rich Ecosystem

While damar trees are clearly dominant in mature gardens, representing about 65 percent of the tree community and constituting the major canopy ensemble (figure 7-6), damar gardens are not simple, homogeneous plantations. They exhibit diversity and heterogeneity typical of any natural forest ecosystem, with a high botanical richness and a multilayered vertical structure,[15] as well as specific patterns of forest dynamics (table 7-1).

Plant inventories in mature damar agroforests have recorded around forty common tree species, and several more tens of associated species, including large trees, treelets and shrubs, liana, herbs, and epiphytes. Important economic species commonly associated with damar are mainly fruit trees, which represent 20–25 percent of the tree community. In the canopy, durian and the legume tree *Parkia speciosa* associate with the damar trees. In the subcanopy ensembles, langsat is the major species with, to a lesser extent, mangosteen, rambutan, jacktree, palms such as the sugar-palm *Arenga pinnata* or the betel-palm *Areca catechu*, and several water-apple species—*Eugenia* spp.—as well as trees producing spices and flavorings

**Figure 7-6** Inside the Damar Agroforest. Damar trees are usually dominant in dmar gardens, but density greatly varies among individual gardens in accordance with individual owners' objectives, which include its use as a property marker, as part of a mixed orchard (producing damar and fruit), and as a main production objective.

(*Garcinia* spp., the fruits of which are used as acid additives in curries, and *Eugenia polyantha*, the local laurel tree). The last 10–15 percent of the tree community is composed of wild trees of different sizes and types, which have been naturally established and are protected by farmers, either because they do not have adverse effects on planted trees or because of advantageous end uses. These species mainly include bamboos and valuable timber species (Apocynaceae, Lauraceae, etc.). Nontree species characteristic of a forest ecosystem (Zingiberaceae, Rubiaceae, Araceae, Urticaceae) have colonized the undergrowth of gardens, where they contribute to the maintenance of a favorable environment for the development of seedlings of the upper-layer trees.

Management of mature gardens is centered around the harvest of resin (figure 7-7) and of fruits. Labor allocated to routine garden maintenance is mingled with labor devoted to resin harvest, and the tempo of harvests is determined by labor requirements for wet rice cultivation. Work in the gardens is postponed at the time of the rice harvest or of ricefield preparation, so that tree gardening never competes with subsistence agriculture for labor.

Once established, the damar plantation evolves with minimal human input. The silvicultural process in damar gardens is not conceived, as it is in conventional for-

**Table 7-1**

Comparison of structural characteristics between damar agroforest and primary forest in Pesisir Krui

| Sample number | 1 | 2 | 3 | 4 | 5 |
|---|---|---|---|---|---|
| Sample plot area (m²) | 600 | 1,000 | 400 | 1,000 | 2,000 |
| **Damar trees >10 cm DBH (trees/ha)** | | | | | |
| Young unproductive trees | 200 | 140 | 200 | 150 | n.a. |
| Mature and old productive trees | 200 | 140 | 250 + 50 | 190 + 70 | n.a. |
| Total stand density (damar) | 400 | 280 | 500 | 410 | n.a. |
| **Total trees over 10 cm DBH** | | | | | |
| Total stand density (all species) | 680 | 300 | 650 | 560 | 500 |
| **Vertical structure** | | | | | |
| Number of canopy ensembles | 2 | 3 | 3 | n.a. | 4 |
| **Distribution of crown coverage** | | | | | |
| Emergent trees | 0% | 0% | 0% | n.a. | 25% |
| Upper-canopy trees | 130% | 88% | 114% | n.a. | 60% |
| Lower canopy trees | 34% | 5% | 8% | n.a. | 33% |
| Undergrowth trees | n.a. | 12% | 12% | n.a. | 13% |
| Immature trees (trees of the future) | 41% | 38% | 33% | n.a. | 45% |
| Total | 205% | 133% | 167% | n.a. | 176% |

1, 2, 3 = damar gardens in Penengahan, Central Pesisir (Michon 1985).

4 = damar gardens in Pahmungan, Central Pesisir (Torquebiau 1984).

5 = primary forest (Laumonier 1981).

DBH = diameter at breast height.

From Michon (1985).

est plantations, as a mass treatment applied to a homogeneous, even-aged population of trees; instead, it aims at maintaining a system that produces and reproduces without disruption either in structural or functional patterns. Natural processes are given the major role in the evolution and shaping of the cultivated ecosystem. Global continuity is ensured through a balanced combination of natural dynamic processes[16] prevailing in the tree population, and the appropriate management of individual trees of economic species. Since the natural decay of planted trees is predictable, farmers can easily anticipate and plan their replacement. The main task of the gardener is to regularly introduce young trees into the garden plot to constitute and maintain an uneven-aged pool of replacement trees. In a well-managed garden, the size of the replacement pool ensures the sustainability of the productive stand.

## Between Plantation Economy and Forest Use: The Economic and Social Value of Damar Gardens

Damar gardens have been established by farmers for commercial production, and their economic management is closer to that of an agricultural smallholder planta-

**Figure 7-7** Damar Resin Harvest.

tion than to that of a forest. However, some functions of the damar gardens still relate to the former harvested forests that complemented rice swiddens in ancient production systems.

### The Damar Garden as an Enterprise

Damar trees represent the main source of household cash income (figure 7-8), and damar collection is far more lucrative than other agricultural activity in the region (Mary 1987; Levang and Wiyono 1993). Resin is harvested on a regular basis: individual trees are usually tapped from once a month to once every two weeks. A single villager can harvest an average of 20 kg of resin a day. In the central subdistrict villages, average harvests are between 70 and 100 kg per family per month. Resin sale represents a regular income allocated to day-to-day expenses:[17] purchase of additional foods, weekly costs for children's schooling. Five days of work in

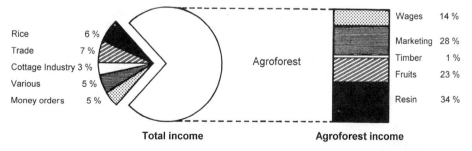

| Rice | 6 % | | | Wages | 14 % |
| Trade | 7 % | | | Marketing | 28 % |
| Cottage Industry | 3 % | Agroforest | | Timber | 1 % |
| Various | 5 % | | | Fruits | 23 % |
| Money orders | 5 % | | | Resin | 34 % |

**Total income**                                                   **Agroforest income**

**Figure 7-8** Origin of Houshold Cash Income in a Damar-Based Village, Pahmungan (Levang and Wiyono 1993).

damar gardens is usually enough to ensure a month's subsistence for the whole family (Levang 1989, 1992). For those who do not own permanent ricefields, the damar income also allows for the purchase of some rice and thus complements dry rice culture where it still exists. However, the damar income is usually not sufficient to raise significant amounts, nor is it enough for hoarding.

The damar activity also generates a series of associated activities: harvest, transportation from the field to the village, stocking, sorting, and transportation to wholesalers in Krui (table 7-2). Harvest, transportation, and sorting are carried out either by the grower himself or by members of his family, or by specialized agents who are paid employees. Independent entrepreneurs ensure resin stocking in the village. These activities raise significant additional income for the village[18] and allow those who do not own a damar garden to benefit from damar production (Bourgeois 1984; Mary 1987; Levang and Wiyono 1993; Nadapdap et al. 1995).

As in many other places in Sumatra, the contribution of the fruit component to household economy has been increasing in recent years because of the growing importance of urban markets and because of recent major improvements in the road network. For the last productive years, marketing of the major commercial fruits, durian and langsat, has allowed the global agroforest income to double (Levang and Wiyono 1993; Bouamrane 1996; de Foresta and Michon 1997). However, because of high irregularities in fruiting seasons,[19] income from fruits cannot be fully integrated into daily household budget planning. It is still used mainly for exceptional or "luxury" expenses.[20]

Damar gardens constitute one of the most profitable smallholder production systems in Sumatra (table 7-3). They globally ensure reasonable quality-of-life levels, including high schools for children (given top priority in most villages of the area). In addition, they can be managed—and used accordingly whenever needed—as a

**Table 7-2**

Main characteristics of the damar resin trade chain inside Indonesia

| Agents | Relative profit margins[a] | | Activities[b] | | | | | |
|---|---|---|---|---|---|---|---|---|
| | 1st trade chain | 2nd trade chain | Harvest | Stocking | Drying | Sorting | Transport | Processing |
| Damar grower | 70% | 70% | xxxx | x | x | 0 | xxxx | 0 |
| Village traders | 3% | 6% | 0 | xxxx | xx | xx | xx | 0 |
| Krui dealers | 1% | — | 0 | xxxx | xx | xx | xxxx | 0 |
| Direct traders | — | 6% | 0 | xxxx | xx | xxxx | xxxx | 0 |
| Krui wholesalers | 13% | — | 0 | xx | xx | xxxx | xxxx | xx |
| Expenses | 10% | 15% | | | | | | |
| Losses | 3% | 3% | | | | | | |

From Bourgeois (1984).

[a]Expressed in percentage of the resin price in Tanjung Karang or Jakarta.

[b]xxxx = principal activity; xx = frequent, x = occasional, 0 = never.

**Table 7-3**

Average production per hectare per year in mature damar agroforest, Pahmungan village,
Central Pesisir subdistrict, April 1995

| Species | Density trees/ha >20 cm DBH | Production | Traded | Labor (family level) | Yearly Income Rp | US$ |
|---|---|---|---|---|---|---|
| Shorea javanica (resin) | 145 | 1,550 kg | 1,500 kg | 50 | 1,500,000 | 682 |
| Durio zibethinus[a] | 25 | 625 fruits | 600 fruits | 10 | 420,000 | 191 |
| Lansium domesticum[b] | 15 | 600 kg | 500 kg | 10 | 250,000 | 114 |
| Parkia speciosa | 8 | 1,200 pods | 1,000 pods | 10 | 100,000 | 45 |
| Baccaurea racemosa[b] | 7 | 200 kg | 50 kg | 2 | 10,000 | 5 |
| Artocarpus cempedak[a] | 6 | 100 fruits | 50 fruits | 2 | 50,000 | 23 |
| Other fruit trees (6 species)[b] | 10 | 200 kg | 50 kg | 3 | 50,000 | 23 |
| Timber (all species are used) | 250 | 5 m³ | 2.5 m³ | 0[c] | 50,000 | 23 |
| Total labor (mandays) | | | | 87 | | |
| Average yearly income | | | | | 2,410,000 | 1,106 |
| Minimum income (no fruiting season) | | | | | 1,650,000 | 750 |
| Maximum income (fruit season) | | | | | 3,570,000 | 1,625 |

From de Foresta and Michon (1997).
[a]Production every 2 years.
[b]Production every 3 years.
[c]No family labor involved in timber harvesting.

safety asset: a garden, or a part of it consisting of several selected trees, can be
"pawned" through special agreements called gadai[21] (Mary 1987; Lubis 1996) that
allow any family to overcome difficult periods without resorting to selling trees or
land,[22] which is considered one of the worst things that might happen to a family.

Indeed, in accordance with an agricultural conception of resource management,
damar gardens also represent a patrimony. Arising from a strategy of land property
creation, the fruit of labor invested for a distant term to benefit future generations as
well as present, the damar garden constitutes an inalienable lineage property (Mary
1987; Nadapdap et al. 1995). In the very particular social and institutional context
of the Pesisir, where families are defined mainly by their land assets, this notion of
lineage patrimony defines the agroforest not only as the source of living for a household, but also as the land foundation of a lineage.

### Damar Gardens as a Useful Forest

Damar gardens also fulfill a role equivalent to that of natural forests in the
economies of forest villages. Wild resources associated with damar trees support a

wide range of gathering activities that are more typically linked with natural forest ecosystems—hunting, fishing, and harvesting of plant products—and provide important complementary subsistence resources for households. These include various noncommercial fruits, vegetables, spices, and firewood, as well as other plant material and timber for housing purposes.[23]

Damar gardens also represent, as does any natural forest, a source of products that are potentially marketable commodities at a larger scale: timber, rattan, and medicinal and insecticidal plants can be harvested for sale whenever needed or if market conditions are considered favorable.[24] As new markets develop, some of the traditional subsistence products have actually emerged as new commodities. Timber presently stands as the major "new" commodity that might even revolutionize the management of damar gardens[25] (de Foresta and Michon 1992, 1994a; Michon et al. 1995a; Petit and de Foresta 1996).

Damar gardens have taken over the essential role traditionally devoted to natural forests in household economy: a place opened to subsistence gathering and extractivism and used to fulfill the family's immediate needs. This forest function also appears in some of the egalitarian social attributes of the gardens, i.e., product exchanges, sharing and donations,[26] and free harvesting rights.[27] This creates important networks of reciprocity that act as a counterpart to mercantile networks created through agricultural activities, and that help maintain a social balance between well-endowed people and those without resources.

In replacing natural forests by damar agroforests, the villagers' aim has been to amplify commercial strategies linked to the forest ecosystem. This is a widespread dynamic all over Indonesia: slash and burn practices are usually not targeted to staple food production, but primarily to the establishment of income-generating agroecosystems (Pelzer 1978; Scholz 1982; Dove 1983; Weinstock 1989). Here—and this is one of the main originalities of the land conversion process in the area—although converting natural forests into a commercial plantation, Pesisir farmers also managed to restore a wide range of economic products and functions originally derived from the forest. Forest conversion did not go along with a radical process of biological simplification; rather, it restored plant and animal diversity through cultivated, preserved, and spontaneously established species. Specialization did not entail economic reductionism; instead, it restored the whole range of economic choices present in a natural, untransformed ecosystem. From the perspective of an integrated conservation and development program, this preservation of existing and potential economic diversity appears as important as that of biodiversity.

## BUILDING A FOREST RESOURCE: SPECIES DOMESTICATION OR FOREST RECONSTRUCTION?

The damar story in the Pesisir constitutes a highly original example of spontaneous appropriation of a forest resource, the damar tree, by local farming communities. It was achieved as the wild resource itself was vanishing (Michon et al. 1995a, 1996). If human history is rich in examples of natural resource appropriation through cultivation to achieve domestication, the originality of the damar example is that, while cultivating this particular forest resource, villagers have achieved the global restoration of a forest in the middle of agricultural lands (Michon and de Foresta 1996). Biologists will argue that the damar agroforest is far from a natural, pristine tropical forest: although close to it, damar gardens cannot totally replace the natural forest *ecosystem*. However, they represent a rather integral forest *resource*, which is much more significant for local people than a natural forest that increasingly eludes their control, the conservation of which is not, for longstanding and external institutional reasons, within their power. Besides the technical success linked to the establishment and reproduction of a large-scale dipterocarp plantation over more than a century,[28] it is this appropriation of the global forest resource through an agricultural strategy, and its integration into farmers' lands, that are worth analyzing.

## From Extractivism to Culture: A History of Resin Harvesting and Production in the Pesisir

### Damar Resin: An Ancient Forest Resource

Resins, which are sticky plant exudates found in various families of forest trees,[29] are among the oldest traded items from natural forests in Southeast Asia: they entered short-distance trade between Southeast Asia islands as far back as 3,000 B.C. and were probably included in the first long-distance exchanges that developed with China from the third to fifth centuries (Dunn 1975). The word *damar* appears in the lists of items traded to China from Southeast Asia in the tenth century (Gianno 1981, citing Ma Huan 1451). The first exports to Europe started only in 1829 and those to America, in 1832 (van der Koppel 1932). Locally, damar served for lighting purposes and for caulking boats. It was traditionally traded as incense, dyes, adhesives, and medicines (Burkill 1935) and acquired a new commercial value by the middle of the last century with the development of industrial varnish and paint factories. Collection intensified for export trade to Europe and the United States, and then to Japan and Hong Kong. After 1945, however, exports dropped rather sharply because of competition with petrochemical resins, which are preferred for most industrial uses.

Nowadays, Indonesia is the only damar-producing country in the world. Damar resins are marketed through both interinsular and export markets. Major end-users are low-quality-paint factories in Indonesia, which use the lowest grades. The best-quality damar is reserved for export, mainly to Singapore, where it is sorted, processed, and reexported as incense or a base for paint, ink, and varnish factories in industrial countries. Other destinations include handmade batik industries and the manufacture of low-quality incense (Bourgeois 1984; Dupain 1994; Lembaga Alam Tropika Indonesia 1995).

In the glorious period of intensive harvesting for export, from the beginning of the twentieth century until World War II, the main damar-producing areas were the natural forests of southern and western Sumatra, as well as West Kalimantan (van der Koppel 1932). Today, West Kalimantan and South Sumatra still produce some damar, but the main producing area is certainly Lampung, the southernmost province of Sumatra.

### Resin Extraction in the Pesisir in the Past

From at least the eighteenth century onward, agricultural economy in the Pesisir combined subsistence strategies (with swidden rice production dominating until the end of the nineteenth century) and market-oriented strategies, thus associating the production of copra along the coast, and pepper, coffee, and clove[30] on the hills, with commercial gathering of forest products, mainly gutta-percha, wild rubbers, rattans, birds' nests, and damar. Chinese traders waiting in all the small harbors along the coast ensured the export of agricultural and forest products northward to Bengkulu or southward to Tanjung Karang, Batavia (Jakarta), and Singapore.

Damar production is reported to have been a major activity in the whole area (figure 7-9). As early as 1783, the British historian Marsden mentions a type of resin "yielded by a tree growing in Lampung called *Kruyen* [one cannot but think about Krui] the wood of which is white and porous . . . and which differs from the common sort, or dammar batu, in being soft and whitish. . . . It is much in estimation for [lining] the bottoms of vessels. . . . To procure it, an incision is made in the tree" (Marsden 1783). Harbor accounts in Teluk Betung, from the middle of the eighteenth century, stress that the trade of damar mata kucing was a source of considerable profit in Lampung, 285 tons being exported in 1843 (cited in Sevin 1989). A map drawn by the Belgian geographer Collet in 1925 mentions damar as one of the three main exports of Krui (Collet 1925). Rappard, a Dutch forester who visited the area in 1936, mentions that damar ranks third in the agricultural exports of Krui, after coffee and copra, but before pepper; the total production in 1936, for the Krui area alone, was more than 200 tons (Rappard 1937).

In villages, there is still a vivid memory of the importance of wild resins, and people can point to old wild damar trees that were protected in the *ladang* (the Indonesian word for swidden) while the forest itself disappeared.

### Building a Tradition: Damar Cultivation as a Heritage of the Past?

When and why did the cultivation of damar trees start? Farmers present it as a "tradition" inherited from their "ancestors"—*nenek moyang*—which roughly means, "That's what we do, what we can do; that's the basis of our lives." However, this tradition is certainly not older than a century or so and probably occurred through a combination of internal factors and external influences. Some villagers trace its origin to the beginning of the twentieth century, the result of a visit by two respected hajjis from Singapore, who were convinced of the damar market's bright prospects and returned to establish plantations. Other informants in the ancient producing area of Pugung assert that, about six generations ago (at least 130 years ago, i.e., circa 1870), villagers came from the central subdistrict to ask for damar

**Figure 7-9** Tapping wild damar trees in natural forests used to be a major economic activity in the Krui area until the beginning of the twentieth century.

seedlings, which were then taken from Batu Bulan forest where natural damar trees were famous (Dupain 1994). Villagers commonly agree that the oldest planted damar trees are to be found in the south, where "you can find huge trees that were planted more than 200 years ago." The only written material is provided by Rappard (1937) who reports that he encountered 70 ha of plantations around Krui, among which several were at least fifty years old, which indicates that the first plantations might have been in 1885. Rappard mentions that 80 percent of the damar produced in Krui in 1936 is from cultivated trees, and that *Shorea javanica* no longer exists in a wild state in the area. He notes that production increases from year to year, with 120 tons in 1935, 201 in 1936, and an estimated 358 for 1937 (Rappard 1937).

Among the imperatives leading to the initiation of a generalized cultivation process, the main one was probably the increasing difficulties encountered in the collection of wild damar, which could closely resemble the conflictual processes regarding access to common-property resources encountered today for other forest products (Peluso 1983b, 1992c; Siebert 1989). At the turn of the century, the high increase in resin prices led to intensive and generalized tapping of trees in natural forests. Overcollection entailing the rarefaction of mother trees blocked natural regeneration, whereas the extension of the cultivated territory entailed the rarefaction of the forest itself. Damar trees were spared in the slash-and-burn process (figure 7-10) and could easily survive in the modified environment of ladang and secondary vegetation, but natural regeneration in these conditions appeared difficult. Some serious conflicts are reported to have occurred between (and within) villages concerning access to the remaining damar trees (Levang and Wiyono 1993).

**Figure 7-10** A Wild Damar Tree Preserved in a Swidden. Damar trees were systematically preserved in converting natural forest to damar agorforests. These trees could continue to be tapped. They were sometimes used as a source of seed for damar nurseries. However, the low density of damar trees in natural forests and their erratic and unpredictable regeneration prevented spared trees from playing a significant role in damar garden establishment, and farmers had to plant the trees to ensure sufficient densities of productive trees in their gardens.

Damar gardens also appeared to be an answer to increasing problems in commercial agriculture. For example, pepper plantations encountered serious difficulties on the western coast of Lampung around 1920, when a serious disease reportedly killed most pepper vines (Levang 1989). The induced disturbance of the balance between subsistence and commercial strategies in cropping systems could partly explain the development of damar plantation after 1930. Colonial administration might have played a role in advising local people to continue their process of domestication. It is also most probable that Chinese traders actively encouraged the diffusion of damar cultivation, as they did for rubber in other parts of Sumatra (Pelzer 1978).[31]

Damar gardens have gradually spread in the Pesisir, and productive gardens presently cover at least 50,000 ha [according to a recent interpretation (December 1997) by the Department of Forestry and the International Center for Research in Agroforestry (ICRAF) of a Landsat image dated November 1994], the main center of cultivation being located around the city of Krui, where the hills are almost totally covered with a mature damar forest. Yearly damar production was estimated to be around 8,000 tons in 1984 (Bourgeois 1984) and reached 10,000 tons in 1994 (Dupain 1994). New gardens are still being established in the northern and southern subdistricts.

## How Cultivation Techniques Evolve into Forest Appropriation Strategies

The reconstruction of a forest by Pesisir villagers was not planned as such. Rather, it appeared a posteriori as the consequence of a particular cropping system that minimized labor input and maximized the use of natural production and reproduction processes of an artificial ecosystem dominated by trees. In that sense, it is the choice of particular cultivation techniques and patterns, more than the initial selection of a given forest tree, that allowed true forest reappropriation (de Foresta and Michon 1997).

### Technical Appropriation of the Damar Resource: Overcoming Biological Constraints and Using Biological Advantages

The main ecological disadvantages of the selected forest species are typical of dipterocarps: difficult natural regeneration due to irregular and only occasional flowering, lack of seed dormancy (figure 7-11), and need for mycorrhizae association. But one important advantage should be noted: unlike many dipterocarp species, *Shorea javanica* appears to be rather light-tolerant, which made it suitable for cultivation in plots already cleared for agriculture.

Villagers solved the regeneration problem through a technology of "assistance storing of seedlings" (Michon and Bompard 1987a,b; Michon and Jafarsidik 1989). The establishment of small nurseries, where the seedlings could be kept for several years and used whenever planting material was needed, allowed growers to overcome fruiting irregularity and lack of seed dormancy. The mycorrhizae problem was

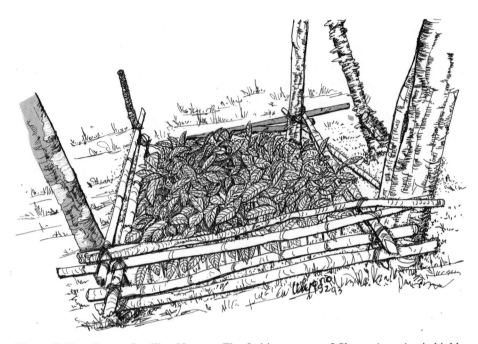

**Figure 7-11** A Damar Seedling Nursery. The fruiting season of *Shorea javanica* is highly unpredictable: seeds are only available ever 4 to 6 years. Also, seeds cannot be stored more than a few days. Local farmers have solved these two problems storing seedlings, using a locally devised technique. This enables them to have access to seedlings whenever needed (for replacement of dead trees or establishment of new gardens). In season, seeds are selected and collected in gardens; they are then planted in small nurseries located in damar gardens, in the village or in the swidden. Density in the nursery is very high, which results in seedlings that do not grow more than 20–30 cm high and that can survive 4–5 years.

avoided through a first phase of direct transplantation of seedlings from the forest to the plantation site.

Among other biological constraints is the long renewability rate of damar as a resource: it takes at least one generation—twenty to twenty-five years—for a tree to attain a minimum tappable size. The economic consequence is that, for the first twenty-five years, a pure damar plantation would be of little, if any, use for the planter. This difficulty has been solved through a strategy of crop succession starting from the ladang and planned over the medium term.

## Integration of a Forest Tree in a Farming System: The "Ladang Way"

Expansion and success of damar cultivation are indeed closely related to swidden agricultural practices (Michon and Bompard 1987a,b; de Foresta and Michon 1994b). It is through the ladang, and through its traditional crop succession structure, that damar trees have been restored to the landscape. In the former dry land cultivation system, ladang were opened primarily for rice production, but some did not directly return to fallow. Instead, they were further transformed into either coffee or pepper plantation.[32] The first damar trees were introduced in these successional ladang gardens, amidst coffee bushes and pepper vines, where they found a suitable environment to establish themselves and further develop. After the abandonment of the coffee or pepper stand, damar trees were strong enough to grow along with secondary vegetation and to overcome competition from pioneers. The subsequent fallow was a mix of self-established successional vegetation and deliberately planted damar trees, which developed fully until reaching a tappable size, some twenty to twenty-five years after the plantation, but no more than ten years after the plot abandonment (figure 7-12). Damar plantation soon became a success story: everyone started to plant seedlings in his own ladang garden. Through this very simple cropping technique, after two decades, a traditional fallow land had changed into a managed tree garden that included damar trees as well as other introduced fruit species and self-established trees, bushes, and vines.

This process of establishment still prevails today in areas that are being converted. Ecologically, the whole development of these successive crop mixtures imitates natural forest succession,[33] with all its ecological benefits: soil protection and microclimate evolution in accordance with successive component needs. Technically, it is similar to a classic agroforestry process of forest plantation establishment—the taungya system—in which young seedlings of economic tree species start to grow in favorable, controlled conditions. Here, maintenance of the coffee/*Erythrina* stand secures good microclimatic conditions, shade and humidity, favoring transplantation success, and provides weed control during the first four to fifteen years following the introduction of seedlings.

Economically, this vegetation succession process is of tremendous importance as it is the basis of a succession of harvestable commercial products, thus reducing the unproductive time span of the plantation to some five to ten years. Costs of labor devoted to damar establishment are mingled with those devoted to rice and coffee cultivation on swidden fields. Cultivation of commercial tree crops does not compete for labor with subsistence agriculture. On the contrary, it allows the maximization of returns on labor inherent to the swidden system—vegetation cutting and field maintenance—successively through coffee and trees.

Pesisir villagers have succeeded in doing what most foresters dream of: establishing, maintaining, and reproducing, at low cost and on huge areas, a healthy dipterocarp plantation. This is still a unique example in the whole forestry world. The best part of the story is that this success is inextricably linked to shifting cultivation, the agricultural system held in contempt by foresters. The acceptance of the wild tree as a cultivated tree crop and the subsequent expansion of the plantation were permitted by the particular structure of the swidden production system, and ladang was at the very heart of this success.

In achieving the switch from the "natural and sometimes protected" status of the damar tree in traditional extractivism systems to its adoption as a new crop in the farming system, farmers have clearly reinvented the common process of resource appropriation through (agri)culture. Indeed the control of the damar resource, based

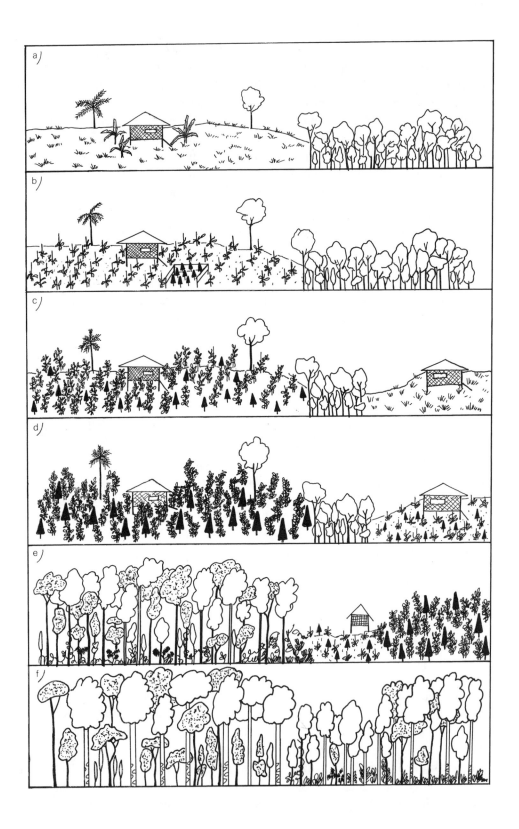

on the mimicry of natural forest processes which adapts the cultivated ecosystem to the plant characteristics, runs counter to conventional domestication processes, which emphasize modification of biological and ecological characteristics to achieve adaptation of the plant to a cultivated ecosystem (Michon and de Foresta 1996).

### Further Appropriation of the Forest Resource: Restoring Biodiversity

The plantation process that usually associates damar with fruit trees and leaves pioneer trees establishing naturally in the ladang garden basically recreates the skeleton of a forest system. But the real appropriation of forest richness and diversity is achieved through the free development of natural processes of diversification and niche colonization. As in any secondary vegetation dominated by trees, the newly maturing damar plantation provides a suitable environment and convenient niches for the establishment of plant propagules from the neighboring forests through natural dispersion. It also offers shelter and food to forest animals. In this natural enrichment process, farmers merely select among the possible options offered by the ecological processes: favoring resources, through introducing economical trees and protecting their development, or tolerating development and reproduction of nonresources as long as they are not considered "weeds." After several decades of such a balance between free functioning and integrated management, the global biodiversity levels are fairly high. As natural forests below 700 to 800 m above sea level have almost disappeared in the Pesisir, damar gardens constitute the major habitat for many plant species characteristic of lowland and hill dipterocarp forests that would otherwise have disappeared (Michon and de Foresta 1992, 1995). The agroforest also shelters many animal species, some of which are highly endangered, such as the Sumatran rhino and the Sumatran tiger.[34]

Seen from the planter's point of view, while the introduction of economic species in the damar agroforest is intentional, global biodiversity reestablishment is "accidental." But it is precisely this "accident"—the establishment of diversified flora and fauna as in any silvigenetic process—which reconstitutes the real forest aspect of the

---

**Figure 7-12** (*on previous page*) From a Mosaic of Swidden Field to a Damar Agroforest. Most damar gardens begin with a rain-fed rice phase, followed by a coffee plantation phase, in the following steps: (a) Year 1: Slash and burn of secondary vegetation (sometimes primary forest) area, and planting of the first rain-fed rice crop, along with vegetables and fruits such as papaya and banana. (b) Year 2: Introduction of coffee seedlings and establishment of the second rain-fed rice crop; damar seedlings in the nursery are ready to be planted. (c) Year 3: No more rice crop on the field; introduction of damar and fruit seedlings between the coffee rows. (d) Years 4 to 8: (but sometimes to year 15): Coffee trees begin to produce significant amounts and are usually harvested for 3 to 5 years (but sometimes managed so that they still produce at year 15). (e) Years 15 to 20–25: The field has been temporarily abandoned after the last economic coffee harvest; a spontaneous component is now developing (trees, lianas, shrubs, forest herbs, epephytes), along with the planted damar trees and fruit trees. (f) Damar trees begin to be tapped; the damar garden continues to develop; farmers' management ensures that the garden produces and reproduces without having to return to a slash-and-burn phase.

agroforest. These combined processes, the intentional and the accidental, are essential for several reasons. They restore resources that otherwise would not have been conserved purposefully because they do not appear to be important economic resources, but they also permit the restoration of biological and ecological processes that are determinants in the functioning and reproduction of the agroforest as a forest ecosystem. In that sense, even those components that are not economic—or potentially economic—resources are not neutral. Nonedible fruit trees in the agroforest help in supporting populations of fruit-eating birds, squirrels, and bats, which are essential natural pollinators and dispersers of economic fruit species. One should not forget those "functional" resources that are not—and will never be—valued as commodities but are nevertheless essential; restoring diversity, either economic or biological, is meaningless if ecological processes are not maintained.

## FOREST REAPPROPRIATION: A CONCEPTUAL AND CULTURAL PROCESS

Appropriation of the forest resource through damar plantation has also involved the modification of traditional perception systems for forest resources, as well as the establishment of a modified system of social, legal, and institutional access to resources.

### Problems of Perception and Representation: A Garden, Not a Forest

The ancient perception and representation systems of natural forests and forest resources are presently quite obliterated, at least in villages that have reached the limits of their territorial expansion. While disappearing from the immediate environment of villages, the natural forest has lost its importance in the farmers' "imaginary," or system of beliefs (note the French *l'imaginaire*).[35] Reference to the ancient myths or to forest spirits and magic is presently very rare. The forest of the past, a source of material wealth as well as spiritual blessing, represented an imaginary world as well a major source of life. Today's forest is neither mythical nor mystical; it is no more than the domain of forest administrators, and secondarily an area of trouble for those who dare enter it too conspicuously.

The agroforest has replaced the forest in the landscape and in the local economy. Allowing the maintenance of a lifestyle that coexists with a forest culture from which it directly evolved, it represents the last witness of an ancient alliance between Pesisir communities and forest resources. All the interactions between people and forest resources presently happen through the agroforest. But the agroforest did not replace the natural forest in local representation systems. It remains first and foremost an agricultural unit[36] (Lubis 1996; Michon et al. 1996), the determining factor being that it results from a plantation process, even though important components have regenerated spontaneously. However, if damar is actually considered a crop, the distinction farmers make between cultivated and managed (or even wild) plants remains somewhat vague and highly subjective. But the agroforest components are never perceived as "forest plants," even though farmers recognize that most agro-forest plants can be encountered in the forest as well.

The agroforest itself will never be mistaken for a forest (pulan), except in some very specific activities, usually linked to natural forests. People ordinarily go deer hunting or collect rattans "in the forest." In that case, *forest* refers not to a given garden or a given part of the agroforest, but more globally to the former communal forest space, stretching between the reserved forest and the villages, that the agroforest has replaced but that also comprises ladang, successional vegetation, and remnant old-growth forest.

This well-established distinction between the forest and agroforest is a logical one. Agroforests result from important initial work and represent long-term investment and years of a process that can be compared with capitalization. Identifying agroforests with natural forest would mean denying this work and time investment, this long-term planning of the "ancestors" for their heirs, which is an obvious tradition to any Pesisir farmer. It would also deny the whole resource appropriation process achieved through destruction of the forest and plantation of trees. Confusing agroforest with the forest is, in that sense, heresy.

## Access Systems: Private Ownership and Individual Decision for Gardens Versus Common Property for Forests?

Institutionally, appropriation of the forest resource has entailed a total reorganization of the traditional tenure system for forest lands and goes along with the increasing importance of land as property and privatization of this property (Michon et al. 1995a,b).

### The Ancient Access System to Forest Lands and Resources

According to the ancient customary tenure system, forest lands and resources were managed as common property by the village community—*marga*—and designated as *hutan marga*, unlike irrigated lands for rice production which were privately owned. Individual claims over economic resources in the hutan marga were acknowledged for certain species and through certain technical processes. Thus, a wild damar could be appropriated by the person who first began tapping it; collecting damar from that tree was then considered his own and exclusive right. However, nobody could claim rights over a piece of unmanaged, pristine forest. Access to land for subsistence and cash cropping was usually gained through clearing a piece of land in the communal forest and cultivating it. Distribution of access rights between the different families of the marga consisted of long-term individual usufruct rights. The land itself remained the property of the marga. These individual usufruct rights were in fact tacitly maintained long after the crops were abandoned, and the same family could recultivate the land after a fallow period without asking permission from the marga. However, customary rights strictly forbade the planting of perennials on these communal forest lands, except for short-lived perennials such as coffee or pepper. Tree plantation was indeed considered a major investment for land development, which was likened to labor invested in irrigation works for ricefields. As it was acknowledged for ricefields, this labor investment would have led to private appropriation of the land itself.[37]

As more people developed an interest in damar cultivation, the assembly of pasirah, responsible for the customary law, formally accepted the removal of the prohibition against planting perennials in the marga lands, which boosted the spread of the plantation movement and led to drastic land appropriation activities by individuals in the former communal forest domain (Levang and Wiyono 1993). However, land property could be claimed only through tree plantation, and the old tenure system—communal property of the land and usufruct rights—prevailed for non-planted plots that were still considered hutan marga.

### Privatization of the Commons? Individual Ownership Revisited

As the plantation process was conceived in a context of the relative failure of common property systems, its success required an assurance that the planter's children would effectively enjoy the right to harvest the trees, which implied not only that property rights be acknowledged and enforced, but also that transmission rights be secured.[38] The consequence is that created land properties never returned to the community, and the commons gradually disappeared. However, the privatization process remained original and unpredictable as it did not entail promotion of individual control nor fragmentation of the agroforestry domain.

The customary law makes a clear distinction between hak milik penuh ("full property rights"), a right more or less similar to Western ownership and one that designates newly created or newly bought land, and hak waris ("inherited rights"), which concerns inherited properties and defines a lineage patrimony. In Pesisir, the owner of a piece of inherited land is bound by traditional restrictions regarding both the transfer of land and the right to use it. Even though hak waris is acknowledged as an individual ownership right, the owner can transfer neither the land nor the trees, nor can he cut productive trees without obtaining permission from the whole extended family.[39]

This rights restriction is more a moral obligation than an enforced regulation. It operates through a strong social control system, in which the customary legal authority over land and resources is transferred from the village community to the lineage community. Transmitting an intact family patrimony to one's eldest son is as important as receiving it, and the property rights system cannot be dissociated from a social system in which many community traditions are still alive. The domestic group largely exceeds the limits of the nuclear family,[40] and the lineage maintains overall authority over the family patrimony. The head of the domestic group, who actually holds the hak waris, although he is the legitimate and real owner of the garden, is socially more the depositary of a patrimony, the continuity of which is under the control of the whole lineage. Hak waris bukan hak milik saya, "My heritage is not my property," summarizes the ethic of the property rights systems in the Pesisir, and this ethic still constitutes, more than any formal regulation, a clear safeguard against total privatization and individualization. Even though the customary law acknowledges individual rights, exclusive control over land or trees does not occur in the Pesisir. This has obvious consequences for the sustainability and the efficiency of agroforest management. The breakdown of the agroforest block into individually owned plots could evolve into a mosaic of fields with different structures and vocations if it is not controlled by a strong social structure. This would lead to a drastic fragmentation of the ecosystem that could greatly endanger the overall reproduction of biological and productive structures[41] (Mary 1987; Levang and Wiyono 1993; Michon et al. 1995a).

### Reestablishing Former Common Property Rights and Values in the Framework of Private Agroforests

As forest resources and structures have been reestablished, common property traditions have been redefined and reinforced in the context of privatization: the balance between formal individual rights and moral obligations toward the "community" also concerns minor resources included in the agroforest to which other members of either the domestic group, the lineage, or the village community can gain access.

Individual appropriation does not concern the totality of the agroforest domain. The degree of any owner's control over resources included in his garden actually depends on the nature of any given resource. Important economic resources such as resin and commercial fruits, as well as land, are effectively individually owned assets, with the traditional restrictions mentioned previously. However, on these private agroforest lands, many resources are still considered common property or open access resources. In fact, the only strictly privatized resource is the damar resin, and taking resin from someone's tree constitutes a real theft. Other resources, such as fruits, sap from the sugar palm, bamboos, special thatching leaves, which are provided by species commonly considered planted, remain at the disposal of the community. But which community may harvest which resource and to what extent varies according to resources, from the family group to the lineage or the village itself. Usually, permission of the owner should be asked before collecting what could be considered "important quantities," and sharing of the benefits usually occurs for those products harvested for commercial purposes, but picking fruits or bamboo for one's own immediate consumption while passing by a garden is considered normal. Resources considered to be pure forest resources, such as rattan, wild vegetables, medicinal plants, and firewood from spontaneous species, in fact those plants that are perceived as wild as opposed to planted, are covered by rules that fluctuate between a very broad sense of common property—firewood for example may be collected in small quantities by anybody from the village community—and open access. In most villages, the customary rule allows not only subsistence hunting and gathering of vegetables and medicinal plants without any restriction concerning the origin of the collector, and over all the agroforest area, but also more income-generating activities such as commercial gathering of rattan.

In the same way that the technical appropriation of the forest resource did not fundamentally change the Pesisir landscape, the institutional reappropriation of the former forest commons through "controlled privatization" did not result in a total institutional revolution that erased old values. This maintenance of the communal philosophy in agroforest management is essential. If the damar agroforest is more a garden than a forest, it is nevertheless not an agricultural field with a short rotation, but a forest plantation in which long-term management—which is more a principle of forestry than a concern of subsistence farmers—is essential. Establishing access to productive structures and resources that will start producing for one's child and be fully productive for one's grandchild constitutes a new logic, in which short-term individual considerations need to be buffered by a community concern for perpetuation of these structures and resources. In the way that former common-property regulations controlled the permanence of the commons, the new property ethics in the Pesisir ensures that trees and land will be integrally transmitted to future generations.

For village communities, the private property legal framework could secure a better bargaining position with external bodies than common property, which is still negatively perceived or easily denied by most state bodies as well as by private companies. Private claims over land are more easily acknowledged, and compensated for, by the Indonesian administration. Privatization could therefore be used as a political strategy for local communities to protect their resources.

## From Extractivism to Cultivation: The Agroforest Strategy as a New Framework for the Reappropriation of Forest Resources by Local Communities

The main objective of the agroforest strategy of Pesisir farmers was clearly to develop a sustainable commercial enterprise that could overcome the weaknesses of the ancient extractive system. Indeed, it halted the process of resource exhaustion from the natural environment, and it allowed the solution of social problems linked to the failure of former common property regimes devised for the control of that resource; it also introduced a new balance in local production systems. The technical, ecological, social, and institutional framework devised by Pesisir farmers, offering both economic success and ecological sustainability while avoiding sociocultural breakdown, proved quite efficient for the sustainable management of forest resources.

However, more than a success for natural resource management, agroforest establishment in the Pesisir does constitute a true revolution in both the forestry and agriculture contexts. As a forest plantation strategy, the damar agroforest model runs counter to the conventional model of timber estates that are presently being developed. While favoring a selected resource, as estates do, the agroforest allows the maintenance of numerous other resources that otherwise would not have been conserved purposefully, and species that are not direct resources to be restored as well. Moreover, the establishment process allows the restoration of integral biological and ecological processes that are crucial to the overall survival and reproduction of the agroforest as an ecosystem. If encompassed in the framework of agricultural plantation strategies for the development of forest lands, extension of the damar agroforest represents a process of forest conversion that does not go along with economic reductionism, which does not irreversibly close economic potentialities linked to the presence of the natural forest. On the contrary, through the restoration of biodiversity in the agroforest, farmers have achieved the restitution of a wide range of economic choices for the present and the future, which appears indispensable in a sustainable development perspective. The agroforest development also represents a successful strategy for agricultural intensification that has helped to set farming system patterns without any disruption in food availability or in living standards, while maintaining intact the productive potentialities of the land itself.

Damar gardens are certainly an interesting example of agroforestry association. But in the agroforestry context, they convey a totally new dimension: that of the association, not between trees and crops as in conventional agroforestry, but between the forest resource and agricultural logic (Michon et al. 1995a). It is, above all, the integration of forest resource management into the farming system that constitutes the success and the originality of the damar agroforest.

Damar gardens offer new insights into the definition of technical and ecological—as well as socioeconomic and institutional—bases for managing forest resources within farming systems. They open new perspectives for reinventing forest common property resources through an original agricultural perspective. They also bring new insights to the open debate about natural resource management by local communities. As a development strategy, the establishment of damar agroforest represents an interesting example of forest product management for commercial purposes, which has entailed a total transformation of the original ecosystem while preserving most of its resources and retaining an important part of its biodiversity. The transfer of forest functions from the natural ecosystem to the agroforest implied not only a transfer of resources, structures, and economic vocations, but also a guaranty of their renewability. In that respect, the agroforest should be considered by foresters to be an ecological model of forest reconstruction of great potential for reforestation and land rehabilitation programs. But the agroforest is more than a biological duplicate of the forest. In the present political, institutional, and socioeconomic context in Indonesia, which appears quite unfavorable to long-term maintenance of the forest itself, the whole process of damar agroforest establishment and development appears to be an extremely original strategy for reappropriation of forest resources by local populations, or, more than forest resources, of the traditional "forest resource" of peasant economies in the forest margins of Indonesia.

After years of conflicts with the Forest Services in the Pesisir, farmers apparently gave up most of their claims over the National Park natural forest, which is henceforth considered more a geographical unit in an administrative landscape than a resource in the village landscape. For farmers, the forest is an exclusive, reserved, and closed domain of the state, and entering it is often an act of exquisite provocation. Agroforest, on the contrary, represents a man-made structure where the forest resources are appropriated and managed in accordance with the farmers' needs, philosophy, and beliefs. Through the agroforest, farmers can claim that they have restored, in the middle of an agricultural territory over which they believe they have firmer control, a privileged space in which their forest resource is protected. And, in that sense, the fact that farmers do not confound their agroforest with a "forest" is essential to consider.

Agroforests are *not* natural forests that have been gradually modified through management. They represent a man-made area that has been created by farmers' communities. They result from the voluntary decision of these communities to reestablish forest resources and to recreate forest structures. Natural forest management in Indonesia, including extractivism, is still a form of exploitation of nature's gift. Agroforest management is beyond that: it is the invention and the achievement of a new form of forest resource management on former natural forest lands.

## FOREST FARMERS VERSUS FORESTERS, PLANTERS, AND THE STATE: THE NEED FOR LEGAL RECOGNITION

The isolated situation of the Pesisir area and the absence of projects until the last few years have protected damar agroforests and damar farmers from the outside world. Since the early 1990s, the acceleration of regional development has clearly shown the limits and weaknesses of agroforests as a strategy of appropriation of forest resources by local people, as long as agroforests are not recognized by the state and as long as this recognition is not given legal status.

Regarding the evolution of damar agroforests' official status, many interpretations can be developed, many positions questioned, many questions raised, many answers debated. However, as this status is presently being discussed at the highest governmental levels after a series of recent and difficult conflicts, we shall here limit ourselves to a basic reading of the events that might indicate the dramatic end of a now famous success story.

Until very recently, the history of relations between damar farmers and state officials and state-supported bodies was dominated by misunderstandings and rights abuses. It may be divided into a first period of exclusive conflicts with forest authorities, followed by a second phase that saw the emergence of unexpected but powerful stakeholders: the regional authorities and the private sector.

### State Forests, Customary Forests, and Damar Gardens: Conflicting Views over Land Status, Land Use, and Land Control

No official map above the village level mentions the existence of damar gardens. The most recent map for the West Lampung district[42] classifies the land occupied by damar agroforests as either "swidden and dry fields," "secondary forest and degraded vegetation," or "plantation area."[43]

According to the official statistics of the Department of Forestry (Dinas Kehutanan Lampung, 1995), 10,000 ha out of the estimated total of 17,500 ha covered by damar gardens lie on state forest lands, classified as either Limited Production Forest or Protection Forest. However, in village statistics and maps, these state forest lands are still considered "customary forest" [hutan marga, as they were called before the Forest Land-Use Master Plan by Consensus (TGHK) published in 1991], which, in fact, is no longer an officially acknowledged status but has led villagers to think that their customary lands were recognized by the state. They were therefore quite surprised when, between 1992 and 1996, depending on the area, they saw forestry employees planting sticks in their damar gardens and telling them, "Beyond this limit, the land belongs to the government." This indeed sounds a bit colonial, although farmers believe that "even the Dutch would not have done this." For the villagers, these border sticks have suddenly brought home the extent and the gravity of forthcoming conflicts: in some villages, the first sticks are located 100 meters from the last house.

The remaining 7,500 ha of damar agroforests are "unclassified" by the Forestry Service; they are not public land and are therefore sometimes called private land, which sounds great but is not: private appropriation by local people is not formally acknowledged, as farmers still do not hold any official land certificate for either rice-fields or damar gardens. The district under whose jurisdiction these private lands fall has indicated that its top priority in regional land-use planning has been given to estate plantation development based on private investments.

Damar farmers are therefore caught between two mutually exclusive administrative mechanisms regarding their lands, neither of them really appearing to hold positive prospects: in both cases, the farmers' legitimate position in regard to the law is dramatically weak. To forest authorities, they are undoubtedly outlaws: conducting any agricultural or harvesting activities on forest lands without permission from the Department of Forestry is constitutionally illegal and implies a penalty. Under a "private" regime, but without any official land title, damar farmers may be considered to be squatters on empty lands that are reserved for regional development. In both cases, they are highly subject to eviction in order to give way to projects.

## Forest Farmers and Foresters: From Conflict to Alliance?

Damar farmers have never been involved in official decision-making processes regarding the planned development of lands they have actually and efficiently managed for centuries. Neither have they been really informed of these decisions that obviously may have profound implications for their future. On the contrary, after having worked hard and believed that they were developing their lands for their children and grandchildren, they suddenly learn, usually through rumors more than through clear explanations, that these lands belong to the state and that the state has in mind "better" projects for their development. Not surprisingly, this chronic disinformation has led to the multiplication of conflicts between farmers and government-sponsored agents, which has commonly translated into cheating on one side and power abuse on the other.

### Episode 1: The Reserved Forest

The forest reserve (BoschWeesen, or BW as it is still known locally) was established by the Dutch administration in 1937. Its borders, decided after consultation with local people, were located far from the agricultural territory of villages. The status of this reserve was upgraded to National Park in 1991 (Bukit Barisan Selatan

National Park; see figure 7-3). As an old constraint on their territorial development, villagers are well aware of its existence and its borders. However, they do not really fully agree with the legitimacy of the park for flora and fauna protection. This disagreement has gained importance in villages where land shortage problems were acute, but it is basically more conceptual than factual. The fundamental grievance of farmers against the ideology—and practice—of conservation forestry is that it values the forest more than humans and will always give preference to wildlife and plants, whether or not this results in serious problems for local people.[44]

As a result of land shortages in several villages, encroachment of ladang and damar gardens in the forest reserve, especially along the Krui-Liwa road, started as early as 1955. In the late 1960s, a tacit agreement was concluded between farmers and the forestry authorities, allowing several dozens of families to open land in the reserve and establish damar gardens (Mary and Michon 1987). But police and conservation guards continued to regularly visit the farmers to receive a "reward" for this agreement. This continuous annoyance led many families to leave the area at the end of the 1970s. Today, in the park where no ladangs have been opened for more than twenty years, the canopy has closed and only the expert eye will distinguish the damar islands in the forest.

### Episode 2: The Production Forest

Between the reserved forest and the Indian Ocean, the government granted concession rights covering 52,000 ha to a logging company, HPH Bina Lestari (Kusworo 1997) in 1981.[45] This company had formal rights to collect timber all over the three Pesisir subdistricts. Damar farmers did not know that their territory was given over to logging, as the company only logged timber in the extreme north and in the extreme south and did not dare harvest timber planted by local farmers in their agroforests; had they done it, this would not have been considered illegal, and farmers would not have had any right to claim compensation.[46]

The HPH left in 1991 as the area was divided, according to the first officially recognized TGHK maps, into conversion forest (7,500 ha) in the extreme south, protection forest in small pieces distributed all along the western border of the National Park,[47] and production forest (about 42,000 ha). The management of the production area was given to the state-run company Inhutani V and rumors quickly arose of an Inhutani project of industrial forest plantation for forest "rehabilitation" with large-scale acacia planting, to start in 1992. Fortunately, this never materialized, and it is not likely to happen since the latest development of the "Krui case" in the Department of Forestry.

Between 1992 and 1996, the Forestry service fixed these maps by measuring the state forest borders and by placing poles. During this period, damar farmers began to suspect that their lands were also being claimed by the state. They were never directly informed of the legal consequences of their land being classified as Production or Protection Forest, and when they asked, the answer was always that nothing was changed, *at least for the time being*.

### Episode 3: Recognition or Reappropriation? Ambiguities of the Forestry Support

Since 1992, new developments occurred that induced changes in the attitude of foresters regarding the damar enterprise of Pesisir farmers. Among the combined forces that pushed these changes are (1) the joined efforts of local and international researchers and nongovernmental organizations (NGOs) to promote the "Krui case" as an outstanding example of reforestation and forest management by local communities[48]; (2) the politically correct switch in the Department of Forestry itself toward allocating more support to forest communities; and (3) the (timid) acknowledgment by regional authorities of potentially serious social troubles induced by repeated rights violations and power abuse. This translated into more serious consideration of the originality and value of the Pesisir system at various levels of forest and regional administration.

However, this recent support may be a double-sided sword for damar farmers. While many foresters in Jakarta acknowledge the value and validity of the damar garden system, they seem unable to acknowledge that it arose and worked perfectly well without them for approximately a century.

## The New Deal: One for All, All for One, Against the Planters?

Forests, as well as nonforest lands, in the Pesisir, represent the last wild frontier in the already overpopulated province of Lampung. Because of the proximity to Jakarta and ongoing road development, it is a tempting invitation for private speculators: estate developers and agro-industries. For the regional authorities, these potential investors represent highly interesting parties. Besides being important taxpayers, which farmers are not, their investments would greatly increase the regional development index, and they would supposedly increase the level of industrial activities in the area (Kusworo 1997).

Since the early 1990s, following completion of logging operations, the district authorities have begun to allocate "private lands," as well as part of the logged-over forest lands,[49] in the three Pesisir subdistricts to two oil palm companies: 24,500 ha to PT Karya Canggih Mandiri Utama in the south, under development since 1994, and 17,352 ha to PT Panji Padma Lestari in the north, starting in January 1996, with an additional 4,500 ha in the south to the same company. Local farmers were not informed of these projects and started asking questions when they encountered field teams measuring land, including their damar gardens and even their ricefields. They were not always given the right answer.

Local authorities specified that oil palm would be planted only on actually "empty" lands, although local farmers could also be invited to join with their own lands if they wish. The local authorities started campaigning to support the project, asking the village heads to speak highly of the economic merits of oil palm planting and to ensure farmers' cooperation. But they also specified that no farmer should be compelled to give up his damar land for the company, and that no damar tree should be felled without the consent of the owner. PT Karya Canggih Mandiri Utama soon applied its own conception of "inviting" farmers to join. After a formal convocation conveyed through the subdistrict head—*camat*—to the village authorities, and given the subsequent lack of enthusiasm from damar farmers, they decided to use fake but positive agreements signed by farmers in lieu of true but negative ones, and they started clear-felling damar gardens under moonlight, as this makes for less visibility than bright morning light.[50] Angry farmers started publicizing this blunt violation of their basic rights to the provincial assembly and to local newspapers.

Farmers in the northern subdistrict, aware of the hidden practices of the companies, began to affirm and publicize their resistance to the arrival of PT Panji Padma Lestari even before it actually started measuring land (1996).

The joint claims of farmers, NGOs, and international research institutions asserting that replacing farmers' damar gardens by oil palm estates was neither ecologically defensible nor socially acceptable, and that the way this replacement was about to happen was clearly a classic case of power abuse by economic and political elites, finally succeeded. In December 1996, the Ministry of Forestry asked PT Karya Canggih Mandirutama to suspend its activities and solve the current conflicts with local damar farmers, while in March 1997, the provincial governor asked PT Panji Padma Lestari to halt its activities.

## CONCLUSION: WHICH JUSTICE? WHICH STRATEGY FOR CONFLICT RESOLUTION?

The Pesisir case addresses many justice issues. The main one concerns civil justice. The basic property and use rights of local people over lands and resources they have not only managed, and sustainably managed, but also developed and enriched over centuries are not fully recognized by the state, in spite of constitutional facilities that accommodate the acknowledgment and legalization of such rights. This issue is not specific to the Pesisir; it constitutes the major confrontation area between the state and forest farmers' communities, while revealing the major impediment to the integration of indigenous communities as groups of fully vested citizens into the Indonesian nation. The closure of the damar lands by the state would constitute not only a violation of basic rights but pure theft. Although granting rights over land to either public companies or private firms does not directly mean seizing the rights of local people, the practical interpretation and the implementation of these concession rights unavoidably erase the expression, and the very existence, of local rights. Replacing damar gardens with estates, either forest or agricultural plantations, or reserving the damar gardens for any project of conservation or production forestry, would obviously constitute a forceful appropriation of not only other people's lands, but also the fruit of other people's labor: i.e., trees and a considerable amount of valuable timber. In trying to defend their legitimate rights and properties, farmers might well lose even their basic human rights, given the fact that private firms as well as public bodies have shown their own conception of policy enforcement in the past decades: fake promises, verbal and physical intimidation, and passivity in the face of violent military intervention.

The second issue is one of economic and social justice. Replacing damar gardens with specialized oil palm or acacia plantation might prove, in the short term and with a partial economic valuation, an economic gain for the region itself. However, it is not certain that this economic gain will be redistributed to the farmers who will, certainly, contribute to this gain through their—underpaid—labor force. In terms of equity, the overall economic characteristic of the damar gardens is that the majority of the benefits they provide go to local people: farmers, wage laborers, local trade entrepreneurs. But the income officially derived from the damar activity by and for the district is almost nonexistent: taxes on the damar resin represent less than 0.1 percent of the district budget. Industrial plantation estates provide much higher profits—but to a far lower number of people—whereas levies raised by the district through the estates and the related industrial processing units are numerous and substantial. Seen from the point of view of regional administrators, the choice is obvious.

The last issue concerns environmental justice. The damar garden system developed by Pesisir farmers has proven to be an almost perfect ecological substitute for natural forests, in fact probably the best possible one for a diversified production system. Destroying damar gardens to make room for specialized oil palm or acacia plantation would constitute an ecological crime, with, among other immediate consequences, the destruction of the specific habitat for many lowland plant species; a significant reduction in the feeding and breeding areas of many endangered mammal and bird species (Sumatran rhino, tiger, tapir, elephant, siamang, hornbills, and rapaces); and a drastic increase in soil erosion, with consequent siltation of the Pesisir coast and of irrigation works in the lowlands, not to mention the increase in ecological risks for people as well as for the plantation. An additional consequence is the uncertain ecological sustainability of monocrop plantation over the long term, which has to be compared to the proven sustainability of the damar enterprise over the last 150 years. Crimes of this sort do not result in immediate punishment, but their long-term costs, for locals as well as for the nation itself, are potentially immense.

Without any doubt, damar agroforests represent a rare and precious example of successful sustainable management of forest resources in the humid tropics. However, for the last few years, the damar success story has been strongly endangered. Pesisir farmers have been facing urgently threatening choices, either to become laborers on their land as their damar agroforests might be converted to oil-palm estates, or to see their rights strongly restricted by zealous foresters who confound damar agroforest with natural forest and thus forget that there are no damar agroforests without damar farmers.

Indeed, culturally, biologically, economically, and socially, damar farmers have succeeded in reappropriating their forest resources. However, what the last few years of threats have shown is that this reappropriation was incomplete, enough to ensure the long-term sustainability of the system but not enough to protect its short-term survival. To ensure against forceful conversion, a fifth element is needed that would translate into legal terms the formal and official recognition of the damar farmers' contribution to overall national and regional objectives.[51] But the agroforest situation does not fit any of the existing legal forest categories. Acknowledgment of this fact may well be the most important output of a recent meeting (in Liwa, 12–13 June 1997), where all Krui case stakeholders were gathered mainly to discuss the problems linked to land and product status in the Pesisir. This means that if damar agroforest lands are to remain as state forest land, a new legal status should be devised to fit damar farmers' needs and to ensure a future for damar agroforests. As this is the new direction that has been taken after the Liwa meeting by the Department of Forestry, under the minister's initiative, there is now hope, after years of doubt, that justice for damar farmers will be respected, and that this may be a first step on the way toward justice for all the other agroforest farmers in Indonesia facing the same kind of problems.

The agroforest framework offers a good opportunity to escape the formal forestry context and to devise new forms of association between farmers, foresters, and regional authorities concerning forest resources. Ecologically, economically, and socially, the agroforest should not be identified with a natural forest, and indeed, as long as this confusion between forest and agroforest is maintained, as long as local practices for management of forest resources in farming systems are ignored, the chances of survival of agroforests as a unique model of integral forest management continue to decrease. Agroforests, once they are recognized, will open a totally new field for negotiations between foresters and local communities, a field favorable to institutional innovations where ancient conflicts might be resolved without one or another party's losing face. In particular, it could facilitate the formation of new alliances between the conventional forestry sector and local communities, and new options for land or resource management and control, without destabilizing the existing forestry legislation. Furthermore, it would be a pity not to take this as an opportunity to rethink, on a real field basis, the whole conventional context of forestry and agriculture.

## ACKNOWLEDGMENT

We wish to thank Wiyono, an artwork-cartography specialist who has been working with the French Research Institute for Development through Cooperation (ORSTOM) (now l'Institut de Recherche pour le Développement, IRD) in Indonesia since 1984. He was also involved in field studies carried out with P. Levang in Krui. Wiyono is presently posted with G. Michon and H. De Foresta at the International Center for Research in Agroforestry (ICRAF) South East Asia Regional Office.

## NOTES

1. While being constantly refined, these laws and regulations show only minor differences from those promulgated during the colonial period, at least in terms of underlying con-

cepts (Peluso 1992b). With the major changes that have shaped the modern history of forests in Indonesia, present laws and regulations are clearly outdated and are more and more unable to fulfill their overall objectives.

2. In a broad sense, this includes some true forest dwellers, but mainly the bulk of shifting cultivators living on the forest margins.

3. Population pressure is a well-recognized cause of this increase in space constraints; other, less recognized but major causes are land appropriation by the state (state forest land) and land designation by the state for "development" projects.

4. For example, the official version of the underlying causes of deforestation states that shifting cultivators are the main, if not the sole, agent of increased deforestation in the country. Another official dogma is that indigenous people are totally unable to manage forest resources sustainably.

5. Watershed protection, forest production, source of livelihood for the community inside and around the forests, flora and fauna protection, transmigration, agriculture, plantation, and cattle raising.

6. However, in relation to indigenous communities, law enforcement is easily achieved through the help of the armed forces.

7. Conservation, mining or logging concessions, transmigration schemes, industrial forest plantations, or any other kind of elite estate.

8. Farmers might try to ask for compensation for their lost resources, but in no way will most of the components of the indigenous agricultural systems—swidden fields, fallowed fields, rattan gardens, or any kind of agroforestry field—be compensated for.

9. Not more than a dozen timber conglomerates control the 56 million ha of production forest—about 30 percent of the total land surface of Indonesia (Gillis 1988).

10. In most of the outer island provinces, forest land borders are being renegotiated in concert with the regional government and forestry agencies. This negotiation process does not involve any representative of the farmers' communities.

11. Foresters definitively lose authority and jurisdiction over converted forest lands.

12. Population density ranges from 100 people per sqare kilometer in the central district, where available space for agriculture has been saturated for more than thirty years, to less than 20 p/sq km in the south, where land can still be easily appropriated. From ancient times until recently, the main communication links with regional centers (Bengkulu, Teluk Betung, Batavia/Jakarta, Singapore) were established through direct maritime connections, and several small harbors were scattered along the coast, with an important market center at Krui, in the central district. Recently, the old road to the east, through the central Barisan range, and through the national park, has been rehabilitated, a provincial road to the north has been completed, and another is being developed to the south.

13. *Damar* is a generic term used in Indonesia to designate resins produced by trees of the dipterocarp family.

14. These are mainly villages whose territory lies on the sandy sediments of the coast and which have specialized in coconut growing, as well as new transmigration villages in the south, and several former "clove villages" in the north (Dupain 1994).

15. Tree stands in damar gardens show a mean density of 245 trees per hectare (from a record of all trees over 20 cm in diameter on eight 4,000-$m^2$ randomly selected plots) and a mean basal area of 33 $m^2$/ha. These quite high figures, associated with a well-balanced diameter class distribution, are really close to structural patterns found in natural forests (Michon 1985; Wijayanto 1993).

16. Pollination, fructification and production, seed dispersion and germination, seedling and sapling development, gap colonization, water, and nutrient cycling.

17. In the eleven villages "less involved" with damar activity, resin production and processing make up 45 percent of the average household's cash income. In the remaining forty-six "damar villages," it represents between 70 and 100 percent. Production in 1993 generated a regional gross value estimated (in Indonesian rupiahs) at Rp6.5 billion (US$3.25 million) for Pesisir farmers from the sale of the damar only, to which Rp5.3 billion (US$3.25 million) have to be added as additional value generated by trade, and Rp2.7 billion (US$2.65 million) for related wages, which makes a total of Rp14.5 billion (US$7.25 million) of regional gross value for all Pesisir villages. To this should be added Rp542 million (US$271,000) of profit margins made by the nine Krui traders (Dupain 1994).

18. The sale value of the resin itself represents only less than half (44.5 percent) of the total income provided by resin production in villages, related activities accounting for the largest share (Mary 1987; Levang and Wiyono 1993).

19. Because of adverse climatic conditions, there was no fruit season in the area from 1992 to 1994.

20. House repair; purchase of furniture, chainsaw, or satellite dish; wedding ceremonies or any festive activity.

21. "Pawnbrokers" (any villager with funds available can become a pawnbroker) may provide loans of several thousand rupiah for one garden for an undetermined period (at least one year). Tree production serves as yearly interest for the creditor, who can use the garden for his own convenience during the entire loan period, except for selling or transforming it. The agreement ends as soon as the garden's owner refunds all the money to the creditor, or when he claims the profits made by the creditor are sufficient.

22. Bank credit is still uncommon and unreliable in villages.

23. Thatching material from palm and *Garcinia* leaves, rattan and other liana, fibers from tree bark, and bamboo. For timber production, damar and fruit trees appear as important as wild species.

24. The most valuable but also less predictable extractive commodity in the damar gardens is rattan. Rattan cane harvest is subject to the profit/failure dynamics of local buyers, and rattan fruits once appeared to be a valuable product. This important economic unpredictability constitutes the main impediment to the development of rattan harvesting into a real garden production.

25. As other sources of timber in the area vanish, the economic potential of damar timber is increasing. However, timber harvesting and marketing regulations, taxes, bribes, and police harassment constitute major impediments to the development of timber production as an integrated production of damar gardens.

26. Poor people, and children (to pay their weekly school expenses), can harvest resin fallen on the ground; they are even allowed to collect resin from the lowest tapping holes. Valuable fruits are traditionally shared by the family, and, in season, distant relatives may come and join for a durian party or leave with a basketful of langsat, which is considered a valuable practice for maintaining family cohesion.

27. Useful garden products, such as firewood, sugar palm sap, small fruits, and medicinal plants, can be collected in privately owned gardens by whoever needs and asks for them.

28. Which, in itself, is quite remarkable as foresters have often been unable to achieve success in the industrial growing of dipterocarps.

29. Several types of resin can be harvested from natural forests in Indonesia. Turpentine

(the pine resin) and copal (*Agathis* resin) used to be known as the major economic resins traded from Indonesia before World War II. Some 115 dipterocarp species, distributed in seven of the ten genera of the family, produce damar (Foxworthy 1922). Because of the dominance of dipterocarps in lowland forests of the region, damars form the most common resin type in western Indonesia but are usually considered of lower quality than copal and turpentine. In fact, two categories are commonly distinguished that largely differ in quality. *Damar batu* are low quality damar of dark color resulting from spontaneous outflows caused by occasional injuries. Large pieces fallen from the bark can be collected by digging the ground around the trees, which usually provides huge quantities from old trees. *Damar mata kucing* are clear to yellow damar of high quality (comparable to copal), obtained by making incisions in the bark. About forty species from the genera *Shorea* and *Hopea* produce damar mata kucing, among which the best are *S. javanica* and *H. dryobalanoides*.

30. Pepper cultivation probably developed very early (i.e., in the seventeenth century) but lost its importance in the 1920s; coffee was developed after the beginning of the nineteenth century and is still intensively planted; clove was introduced in 1930 and fully developed in the 1970s, but it almost totally disappeared as a result of a serious disease which hit Sumatra after 1975.

31. Rappard (1937) mentions that extension of the damar plantations around Krui was helped by the former governor Helfrish. Dupain (1994) notes an astonishing coincidence between the oldest centers of damar cultivation and the former centers of residence of *pangeran* (representatives of the *marga*, which corresponded to territorial units reinforced by the Dutch; the pangeran were an important link between colonial authority and the villages). Dupain formulates the hypothesis that, had the pangeran been more "informed" or "controlled," they could have facilitated the extension of damar cultivation. But these centers of pangeran residency were also the place of residence and activity of Chinese traders (Pelzer 1978).

32. Coffee or pepper and *Erythrina* as shade trees or living poles were coplanted with dry rice and vegetables; the productive plantation was maintained for four to six years for coffee, and up to fifteen for pepper, and then returned to fallow.

33. Rain-fed rice as the first grassy phase, coffee/pepper as the early pioneer tree phase, subsequent secondary formation with young damar and fruit trees, and damar/fruit trees related to various wild trees as the mature phase.

34. To assess biodiversity levels, comparative studies have been conducted between agroforests and related primary forests for several fauna and flora groups, including higher plants (from ferns to dicotyledons), birds, mammals, and soil mesofauna. Soil mesofauna diversity levels are quite similar in forest and agroforest. None of the numerically important species of the forest population are absent in the related agroforest; however, because many species in that large group are rare species, results do not prove that all forest species exist in the agroforest (Deharveng 1992). Bird richness in damar agroforests is 30 percent lower than in primary forest: 96 and 135 species have been recorded in those two ecosystems, respectively. About 57 percent of the bird species found in the forest have not been encountered in the agroforest, whereas 40 percent of the agroforest species were not present in the forest surveys (Thiollay 1995). Reduction of bird diversity can be related to biological factors (simplification of composition and vertical structure from forest to agroforest), but it is probably mainly the result of hunters (birds are caught for food, but they are also often kept in cages in the village or sold to outsiders, as bird keeping is more than a hobby in Indonesia). As far as mammals are concerned, almost all forest species are present in the agroforest. Densities of primate populations (macaques, leaf monkeys, gibbons, and siamang) in the agroforest are

quite similar to those observed in natural forests. Footprints of the rare Sumatran rhino have been recorded in the agroforest, less than 2 km from villages. This represents the first record of rhino in this part of Sumatra and allows us to draw hypotheses about the usefulness of agroforests for the conservation of endangered animals, an important adjunct to protected forests (Sibuea et al. 1993). Global flora richness is reduced to approximately 50 percent in the agroforest. However, results have to be dissociated by biological groups, as they can be very different from one group to another. The largest loss occurs for trees (agroforest diversity represents merely 30 percent of original diversity levels), which is quite understandable as economic intensification, and therefore selection operates mainly on trees. Epiphyte and liana richness in the agroforest is at least 50 percent of forest richness, whereas for undergrowth plants, our samples show the percentage to be much higher in agroforests than in natural forest (two to one), which should be related to the common abundance of this group in secondary forests as compared to primary forests (Sibuea et al. 1993; Thiollay 1995).

35. Among the explanations put forward for such a mutation, the religious evolution toward an "adult" Islam comes at the right time. Islam has shown the errors of former beliefs in which spirits and magic held a determining role. This feeling of returning to the straight and narrow path is totally in line with the first principle of the political ideology of Indonesia, known by all Indonesian citizens since elementary school, which gives full support to the great religions in their holy wars against animism.

36. It is commonly called darak, equivalent to the Indonesian ladang, a generic term that defines any field opened in the forest, or repong, which formerly designated privately owned fruit gardens surrounding villages. Pesisir farmers also often describe it with the Indonesian word *kebun*, which means garden or plantation.

37. Such systems still exist in other parts of Sumatra (Levang 1989).

38. Transmission of rights to the damar gardens commonly follows the traditional patrilineal tenure regime formerly devised for irrigated ricefields and fruit gardens. A piece of land, once acknowledged by the community as the private property of the individual who "created" it, remains in the lineage of the "creator" through its inheritance by the eldest son.

39. Sharing landed properties usually occurs after the birth of the first male child of the eldest son. The newly endowed heir becomes the *kepala keluarga*, head of a family unit that comprises his children and his parents, as well as his unmarried youngest brothers, his unmarried sisters, and sometimes the children of his married brothers. As the exclusive heir of the family properties, the eldest son is in charge of housing and feeding all this extended family group. This heavy responsibility, according to the heirs, largely compensates for the inequality of the transmission system, but it does not seem to negatively affect individual incentives for production and investment.

40. Usually parents, if present, and direct uncles and brothers.

41. This situation of ecological and economic collapse happened in some villages, with the introduction of clove. The collapse of clove gardens and the related extension of bush and grass vegetation has led in some areas to fire problems that threaten to destroy the remaining patches of damar gardens.

42. Dated 1995 but available in early 1997, this map was recently shown by some of us in one Central Pesisir village and quickly rolled up, following the advice of the village chief, who explained that villagers would riot if they knew how the result of their plantation work is treated by officials.

43. Already mapped as an oil palm estate plantation in South Pesisir.

44. Wildlife, especially elephants, constitutes another source of conflicts between villagers and the directorate general for Forest Protection and Nature Conservation (PHPA). For several years—in fact, since the remaining production forest was logged over; the related opening of roads attracted migrants, who cleared large areas in the logged-over forest for coffee and pepper growing—elephants have frequently come out of the national park, destroying the crops and attacking farmers in their *ladang*, and now even in villages. Villagers require the right to carry guns to protect themselves, which is of course not accepted by the conservation guards.

45. The Pesisir area is known in the Department of Forestry as Bina Lestari's forest.

46. Under the same circumstances in the province of Bengkulu, private foresters have deliberately logged damar gardens belonging to local people.

47. These pieces of protection forest have no rationale other than as compensation for other areas in the province where previously designated protection forest has been declassified to make way for public and private development projects.

48. As a result of this joint effort, the prestigious National Kalpataru award for the environment was given by the president of Indonesia to the "customary community of damar farmers in Pesisir, Krui" on June 5, 1997.

49. Using the process of revision at the district scale of the TGHK as a way to declassify the targeted forest land.

50. Farmers actually woke up and noticed that all their damar trees had been cut overnight.

51. This need was recognized quite early by the minister of forestry, who, after having visited Krui in 1994, has always been quite positive about damar farmers and their contribution.

# CHAPTER 8

## Tropical Forests Forever? A Contextual Ecology of Bentian Rattan Agroforestry Systems

*Stephanie Gorson Fried*

> Rattan income has already made a substantial contribution to the Bentian people. They school their children with rattan money, they build houses with rattan money, they buy food and many other goods with rattan money. . . . The goal of development itself is to increase the standard of living of the people, not to destroy it and certainly not to destroy the commodities, like rattan gardens and land which in reality make a large contribution to the [welfare of] people.
>
> *Bentian oral historian[1] and author, rattan farmer, forest manager (1991)*

> The loggers say, "Who owns this land? No one because there are no [sign]posts." But families and neighbors know. There is no "proof" like the government wants, but farmers A and B know their boundaries. [Sign]posts are regulations of down-river city people. It's not like the city here. In our forest gardens, fruits are free for a hungry visitor. Eat them in the forest if you are hungry but do not take someone else's fruits back to the village, or to sell.
>
> *Bentian oral historian and grandfather, rattan farmer, forest manager (1991)*

Rattan, as climbing, spiny vines belonging to the palm family are often called, is considered to be the most important nontimber forest product in Indonesia (Weinstock 1983; Peluso 1983a). Eighty to 90 percent of the world's rattan supply comes from Indonesia's tropical forests (Menon 1989; Priasukmana 1988). My original intention was to study the unusual sustainable forest management and rattan production system of the Bentian Dayak[2] people of Kutai Regency, East Kalimantan (figures 8-1–8-3). However, armed bulldozer crews associated with private-sector logging concessions and plantations moved into the Bentian region and, in

**Figure 8-1** Indonesia.

the name of "development," proceeded to destroy the remarkable Bentian rattan gardens that were the object of my research and that conducted by the Indonesian Ministry of Forestry. Thus it became necessary, in addition to providing a basic description of the rattan system, to examine the process by which a productive, sustainable, foreign-exchange-earning agroforestry system, and the people who designed and implemented it, could be purposefully and brutally eliminated in the name of development.

This study, therefore, presents the Bentian rattan agroforestry system in the larger context of state and military control over access to natural resources, and of the increasingly frequent battles over the rights to use, own, and conserve Indonesia's forests.

The Bentian, their forest management system, and their struggles for full national citizenship rights are not unique. Tropical and temperate forest ecosystems throughout the world are frequently the sites of conflicts over access, ownership, and management rights. Areas of natural resource abundance often lie on the economic frontier between an expanding global economy and traditional societies that have managed these resources for centuries.

Such conflicts occur in the context of accelerating resource extraction in remote, resource-rich regions and are commonly accompanied by environmental as well as cultural destruction. Many of these conflicts go unseen and unrecorded by those not party to them. They commonly bring substantial suffering for the inhabitants of remote communities who find themselves plunged into poverty and despair as their resource base is destroyed and their communities are uprooted. The very existence of these conflicts is often vigorously and publicly contested by the government officials and private-sector parties engaged in them.

**Figure 8-2** Kalimantan, Indonesia

Conflicts usually occur in isolated and relatively marginalized resource-rich areas including coastal fishing villages and remote forest communities. They are often played out between literate, well-educated members of state bureaucracies, national and multinational resource extraction companies, and armed security forces, on the one hand, and skilled but less literate resource managers from traditional societies, on the other.

In the past, records and descriptions of traditional resource management techniques and conflicts over resource use and appropriation have been recorded either by state-sanctioned researchers or by parties to the conflicts. There are usually few outside witnesses to the brutal and brutalizing processes of land seizure, dispossession, and forest destruction. Official and corporate documentation is usually stored

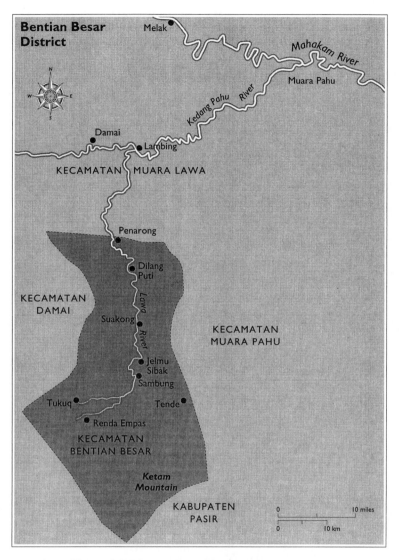

**Figure 8-3** Bentian Besar District, East Kalimantan.

in the files of government bureaucracies, university offices, and resource extraction companies, remaining, for the most part, unavailable to a wider audience.

In recent years, however, responding to modern high-speed incursions of the state into the lives of citizens who were formerly marginal to the nation-state enterprise, a number of traditional or indigenous societies inhabiting tropical forest ecosystems from Brazil to Indonesia have begun to document their own resource management techniques and land tenure practices, as well as their own accounts of conflicts over access to natural resources.

By recording their accounts on film and in writing, otherwise marginalized citizens, such as the Kayapo of the Brazilian Amazon and, more recently, the Bentian of Indonesian Borneo, have been able to strengthen their claims to legitimacy as resource managers, property owners, and citizens of modernizing nation-states. They have also been increasingly able to affect national and international discourse on development, conservation, and resource management.

As modernizing nation-states such as Indonesia move into new forms of "hypermodernity" or "fast capitalism," social relations are torn from the hold of specific locations and recombined across wide time-space distances (Giddens 1991:2). This has led to a tremendous growth of what some have called "placeless power," that is, power freed from the constraints of local relationships, customs, and norms, with a corresponding increase in "powerless places" (Pred and Watts 1992).

In the case of Kalimantan, indigenous peoples are attempting to capture some of the "placeless power" that otherwise threatens to destroy them by speaking and writing about their forest management practices, their knowledge of ecology, and the challenges they face.[3] This chapter presents ecological aspects of the sustainable Bentian agroforestry system, set in the context of the ongoing destruction of Bentian forests by private-sector logging interests.

## BACKGROUND: RATTAN CULTIVATION

Rattan vines have been harvested and utilized throughout Southeast Asia for centuries, if not millennia. Typical uses of rattan include "cordage, baskets, mats, fishtraps and snares; [lashing for] house-poles, ladders, carts, masts, water-wheels, and other such constructions [obviating] the need for nails; sunblinds, halters, tethers, guy-, mooring-, and tow-ropes; [and] perilous swinging bridges over gorges" (Purseglove 1972).

In addition to its widespread local uses, rattan has played an important role in international trade for centuries. The Portuguese, who conquered Malacca in 1511, brought the rattan trade to Europe (Purseglove 1972). The term *rattan* entered the English language, slightly modified from the Malay *rotan*, in the seventeenth century (Oxford Dictionary of English Etymology 1966). The island of Borneo, home to over 300 varieties of rattan, has historically played an important role in the rattan trade.

In the early 1970s, world demand for rattan rose dramatically as wicker furniture grew increasingly popular in Europe and North America, and as rattan floor mats gained popularity in Japan. Although long considered a minor forest product, by the late 1980s unprocessed rattan generated $600 to $800 million in annual revenues on the world market (Menon 1989). The trade in finished rattan products, primarily furniture and woven mats, produced more than $15 billion for secondary processors and manufacturers located, for the most part, in Taiwan, Singapore, and Hong Kong.[4]

Since the 1970s, Indonesia has been faced with a rapidly diminishing supply of wild rattan as a result of logging activities, forest conversion, over-harvesting, and forest fires. In 1988, because of the high potential for the rattan industry to have a substantial positive impact on national economic development, small-holder incomes, and the ecological stability of forest lands, the Indonesian government tar-

geted 1.25 million hectares of land to be brought into rattan production over the following decade (Menon 1989; Dransfield 1988; Godoy and Feaw 1988; Tardjo 1982).
Approximately 9.3 million hectares of forest land in Kalimantan, Sumatra, Sulawesi,
and Irian Jaya have been identified as potentially productive rattan forest (Menon
1989). The province of East Kalimantan contains over 4.3 million hectares of wild
and cultivated rattan, making it one of the largest rattan production areas in Indonesia (Peluso 1983a).

Despite the economic importance of rattan, there is little detailed information
available on the manner in which rattan production is already successfully implemented and integrated with the agroforestry systems of small-holders cultivating the
crop, or on the extent to which diverse groups of agriculturalists and nonagricultur-
alists depend on rattan for their livelihood. For hundreds of years, forest and riverbank dwellers in Kalimantan have cultivated rattan on a commercial basis (Mayer
1989; Peluso and Jessup 1985; Weinstock 1983). Studies by Mayer (1989), Dransfield (1988), Godoy and Feaw (1988), and Weinstock (1983) suggest that the social
organization of various small-holder rattan agroforestry systems allows for economically viable and ecologically sustainable production in a manner compatible
with the maintenance of social cohesiveness and cultural identity. Despite the value
of rattan to national, regional, and local economies, rattan management and harvesting practices that have proven sustainable for centuries have not been documented in detail.

## THE BENTIAN

The Bentian of East Kalimantan and thirteen other groups identify themselves with
the Lawangan Dayak family. Approximately 250,000 Lawangan occupy a territory
of one-quarter million square kilometers from the middle section of the Barito River
in Central Kalimantan to the Middle Mahakam River region in East Kalimantan
(Weinstock 1983).

Bentian oral historians and *adat*, or traditional leaders, trace their migrations
from Central Kalimantan to East Kalimantan approximately twenty-four generations ago. The Tementa'ng inhabitants of East Kalimantan did not permit the newcomers to settle in their area until they had fulfilled a *penyua*, or gift obligation, of
100 Chinese urns, 100 water buffaloes, and a slave. In return, they were granted the
rights to an area called Nine Rivers, which encompasses the current administrative
district of Bentian Besar (Nasir 1991). The new groups eventually came to be called
Bentian.[5]

The Bentian, like other Lawangan peoples, are swidden agriculturists who practice a form of rotational agriculture suited to the poor and easily eroded soils of the
region. Clearing patches of old-growth secondary forest to plant rice with dibble
sticks, Bentian farmers, unlike many other swidden cultivators, also sow large quantities of rattan[6] seeds and seedlings in their fields. These seedlings mature after seven

to ten years into productive rattan vines, their weight supported by thick secondary forest regrowth in the fallowed fields. The Bentian continue to harvest rattan for up to thirty years or more after planting. Rattan is currently a staple commodity for the local inhabitants of the region, as it has been for well over a hundred years.

Norwegian naturalist Carl Bock, during his 1879 visit to the Bentian region, noted that rattan,

> besides forming the chief article of trade in its raw state, . . . furnishes the material for the manufacture of an endless variety of useful objects. . . . Take away his rattan, and you deprive the Dyak of half the articles indispensable to his existence. What crochet-work is to the European lady, rattan plaiting is to the Dyak housewife. She is always manufacturing either sleeping-mats, sitting-mats, sirih boxes, baskets of all shapes and sizes, and for all kinds of uses, besides long pieces of plait to be used as cords, ropes, or threads, in dressmaking, house-building, raft construction, and the hundred-and-one other purposes of daily life in the forest.
>
> Bock (1985:204)

Currently, rattan gardens in the Bentian region function very much like bank accounts for their owners. When rattan prices are low, or when farmers choose not to sell for other reasons, the clumps of constantly growing rattan shoots may be left in the fields, growing longer and more valuable (adding "interest") every day that the farmer does not sell the crop. In the case of a family emergency, a withdrawal may be made from the bank account. Even if rattan prices are still low, the harvest of a ton or two will cover hospitalization costs, school expenses, or other urgent needs. Additionally, small amounts of rattan are harvested every month, regardless of the market price, to provide the income necessary for the purchase of goods such as cooking oil, batteries, and soap.

The Bentian strategy of a diversified and flexible household economy, which involves rattan production for export markets as well as fruit, vegetable, and grain production for local consumption, has proven successful enough to see them through the first few years of a traumatic drop in rattan prices. This flexibility makes Bentian swidden agriculturalists easily competitive with the large, single-commodity-dependent commercial plantations currently favored by the Indonesian central government and international financial institutions such as the World Bank and the Asian Development Bank. Such externally financed monoculture plantations, when exposed to severe price fluctuations, face a substantial likelihood of worker layoffs, social unrest, and bankruptcy.

## INVISIBLE GARDENS

> If you find clumps of fruit trees, *rambutan* together with *cempedak* and *nangka,* for example, this means that in the past the ancestors planted them. Descendants of the planters know the borders of their fields, usually hills, rivers and streams.
>
> Bentian oral historian and village government official,
> rattan farmer, forest manager (1992)

To the untrained eye, Bentian rattan gardens often appear to be "natural" secondary forest filled with looping, spiny rattan vines. The gardens, indeed, do imitate the canopy structure of the forest. To the Bentian eye, however, rattan gardens are easily recognizable and are a sign of human presence and land ownership. The gardens are clearly demarcated by borders of fruit trees, small streams, hills, honey trees, and other obvious planted or natural features. Much of the Bentian forest management system, however, remains incomprehensible to and unseen by outsiders. This is often a source of conflict, as outsiders fail to consult with the Bentian before rushing in to declare Bentian lands "unoccupied" or "abandoned" before clear-cutting them.

Both social and ecological factors are of crucial importance to the sustainability of the Bentian agroforestry system. The Bentian possess a detailed knowledge of Kalimantan forest ecology. This, combined with a socially regulated system of agricultural and forestry management practices overseen by *hukum adat*, or customary law, has safeguarded their forest territories. Ironically, this has also made Bentian lands inviting targets for logging operations.

## THE SOCIAL AND ECOLOGICAL CONTEXT OF BENTIAN RATTAN PRODUCTION

According to Bentian oral historians, farmers who had always cultivated small amounts of rattan for household consumption were asked to produce rattan for export by Aji Imbut Sultan Muhammad Muslihuddin, also known as the Rose Water Sultan, who was the ruler of the Kingdom of Kutai Kertanegara in the mid eighteenth century. Successive Kutai sultans, responding to Dutch colonial demands, also requested taxes payable in the form of rattan from the Bentian and neighboring groups.

In more recent years, as supplies of wild rattan have been depleted as a result of overharvesting in the Philippines, Malaysia, and parts of Indonesia, the rattan cultivating skills of peoples such as the Bentian have taken on great significance.

Prior to a discussion of Bentian agroforestry techniques, however, it is necessary to explore the social and environmental context of the Bentian system, that is, the close link between sustainable rattan cultivation and swidden agriculture, and the central importance of adat and adat law.

## SHIFTING CULTIVATION

Animals live in our protected forests. If we had permanent rice paddy fields, the animals could not live. When we fallow our fields, the animals use them. When we move, they have a place. After the harvest, they come, get fat, and have offspring. If we need them we look for them with dogs.

Bentian oral historian and trader, traveler, rattan farmer,
and forest manager (1991)

Bentian farmers, like most successful agriculturalists in East Kalimantan, practice a form of rotational agriculture, sometimes referred to as shifting cultivation (*ladang berpindah*) by non-Bentian Indonesians.[7] The Indonesian[8] terms *ladang berpindah* or *ladang berpindah-pindah* are used in official discourse to refer to all types of swidden agriculture. *Ladang* means field and *berpindah* means moving, or changing location. *Berpindah-pindah* means moving around or moving here and there and implies a careless type of uncontrolled nomadism. *Slash and burn* is probably the closest English language equivalent to *ladang berpindah-pindah*, which evokes the illogical, destructive, and dysfunctional use of natural resources.

The Bentian prefer to use the Indonesian terms *ladang bergilir* and *ladang daur ulang*, or "rotational agriculture" and "recycling agriculture" when describing their intergenerational, managed fallow system to outsiders, since the term *ladang berpindah-pindah* is commonly used to describe destructive farming practices that are the target of official eradication campaigns.[9]

Bentian farmers, like many other upriver farmers, operate within a "perennial" as opposed to an "annual" agroforestry paradigm, one that stretches across decades, if not centuries, of forest management. Their concepts of forest and field management reveal cross-generational planning and are reflective of what has been called perennial polyculture. Contrary to official beliefs, Bentian and other upriver farmers often plant slow-growing, highly prized, and commercially valuable timber trees such as meranti (*Shorea* species), trees that are often felled indiscriminately by private-sector logging companies.

## ADAT AND ADAT LAW

In modern Indonesia, an archipelago of thousands of islands and hundreds of different cultures, *adat*, meaning custom or tradition, is a term of great significance. It is most often used in reference to the different customs and practices of Indonesia's myriad cultural groups. Each cultural group is said to have its own specific form of adat, which includes marriage, birth, and death customs, methods of conflict resolution, rules for resource ownership and utilization, regulations for permitted and prohibited behaviors, and processes, including sanctions, for responding to violations of adat. Local methods of conflict resolution and processes for addressing adat violations are often referred to as *hukum adat*, or "adat law." Bentian forest management adat and adat law practices concerning natural resource access and utilization have ensured the success and sustainability of the Bentian agroforestry system. Bentian adat regulates the utilization of forested areas for agricultural purposes, the harvesting of rattan, fruit, honey, and other agricultural and forest products, the inheritance of previously cultivated lands, and the ownership of trees and rattan clumps. The Tatau, or Bentian adat leader, and a council of village elders adjudicate disputes that arise in the region. As early as the first half of the seventeenth century, the ruler of the East Kalimantan Kingdom of Kutai, Aji Pangeran Sinum Panji Mendapa ing Martadipura, officially recognized the adat of upriver Mahakam peoples and bade his followers to respect Dayak adat (Widjono 1991).

Bentian adat law is primarily a law of fines (Pantir 1990). Parties found guilty of adat transgressions must pay fines in traditional Bentian currency: livestock (water buffalo, pigs, chickens), antique Chinese urns, rattan mats, bolts of cloth, hunting spears, machetes, brass gongs, and white plates (Dingit et al. 1994; Pantir 1990; Bentian oral historian 1992). The existence of explicit adat regulations concerning the harvesting of different types of forest trees, the care and ownership of rattan gardens, and other aspects of natural resource management, has ensured that the Bentian have implemented sustainable resource management practices over the centuries. One of the most important aspects of Bentian customary law is the explicit delineation of use rights and ownership of forests and forest gardens.

In the Bentian region, land ownership claims are not usually documented on paper. Instead, these claims are validated by village members chosen to act as ownership witnesses, or *saksi*. For each field owned by a Bentian farmer, at least one other villager (preferably an older one) acts as a *saksi* to the ownership of that particular field.

> This Witness is ready to be called on at any occasion to testify about the truth of the ownership claim. . . . Before testifying, the Witness is sworn in first by the Tatau with the Ineq Rodot oath. Whosoever bears false witness . . . will die within at most one month, eaten by Ineq Rodot. Ineq Rodot is the name of the Tiger God. The Tiger God oath is very honored and feared in the Bentian Besar [region]. This oath is still valid[10] and the apparatus for it is still cared for[11] by the Tatau. . . . Because of the extraordinary power of this oath, very rarely do people want to bear false witness. Because of this, if the witnesses who are called feel hesitant about the truth of their statements, they have the right to refuse to be called as a witness before being sworn in with the Tiger oath.
>
> Nasir (1991)

Disputes about the ownership of property and land in the Bentian region, then, are played out not through official land-ownership certificates and written processes but rather through the medium of living witnesses, sworn to tell the truth on the threat of painful death from supernatural causes.

Severe damage to the Bentian natural resource base occurred recently when outsiders, unaware of and apparently uninterested in local ecology, local resource ownership and utilization patterns, and Bentian adat regulations, have misused and damaged Bentian forests and waterways (Zerner 1992; Fried 1995; Triwahyudi 1992; Dingit et al. 1994). The detailed knowledge possessed by the Bentian of their surrounding forest ecosystem, their individual and collective ownership of specific forested regions, and their adat regulations on resource utilization and ownership provide a strong foundation for Bentian forest management practices. The recognition of Bentian adat processes and regulations concerning forest management and ownership is, therefore, a prerequisite for the maintenance and development of sustainable rattan cultivation in the region.

## BENTIAN AGRICULTURE AND FOREST MANAGEMENT

> In fruit season, plant as many trees as possible. Make a nursery in the field for
> nangka, cempedak, durian, and others. Choose seeds from the biggest fruit. Take
> the seeds carefully from the tree. For keramuk, choose the wide fruits—these are
> female—they will grow and fruit again. Narrow fruits are male.
>
> Bentian oral historian and rural government official,<br>rattan farmer, forest manager (1991)

> For the annual harvest ceremony, Tak Dinas's house is transformed into a fantastic
> vegetable world. Over the next eight days, as she calls visiting spirits down from
> the heavens, a balai is constructed in her main room. The balai is a miniature ver-
> sion of a Bentian field house, built of ritually and ecologically important forest and
> agricultural plants. On the fourth night of the ceremony, the balai is transformed
> from bare bamboo poles into an enchanted plant- and cloth-covered structure. Bird
> carvings top the house poles and the balai is festooned with red pangir flowers. Of-
> ferings of rattan, ironwood, and ngangsang are made. Branches of these and other
> forest trees are requirements for balai construction as are the aromatic and colorful
> telasi, kumpai lati, and geronggong flowers. These flowers are sharply fragrant, and
> appear similar to flowers found elsewhere throughout the tropics which play a role
> in reducing insect pests in agricultural fields.
>
> Fried (1992)

The Bentian system of rattan production, like that of the related Benuaq and the
Pasir peoples, is closely integrated with their swidden agriculture cycle. Bentian
farmers, understanding the structure and nutrient content of the poor soils of the
region, generally cultivate rice and vegetables for two consecutive years on a particular plot of land. After two years, to ensure the integrity of the soil structure and
the maintenance of soil fertility, the land is left fallow and is allowed to revert back
to a forested state, enriched by the planting of fruit and timber trees and climbing
rattan vines.

In any given year, a Bentian farmer will maintain most of her hereditary field
sites in standing forest. Each year only two small plots, totaling perhaps 1.5 to 2.5
hectares (depending on labor availability and the number of family members supported by the field) will be planted to rice and vegetables. One of these plots will be
in its first year of rice and vegetable production and the other will be in its second
year of rice and vegetable production. Other small plots, planted to rice and vegetables in previous years and now fallowed, will be in various stages of secondary forest regrowth.

## BENTIAN FOREST VOCABULARY

The Bentian, like most upriver Kalimantan peoples, are not only farmers but also
experienced forest managers. As such, a brief study of their language reveals a

wealth of indigenous terms concerning forests and forest management. For the purposes of this chapter, we will examine Bentian concepts of forest regeneration. The Bentian language reveals a series of specific terms for stages of forest growth, unparalleled in the English or Indonesian languages which refer broadly to "primary" (untouched) and "secondary" (regrown) forest. The Bentian language contains at least seven terms for different stages of forest growth, ranging from *boaq*, the immediate regrowth that occurs in a former rice field during the first years of fallow, through two stages of *kelewako* (*ureq* and *tuhaq*, or young and old) occurring up to five years after the last rice harvest. The *kelewako* stages are dominated by pioneer tree species that are still small enough that they may be easily felled with a machete. This is followed by the *bataqng ureq* and *bataqng tuhaq* stages. *Bataqng ureq* marks the beginning of the die-off of the pioneer *kelewako* species and is the last stage of forest growth in which trees are still small enough that they can be felled with a machete. Then comes *bataqng tuhaq*, a transitional stage of forest growth occurring between twenty and forty years after the last rice harvest, characterized by the replacement of pioneer tree species by trees commonly found in the mature forest. The diameters of the trunks of trees in *bataqng tuhaq* range from thirty to fifty centimeters. Trees of this diameter are often considered suitable for logging by plywood, paper, and pulp companies. The *bataqng tuhaq* stage is frequently misidentified by non-Dayak visitors to the forest as primary or untouched forest. As such, it is often logged by outsiders unaware of or unconcerned with Bentian ownership rights.

Later stages of forest succession include several categories of *alas*, such as *alas kereroyan*, which occurs between forty and seventy years after the last rice harvest. Alas kereroyan is dominated by what the Bentian call jungle trees, including ironwood (*Eusideroxylon zwagerii*) and commercial Dipterocarpaceae such as *meranti* (*Shorea* spp.). The growth of such trees is actively encouraged by the Bentian, who, in addition to planting them, tend wild seedlings as well by clearing brush and strangling creepers away from them as the forest grows back. In alas kereroyan, tree diameters may reach a meter or more. This stage of old secondary growth is, again, often thought to be "primary" forest by outsiders. More and more frequently, such forests are being logged, without the permission of their Bentian owners, by chain saw crews working for large-scale logging concessions and industrial estate plantations using heavy trucks and equipment.[12]

The Bentian identify not only *alas kereroyan*, but older stages of *alas*, including *alas mentuqng*, or true primary forest, which has not been previously utilized for agriculture. Using their detailed knowledge of local forest ecology and the social history of the region, the Bentian are easily able to differentiate between old-growth secondary forest and such primary forest. Since Bentian farmers plant identifying groups of trees or rattan clumps on their agricultural lands, they know that certain recognizable complexes of rambutan, durian, jackfruit, langsat, and other fruit trees found in a mature forest are indicators of ownership and cultivation rights. The same is true for clumps of certain kinds of rattans, as well as ironwood grave markers and house posts (which may last well over a hundred years in the humid forest). There are also certain tree species known to the Bentian that die off after seventy or seventy-five years. Finding groups of these trees in one area, especially if typical Bentian fruit trees and rattan are also found, indicates that the area was very likely a rice field at some point up to seventy years ago.

In addition to their detailed knowledge of forest ecology, the Bentian are skilled oral historians and have preserved oral accounts of generations of agricultural activities. For example, the Paramount Adat Leader, or Tatau Solai, can recite the names of twenty-four generations of Bentian leaders and the locations of their agricultural lands. This detailed knowledge of the social history of their region means not only that the Bentian are able to recognize that a certain area of forest was farmed seventy years ago, but also that they can identify the exact farmer/owner and her or his descendants. This information is not posted on signs, fences, or markers in the forest, and it is therefore invisible to outsiders if they do not take the time to meet with the Bentian and ask them about their surrounding forests and fields. Most outsiders, especially those seeking wealth in the form of timber, coal, or gold on Bentian lands, do not make this effort.

The majority of visitors to the Bentian region, including highly trained forestry professionals, are unfamiliar with Kalimantan forest ecology. Having been trained elsewhere, they are usually unable to identify even the basic ecological differences between carefully managed, privately and collectively owned Bentian forests and the surrounding and largely unmanaged forest. The tendency to equate "big trees" with primary forest is often coupled with an unwillingness on the part of forest professionals and logging company officials to consult those who reside in forested areas to determine not only the nature of the forest (i.e., primary or secondary) but also the ownership rights in land and trees currently existing in the region.

## RICE FIELD SITE SELECTION: SOCIAL FACTORS

As Bentian farmers begin their rattan agroforestry cycle, social and environmental factors play important roles in determining the location of rice fields and rattan gardens. According to Bentian *adat*, a young married couple has the right to make their annual rice fields on or near sites in the reforested former rice fields previously farmed by the couple's parents and maternal and paternal grandparents. Each married couple, therefore, has access to eight separate areas in which to start their rice swiddens and rattan gardens.

Traditionally (but not always), soon after marriage, a couple lives with the wife's family and is, thus, more likely to farm first in areas previously cultivated by the wife's parents and grandparents. After several years, the couple is likely to move to the husband's natal household, farming lands owned by the husband's parents and grandparents. After several more years, the couple may switch back to the maternal grandparents' sites, continuing to rotate among paternal and maternal sites over the years, planning and negotiating with grandparents and parents as well as siblings and cousins who also have adat rights of access to the same field sites. The planting of rattan or fruit trees on a field confers ownership to the planter.

The choice of a specific field site for a given year depends on a number of factors. Is the land legally available, under adat law, to the women or men wishing to farm it? That is, are they descendants of the original farmers? Or, if not, have they made formal arrangements to borrow the land? Bentian adat specifies the manner by which outsiders, newcomers, or persons not directly related to the original farmers are allowed to borrow forested farm land (Nasir 1991; Fried 1995). Potential borrowers must first meet with the adat leader, or Tatau, and explain which plot of land they would like to farm. If the land is not owned by the Tatau, he will approach the descendants of the original farmers who have the rights to that area. If the descendants agree, the land is lent to the person or persons who requested it. Often, the borrowers will present a white plate (a common item in the traditional Bentian exchange economy) to the owners of cultivation rights as a sign of the purity of their intentions. Sometimes, a borrower will give the owners a small, primarily symbolic portion of the harvest, but this is not necessarily expected. A person farming on borrowed land is not allowed to plant rattan or trees of any kind, since rattan and trees belong to the planter and are signs of land ownership. If a borrower were to plant trees on such land, this would lead to conflict at a later date over the ownership of the trees and the land. If the descendants of the original farmers do not wish to lend their land, the Tatau has no power to force them to do so. Such decisions remain firmly in the hands of the descendants of the farmers who first farmed that land. Common reasons for borrowing land include marrying into a village, the need to share moist and fertile riverbank lands during times of severe drought, and the borrowing of lands close to the village health clinic, schools, and shops by families with pregnant farmers or young children.

After the set of potential field sites for a given farmer or farmers has been identified, the process of negotiating with siblings and cousins, who also have adat law access to the same parcels, begins.

## RICE FIELD SITE SELECTION: ENVIRONMENTAL FACTORS

Environmental factors are also crucial to the choice of field site: Is this year likely to be a year of drought or flood? Is there a sufficient water supply for drinking, cooking, and bathing near the field? What is the stage of forest growth and the level of soil fertility at the various sites?

### *Bentian Soil Fertility Indicators*

In addition to having detailed knowledge of stages of forest succession, the Bentian are also careful observers of soil fertility. Indicators of soil fertility and suitability for agriculture and agroforestry are varied and range from the identification of indicator plants to tests of the clay and sand content and water-holding capacity of the soil. Indicator plants such as *parap'm*, a sign of good soil, or *nagaq* trees, whose presence makes the land unsuited to rice cultivation,[13] are used by the Bentian in their decision-making process. According to a Bentian farmer,

To see if the soil is good for planting . . . test it with a machete. Stick the machete in the ground and pull it out. If soil sticks to the machete [indicating sufficient clay content], this is a good location. If not, it means this is sandy soil, poor, infertile. The soil will heat up too much for the rice plants. . . . The soil should feel soft to the touch; you should also see very green plants in the area. Plant in dark soils, never in yellow soil.

Bentian oral historian and village official, rattan farmer, forest manager (1992)

These simple tests and observations actually reflect a detailed understanding of soil characteristics and their implications for agricultural production and reforestation. The machete test provides a quick indicator of the clay or sand content of the soil, a crucial factor in the moisture retention ability of otherwise drought-prone soils. The presence of healthy green plants serves to indicate good soil nutrient status. Dark-colored soils of the region are often richer in humus than the heavily leached yellowish soils. These tests, combined with a knowledge of specific indicator plants, allow Bentian farmers to be sophisticated judges of the suitability of their surrounding soils for agriculture, agroforestry, and reforestation efforts.

## LAND "CLEARING," TWO VERSIONS

The Bentian agricultural year begins and ends with the rice harvest, which occurs from February to March. In the months leading up to the intensive activities associated with the rice harvest, households begin the process of choosing the site for the new rice fields that will be established to replace those fields that are in their second year of rice production. In most cases, the fields that have produced rice for two years will be fallowed and allowed to revert to forest, enriched by Bentian rattan and fruit tree plantations.

For a period of a few weeks between February and June (exactly when depends on many factors including weather, the time of the preceding rice harvest, the scheduling of nonagricultural activities, and the availability of labor), Bentian farmers begin to clear the brush and small trees from the forested areas that they have designated for rice cultivation.

It is important to note that, in the Indonesian context, the term *clearing* or *land clearing* is commonly utilized to describe widely different activities—activities that differ tremendously not only in their methods but in their impact on forest ecosystems. The English-language term *land clearing* is often utilized in Indonesian documents and discussions to refer to the initial steps in the establishment of monoculture industrial forest plantations (HTI, *hutan tanaman industri*) or transmigration housing projects. In this case, *land clearing* describes a process in which chain saws are used to fell all standing timber, regardless of tree species, size, or ownership rights—in other words, clear-cutting.

Commonly, timber that is clear-cut during the "land clearing" stages of forest plantation (HTI) establishment is utilized directly by plywood or by paper and pulp industries. What is not used may simply be burnt. After clear-cutting, the land may be leveled by bulldozers, not only removing the remaining soil cover (brush, leaves, etc.) but exposing and often removing the topsoil. After a few heavy rains, much of the topsoil washes away, leaving infertile soils with damaged soil structure. This has essentially the opposite environmental effects of the Bentian method of "land clearing."

## Tropical Forests Forever?

The subject of clear-cutting is a sensitive one in Indonesia. Since at least the early 1990s, the Indonesian loggers association, Masyarakyat Perhutanan Indonesia (MPI—literally, Indonesian Forestry Community), directed by Haji Mohammed "Bob" Hasan, has repeatedly attempted to convince the world that clear-cutting is not permitted in Indonesia and that loggers are carefully managing Indonesia's forests. During their "Tropical Forests Forever" publicity campaign, MPI stated as much in their colorful brochure (published in Oregon, undated, but approximately 1990): "Clear cutting, or removing an entire block of timber, is not permitted" (p. 13).

A glossy clasp of an "endangered orangutan" and comments about Indonesia's "very strict" forestry regulations are followed by other observations, including the "fact" that "shifting cultivators are the major cause of deforestation of tropical forests." This is followed by the somewhat ominous statement, "Family planning programs and creation of new jobs will also alleviate the problem of shifting cultivation" (p. 17).

The brochure continues, "Indonesia is converting 20 percent of its forest lands to farms and forest plantations . . . [HTIs]. These new farm lands are one step in Indonesia's efforts to reduce the loss of forests to shifting cultivators" (p. 17).

Masyarakyat Perhutanan Indonesia also aired commercials on European and American television that claimed that the clear-cutting of forests was not permitted in Indonesia and that 79 percent of Indonesia's forested land has been made a "vast permanent forest." In 1994, in response to consumer complaints of false advertising, the British Independent Television Commission, a governmental regulatory body, banned MPI's misleading representations about clear-cutting and other Indonesian forestry operations from British airwaves.

## Land Clearing, Bentian Version

The Bentian concept of "clearing" the land involves cutting brush (*nokap*) and felling trees (*noweng*) in a small area of inherited, borrowed, or otherwise negotiated forest. The "cleared" brush, tree trunks, branches, and leaves are left on the soil surface. Bentian agricultural techniques ensure that the soil surface is covered at all times— if not by forest growth, then by the mulch that is formed during *nokap*. In addition, the "cleared" patch of land is surrounded by standing forest that acts as a seed bank and ensures the reforestation of the cleared field after it is fallowed. Bentian adat holds that "mother trees," mature and healthy seed-producing trees, should be left standing in swidden fields to assist the reseeding process when the field is left fallow to revert back to forest.

In June and July, after the brush has been chopped, the larger standing trees in the new field are felled. This process takes a few days if a chain saw[14] is used, and up to a month if an axe is used. The Bentian, like most Dayak peoples, follow adat as well as common-sense rules about what kind of trees may or may not be felled during land clearing. For example, one must not cut down productive fruit trees or rattan clumps. It is also forbidden to fell the strikingly large "honey trees" (referred to as *tanyut*, a term that encompasses many species, including *Canarium opertum* and species from the genera *Koompassia*, *Dipterocarpus*, *Drybalanop*, and *Shorea* (Burkill 1935), where wild bees gather to produce their honeycombs.

## HONEY TREES

In the Bentian region, honey trees are owned by the descendants of those who originally discovered them. In 1992 and 1993, centuries-old honey trees were illegally felled by loggers subcontracted by the Kiani Lestari/Kalimanis Group and the Timberdana/Kalhold Utama concession. The Bentian are currently seeking damages under both adat law and national law for the destruction of economic crops, sacred trees, and burial sites.

Felling a honey tree or other trees and rattan clumps without permission of the owners is considered a serious offense since "the destruction of [trees] owned by the [Bentian] people without the permission of the owners or their heirs is the same thing as cursing (*sumpah*) . . . the owners or heirs as if they are dead when, in fact, they are still living" (Dingit et al. 1994)

In other words, not asking for permission to use a privately owned resource (i.e., felling or destroying trees without permission) is the same thing as assuming or declaring that the owners of the resource are dead, in effect, cursing an oath of death upon them. In the Bentian region, the use of feared and supernaturally charged curses are thought by the Bentian and their neighbors to bring death and destruction to their targets or, if wrongfully uttered, to their instigators.

Under the Bentian forest management system, social sanctions for the felling of honey trees and other "mother trees" are combined with the control of soil erosion and fertility loss through the continuous maintenance of ground cover. These two aspects, that is, the continual protection of the soil surface and the ready availability of reseeding material, are perhaps the most crucial factors in guaranteeing the successful reforestation of Bentian agricultural fields.

By contrast, however, the activities of highly trained forestry professionals, including those who work at some of the largest and most technically sophisticated Indonesian logging concessions, indicate that they are either unaware of or unconcerned with the need to maintain sufficiently large standing stocks of seed or "mother" trees or the importance of soil cover in maintaining fertility and preventing erosion from thin, acidic Kalimantan soils.

The following is an account of a typical East Kalimantan occurrence that happened on Bentian territory.[15] In 1982, the United States–based Georgia Pacific Company entered Bentian territory, evicted local residents from the banks of the Anan River, and destroyed a portion of their forest property, including grave sites, in order to set up a logging base camp. This set the stage for over a decade of environmental destruction and escalating regional tensions. In the mid 1980s, Georgia Pacific sold its Indonesian operations to an ethnic Chinese businessman named The Kian Seng, commonly known as Haji Mohammed "Bob" Hasan. Mr. Hasan, a billionaire and a close associate of General Suharto, heads the Indonesian Wood-Panel Exporters Association (Apkindo), a cartel that regulates $3.7 billion worth of plywood exports and controls over 2 million hectares of Kalimantan logging concessions. Mr. Hasan is or has been the chair of most Indonesian forest industry associations, including MPI (as previously mentioned) and its member associations, which include Asosaisi Pengusaha Hutan Indonesia (APHI) and Apkindo.

> An ethnic-Chinese of Hokkien descent, Hasan has had a close business relationship with President Soeharto since the latter was a colonel in the Diponegoro Division in Central Java in the 1950's. Since the inception of Soeharto's New Order regime in 1966, Hasan has drawn on Soeharto's patronage to construct a billion dollar corporate empire, with holdings in shipping, steel, telecommunications, financial services, industrial manufacturing and contracting, pulp and paper, automobile distribution, tea plantations, and recently, publishing. The most prominent of his business interests, however, have been in the timber sector, where he has used his political connections to obtain vast logging concessions and subsidized financing for his wood processing operations. Playing golf with Soeharto twice a week and managing many of the President's personal business interests, Bob Hasan has had virtually unparalleled access to the center of state power, and it is widely believed that this has given him greater influence over timber sector policy than that held by even the Minister of Forestry.
>
> Barr (1997)

In addition, in April 1997, Mr. Hasan was invited to the White House to attend a special "recognition ceremony" for his forestry conglomerate's proposed role in a proposed project to preserve Kalimantan's forests. Mr. Hasan did not attend the ceremony, sending instead the president of Kalimanis Group.

Although Georgia Pacific had turned its former Indonesian holdings over to Bob Hasan, some of its key American employees continued to act in management and advisory capacities for the various plywood factories operated under Hasan's Kalimanis Group/Kiani Lestari conglomerate. The base camp on Bentian territory was to be operated by the P. T. Kalhold Utama Company, a member of the Kalimanis Group, and a subcontractor to Kalhold, the Timberdana Company. Although Georgia Pacific had officially withdrawn from logging concession operation in Indonesia, 1990/1991 plywood import figures for the United States show the company as a major importer of Indonesian plywood, much of this from its former logging concessions such as P. T. Kalhold Utama.

As a result of increasing publicity about the negative impact of logging concessions on surrounding communities, the Indonesian government passed a law requiring each concession holder to implement successful Logging Concession Village Development Projects (HPH Bina Desa) in order to continue to receive governmental permission to log a region.

In response, logging companies designed village development projects with the sole objective of "eradicating shifting cultivation," which they considered a threat to their logging operations. In the case of many HPH Bina Desa programs, the eradication of indigenous agricultural practices is equated with "development." According to the Indonesian Ministry of Forestry and the World Bank, over the years that the HPH Bina Desa program has operated, logging companies, for the most part, have carried out their village development duties in a pro forma and often extremely damaging manner (Departemen Kehutanan 1988; World Bank 1994; Zerner 1992).

Unfortunately, P. T. Kalhold Utama would prove to be no exception to this rule. For one of their village development projects, forestry advisers at the Kalhold base camp on the Anan River decided to "convince" the Bentian to stop "shifting" their agricultural fields and to "convert" to sedentary agriculture.[16] The idea of this particular HPH Bina Desa program was to introduce the Bentian to "settled agriculture" and the use of hoes so that they would stop their rotational system of farming, which was felt by logging company officials and foresters to be both "irrational" and a threat to the forests that had been claimed by the logging concession. Many of these forests, however, were actually Bentian *bataqng tuhaq* or *alas kereroyan* properties, which had been de facto seized from the Bentian by the powerful logging company.

In 1991, P. T. Kalhold officials (who were Javanese and who, when interviewed by Kalimantan-trained foresters, were strikingly unfamiliar with Kalimantan ecology and climate) insisted that Bentian farmers from the village of Jelmu Sibak clear (literally) a steep hillside, removing all trees, brush, and ground cover, then hoe the bare soil (as is the fashion in Javanese agriculture). The Bentian refused to do this, citing the irreparable damage to the soil that was likely to occur as a result. Failing to entice, convince, or coerce Bentian farmers, the company then hired Javanese transmigrants for the task. Such transmigrants are often available as a cheap labor force, desperately seeking cash income when their attempts at agriculture using Javanese (intensive, "permanent" agriculture) techniques on the poor soils of the region fail. In East Kalimantan, it is not uncommon for teams of such transmigrants, having abandoned their own failed transmigration sites, to be hired as laborers for the land clearing and bulldozing phases of new, equally doomed transmigration sites.

Within a few days of the clearing and hoeing of the steep hillside, typical Kalimantan rainstorms occurred, washing away most of the exposed topsoil and leading to severe gully erosion. The experiment was a total failure that not only destroyed the soil structure and fertility, but also immediately increased the soil sediment load of the ponds and streams at the bottom of the hill, leading to a deterioration of water quality. This demonstrated (to the Bentian, at least) the superiority of Bentian knowledge of the local forest environment compared to that of academic forestry "experts" and the forestry staff of one of the most powerful and highly capitalized logging concessions in Indonesia.

Meanwhile, the 1991 Environmental Impact Assessment of another large Kiani Lestari/Kalimanis Group logging concession in Batu Ampar, East Kalimantan (also a former Georgia Pacific concession) indicated the following:

> The activities of [concession]-holder, P. T. Kiani Lestari have brought and will bring a negative impact on all environmental components, except those components of a social-economic-cultural nature. Physical, chemical, and biological components of the environment are also adversely affected by the company's activities save its enrichment planting. The negative impact on the physical components covers such sub-components as soils, hydrology, climate, air, water quality, vegetation, wildlife and aquatic organisms. The study shows that soil erosion increased by 800 times [e]specially at the skidding area. . . . There was an accelerated loss of protected floras such as *Eusideroxylon zwagerii* (Borneo ironwood). On the sub-component of wildlife the study recorded a destruction to their natural habitats, particularly with respect to *Pongo pygmaeus* (orangutan), *Hylobates muelleri*, and bird species; . . . and a decrease in mammal population and ornithological population by twenty-one to one hundred percent and fifty-eight to seventy-nine percent respectively.

By this time, the Indonesian loggers' association, MPI—chaired by the owner of the Kiani Lestari/Kalimanis Group—was in the midst of their global "Tropical Forests Forever" publications campaign—displaying glossy orangutan photographs, proclaiming the illegality of clear-cutting, and targeting "slash-and-burn" farmers as destroyers of Indonesia's forests.

## RATTAN SEED COLLECTION AND PLANTING

As larger trees are felled in the new field, brush and weedy growth are chopped down in the old field from which one crop of rice and vegetables has already been harvested.[17] This field is then prepared to receive its second crop of rice and vegetables. During the period of tree felling and the time when farmers wait for the felled trees to dry (usually between August and October—exactly when, again, depends on many factors, weather and labor among the more important), farmers gather ripe rattan seeds from their mature rattan gardens (often fifteen to forty years of age, or older, no longer productive gardens). Like all good plant breeders, the Bentian follow careful seed selection procedures. They select rattan seeds based on observations of seed coat color, taste (a relatively sweet taste indicating ripeness), size, and seed condition. These seeds are either planted directly in the field that is to be prepared for the first rice crop, or they are grown out in a nursery to provide seedlings for planting the following year.

The Bentian pattern of rattan planting bears some comment. To the eye of the professional forester or agronomist trained to delight in neat and orderly rows of crops or trees, Bentian rattan plantations often seem illogical, untutored, disorderly, and bizarre (see Fried 1995; Tsing 1993). The rattan in a Bentian farmer's field appears to have been planted in a haphazard manner, conforming neither to linear aesthetics nor to a sense of regularity, repetition, or pattern. A closer examination of rattan fields, and discussions with farmers, however, reveal the management strategy involved in the irregular spacing of these plantations. The Bentian specifically look for microenvironments of suitable soil fertility, and they plant their rattan there. They prize areas lying adjacent to felled tree trunks that release nutrients as they decompose. They plant rattan on the east and west sides of the large roots of standing tree stumps. Early in the day, the seeds and small seedlings on the west side of the large roots are protected from the direct glare of the burning sun. Later in the day, those seeds and small seedlings on the east side are protected.

Rattan farmers also look for depressions in the soil filled with a substantial layer of humus as well as areas with underground streams lying close to the surface. These microregions of increased soil fertility occur irregularly in a given field. The pattern of Bentian rattan plantation mirrors the irregular nature of the pockets of increased soil fertility. This irregularity, if not aesthetically pleasing to the forestry or agricultural professional, is eminently logical and reflects a detailed understanding of the local environment.

Uninformed attempts to alter the Bentian system of rattan plantation have been disastrous. In 1990, foresters from P. T. Kalhold Utama, the Kiani Lestari/Kalimanis logging concession operating on Bentian territory, decided to teach Bentian farmers from the village of Jelmu Sibak how to grow sega rattan (*Calamus caesius*) as part of another Logging Concession Village Development Project required by Indonesian law. Having observed that rattan was harvested from the forest, Javanese-trained foresters, unfamiliar with either Kalimantan forest ecology or rattan cultivation, ordered the Bentian to plant rattan neatly in rows in standing forest. This represented another attempt to convince the Bentian to "stop shifting" their agricultural fields. Instead of asking for advice from the Bentian, who may rightly be considered among the world's leading experts on rattan cultivation and Kalimantan forest management, they treated local rattan farmers as ignorant farm laborers.

According to Bentian farmers, however, sega rattan requires substantial amounts of light during its seedling stage. In fact, several farmers had previously conducted experiments to determine the efficacy of planting rattan in brush and young secondary forest compared to planting it in their open fields. They found that it did not grow as well as did rattan planted in an open field (Nasir 1991). For this reason, Bentian farmers continued to plant rattan in their open rice fields.

For P. T. Kalhold Utama's village development project, the Bentian were told not only to plant rattan in straight lines (regardless of soil fertility) in the dark forest, they were also ordered to add commercial fertilizer to the rattan seeds. Earlier Bentian experiments had indicated that commercial nitrogen fertilizers led not only to vigorous rattan leaf growth but also to a rapid thickening of the rattan stems, which severely decreased their commercial value.

Bentian farmers initially refused to participate in the experiment for two important reasons. First, they feared that since rattan plantations were, according to adat, indicators of ownership rights, Kalhold would then claim ownership of the adat lands on which the company wanted the farmers to plant rattan. Second, the farmers predicted that crop failure would result if they followed the company's planting instructions. The company, however, eager to appear to comply with Indonesian law by carrying out a village development project, used threats, coercion, and bribes of village leaders to form a makeshift farmers' group, which was then ordered to plant rattan in standing forest. Alarmed and mystified, but assuming that such highly educated foresters probably knew more than did uneducated villagers, the hastily formed farmers' group, which was never asked for advice, planted the rattan neatly in rows in standing forest. By 1991, when the author and Dr. Sardjono from Mulawarman University's Forestry Faculty observed the P. T. Kalhold Utama rattan plots, we found, not surprisingly, that the mortality rate of the rattan seedlings planted under Kalhold Utama's Village Development project was approximately 90 percent.

## BENTIAN RATTAN EXPERIMENTS

One of the reasons the Bentian are so knowledgeable about rattan ecology is that they continually conduct experiments on most aspects of rattan cultivation. For example, during the years preceding my period of research, farmers had attempted to determine the optimal time for the planting of rattan seeds with respect to the burning of the fields. They experimented by planting the seeds before the brush covering a new field was burned, and after the burn.

Before the burn, seeds were planted in holes made in the ground in relatively protected areas. They were placed, for example, near the foot of a felled tree or in fertile land not too thickly covered with dry leaves. The seeds were then covered with soil. The farmers found, after the burn, that there was some risk that such seeds would be destroyed by fire, but the ones that were not damaged grew very quickly after the first rains. Seeds planted after the first burning of the fields appeared to grow more slowly than those planted before the burn, but more of them appeared to germinate since they had not been damaged by fire.

In the early 1990s, Bentian farmers were also avidly experimenting with new methods of rattan seedling nursery design. Having observed the use of plastic seedling bags ("polybags") for early seedling growth on tree plantations, Bentian farmers came up with their own cheaper and biodegradable version of the plastic bags—bamboo sections. Rattan seeds were placed along with soil in a tube formed by a bamboo section. When the seedlings were large enough, the bamboo sections were carried to the field and the seedlings removed and planted. Other Bentian and Benoaq innovations include various types of seedling trays, designed to increase the ease of transporting the mature shoots to the field and to reduce root stress during transplanting.

## THE PLANTING OF RATTAN SEEDLINGS

Between August and October (the preferred time is closer to August, weather per-mitting), the felled trees and brush in both old and new fields are burned, releasing nutrients into the soil. For fields in their second year of rice cultivation, rattan seedlings that were planted the year before must be protected from the fire. This is done by covering the seedlings with the moist inner sections of banana plant stems, or by digging small trenches next to the rattan seedlings, bending the seedlings into the trenches, and covering them with soil prior to the burn.

Between November and January (or whenever the rains appear), twelve- to four-teen-month-old rattan seedlings, either gathered from mature rattan gardens or grown out in a rattan nursery, are planted in the new rice field. According to farm-ers, the rice should be approximately three months old (or thirty centimeters high) to provide sufficient shade for the rattan seedlings. From November until the rice harvest, the rice and rattan plants are weeded as necessary. Bentian farmers report that, for good early growth, rattan must be weeded well during the first two years after planting. As the rice crop matures, the farmers protect it against birds, mon-keys, wild pigs, and deer if these animals appear likely to cause destruction. It should be noted that, in Bentian fields, agricultural pests are more likely to be mammals or birds instead of insects. As the rice grows, plans are developed for the location of the new field for the next year. In February, the rice harvest begins, marking the end of the annual cycle.

## RATTAN HARVEST

Unlike the rice harvest, which occurs at approximately the same time every year, rat-tan has no fixed harvest date. This makes it an ideal smallholder export crop since farmers are less vulnerable to the price fluctuations that often accompany crops with fixed harvest seasons.

Rattan vines are mature and ready for harvest seven to ten years after planting. Each rattan clump may contain between forty and sixty stems, a portion of which will be mature at any given time (estimates range from 10–40 percent). Bentian adat and knowledge of rattan ecology govern the rattan harvest. To prevent infection of the remnant of the harvested rattan stem and the resultant death of the rattan clump, Bentian harvesters cut rattan stems beginning at a height of approximately one me-ter from the ground (and one meter away from the clump). In addition, they attempt to harvest only the mature canes, ensuring the continued growth and productivity of the clump.[18]

After the first large harvest, succeeding large harvests may occur every two or three years for the next twenty or twenty-five years, depending on the productivity of the rattan garden and the needs of the farmer. In addition to large harvests, small monthly harvests are made, usually of ten to twenty kilograms. Money from the small harvests is used to pay for goods such as cooking oil, fuel oil, salt, sugar, soap,

batteries, and cigarettes. Larger harvests are made to cover costs of weddings, fu-
nerals, the purchase of boat motors, education for children, or emergency costs.

## THREATS TO DAYAK AGROFORESTRY SYSTEMS
## AND KALIMANTAN FORESTS

> There is medicine, food, fuel, and feed for animals in our forests. There are fish in
> the streams and game animals in the forests. Look at ITCI [the former Weyerhauser
> logging concession, now run by a military foundation]. They get water from a
> water tank, fish are brought in. The environment can no longer fulfill their needs.
>
> Bentian oral historian and urban resident, provincial government official (1992)

> The Gunung Putih Indah rubber plantation (HTI) did not ask permission. They just
> came in and cleared places to plant. Thousands of hectares of our lands were taken.
> They said the Bentian and Benoaq people could not reprimand them because the
> company had permission from Jakarta. They said, "Just try to see our permit letters.
> If you force the issue, we'll show you."
>
> Benoaq oral historian and village resident, community leader,
> rattan farmer, forest manager (1991)

Unfortunately, the destruction of thousands of hectares of productive rattan,
fruit, and rubber gardens in Bentian and neighboring Benoaq and Pasir regions has
recently occurred—either as a result of errors of judgment on the part of forestry
professionals unfamiliar with Kalimantan ecology and unwilling to work in partnership with knowledgeable forest dwellers, or as a result of purposeful and destructive plans aimed solely at maximizing profits for developers regardless of the
social, economic, and ecological costs to be borne by the surrounding communities
and their forests. According to East Kalimantan press reports, additional hundreds
of thousands of hectares of such lands are currently slated for destruction as well.

In 1992, for example, the author encountered a forestry team scouting Bentian
forests for Alas Kusuma, one of Indonesia's larger logging concession groups, hoping
to carve out a concession for a paper and pulp plantation on a portion of Bentian lands
already claimed by two other major logging concessions. The Alas Kusuma survey
team and its European consultant, a professional forester for Jaakko Poyry, one of the
world's leading forestry consulting firms (based in Finland), walked and drove through
thousands of hectares of Bentian rattan gardens without speaking to a Bentian farmer,
without asking permission from Bentian authorities, and, strangely enough, without
even noticing the extensive rattan plantations that covered the region.

The team explained that, while assessing the region's timber resources to determine the feasibility of their plans to clear-cut 200,000 to 300,000 hectares of Bentian
forests in the "land clearing" phase prior to setting up a paper and pulp factory, they
had been in only "primary" or "secondary" forest. As far as they were concerned,
primary forest was "wild" and unowned, and secondary forest was "abandoned" and

unowned. They had not noticed any rattan gardens. "What does rattan look like?" a forester trained only to recognize industrial timber species asked me. I described the painfully spiny vines that loop up and around trees all over Bentian forested land. "Yes," recalled their Indonesian team leader, a graduate of Mulawarman University's Forestry Faculty in Samarinda, "there were a lot of annoying thorny vines," which the team had chopped down to make their timber survey easier. "But those vines didn't belong to anyone," another forester said. "They were just growing wild in the forest." "Birds probably planted them. You know, bird shit [tahi burung]," he added grinning broadly.[19]

In this case, as it turned out, the Bentian were temporarily lucky. The logging concession responsible for the survey was no match in Jakarta for the other giants who were already engaged in logging Bentian land. The plan to clearcut 200,000 to 300,000 hectares of Bentian territories was withdrawn after pressure had been exerted in Jakarta by the concessions already in the area. The Bentian were unaware that these concessions, however, had plans to clear-cut and bulldoze their region, as well.

In 1993, after the repeated failure of Kalhold Utama Logging Concession Village Development Projects (HPH Bina Desa) in the Bentian village of Jelmu Sibak, subcontractors for Kalhold Utama began to bulldoze the village's rattan gardens, fruit gardens, honey trees, and even grave sites as part of the "land clearing" phase in the establishment of an HTI. When Bob Hasan, the company's owner, was questioned by reporters in Jakarta about the seizure and bulldozing of Bentian adat lands being carried out by his company and subcontractors, he responded, "What profit would I gain by seizing adat lands and evicting the people? If my people dare to do that, I will fire them" (*Kompas*, 10/93). Unfortunately, however, after Mr. Hasan's public announcement the bulldozing and destruction actually escalated, as did the climate of terror throughout the region.

The establishment of Kalhold Utama's HTI was required by Indonesian law as part of a national reforestation effort. Not only did the company intend to clear-cut at least 5,000[20] hectares for a plantation, but it planned to settle 250 families of Javanese transmigrants on the bulldozed and exposed soils of the former Bentian rattan gardens (Department of Transmigration, Kutai Regency Office, 1993). Indonesian law, however, also stipulates that such plantations must be established only on marginal and unproductive lands in an attempt to rehabilitate them, and not on productive and privately owned lands.

The thirty-five families initially affected by the bulldozing (which was conducted under armed guard) recorded the destruction of over 10,000 productive clumps of rattan and 2,000 fruit trees and honey trees in direct violation of both adat and national law. Ironwood grave markers and grave sites were bulldozed and burned as well. Despite the ongoing dispute and violations of law, the logging company hired teams of Javanese laborers, desperate transmigrants from failed transmigration sites, to construct tiny shacks for 250 transmigrant families that the company was planning to bring from Java as the labor supply for its plantation on Bentian lands. Given the ecology of the region and the clear-cutting and bulldozing of the forest, it is difficult to imagine that such a plantation will ever produce a yield sufficient to replace the lost income that would have been produced for generations to come for the surrounding Bentian communities as a result of their careful forest management, or even simply the value of the timber seized from Bentian lands by the company during "land clearing."

During the first few years of operations, it is likely that the company will provide a substantial subsidy to the transmigrants in an attempt to have them remain in place as long as possible in their highly publicized site. After the destruction of the Bentian resource base, however, the transmigrant laborers will most likely be forced to join their desperate brothers and sisters in the growing, marginally employed labor force in the cities of East Kalimantan. As for the Bentian, after the seizure and destruction of their forested homelands, a similar fate is likely to be theirs as well.

Events such as these are not unusual in East Kalimantan. They occur in most remote regions where logging, plantation, and mining companies carry out their operations in violation of both national and adat law. In the Bentian and neighboring Benoaq areas of Dilang Puti and Penarong, for example, a rubber plantation company has been clear-cutting and bulldozing thousands of hectares of productive Bentian and Benoaq rattan gardens. The company silenced the protests of the horrified villagers—frantic as they watched their rattan "savings accounts" being destroyed—by a typical heavy-handed combination of coercion, bribery, and the use of uniformed on-duty government police forces to prevent villagers from protecting their lands.

As of 1993, the rubber company announced plans to expand its land-clearing operations to additional tens of thousands of hectares of Bentian and Benoaq rattan lands, further devastating the economic base of the region. This occurred despite the fact that company management had been repeatedly informed by local citizens that they were operating on privately owned Bentian and Benoaq territories, against the wishes of the Bentian and Benoaq inhabitants.

Even easily accessible areas such as the touristic Tanjung Isuy region, a short boat-ride from the provincial capital, Samarinda, do not escape the ravages of destructive logging and plantation establishment. In 1993, Benoaq Dayaks reported the bulldozing of their fruit and rubber gardens as well as their grave sites near Tanjung Isuy by logging companies and industrial forest plantations. These activities are, clearly, against both Indonesian and adat law. Strangely enough, however, the rationalization that company officials provided to angry and astonished farmers was that the "land clearing" was being done to comply with forestry laws mandating the "reforestation of marginal lands" and the "development" of local communities.

Since many resource extraction companies operating in Kalimantan appear to comply with national laws in a pro forma manner only, if at all, they continue to destroy the very resource base that the forestry and environmental laws were meant to conserve. As the resource base of the region is increasingly destroyed by such logging companies and plantations, once-independent and relatively well-off Dayak farmers such as the Bentian find themselves increasingly impoverished and reduced to plantation labor at salaries lower than the cost of living.

## CONCLUSION

The Bentian system of rattan production represents a successful example of locally initiated sustainable development and resource management. The Bentian agroforestry system, which generates foreign exchange, protects the resource base, and provides substantial local income, is a fitting model for sustainable rattan production elsewhere. In addition, lessons learned from the Bentian system, which so neatly combines ecological and social aspects of resource management, should be utilized in the design of other types of sustainable resource management efforts.

The success of the Bentian agroforestry system is based not only on its relevance to the world market but also on a detailed knowledge of forest ecology and the desire of the Bentian to protect and conserve their resource base. These factors are supported by Bentian adat practices, which guarantee ownership of resources to those who invest their labor in forest cultivation, while also ensuring proper resource management techniques and providing sanctions for the improper use or purposeful damaging of natural resources. Thus, as we can see, the success of this export-oriented agroforestry system consists of both a detailed knowledge of a specific resource base and a social system that provides clear resource ownership rights and acts to support sustainable management techniques.

Before, however, the Bentian system and others like it can be promoted as examples of sustainable development, such systems must first be protected from destruction by misinformed or, less charitably, careless and destructive private-sector logging concessions, industrial forest plantations, and other resource extraction industries.

## ACKNOWLEDGMENTS

Institutional support for this research was provided by the Research and Development Institute of the Indonesian Ministry of Forestry, the Indonesian Institute of Sciences (LIPI), Cornell University, the East-West Center, the Ford Foundation, and Mulawarman University in Samarinda. I owe thanks to my current employer, the Environmental Defense Fund, for being supportive of the final editing process. The research was funded by a Fulbright-Hayes Dissertation Fellowship, with additional assistance from the GTZ Forestry Project at Mulawarman University. Many individuals have contributed to the thought processes and data gathering leading to this article. I cannot name them all here for various reasons, including, on the part of some, their wish to maintain their privacy. I owe a great debt of thanks to my Bentian hosts and friends for their ideas, their patience, and their contributions to my research efforts. I owe thanks also to Ministry of Forestry officials, including Dr. Johannes S. H., Dr. Priasukmana, Dr. Tantra, Dr. Kosasi, Mr. Amblani, Mr. Endang, and the rest of Litbang Kehutanan Samarinda staff; to Nancy Peluso, Chris Barr, Joe Weinstock, the Jl. Pertahanan regulars, to M. Scharai-Rad at GTZ, Samarinda, and to Michael Dove. I also thank Martijn van Beek for providing substantial insights into theoretical aspects of my data and Charles Zerner for his enormous patience and editorial encouragement.

## NOTES

Portions of this chapter were presented at the Fourth International Conference on Ethnobiology, Lucknow, India, in November 1994.

1. Throughout this paper, there are citations referring to Bentian oral historians. These are the men and women of the Bentian region who verbally articulated the history of their people in my presence during interviews and religious ceremonies. I have not yet been able to ascertain whether these historians wish to have their names recorded in print, or whether, because of the complexities of village relationships and local and national politics, they wish to remain anonymous. For now, I have decided to err on the side of anonymity. At a future date, if I am permitted to do so, I will publish the names of those who so patiently and so vividly described their history and culture.

2. The term *Dayak* has, at times, had derogatory connotations. Currently, however, members of groups labeled by outsiders as Dayak, such as the Bentian, the Benuaq, and the Tonyoi, refer to themselves as Dayak out of a sense of pride and unity. It is in this spirit that the term is utilized in this paper.

3. For a more detailed account of materials written by the Bentian, see Fried (1995).

4. In 1986, in an attempt to stimulate the development of domestic rattan-based industries, the Indonesian government banned the export of raw rattan. This was followed in 1988 by a ban on semiprocessed rattan exports. Less than one year after the ban on the export of semiprocessed rattan, the manufacturing capacity of the Indonesian rattan industry increased by more than 600 percent (Menon 1989). Nonetheless, this increase was still not sufficient to make up for the relative lack of domestic manufacturing capacity. By the end of 1989, rattan prices remained drastically depressed as a result of the glut of the rattan market.

It is beyond the scope of this paper to present a detailed analysis of the impact of the rattan ban on Bentian communities. Clearly, however, the ban has resulted in substantial economic losses to Bentian rattan farmers. My field research also documented a remarkable increase in the smuggling of unprocessed rattan from most rattan-producing regions of East Kalimantan to Malaysia and Singapore as a result of the ban.

5. The name *Bentian* appears to be a rather recent appellation. Carl Bock, in his 1879 visit to what is now called Bentian villages, never uses the term and refers instead to "Dyaks of village X." The Bentian say that they were named by a Kutai sultan during one of the annual pilgrimages made by Dayak leaders to pay their respects to the sultan in the royal city of Tenggarong. Leaders of the Lempenai, Teriek, and Jorent groups, who were descendants of those who had participated in the *penyua* with the Tementa'ng king, visited the sultan to report, as requested, on "the conditions of life and the livelihood of the people under their leadership." They reported that their rice fields had all been attacked by ricebirds ("Bentian" birds) from a nearby cave. The sultan declared that if this was so, "then from now on the name of your group will be Bentian since your place of residence is in the Bentian cave area." Little by little, the groups that were originally called Lempenai, Teriek, and Jorent all became known as the Bentian people (Nasir 1991).

6. *Calamus casius*, or sega rattan, is the most commonly cultivated commercial rattan

species in the Bentian region. More than thirty other species are also present in the region and are either cultivated, semicultivated, or wild.

7. For detailed descriptions of Indonesian policy toward shifting cultivation, and official responses to the Bentian agroforestry system, see Dove (1983, 1985b, 1986), Zerner (1992), Fried (1995), Triwahyudi (1992), Dingit et al. (1994), and Pantir (1990).

8. Indonesian, the national language of the Indonesian archipelago, is utilized in governmental discourse and trade. Many Bentian speak fluent Indonesian and use the language with outsiders as well as among themselves.

9. In Indonesia, as in other parts of Southeast Asia, shifting cultivation is often carried out by ethnic minority groups inhabiting remote, resource-rich regions. Governmental and private-sector campaigns to "eliminate shifting cultivation" often also appear to aim at the elimination of the cultural identities of the peoples engaged in shifting cultivation, if not the outright elimination of their communities. It is as if "eliminate slash and burn" campaigns represent a frontier form of "ethnic cleansing."

10. "Still valid," that is, still utilized even though the area is nominally Christian. North America–funded missionaries have insisted that conversion to Christianity requires the Bentian to burn all "pagan" idols, charms, and magical oils. In addition, the missionaries forbid the use of supernatural oaths (*sumpah*).

11. Cared for, or *dipelihara*, probably refers to the fact that the magically charged apparatus, in this case possibly consisting of a sacred tiger's tooth necklace and other magically endowed items, must be regularly "fed" offerings such as a few grains of cooked rice, tiny cigarettes, and blood (usually of sacrificial animals, although in former times human blood from a freshly killed victim was a requirement for the "feeding" of certain talismans).

12. To the dismay of Bentian farmers, the use of heavy machinery in their forests has not only injured the remaining trees but also has led to severe soil compaction, rendering these areas unsuitable for further agricultural or reforestation efforts.

13. Nagaq trees have dense root masses at the soil surface. These root masses hinder the growth of rice seedling roots and thus have a negative effect on rice plant growth and the final yield.

14. In the last thirty years, the introduction of chain saws, boat motors, and, to a lesser extent, modernized forms of blowguns (*sumpit senapan*) have had a significant impact on rural Kalimantan life [for more detail, see Colfer (1980), Peluso (1985), and Fried (1992)]. Farmers may now decide to substitute capital (chain saw rental or ownership costs) for the extensive labor necessary to clear forest plots with a hand axe. This does not seem to have led to an increase in swidden field size, however, since there are still labor bottlenecks for all other phases of agriculture (planting, weeding, and harvest).

15. Portions of these events are also recorded in a World Bank Forestry Sector Report (1994), in Zerner (1992), and in Triwahyudi et al. (1992).

16. The description of P. T. Kalhold Utama's HPH Bina Desa project is based on my field notes, on Zerner (1992), and on Triwahyudi et al. (1992).

17. For further analysis of Dyak rattan cultivation systems, see Iskandar (1993).

18. During the "rattan rush" of the mid 1980s, however, an influx of rattan harvesters from other regions led to the severe overcutting of many Bentian rattan gardens. In the Bentian area, rattan harvesters who are not garden owners receive 50 percent of what they harvest. Garden owners receive the remaining 50 percent. Compared to other agricultural commodities on other islands of Indonesia, where sharecroppers may receive as little as 30 percent of the harvest that they themselves have planted on borrowed or rented land, rattan

harvesters who neither own gardens nor have invested time in cultivation receive a relatively high share of the profits of their labor.

19. The tahi burung, or bird shit, theory of the origin of the massive rattan plantations found in the Bentian region is frequently proposed by logging concession employees to explain (without admitting Bentian ownership rights to the surrounding forests) the presence of plantations of commercially valuable sega rattan (*Calamus caesius*) in the region. This theory fails to explain, however, not only the detailed Bentian oral histories documenting their forest use and ownership, but also the appearance of large (often two to three meters tall) ironwood house posts and carved ironwood grave markers in and around Bentian rattan gardens—unless, of course, birds of a size unknown to modern science and with a taste for ironwood posts are postulated.

20. Some newspaper accounts indicated that 80,000 hectares of Bentian lands were to be utilized for the plantation.

# CHAPTER 9

## Global Markets, Local Injustice in Southeast Asian Seas: The Live Fish Trade and Local Fishers in the Togean Islands of Sulawesi

*Celia Lowe*

> The bourgeoisie was to leave to itself the illegality of rights: the possibility of getting around its own regulations and its own laws, of ensuring for itself an immense sector of economic circulation by a skillful manipulation of gaps in the law—gaps that were foreseen by its silences, or opened up by de facto tolerance.
>
> *Michel Foucault (1975)*

With strong words, a Sama[1] fisher exposes his fatalism, marking therein the anxieties and paradoxes of a new export trade in live reef fish: "If people were using poison and my take dropped to only a little, I would accept it," he said. "But I feel heartsick that people have used cyanide here and then I catch nothing at all. I have not caught a big fish in a month so there's no point in going fishing this afternoon. There won't be any results." How is it that experienced fishers in the Togean Islands of Sulawesi, Indonesia, no longer want to fish? Only in the past few years has this fisherman taken up live fishing himself. At first he would go out to the reef, paddling his canoe two hours in the morning as the stars faded around him, count out twenty arm spans of nylon line and drop it over the side, then look deep into the water, waiting, forearm resting on the canoe's edge, the heavy line wrapped three times across his palm. But he has had competition: younger men who motor along the reef's edge, also on the lookout for live fish species, who do not use lines, hooks, or baits—only chalky clouds of poison that erupt out of their plastic squirt bottles to stupefy and tame otherwise wary fish. This fisher is torn; he also wants to be able to catch live fish to sell to the fish camps, which will export them to Hong Kong and Singapore and pay him very well for his effort. He maneuvers, in our conversations, to protect and perpetuate the industry, defending this camp, or that, as "clean," not supporting poison. Yet, he also recognizes that live fishing has brought his community to a difficult place: because of cyanide use, there are fewer fish for people to eat.

This chapter is about the bind Togean people are in, relative to the live fish trade, a trade that has quickly proven harmful for the majority of fishers, for their communities, and to coral reef environments. Radiant wild reef fish, which provided fishers with a few extra rupiah[2] through longstanding markets for salt fish, and which were once reliable as a source of food, are becoming rare in Togean waters. Yet not only fish and corals are affected; in addition to the biophysical damages are the human ones. These consequences are not logically explained as the independent acts of a few misguided fishers, though. Togean people are caught within the matted fibers of market, law, bureaucracy, and identity that determine the patterns of who will fish with cyanide, who will profit the most by it, and who will suffer the consequences. These questions then emerge: Why do conservationists and bureaucrats stigmatize Sama communities more than others as responsible for cyanide use? How do market ties to traders bind individual fishers to destructive practices that some of them would rather be free of? In what ways do legal frameworks supporting business practices over fisher interests make Togean people vulnerable to police extortion and the enforcement of unjust regulations? Answers are illuminating: these brachyating features of live fishing are what disable Togean people's control over reef resources while allowing a few local bureaucrats and outside entrepreneurs to become wealthy.

I am interested here largely in one set of imbricated facts: while most live fish are caught with cyanide, not all cyanide fishers are ethnically Sama, and the majority of Togean fishers, of any ethnicity, use traditional handline techniques. These ratios are disguised within the current discursive articulations of causality formulated by many conservationists who condemn the live fish trade as inherently destructive and who blame local people as poor conservators of nature. During twenty-four months of ethnographic work between 1994 and 1997, I learned how the live fish trade developed in the Togean Islands, and in Sama villages in particular. In an interconnected scenario, traders targeted Sama people as suppliers to fish the resource, while other outsiders—conservationists and government officials—began to associate Sama fishers in an overly simplified manner with the illegal cyanide use sometimes encountered in the capture of live fish. In querying perspectives and practices of bureaucrats and conservationists from Sulawesi, and from the nation's capital, Jakarta, and in trying to expose translocal contours of a Togean political ecology (Bryant 1992; Moore 1994; Peet and Watts 1996), it became clear that the views of very well intended conservationists overlap with those of less altruistic entrepreneurial bureaucrats. While local fishers attract a disciplinary gaze from both sides, neither pays critical attention to new market and jural regimes of accumulation, which are changing land and seascapes around Indonesia, as elsewhere.

## THE TOGEAN ISLANDS IN THE LIVE FISH TRADE

The Togean Islands are a small archipelago[3] in the middle of the eastward-facing Gulf of Tomini harboring a volcano and six elongated raised limestone islands (figures 9-1, 9-2, 9-3). Smaller, craggy, thinly soiled islets, bordering the shores of the main islands, create hideaways, anchorages, mangrove-lined boat passages, and resource collecting sites. Sama villages are normally found on outlying islets off-shore from the larger islands. In these, and in other village communities dispersed along the perimeters of islands, houses built up from cement, wood, and forest materials lie at the edges of the land, or on stilts over the fringing coral substrate. Oil palm and vegetable gardens spread from coasts into the interiors, and forests, in the midst of these encroaching cultivations, supply residents with canoe timber, rattan, medicinals, and other timber and nontimber products. Surrounding Togean waters, reflecting a violent equatorial sun, contain fringe, barrier, patch, and atoll-shaped reefs where people collect subsistence and market-oriented marine goods, of which fish is only one among many. Beyond the reef, in deeper waters, pelagic fish school, drawing local fishers and commercial boats from the mainlands of North and Central Sulawesi. Conservationists, who perceive a natural abundance they articulate in terms of terrestrial and marine biodiversity, proposed the Togean group as a national park in 1982 (Salm et al. 1982). The bureaucratic realization of a park is still in process, although conservationists' efforts to limit land and marinescape modifications are ongoing.

In 1995, the population of the Togean Islands was roughly 29,000 people. Five major ethnic groups in addition to Sama share the islands, with smaller numbers of people of several other ethnicities. Some Togean villages are identified as multiethnic, and some by a dominant ethnic group; Sama villages and village subdistricts are demarcated by their ethnonym, although there is no such thing as an ethnically homogeneous Togean locality. Sama and other islanders work at small-scale resource harvesting, gardening, oil palm farming, and petty trade. Sama people, in particular,

**Figure 9-1** Indonesia.

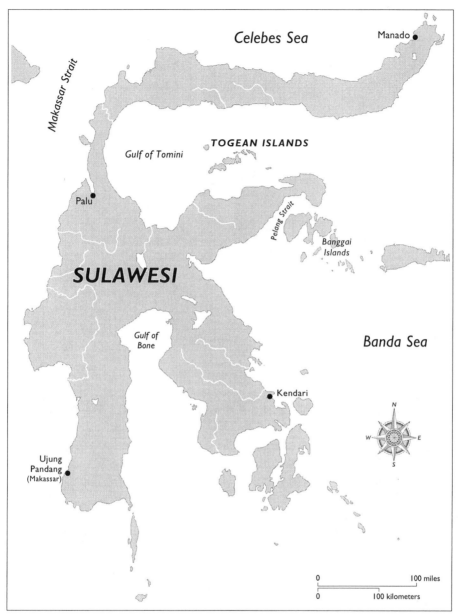

**Figure 9-2** Sulawesi.

are stereotyped by their collection of marine products, although their activities are more nuanced than fishing alone, and they too farm and trade. Average earnings for Togean fishers who do not work any oil palm–producing land are $50 a month, placing them among the least affluent of Togean residents, and of Indonesians in general.

In this context, Togean people have, by and large, welcomed the live fish trade. Profits from live fish have allowed them to improve their homes with cement and wood that they could not otherwise have afforded. Some fishers have used proceeds

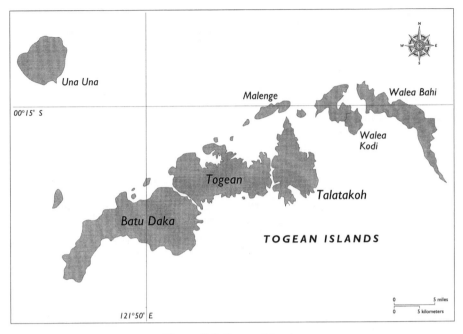

**Figure 9-3** Togean Islands.

to buy consumer goods such as watches and gold jewelry. All would concur that live fishing has brought at least a momentary economic boom to the Togeans, and often Togean people are willing to invest their time less profitably in live fishing, rather than in other economic activities, with a gambler's eye toward catching "the big one." Yet, while live fish are profitable locally, they are even more so beyond the Togeans. Fishers say that one large Napoleon (humphead) wrasse (*Cheilinus undulatus*) sold by the "bosses" in Hong Kong brings wealth enough to buy a new car. A fish buyer, in another instance, told people that rather than have even one Napoleon wrasse escape, he would prefer his outboard motor to sink to the bottom of the sea. Stories fishers tell highlight the economic value of live fish in the Togean economy and in fishers' lives, while pointing out the social and economic divides between the live fish harvesters and fish buyers, and between suppliers and the people who eat live fish in expensive restaurants overseas (table 9-1).

Despite initial enthusiasm, the negative effects of the live fish trade on fishing communities, fish species, and coral reef ecosystems is of sudden and intense concern for local people, community activists, and some government officials, and it is high on the agenda of most marine conservation programs in Indonesia. Everyone's worry is that live fish are sometimes caught using sodium cyanide (NaCN), which causes a high mortality rate in the fish and, more important, bleaches white the surrounding reef, killing the coral habitat. More of the reef is ruined as fishers tear apart the coral to get at stunned fish hiding in rocky crevices and nooks. While local people note general declines in fish abundance, conservationists are additionally concerned for the survival of individual species seen as endangered, most particularly

**Table 9-1**

Live Fish Prices per Kilogram (Mid-1997)

| Fish Species | Price Given Fishers | Price Fish Camps Receive | Price in Hong Kong Restaurants (1995)[a] |
|---|---|---|---|
| Napoleon wrasse (*Cheilinus undulatus*) | $7.50 | $125 | $180 |
| Polkadot grouper (*Cromileptes altivelis*) | $9 | $90 | <$180 |
| Coral trout (*Plectropomus leopardus, P. maculatus*) | $6 | $25 | — |
| Flowery cod (*Epinephelus fuscoguttatus*) | $2.50 | $25 | — |

*Source*: Johannes and Riepen (1995:9).

[a]These 1995 prices are certain to have increased by 1997.

the Napoleon wrasse. This fish, the most sought-after species in overseas restaurants, very quickly became rare after the introduction of the live fish trade, which began in Indonesia in the early 1990s, and in the Togean Islands around 1994. Marine biologists predict that the species involved in the trade will be commercially, and in some cases physically, extinct in Indonesia by the year 2000. Togean people, as we have seen, already speak of the fruitlessness of their own subsistence fishing under current conditions.

Togean Island markets are only one instance of a new form of commodification that has spread throughout Southeast Asia and much of the western Pacific. According to the detailed investigative work of Robert E. Johannes and Michael Riepen (1995)[4] on environmental, social, and economic implications of live fishing, markets were first developed in the late 1960s with fish stocks coming from reefs near Hong Kong. The market proved lucrative enough that Hong Kong fishers soon began to move further afield. The Philippines became the next main fishing grounds for the trade in the mid 1970s, and in the 1980s and 1990s live fishing spread into Palau, Papua New Guinea, the Solomon Islands, the Maldives, and Indonesia. Currently, half the live fish market is supplied by Indonesia. At first, live fish were caught by foreign fishers who fished illegally in Indonesian waters. More recently, the Togean pattern of small collecting companies employing local fishers has taken shape. Johannes and Riepen approximate the total annual market for live fish at 25,000 tons a year, some of that being cultivated fish, although accurate figures are not usually available.

In the Togean Islands, centrally located fish-buying camps, which function as collecting and transfer points for live fish, purchase ten different species (table 9-2) and keep them in pens where they are fed and medicated while awaiting shipment overseas. The fish are picked up by live fish transporting vessels after being stored as long as two months. Live fish in holding pens often die very quickly; mortality is highest for coral trout (*Plectropomus leopardus*), where 50 percent of the fish die

**Table 9-2**
Species Bought in the Togean Island Live Fish Trade

| Sama | Indonesian | English | Latin |
| --- | --- | --- | --- |
| Langkoe' | Maming | Napoleon wrasse, humphead wrasse | *Cheilinus undulatus* |
| Kiapu tikos | Kerapu tikus | Polkadot grouper, panther fish, barramundi cod | *Cromileptes altivelis* |
| Sunu | Sunu/super | Coral trout, leopard grouper | *Plectropomus leopardus* |
| Sunu | Sunu/super | Polkadot cod | *P. areolatus* |
| Sunu | Sunu/super | Bar cheeked coral trout | *P. maculatus* |
| Sunurang | Sunurang | — | — |
| Sunu macang, kiapu macang | Sunu macam | Highfin grouper | *P. oligocanthus* |
| Tembolang | Kerapu tembolang | — | — |
| Gomez | Kerapu gomez | Flowery cod | *Epinephelus fuscoguttatus* |
| Gomez pipi' | Kerapu lumpur, kerapu gomez | Queensland grouper | *E. lanceolatus* |

before transport, and for a grouper species known locally as *tembolang*, which is no longer purchased by many camps for this reason. Moreover, the death rate for coral trout has increased a further 20 percent in less than a year, and the reason is not yet known. In 1996, the four camps in the eastern Togeans subdistrict were each exporting 300 tons of fish a month; in 1997, this figure had fallen to 200 tons per camp. Rising fish prices are one indication of the decreases in productivity and availability of fish, and prices for all species are on the rise.[5] Competition for the catch fishers haul is also intense, and new buyers are still entering the Togean market despite declining numbers of fish.

Sodium cyanide (I: *bius*, S: *mius*)[6] was first used extensively in the Philippines, where reefs and fish stocks were negatively affected by the mid 1980s.[7] Cyanide fishers are now seen daily in the Togean Islands, and fish are similarly depleted. Using cyanide does not require deep ecological knowledge on the part of fishers. A cyanide fisher swims along, towed in the water by a small outboard-powered canoe, looking side to side for live grouper and Napoleon wrasse. In the boat is a jerry can filled with one tablet of cyanide for every five liters of sea water, and the fisher himself carries a small bottle filled with the mixture. When he spots a likely fish, he chases it into a coral hiding place (either holding his breath or breathing from a hookah rig attached to an air compressor) and squirts cyanide in each of the holes going into the rock. The fish, having nowhere to go, may come out "drunk" (I: *mabuk*), or the fisher may tear apart the coral with his hands or a crowbar to bring the stunned animal out of hiding. Once in the hands of the cyanide fisher, the fish is guided quickly to the surface and put in the canoe's water-filled holding compartment. Breaking apart the coral is one outcome that worries biologists; another is the spillover of the cyanide onto fish roe and fingerlings, which, along with coral polyps, can be directly killed by it. The chemical's longevity in the ecosystem is a further concern for scientists. Local Togean people are also worried about the effects of cyanide. I shall discuss some of these shared concerns.

## CYANIDE USE: FISHER AND BIOLOGIST EXPLANATIONS

Fishers and biologists construct cyanide use in culturally different ways: they muster different tactics to support their empirical evidence; they employ different descriptive aesthetics to bolster their explanatory frameworks. In a laboratory experiment, for example, on two freshwater food fish, *Tilapia mossambica* and *Cyprinus carpio*, a marine biologist at Universitas Sam Ratulangi in Manado, exposed fish to sodium cyanide at a concentration of 1.5 parts per million. The scientist described the symptoms of the poisoned fish, which died within twenty-four hours, in the following way: "The fish breathed on the surface water, jumped out of the water, imbalancedly [sic] swam, put down on the bottom, the fins stood, the body stiffed [sic] and the mouth was widely opened then the fish died and the skin released much mucus" (Pratasik 1983). Sodium cyanide, he explained, converts to hydrocyanide (HCN) in the body and absorbs the blood's oxygen. This experiment took place in enclosed tanks where the fish could not escape to uncontaminated water.

In the sea, fish encounter smaller doses of poison, which stun without killing them. In this environment, fishers too conduct their experiments. A fisher describes the behavior of a cyanide-exposed fish using his hands to trace its dizzy path: "The fish is drunk, it spins and spins without knowing where it is going." Fishers laugh at the antics of the drunken fish and take tactile pleasure in catching wild fish with bare hands, but they do not want to see it die, and, if a fish is really having trouble, they will fan clear water through its gills to revive it. Scientists and fishers alike explain that cyanide is not harmful to humans who eat the fish because the poison rapidly works out of the fish's system when it "comes to." (The availability of cyanide in the Togean Islands, though, has resulted in human mortality: a few people have committed suicide by deliberately ingesting the poison.)

Fishers document the biological effects of the live fish trade by observing fish rarity, changes in fish behavior, coral bleaching, and systemic effects of cyanide. In contrast with the more "experience-far" concerns of scientists, fishers express their worries in relation to regular interactions with the coral reef environment. A woman fisher, for example, demonstrated the effects of cyanide for me by holding up two trigger fish (S: *pogo'*) she had caught, one fat and the other very thin. The thin one, she said, had encountered cyanide, which caused it to eat lethargically. Fishers explain that cyanide causes a fish to lose its appetite, which further means that it will not bite their fish hook. They use metaphors of flow to depict the travels of cyanide through the water column: "Cyanide is like expensive cigarette smoke or perfume: everyone in a room comes under its influence." Or negatively expressed: "It is as though someone hasn't bathed and everyone around has to smell him." The poison makes the fish wary and sluggish, they observe. Formerly, they could catch large Napoleon wrasses close to shore, in the open near the mangroves, and in only a meter of water, right under people's boats. "The fish know," I was told. "They hear an outboard motor now and they run, they run deep out to sea—they hide from us."

While many fishers share the concerns of conservationists, a productive collaboration between conservationists and local people has not always emerged. Local fishing communities dotted along the coasts of Indonesia's more remote islands and reefs are experiencing sharp criticism, surveillance, and enforcement activity, and they are impugned for the coral bleaching and ecosystem damage associated with cyanide use in live fishing. Both the government enforcement apparatus in Indonesia, and conservation nongovernmental organizations (NGOs), through imbricated logics, focus on intervention at the community level. Local fishers are threatened with fines and incarceration; they experience the state through threats of violence and extortion. Conservationists, tourists, and other well-meaning outsiders query, "Why do fishers destroy their own resources?" Unfamiliar with local practices, they assume that local people are ignorant of biological processes and are too poor to find alternative recourse. Whether through government campaigns or in NGO extension work, fishing communities are stigmatized as peopled by fishers who harm the marine environment, while the complexity and specificity of who is doing what and why is glossed over. Moreover, structures of significant state complicity, which will be discussed further, continue to go unrecognized.

Paradoxically, the live fish trade is generally thought by biologists to be a sustainable industry that could have widespread positive effects on the incomes of some of the most economically vulnerable Indonesians.[8] Many fishers know how to use, and do use, handlines, with which they can extract fish at lower, arguably sustainable, rates of harvest without poisoning reefs. Most local people are against the way the trade is conducted but feel helpless to oppose it. Thus, in seeking to shift the terms of the debate away from blaming local communities, I ask, How do market and jural regimes serve to alter marinescapes? And, how might the majority of fishers using traditional fishing techniques be empowered to oppose the few who jeopardize the resource and industry for all? These markets and legal frameworks are the focus of the rest of this chapter.

## PATTERNS OF ETHNICITY AND IDENTITY: WHO USES CYANIDE, AND WHO DOES NOT?

The live fish trade in the Togeans is a multiethnic and economically stratified project. Live fish businesses are owned by wealthy local and Jakartan Indonesians of Chinese descent who operate through established connections with foreign Chinese buyers. This pattern has historical roots in other marine resource trades. Sea cucumber (*teripang*), for example, has been exported to China from Sulawesi for at least 400 years and possibly longer (Warren 1981). At the local level, the trade is managed by fish camps run by non-Togean Indonesian bosses. Camp bosses, coming from far-away Java and Kalimantan, fit relatively comfortably into village life and sometimes marry locally. Togean villagers, who are positioned as harvesters in the trade, are also occasionally hired to do physical tasks at the camps. Of the people who fish, all Togean ethnicities are represented and it is certainly part of local experience, Sama and otherwise, to be at the tail end of extractive commodity chains. One reason is that the costs associated with operating a camp are considerable. They include purchase of fish holding pens, camp buildings, and short-wave radios; salaries; and especially permits and bribes—all expensive by the standards of the local economy. "It's lucky the camps have deep pockets," a villager told me. "Poor people like us could never get such a thing going."

Ethnicity is the most frequent lens through which non-Togean people understand marine resource use. Sama peoples are known in Indonesia through an essentialized identity as "sea peoples" (I: *orang laut*), and this gets them into all sorts of tight spots. Togean reefs show evidence of blast (dynamite) fishing, cyanide use, and decreases in numbers and abundance of commercial marine species, and Sama people are the usual suspects in the use of illegal poison and other harmful activities. The connection of Sama people to cyanide has come about partly because fish-camp managers strategically placed their camps near Sama villages and Sama participation was actively recruited. Thus, Sama do participate in the trade in all its forms. But the notion that Sama people damage the marine environment also relates to the general perception of Sama as *suku terasing*, a term meaning an ethnic group for whom the process of national modernization is still foreign. Sama villages in the Togeans are all part of the Program for Left-Behind Villages (I: *Inpres Desa Tertinggal*), which is a poverty-alleviation scheme for putatively backward villages. A consequence of typologies characterizing Sama communities as backward or alien is that Sama people are not understood to be the same kind of citizens as wealthier, or more educated, Indonesians. The view, held by urban Indonesians, of Sama as pirates plundering reefs and coastal seas, or as maritime "primitives," has relevant consequences. Sama have a weak negotiating position in relation to the national rhetoric of development and in relation to bureaucrats and conservationists as national developers. They are ridiculed for their cultural and fiscal impoverishment and at the same time prodded to come up with means for their own financial advancement. As some of them, through the live fish trade, have found the means to "develop"—build new houses, wear new clothes, own motorized transport—they are then chastised for being environmentally destructive.

Ethnic and class-based hierarchies in the live fish business that place Javanese, Chinese, and urban elites at the center of lucrative extractive economies, and "marginal" rural people at the periphery, are not coincidental: they mirror the way most natural resource projects and commodity productions are operationalized throughout Indonesia (Peluso 1992b; Robeson 1986; Tsing 1993). At the same time, ethnic economies articulate with bureaucratic ones when, as often happens, government workers also become entrepreneurs (Stoler 1985). The patterns of the live fish trade differentiate it from earlier marine resource trades (such as sea cucumber), because in the former, the contemporary bureaucracy is instrumental in taming the populace, domesticating it through rhetorics of inclusion and exclusion in the national polity, and demanding compliance as a means of its own resource control. These partnerships between bureaucrats and internationally focused extractive elites Anna Tsing (1997) calls her "nightmare scenario" of the wrong kind of collaborative work.

Yet, in contrast to outsiders' perceptions and rhetorics, most Sama people oppose cyanide use. In better times, they told me, the walls of their fishing houses, and all the space on the decks of live-aboard fishing boats, would be layered with fish drying, and the air would reek of fish around them. This is no longer true, and people blame cyanide for the change.[9] Although a greater quantity of live fish are caught using cyanide, the large majority of fishers fish with traditional handline techniques. This pattern is quite salient. Although under current conditions in the Togeans, poison produces a higher yield,[10] most people choose to avoid it. Local views and practices concerning cyanide are not uniform within communities, yet whole communities—especially Sama ones—tend to be stigmatized as unvariegated spaces where inhabitants ruin the marine environment. Although Sama people are often cohesive in the face of outsiders, over the issue of cyanide they will sometimes complain about a fellow villager. For example, a Sama person called out to a non-Sama fisher passing by, "Don't bother fishing here, they won't eat your hook. Someone from my village was using *bius* here this morning. You should get your village head to report him." Another villager, spotting a person using cyanide, said, "Tie him to a rock and dump him in a deep spot," and cyanide fishers are frequently insulted as "rock heads" (I: *kepala batu*). Fishers who use traditional hook and line techniques to catch live fish are one constituency who oppose the use of poison; many report their fish take has dropped to nothing. Ordinary fishers are angry at the ones using "techniques," and fissures in social networks open up around cyanide use. Cyanide fishers, they say, are the people who have outboards and are making lots of money, while everyone else's take of reef fish—for both trade and food—disappears.

A closer look at the practices of Sama fishers indicates they may actually be less likely to use cyanide than fishers from other ethnic groups. While bureaucrats and conservationists polemically blame the Sama for degrading their environment, one rarely hears that Sama people's experience with the sea could make them important marine conservators in Togean and other Indonesian settings.[11] While cyanide use on coral reefs is attributed to fisher ignorance, the biological knowledge of Sama fishers concerning the marine world, in contrast, is quite extensive and directly relevant to the capture of live fish. Handline fishing, for example, requires intricate knowledge of species, currents, locations, equipment, and baits, knowledge that constitutes a reservoir of natural history and environmentalist thinking largely unexplored by conservationists.

When I interviewed a Javanese migrant fisher about the presence of cyanide-caught live fish in fish-camp holding pens, for example, he responded, "All

Napoleon wrasse are caught using cyanide." Yet, whenever I fished with Sama handline fishers and observed them fishing selectively for Napoleon, they employed an ecological knowledge of habitat and bait that narrowed the territory of catch locations and enabled fish catch without poison. While cyanide fishers swept the seas attempting to harvest every possible fish from every imaginable reef, Napoleon wrasse fishers were sedentary, waiting for the mobile fish to swim along a favored path. Sama fishers know that all varieties of squirrelfish (S: *babakal*) are caviar to a Napoleon. Experienced Sama live-fishers can name eleven separate subspecies of squirrelfish in a Sama taxonomy that are appropriate as Napoleon bait. A fisher will first catch squirrelfish with a speargun over the shallows before paddling out to a reef precipice to wait for the Napoleon. Non-Sama fishers, on the other hand, who have shallower marine ecological knowledge, are more inclined to use cyanide because they do not recognize the appropriate bait to use, or the appropriate spatial tactics for Napoleon fishing. As new entrants into the market, they are less inclined to take up these practices and more likely to go directly to easier and more destructive ways of fishing.

Togean communities have resource-use divisions both between and within ethnic groups, and ethnicity may not be the most useful way to learn who uses cyanide and why. All Togean ethnicities are represented in cyanide use, and patterns of cyanide practice can be more accurately explained by examining factors such as age, social standing, and gender. Since women fishers never use poison to catch live fish, gender can be used to divide cyanide fishers from non-cyanide users. Women too participate in the live fish trade and they know what is going on. One day, when I asked why everyone was trying to catch a small sardine (S: *solisi*), a woman I was fishing with answered, "We are looking for solisi because all the big fish have been poisoned." Women fishers are doubly affected by cyanide use: they lack fish to catch when poison is used, and they lack food to feed themselves and their children at home. Throughout the Togean Islands, women and children eat the food left over when husbands and fathers have finished eating. When there is a shortage of fish, women are hungry and children are screaming. Thus, while to protect their families women will defensively cover up for a spouse or son using cyanide, it is also women—and women fishers in particular—who most openly criticize the destructive fishing practices they know to exist.

Contrary to many informal village and family power arrangements, the Indonesian state, religious institutions, and conservationists teach that men head families. Government bureaucrats in the Togeans, for example, order people to paint their fences in gender-coded colors: against a white background of vertical pickets, two low blue stripes symbolize the number of children allowed in the government's family planning program; a bar near the top signifies their mother; and a blue fence cap represents the paternal rule that binds the family together. When government officials and conservationists discuss cyanide use (or almost any issue besides cooking and family health), they direct their comments to congregations of men. By officially sanctifying male authority, however, these efforts bypass concerted female interests and structures of political authority that could be efficacious in opposing cyanide use. Through their gendered gaze, they overlook women's habits as fishers and community members vested in environmental outcomes.

Additionally relevant to a social structure of cyanide use is the age of cyanide users. Nearly all cyanide practice is carried out by young men, for several reasons. Older people have trouble diving the way younger folks can; it is a physically strenuous activity and older men complain of the cold. High live-fish profits through cyanide use are a way for young men to build houses and establish new, independent families. Cyanide also has a status that is most appealing to younger people. Cyanide fishers peacock their wealth and modernity by controlling outboard motors. They also have the money to smoke expensive cigarettes and wear new clothes. That the activity is illegal further demands their daring and indicates their tightness with officials who will protect them from prosecution. In short, cyanide fishing is where it's at—what's happening—and this makes it a game the young guys want to play.

While young men catch fish with cyanide, some older men participate by using the attachments to bureaucracy and bureaucrats they develop as village leaders. The Indonesian state bureaucracy, which percolates down to the village level and radiates out into villages through kinship connections,[12] is the diacritic most tightly correlated with illegal trade in natural resources. The anatomy of an enforcement event in the Togean Islands will help lay bare these aspects of cyanide activity.

At one time, a regional official, under pressure to appear to be enforcing cyanide laws, ordered "sea operations" (I: *operasi laut*), also called sweeping (I: *swiping*), where village officials were instructed to go out and "clean up the ocean." I went along with a party of five "sweepers," all wearing khaki uniforms, who set off with their boat drivers in three directions. Only our boat had any "success": we caught five skinny boys, none of them older than ten, using *bius* to catch anemone fish. Children like to play with the fish that symbiotically inhabit sea anemones by making them fight each other in a small container of sea water. The boys, all yelling at once, begged us not to report them. The police would be angry, and it is common knowledge that angry police are physically violent. A solution emerged: the boys would deliver edible anemones to the village official's house. Conspicuously, their poison was not confiscated and they were left to resume their activity.

From there, we proceeded to a less affluent part of the village, far from where any officials live, and it was mentioned to the parents that their children had been caught using *bius*. In the same breath, the khakied bureaucrat made a casual inquiry as to whether there were any ripe mangoes. We were soon sitting on the porch, chins dripping with mango juice. He asked again for fried sago (don't forget the coconut!), which they procured with ingredients quickly borrowed from a neighbor. Coffee with tablespoons of expensive sugar was served after the mangoes; "gifts" of limes and chilies were taken before leaving. Our group had imposed on the hospitality of a family clearly struggling to meet its own food needs. Conversation between the high-status village officials and their subordinate fellow villagers had been smooth, never strained, polite. The threat was always implied, wound around, sweet and hot, in and out of discussions of mangoes and chilies.

Fishers involved in cyanide use, if not immediate family members of bureaucrats, are at least closely related. Top officials provide protection against prosecution for their relatives and workers. "He uses a code when he directs us not to use *bius* which indicates that in his heart he will not really be mad if we do," says one fisher about the village head. Ties to bureaucrats are also relevant as to who pays bribes or is prosecuted. The children collecting anemone fish were from families without strong ties to village leadership and thus vulnerable to demands for payment. Local leaders of cyanide operations work intimately with fish camps and tend to channel financial opportunities and protection benefits to family members whom they can both trust and control. Outside this circle, fishers, even small boys, use poison at their own risk. Subsequent to the "sea operation," a friend who is part of the village faction opposed to cyanide use complained to me that the manner of our operation was all wrong—not *really* designed to catch anybody. It was conducted at the wrong time of the day, and not where people really fish. More important, he said, the people involved in the operation are heavily implicated in cyanide fishing themselves. Village-level bureaucrats, nested tightly in live fish procurement networks, work closely with fish camps that supply cyanide and buy the cyanide-caught fish. Our "boat drivers" had steady work as cyanide fishers in the employ of these leaders. In other words, we had been out only to catch ourselves.

It would be easy to be swept up by the theory that local officials who profit from cyanide are just bad people, but these networks originate, and are patterned on, an entrepreneurial culture that starts at the top of the Indonesian leadership hierarchy and that, even beyond that, facilitates the interests of an international globalized business community. Local village leaders are recruited and inducted into this culture by members of the regional bureaucracy. In fact, they would not be permitted to keep their jobs if they were unwilling to participate in bureaucratic entrepreneurship. Village officials may reinvent, but they surely do not invent, these ways of organizing economic life. Moreover, contemporary neoliberal cultures of economy concur with these business practices and powerful northern nations have acted as guarantor of an Indonesian government bureaucracy that enables international trade while subverting political opposition. And this officially choreographed subversion of opposition is efficient in Indonesia right down to the village level. This is an important alternative context for understanding the rather limited complicity of some local people in cyanide fishing.

## MARKETS AND FISHERS

I have been discussing live fishing as a localized epiphenomenon with ties to a wider entrepreneurial bureaucracy in Indonesia, but we should remember that markets for these fish are international in scope and stake claims to locality in many different places. Wild reef fish from some of Indonesia's most inaccessible regions, hearts stopping as they hit the wok or steam pot, are consumed by tourists and wealthy residents in the luxury restaurants of Singapore and Hong Kong. In what we are calling the era of neoliberal globalization, where markets for high-priced consumer goods tie global suppliers into ever-expanding networks of international buyers, live groupers and wrasses caught by subsistence fishers on isolated Indonesian coral reefs fit into stories of the new, the fast, and the for sale. Thus, goods become meaningful within different contexts of culture and economy, and markets are uniquely local, despite the nomadic travels of a commodity (Appadurai 1995a; Bourdieu 1984; Kopytoff 1995).

Markets for live fish, linking some of Southeast Asia's least affluent rural people with wealthy global elites, have developed hand in hand with rising incomes in urban Asia. In Hong Kong, for example, the well-to-do are known to spend much of their disposable income on food. Johannes and Riepen, who have traced the live fish commodity's migration from capture to consumption, write that meals at celebrations and religious festivals in Hong Kong often focus on fish as symbolic and literal representations of wealth. The traditional preference for wild-caught, freshly killed fish is based on the consumers' belief that taste, texture, and healthful properties fade immediately after the fish dies. International demand for live fish is relatively elastic with respect to price, and Johannes and Riepen suggest the Hang Seng Index is a good indicator of overall demand for live fish, although the endangered status of the Napoleon wrasse, they claim, has also increased its value.[13]

This link between rarity and price is only an obscure gap, however, in the otherwise seamless way the commodity effaces its history. In the forging of status identity in Hong Kong, one imagines little awareness of the genesis of the grouper that sits on the plate. On both sides of the South China Sea, the contexts of live fish consumption and live fish production are mutually obscured.[14] While consumers would have a hard time imagining the conditions of live fishing, fishers can only fantasize the extravagant cultures of foreigners who eat live fish. But, whereas the sins and seductions of eating live fish are opaque for consumers whose high demand motivates live fish production, the social and ecological relations of that production cannot be as easily evaded in the Togean Islands. The localization of international markets in the Togeans occurs through the links local fishers have with fish-buying camps. Fishers too are motivated to catch live fish by their own economies of luxury. But in fisher calculations of value, it is not the fish that have worth, but rather the incentives they are given to catch them. In addition to the purchase price of the fish, the camps give their fishers gasoline for outboards, cigarettes, and sometimes small loans. But the participation of fishers is also more complicated than mere willful consumerism.

Fishers, even those using traditional methods, become tied to the fish camps through the purchase on credit of outboard motors. Outboards are both eminently practical and a status commodity in the Togeans. In a small archipelago where sail and paddle are the most prevalent means of moving about, outboards are highly desired, though financially out of reach for most folks. In the four years since the live fish trade has entered the Togeans, outboard motors have become available to 10 percent of households, largely through loans from fish camps. The camps give fishers five-horsepower longshaft motors (I: *ketinting*) that cost US$450 when they are bought on credit from the camps ($300 in towns on the mainland), which the fishers pay for with irregular payments (I: *uang cicil*) taken out of their live fish sales.

The camps maintain the outboards, changing oil and spark plugs, as long as the motor is being paid off, although once the payments are made, the outboard will have little life left in it.

Fishers are required to put down some cash each time they sell a fish but can decide for themselves how much to deposit. A look at fishers' accounting books shows that they make random payments up to four times a month of amounts between $1 and $6. Fishers pay off a quarter to a third of the debt in a year and are frequently optimistic about the amount they have already paid. Although the implicit rate of interest on the outboard is not onerous (roughly 12.5 percent), and they like owning the outboard, fishers tend to rue the ties they have to fish camps. One informant, who grows chocolate and coconuts in addition to trading in live fish, explained for me the tie he has to one of the buyers. Every morning he fishes, and in the afternoon he comes home to work in his garden. He is tired of fishing, he says, and would like to spend more time farming. But he has to keep going or the camp will take his outboard away; the camp boss notices if he is not bringing fish around. Another informant, who is a full-time fisher, was using his outboard to fish for pelagic tuna. He needed his outboard to get to the offshore fish concentrators where the deep-water fish school. Yet, he was told, his outboard would be confiscated if he did not start catching live fish again for the camp. The camp's profits are not in the loan, but in guaranteeing its supply of fish by ensuring that fishers are forced to keep looking for live fish and forbidden to sell fish to any other buyer. So if a fisher wants to pay off the outboard before it gets repossessed, he must continue to fish on a regular basis. As the supply of fish is becoming tighter, one can only expect fishers to be squeezed by camps even further.

The relationships fishers have to fish camps are significant for another reason: live fish camps are centrally implicated in the procurement of the cyanide. Fishers report that poison was not used in the Togean Islands before the arrival of the live fish industry, when fish traders came with cyanide and taught fishers how to use it. They then supplied it at no cost. "You shouldn't teach people how to use stuff like this; then it wouldn't be used," one handline fisher accurately observed. Traders working with the camps also continue to supply the compressor rigs used to catch live fish in deeper waters and in more difficult locations. Compressors and hookah rigs go hand in hand with cyanide use, as there is no other way for a diver using a hookah to catch live fish. Conservationists have observed that it is these techniques that allow some live fishers to target literally every single sizable grouper and wrasse on a given reef.

## LAWS FROM ABOVE, ENFORCEMENT FROM BELOW

If cyanide fishing is illegal, and many people oppose its use, how does it then come into common practice? An examination of Indonesian legal frameworks provides one answer. Laws, which may elucidate overt intentions, also bear subtexts of unstated agendas; the design and implementation of laws reflect cultural norms and dominant ideologies as well as political capacity and will. Indonesian resource law, which I will explore in relation to Napoleon wrasse regulation, reveals a consistent support for large business interests and a structure that is suspicious and punitive at the community level. These laws differently regulate fishers, businesses, and government agencies determining the calculus of gain and loss in live fishing.[15]

In official discourse, Indonesia calls itself a "legal state" (I: *Negara Hukum*) and, at one level, it does have a legal framework generous in its protections for both ordinary people and the resources they depend on. Environmental laws, for example, prohibit the use of destructive technologies, the harvest and export of endangered species, and the penetration of foreign fishing vessels into Indonesian waters. The basic framework for environmental protection in Indonesia falls under the Law of Living Natural Resources and their Ecosystems.[16] Embedded in a philosophy that nature is God's creation, the law states that the natural environment should be preserved for the common welfare of Indonesian citizens specifically, and for humankind more generally. At a biological level, it acknowledges elements in an ecosystem as interdependent, and ecological processes as part of life support systems. At a social level, it proclaims that ecosystems and species should be utilized in a manner protective of the ecosystem, its flora, and its fauna. The Basic Fisheries Act[17] (see Warren and Elston 1994) is one example of a law rhetorically focused on popular well-being. Its aims include improving the lives of fishers and preserving the fish resource. Environmental legislation broadly argues that conservation and development are both inseparable and necessary for human welfare; thus, environmental protection is considered the responsibility of all citizens and of the state.

Yet these rules, which in ideology are dually biocentric and anthropocentric, fail on both accounts and are, instead, substantively centered on facilitating trade. In a country where "development" is a theology (see George 1996; Hefner 1990; Keane 1997; Steedly 1993; Tsing 1993), environmental legislation in Indonesia creates a protected environment for business while focusing conservation responsibility, enforcement, and blame onto communities. Bracketing momentarily the argument that Indonesian legislation is not expected to fulfill populist rhetorics, we can see how official structures move from protecting people and ecosystems to protecting the interests of bureaucrats and traders by looking at the legislation for Napoleon wrasse. The decrees titled "Ban on the Napoleon Wrasse Fish Haul" (Government of Indonesia 1995a) and "Ban on Export of Napoleon Wrasse Fish" (Government of Indonesia 1995c)[18] appear, in title, to insulate this species from catch and sale, since markets are almost wholly foreign and export is "banned." The law on export states, "The hauling of the Napoleon wrasse has been conducted using manners that may be harmful to coral reef ecosystems and other marine biology," and, "In the framework of development and conservation of fish resources and coral reef ecosystems, it is deemed necessary to establish a ban on export of Napoleon wrasse fish."[19] Both laws, however, contain kernels of exception that actually *facilitate*, rather than hinder, the catch and export of the fish.

Fish camps, exporters, and government officials are the direct beneficiaries of current legal frameworks. Fishers are required by law to sell their catch to a "collecting company." Even though in practice these companies, as employers, control capture methods through the provision of equipment and incentives, the law does not speak to this role, and they are hardly regulated or hampered in their ability to sell and profit from the Napoleon wrasse. For example, article 8 of the ban on haul states that "fish shall weigh not less than one kilogram and not more than three," while article 9 says that fish "weighing more than three kilograms or those weighing less than one kilogram will be allowed to be sold locally to a marketing entrepreneur." Thus, although the law formally "disallows" export, fish that weigh too little or too much may legally enter the hands of traders whose only intent is to sell fish abroad. Napoleon laws also allow catch for research purposes, but collecting companies—fish camps—are not set up as research stations, and no research is facilitated through any Togean Island fish camp. Local markets for live fish are minimal and there are no good reasons other than export to purchase Napoleon wrasse from fishers. It is unrealistic, therefore, to believe that the fish camps' purchases will not be exported (or that large fish will shrink to permissible export sizes!).

Laws that enable trade in live fish, simultaneously attenuate bureaucracies and enrich individual government workers. This is operationalized through the government's reporting, evaluating, and permit-granting roles outlined in live fishing laws. Government agencies grant permits for the haul of fish and require other permits to export live fish. Each Napoleon wrasse, for legal export, also needs an official "letter of origin." The provincial fisheries department is further obliged to oversee the biological prosperity of the fishery; it "shall determine the fishing ground by evaluating the resource and its environment." Despite all this oversight, the Napoleon wrasse resource is well on its way to extinction, indicating a deficit in the will, funding, expertise, and even intent of bureaucrats to carry out their role as resource guarantor. This is because fees are collected for permits and "services" ensuring that the government's oversight practices create conditions for maximal exploitation and minimal protection of live fish and other natural resources.

Every government agency has its agents. In Indonesia, bureaucracies need to secure their own funding for all but the most rudimentary operations, and the personal incomes of government workers are rarely dissociated from office income. Granting permits and facilitating trade are the ordinary profit-making activities of many branches of government. Permits for fish camps in the Togean Islands reportedly cost $1,000 in "official money," implying that the unmeasurable hidden fees surpass this figure. As they are rational people, it is in the self-interest of bureaucrats to grant permits, not to restrict access to natural resources. Johannes and Riepen report that exporters without proper permits call Napoleon wrasse "grouper" on customs forms and pay officials not to inspect their shipments.[20] Likewise, laws regulating fishing techniques allow enforcement in villages, which generates illicit income for multiple officials and regulating agencies. In the case of the live fish industry in Sulawesi, permit requirements that provide personal income for officials are the rule, not the exception. There is no evidence to support the idea that "enforcement" limits, or is even intended to limit, cyanide use.

Unlike traders and bureaucrats, who are structured to profit without liability, risk, or blame, fishers are vulnerable to prosecution and extortion, and they live with the material consequences of weakened ecosystems. In another regulation, the decree of the director general of fisheries regarding "Size, Location, and Manners of Hauling Napoleon Wrasse Fish" (Government of Indonesia 1995b), we see the role fishers are supposed to play in the interstices between traders and bureaucrats. This law allows catch and trade by "traditional fishers," defined as persons or groups whose means of livelihood is catching fish using nonmotorized craft or small outboards, and "which utilize fish catching devices and substances that shall not harm the fish resource or its environment." Different types of rules and responsibilities within the decree apply to fishers, businesses, and government agencies, and the state's positioning of fishers as responsible for cyanide use is transparent in the text. Rules pertaining to fishers focus on technique and equipment: fishers may use lines, traps, and nets to catch Napoleon wrasse. An emphasis on catch method implicates fishers as the party responsible for how fish are caught. "Scientific" expertise is attributed to collecting companies—not to fishers.[21] Yet, since cyanide is distributed from fish camps to fishers, the law unjustly implicates only the most vulnerable party. Moreover, as I have argued, part of fishers' vulnerability is located in their inability to pay the fees and fines that grease the wheels of extralegal economic activity.

In the Togean Island fishery, a story of the "enforcement" of a law prohibiting the use of air compressors to catch fish reveals the manner in which this *Negara Hukum* (this legal state) functions on the ground. Representatives from the police, navy, and fisheries departments descended on the Togean Islands during my research period to perform what they called a "secret operation" (I: *operasi rahasia*). They were looking for air compressors, the possession of which normally requires permits. A local village leader was first told to notify the owners of the compressors of their permit violations. Indexing his own involvement in illegality, he answered that he had "influenza" and could not leave his house. Unlike the "sea operation" run by the low-level village bureaucrats, however, these outsiders were successful in finding "culprits," and three compressors were temporarily confiscated.[22] A rumor floated briefly around the village that the equipment owners would be taken to Poso, the regency capital, to face charges. Clearly, the way out of the difficulty involved cash, and the sooner the problem was dealt with, the less expensive it would be. People told me if money was fast flowing (I: *kencang*), the problem would be put to rest, and this is, in fact, what happened.

Local people consistently express the idea that it is only poor folks who end up in trouble with the law; those with means can pay their way out of difficulties. One fisher said, "I want to help [those arrested], but to help with money—there isn't any money. To help with advice—I don't want to seem like I go along with the position of the police. What do I do? Poor people just don't have the capacity to escape these things." The average savings ordinary Togean Islanders can lay their hands on would be $50 to $100; everyone seems to know (and fear) that getting out of jail would cost the impossible sum of $5,000 should matters progress that far. This forces local people who do become caught in enforcement webs to turn to village officials and entrepreneurs who trade immediate cash and protection for future illegal resource harvests. Ironically, while poor people are the first to suffer penalties, and they assume the greatest risks, they also are excluded from the highest live fish profits and the greatest rewards. As villagers all know, laws, as they are written, interpreted, and enforced within an entrepreneurial bureaucracy, enrich bureaucrats and their organizations while failing to protect either species or citizens.

Even when village people are not the perpetrators of cyanide use, they find that laws are not meant to be employed by people like them. Villagers usually find it impossible to protect their legal interests when confronted by men in uniforms. An example: In North Sulawesi villagers once tried to arrest the captain of a boat belonging to a "cartel" that was in Indonesian waters illegally and using cyanide on local reefs. The captain of the fishing boat was from the Sanghir Islands, a small Indonesian island group near the Philippine border. All of the crew were Filipino. Cyanide fishing was first perpetrated in Indonesia by Philippine fishers coming illegally into Indonesian waters once they had fished out their home reefs. These boats are less frequently encountered now that Indonesia has developed its own live fish businesses, but when they do appear they have police and military ties that protect them from prosecution. A local villager told the boat captain, "The problem is that the villagers here are small fishers. There is nothing for them if you take their fish." The captain, with great bravado, replied that he could do as he pleased, since he had friends in the police all over the province.

Two weeks later, villagers were summoned to speak with a policeman who, sidearm over his shirt, bullets lined up across his chest, chastised them for their action. To lessen its meaning in the face of the police, villagers protested that what they did was not really an "arrest." The policeman said, though, that the villagers had been wrong and the captain might have to be compensated with village funds for lost revenues—clearly an impossible burden. In the end, the villagers were warned, the policeman was paid, and the captain was free to use cyanide where and when he wished. "The village has the right to regulate its own affairs, but sometimes the police want to mix in," somebody grumbled. Thus, even when communities are right, they are wrong from the perspective of Indonesian enforcement. Marine conservation experts in Indonesia claim that one of the most important steps needed to protect Indonesia's reefs is for local people to be able to defend their territories from outsiders who use harmful methods to meet short-term interests. This is quite a task, given the current discursive positioning of villagers in relation to the state. As a fisher I often accompanied explained to me, "Clearly there is no longer a role for the people."

## ON SOCIAL JUSTICE, NEOLIBERALISM, CONSERVATION, AND POSSIBLE FUTURES

The story I have presented is of a newly commodified resource that could be used to benefit a large number of Indonesia's fishers over the long haul but, instead, is organized around making a small number of well-connected officials and entrepreneurs wealthy at the expense of coral reef ecosystems and the local communities that depend on them. I have argued that, rather than condemn the industry and its fishers, we should understand who participates in the destructive aspects of live fishing, what structures facilitate participation, and how and why they have come to exist. In this inquiry, it becomes obvious that destructive fishing practices should be contextualized in a variety of ways that take us well beyond fishing communities; they are not just a consequence of the malintentioned actions of fishers.

Live fishing articulates well with a transnational neoliberal paradigm that seeks to deregulate international trade and facilitate the flow of commodities and profits. This neoliberal template, put succinctly, proposes quantum reductions in government regulation, freeing up Adam Smith's "invisible hand" to calibrate economies. Competition and reduced trade barriers, in this tale, enlarge aggregate wealth through a system of "comparative advantage," whereby regions or countries each produce what they are able to most "efficiently."[23] Within this narrative, Indonesia has been considered a neoliberal success story by the international economic community.[24] Yet it appears at once to present a paradox: Indonesia has achieved its market success not through a contraction of government apparatuses, but by an expansion of bureaucratic networks into all sectors of the export and domestic economy. In forming a collaboration—and in some senses an identity—between bureaucrats and entrepreneurs, the Indonesian state has used the market to constitute and reproduce itself, rather than to limit its internal effects. The Togean Islands are a clear example of this: administrative, police, and fisheries bureaucracies are all parasitic on the live fish trade and work to ensure maximal short-term profits from the resource.

As the Indonesian bureaucratic skeleton continues to attenuate, it appears as a corporate body, not a body with corporality. The apparent paradox of bureaucratic expansion under neoliberal conditions becomes something less of one, however, when the emptiness of the Indonesian bureaucracy's power-balancing and regulatory functions (at the heart of what neoliberals want to reduce) is taken into account.[25] While ineffective at checking abuses of political power (in fact, it is not really clear what an "abuse" might be in this context), the bureaucracy is hyperefficient at distributing capitalist opportunity within its ranks.[26] In this limited sense, the Indonesian bureaucracy fits the model of reduced oversight in a way consistent with neoliberalism, and in a way that readily accommodates international and domestic capital expansion into an Indonesian resource arena. Thus, where the bureaucracy is constituted not only as the handmaiden of capital, but also as the incarnation of capital, it becomes a key site itself for the production of economic and social inequality.

The live fish industry in the Togean Islands, and in Indonesia in general, developed very quickly. As noted previously, marine biologists tell us the trade in live food fish in Indonesia will be moribund sometime around the turn of the century, and this is the story Togean fishers already relate in their commentaries. Thus, everyone concerned with the effects of this industry on communities and coral reefs is struggling to understand and respond to a phenomenon that will have passed before realistic solutions appear. What may be gleaned from the story of live fishing, then, are some general notions of who wins and who loses when natural resource extraction is organized around an alliance between bureaucrats and traders. The significance of this pattern is not just an academic proposition: Indonesian conservationists have predicted that a trade in ornamental aquarium fish will enter in the wake of the one in food fish. Accessing the same infrastructures, ideologies, incentives, and repressive techniques, traders and bureaucrats will again employ fishers already knowledgeable in using cyanide to catch these fish. The textures of the live fish industry are set to perpetuate themselves.

That social justice is not woven into the fabric of the neoliberal model is empirically evident.[27] On the other hand, wealth inequalities that spin off from neoliberalism are place based and no more exclusively organized by universal dictates than other structuralisms—modernization, development, colonialism, and the like (see Lowe and Lloyd 1997; Moore 1994). As when all such propositions are thrown up against everyday practice, situated neoliberalism is subject to its own contradictions, disorders, ellipses, and gaps. While it is challenging to imagine a world unfettered by this form of economy, it is these localized gaps, these elisions of structure, that present hopeful opportunities for social justice. One such space in the Togeans Islands, certainly, is the potential that conservationists will come to work productively with Togean people to empower them against an omnipotent and greedy bureaucracy. Up to this point, conservationists have taken the government's bait; they have swallowed the official red herring, as it were. They have been overly influenced by the state's progressivist ideology and are themselves embroiled in trying to order and discipline local people. But, on the other hand, as elite Indonesians and foreigners, with ties to prestigious universities, donor agencies, and politicians, and as experts who have facility with the powerful discourses of science, conservationists may be the only group with enough "symbolic capital" to build, with local people, alliances that are able to contend with the state bureaucracy and its rhetorics.

The injustice inherent in Indonesian state development ideology and practice is certainly not a new observation, and to hope that this will all be transformed in time to address the country's many pressing environmental concerns is probably fantasy. What is much more promising is the hope that conservationists will address issues of natural resource use with a fuller understanding of how to work with, not against, local people to set and achieve goals of social equity within their conservation agendas. The situation is hopeful because it is what conservationists themselves claim to want, and, in fact, in the absence of this approach, many conservationists will willingly acknowledge there really is no viable conservation agenda. Further, by recognizing that the most substantive ecosystem abuses are not organized locally, but rather underwritten by a ramifying bureaucracy and business community, conservationists may find a basis for alliance with village people. They also might find they have success helping local people combat cyanide use by coming out fully in *support* of live fishing as the sustainable industry that it potentially is. Thus, the question would become not one of how to prevent local people from doing x or y—that is, how to restrict local people's activities—but rather how to help local people respond to the power dynamics that reward gluttonous resource extractions in the name of development. This is an avenue that could lead to a future with which local people, conservationists, and fish all might be able to live.

## ACKNOWLEDGMENTS

I would like to thank Joseph Errington, Emily Harwell, Nancy Peluso, Hugh Raffles, Anna Tsing, and our editor Charles Zerner for reading and providing extensive comment, sometimes on short notice, on various versions of this chapter. Its shortcomings and omissions are, of course, my own work.

## NOTES

1. *Sama* is the name of an ethnic group whose members live in the southern Philippines, Sabah, Malaysia, and eastern Indonesia. Sama is how they refer to themselves, though non-Sama usually call them by the exonym *Bajau*. An excellent description of the differences between these terms can be found in Charles Frake (1980). Of additional interest is the elegant and detailed ethnography of Sama/Bajau people in the Semporna region of Sabah, Malaysia, by Clifford Sather (1997).

2. The rupiah is the Indonesian currency.

3. The Togean Islands are 100 kilometers wide from east to west, and 50 kilometers in a north-south direction.

4. For background information on the live fish trade, I owe a significant debt to the report of Robert Johannes and Michael Riepen (1995), which they submitted to the Nature Conservancy and the South Pacific Forum Fisheries Agency. Their work should be consulted for a complete overview of the live fish trade in Southeast Asia, including substantial case materials from three importing and six exporting countries.

5. The price fishers received for Napoleon wrasse rose to $7.50 from $6 over a six-month period between 1996 and 1997.

6. Because of the potential confusion when translating the Indonesian and Sama languages, I have used an *I* to indicate Indonesian and an *S* for Sama.

7. Johannes and Riepen (1995:35).

8. Johannes and Riepen (1995:79). This statement is also based on conversations with conservationists working for international nongovernmental organizations in Jakarta. They would prefer to remain anonymous.

9. Christoverius Hutabarat (1995) has written on a similar point. See also the work of Hutabarat, Pramono, and Yulati in the same edition of the journal *Tangkasi*.

10. At high fish densities, line fishing may be competitive with cyanide capture, although as numbers of fish decrease, cyanide catches do not dwindle as fast as hook-and-line catches (Johannes and Riepen 1995:72).

11. See Djohani (1996:162) for a very noteworthy exception to this.

12. Words of the Indonesian comic strip character Djon Domino, as referenced in An-

derson (1990), are apropos here: "Djon and his friend watch a luxurious official limousine roar by. Djon: `Hey there big shot, have a thought for the fate of the people. Don't just think of commissions and young concubines!' Friend: `How can he possibly hear you Djon?' Djon: `Think I'm crazy? If he could hear me he'd clout me a good one! After all, I'm a relative of his!'"

13. Johannes and Riepen (1995:17).

14. Following Appadurai (1995a), I have divided commodity knowledge into categories of production and consumption. Using the example of birds' nests collected by people who live in forests in Borneo, he notes the gap producers have in their knowledge of market destinations, claiming it allows traders high profits and is dependent on the subordinance of the producing region relative to the consuming one. Kopytoff (1995) argues that the knowledge gap also prevents the producing and consuming cultures from having their ideas of the commodity challenged. These remarks suggest an effect relevant for live fishing: the lack of the knowledge of production conditions on the part of consumers allows them to avoid even a hint of the moral and political implications that a transparency in the social and environmental conditions of production would encourage.

15. I am arguing here that Indonesian law is flexible not only in its application, but also in its initial formulation. For an excellent description of the contingent nature of law: see Sally Falk Moore's (1986) discussion of "exact" rules and flexible outcomes in Chaga law on Kilimanjaro. She explains that without an understanding of jural practice, formal laws are as decontextualized as the pottery shards archaeologists use to reconstruct civilizations. But it is not just after-the-fact law for which practice is relevant; rather, law-making itself is a practice, and one also open to strategy and contingency.

16. Law 5 of the Government of Indonesia (1990). For a further elaboration of environmental laws in Indonesia, see Koesnadi Hardjasoemantri (1995), and Carol Warren and Kylie Elston (1994). Warren and Elston argue, as do I, that Indonesian legal regulation is "studded with 'special dispensations' and selective application according to `vulnerability and political value' " (p. 8). I go beyond their argument, however, to claim an element of intentionality in the loose wording of environmental regulation, which allows commerce to prosper despite regulation.

17. Law 9 of the Government of Indonesia (1985).

18. This decree falls under the legislative aegis of the director general of fisheries within the Ministry of Agriculture.

19. See also Government of Indonesia (1995c) parts a and b in this context.

20. Johannes and Riepen (1995:40).

21. Article 6 of the decree of the director general of fisheries (Government of Indonesia 1995b) states, "The collecting company shall provide means for cultivation at the stated collecting location and possess a staff experienced in fish cultivation." This is an unworkable parameter, however, since the Napoleon wrasse can not be spawned successfully in captivity.

22. Note that with the right permit, compressors are not illegal, even though they still would be most productively employed, from an owner's perspective, in catching live fish with cyanide.

23. A radically different take on this same pattern is found in "world systems" and "dependency" theories. Pushed through these frameworks, live fishing is a predictable outcome of the exploitation of a dependent periphery by capitalist urban centers. While certainly more

credible at one level, these models also do not allow for the production of outcomes through human agency and resistance.

24. This was certainly true until the collapse of the rupiah in mid 1997. Immediately after its devaluation, however, Indonesia suddenly was viewed by northern nations as a rogue state in the international economy, one both unresponsive to the dictates of the International Monetary Fund, and "surprisingly" riddled with cronyism and corruption. Indonesia itself clearly had not changed; what was new was the insecurity that the northern nations felt about their ability to control and manipulate this system for its own ends.

25. Dan Lev (1998) has convincingly argued that there is no "there" in the Indonesian bureaucracy. Using the example of courts, he claims the legal bureaucracy is incapable of independent judicial oversight. This is consistent with the contention of Warren and Elston (1994) that environmental regulations are unenforceable in Indonesia.

26. This expansion, or intensification, of the bureaucracy through resource extraction in Indonesia is one that Stoler (1985:173), using material from Sumatran plantations, argues commenced almost simultaneously with the present administrative regime in 1965. She says that government-owned plantation estates were taken over at that time by "one of several groups at the top of the bureaucratic capitalist elite, who derive their wealth and power not so much from their high-salaried jobs as from their strategic positions in the flow of goods, services, and contracts that come under their ostensible control."

27. This is true unless, of course, one believes that economic incentives provided to the wealthy will "trickle down" to the poor, or that redistributive mechanisms, such as taxes, will compensate for a compression of wealth at the top. Helen Shapiro is responsible for alerting me to these exceptions. See Harvey (1990) for a more thorough discussion of these themes.

# CHAPTER 10

## Exploitation of Gaharu, and Forest Conservation Efforts in the Kayan Mentarang National Park, East Kalimantan, Indonesia

*Frank Momberg, Rajindra Puri, and Timothy Jessup*

### THE GAHARU RUSH IN EAST KALIMANTAN

The valuable wood gaharu is disappearing from Indonesia's forests at an alarming rate, much to the concern of conservationists and at great cost to local communities. Is this another tragedy of the boom-and-bust frontier economy of the "wild East," or can something be done to save the forest along with the trees? The experience of local communities and the World Wide Fund for Nature (WWF) in the Kayan Mentarang National Park, in East Kalimantan, illustrates the problem but also gives some reason for hope (figures 10-1, 10-2).

Gaharu, a fragrant wood that occurs in some but not all tree in the genus *Aquilaria*, has been widely collected and traded in Southeast Asia since ancient times. Now, in the 1990s, this valuable resource is being stripped from Indonesia's forests at an alarming rate. Large amounts of gaharu have been extracted in recent years from remote parts of Kalimantan and other islands across Indonesia with little regard for the biological impact on forests and the economic cost to local people, many of whom depend on gaharu and other forest products for their livelihoods. The companies financing this latest "gaharu rush" are based in cities such as Samarinda, Pontianak, Surabaya, and even Singapore. They organize and move collectors into remote areas and supply them by boat and aircraft. These professional collectors cut down virtually all the potentially gaharu-containing trees they find in an area, although most contain little or none of the valuable wood.

People who live by the forests where gaharu occurs, on the other hand, are more apt to use their knowledge of ecology and the growth of the trees to identify those that really do contain gaharu, rather than cutting down every *Aquilaria* tree they come across. In the past, when these local collectors were often the primary source

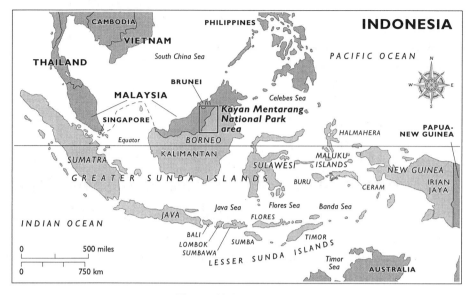

**Figure 10-1** Indonesia.

of gaharu bought by traders farther downriver, people were more likely to inspect trees first, by partially cutting into the stem, and then felling them only if signs of the fragrant resin were found. Trees that contained little or no gaharu were left to be checked again and possibly harvested later. At least, that was an option for communities who chose to conserve resources for their future and their children's future.

Now, however, gaharu is under heavy pressure from a collecting "boom" following recent increases in price: up to nearly Rp2,000,000 per kilogram in coastal towns of Kalimantan in the mid 1990s from a few hundred thousand rupiah just several years earlier. In the absence of effective control by communities over the forest resources they traditionally considered their own, local collectors have largely been forced to abandon any such conservation they may have practiced in the past. For what is the use of saving trees for the future if others will simply come and cut them down in hope of a quick profit? Better in that case to join the rush and get what one can before the trees are all gone. This is the economic logic and the ecological reality of gaharu in the 1990s.

### Indigenous and Exogenous Collectors in Kayan Mentarang

The tropical rain forests of the Kayan Mentarang National Park in the remote, mountainous interior of East Kalimantan help to sustain the livelihoods of some 10,000 indigenous people who live in and around the park (figure 10-3). Forest products such as gaharu, birds' nests, rattan, and cinnamon all contribute to local incomes, but gaharu has been the major source of cash for most households in recent years. The Kenyah, Kayan, Lundaye, and Penan people living in the area now base their

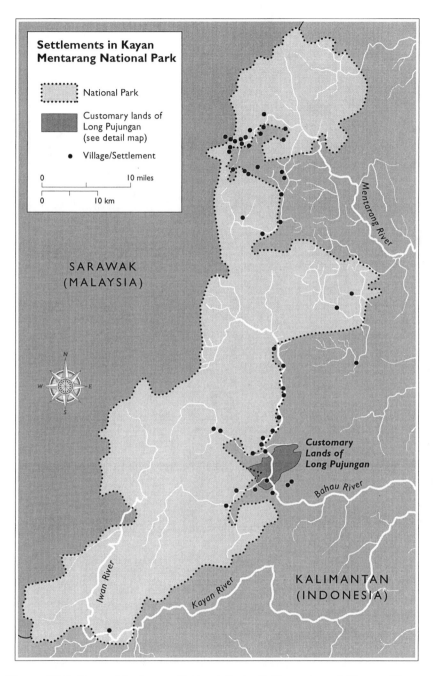

**Figure 10-2** Kayan Mentarang National Park: Settlements and Rivers. (From Geographic Information System, World Wide Fund for Nature, Indonesia Programme, 1997.)

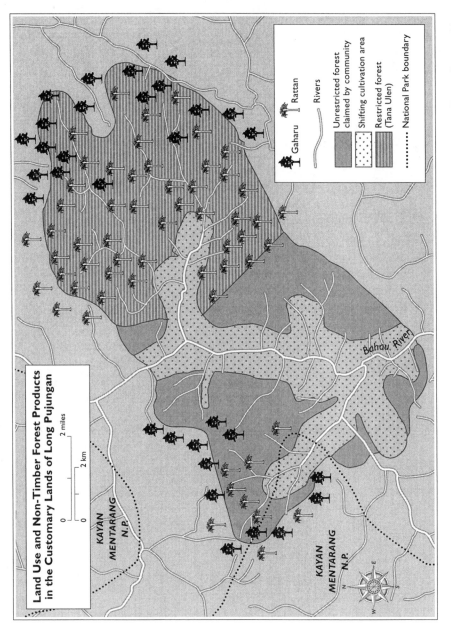

**Figure 10-3** Land Use and Nontimber Forest Products in the Customary Lands of Long Pujungan. (From Geographic Information System, World Wide Fund for Nature, Indonesia Programme, 1997.)

claims to these forest products on a centuries-long tradition of use under *adat* (customary) law. In principle, the government respects these rights and has offered to work with the people to establish good forest management and conservation in and around the new park, which was legally established in 1996 by the Indonesian government with support from the WWF.

Local people have attempted to protect gaharu and other forest resources by excluding nonresident collectors from their community forest lands, both through traditional legal means such as adat councils and local edicts, and by asking government to support their claims and regulate the outside collectors. But these large, remote areas are difficult for anyone to control, and all too often traders are able to bring in large teams of outside collectors rather than working with the local communities.

Although organized gaharu collecting financed by traders and trading companies occurred in interior Kalimantan at various times in the past, the boom in the 1990s has been different because of the extensive use of aircraft to move collectors in and gaharu out. For example, at the southern end of the park in the remote Apo Kayan, an area where the resident population numbers only a few thousand, some 500 professional collectors were brought in by helicopter during just six months in 1994. Helipads were constructed both inside and outside the national park (then a strict nature reserve), and many tons of gaharu, worth many hundreds of millions of rupiah, were removed from the forest and for the most part from the local economy. (Some local collectors participated in the gaharu rush, although they were often paid lower prices by the company contractors running the outside teams. Also, some local people made money for a while selling supplies to the collectors.)

Local community leaders and conservationists protested but could take little direct action to stop the exploitation. Eventually, the gaharu airlift from the Apo Kayan came to an ignominious end after two chartered helicopters crashed in the mountainous terrain and the government issued edicts restricting the outside collectors' activities. By then, however, most of the gaharu, along with the mother trees that might have produced more in the future, had been exhausted.

## Local Efforts to Protect Gaharu and the Future of Forest Conservation

Customary, or adat, rules of forest protection among communities in and around Kayan Mentarang have a long history, and many variations are still in practice today. Specific areas of forest claimed by members of adat communities are protected from clearing, and use of forest resources is restricted in various ways. Outsiders are either excluded outright or are required to pay entry fees to the communities or their leaders.

The system of rules governing forest use in the Kenyah village of Long Pujungan is a good example. The rules apply to an area of several hundred hectares (ha) of common land traditionally belonging to the community. (This strictly protected area of forest is called *tana' ulen*, literally "restricted land" or "prohibited land." Less onerous rules govern a larger territory also claimed by the community under adat law.) Decisions about use of the protected forest are made by the village assembly, while enforcement of regulations is overseen by a committee of village elders. It is forbidden for anyone to make clearings in the protected forest, and outsiders are further excluded from cutting or collecting forest products. Hunting is restricted to protect important wildlife food resources, particularly the bearded pig and sambar deer. Rattan collecting is restricted to a limited season when cash is in high demand, such as the few weeks just before Christmas.

Although rules such as these may have been more effective in the past, and while the present arrangements may still work in a local context, they suffer from inadequate enforcement against powerful outside interests. Even when the government has offered some support to local communities on behalf of forest protection (which is not always the case), well-equipped and well-connected outsiders—and, it must be admitted, some insiders from the communities as well—can find ways to evade the regulations, as illustrated by the example of airborne gaharu collecting in the Apo Kayan.

If conservation is to succeed in places like Kayan Mentarang, where local people depend for their livelihoods on the forest and where protection of the forest depends on local participation, it must be based on three things. First, there must be a consensus among parties with legitimate interests in the forest—local communities, government, business, and conservationists—that all must abide by a clear and agreed-on system of environmental regulations. Second, the regulations must provide a fair degree of security for those holding forest resources or rights to exploit them. Third, the regulations must be enforced in a fair and consistent manner. Otherwise, the fate of gaharu in the 1990s will be the tragic future of Indonesia's forests in the decades to come.

## LINKAGES BETWEEN EXTRACTION AND CONSERVATION OF FOREST RESOURCES

As conservationists, we are concerned with the connections between, on the one hand, the extraction of a valuable forest resource, gaharu, by various groups of people and, on the other hand, attempts by some of those people to protect the resource. The extraction of nontimber tropical forest products by rural people is believed to be compatible with and potentially supportive of nature conservation efforts because of its low impact on the environment (Godoy and Bawa 1993) and its economic viability (De Beer and McDermott 1989; Peters et al. 1989), which may provide sufficient incentive for users to conserve the resource, *provided* they hold secure rights of tenure. Recent quantitative economic and ecological valuations of potential extractive reserves (Godoy and Lubowski 1992; Godoy et al. 1993; Fearnside 1989; LaFrankie 1994; Lawrence et al. 1995; Paoli et al. 1994; Salafsky et al. 1993; Siebert 1993) are insufficient by themselves to demonstrate incentives to conserve, because they do not address the wider context of economic and political pressures for forest exploitation, government policies and their implementation, ethnic and other social affiliations, and the vagaries of local trade (see Dove 1993b), all of which can affect de jure and de facto tenure. Resource tenure and the recognition and enforcement by government of legitimate local claims to forest resources are also at the heart of the justice issue that is the focal theme of this volume. As conservationists, we support a greater degree of local control over forest resources, not only because of the economic incentives for conservation it seems to offer, but also because we believe any attempt at conservation based on the unfair taking of people's property without due compensation is bound to fail.

There are also technical reasons for conservationists to wish to see resource tenure and management devolved to the local level, provided adequate safeguards are enforced. Government is too often ineffective in its attempts to conserve natural resources from a centralized, national level, because of conflicting interests, lack of relevant information and expertise, and institutional distance from local conditions. It is believed that people who use or directly benefit from local resources generally have the greatest level of interest, knowledge, and expertise necessary to effectively manage them (Caldecott 1992; Padoch and Peters 1993; Poffenberger 1990). Small-scale communities in particular are likely to be most successful at collectively managing resources held in common for various reasons, as detailed by McCay and Acheson (1987:23):

> Visibility of common resources and behavior toward them; feedback on the effects of regulations; widespread understanding and acceptance of the rules and their rationales; the values expressed in these rules (that is equitable treatment of all and protection of the environment); and the backing of values by socialization, standards, and strict enforcement.

Thus, sustainable resource management requires that authority, responsibility, and capability be allocated to the appropriate levels, particularly to the local communities that directly benefit from the use of those resources.

Government has a more appropriate role to play in certifying, regulating, and enforcing local arrangements, and in dealing with threats beyond the scale of local management, rather than in directly managing resources on a local scale.[1] Furthermore, a prerequisite to any form of sustainable resource management, including conservation, is the clear acknowledgment and enforcement of just and efficient tenurial arrangements. Community-based management of natural resources requires official recognition of principles and institutions that regulate tenure, access, and use, such as *sasi* marine and terrestrial systems in the Moluccas of eastern Indonesia (Bailey and Zerner 1992), and community-based forest protection in Kalimantan as discussed in this chapter.

## Evolution of Customary Tenure from "Restricted Lands" to "Community Forests"

Strategies by which the present inhabitants of the Kayan Mentarang area attempt to regulate the use of land and forest resources, including gaharu, draw inspiration and authority from traditions or cultural precedents but are no longer tightly bound by custom—if, indeed, they ever were. Control over forests throughout central Borneo in past centuries was established through warfare or negotiations, or both. There is considerable variation in land and resource tenure systems within and between communities belonging to different ethnic groups (Appell 1993; Jessup and Peluso 1986; Rousseau 1990). In stratified societies, such as the Kenyah in the Bahau, resource use was traditionally more strictly organized and controlled to ensure the wealth and prestige of the aristocratic leaders, such as the Kenyah *paren*, who in the upper Bahau maintained control over the tana' ulen (described previously). Tana' ulen were either declared by aristocrats for their exclusive use or granted by them to particular families, such as those who had suffered in tribal wars. Commoners (*panyen*) from the aristocrats' own communities, as well as outsiders, had to ask permission to collect forest products on these restricted lands and also had to pay some portion of what they took as tribute to the aristocrats. Violators of these restrictions could be fined, or worse.[2]

The Kenyah aristocrats of the Bahau area, officially a defunct social class under Indonesian law, still maintain some degree of control over these village lands, now known as community forests, but access for local residents is less restrictive than before. Now, in new roles as school teachers, clergy, village heads and secretaries, and government officials, descendants of the aristocrats are still responsible for major decisions concerning land use in the villages, such as the annual selection of sites for making rice swiddens. Most of the customary restricted areas are still protected from agricultural clearing, and forest resource use is restricted to ensure a steady income for the leadership. Outsiders are either excluded outright or have to pay user fees (Indonesian: *cukai*) to the village treasuries.

The customary restricted forest of Long Pujungan, a village of around 500 people and the capital of the subdistrict, became a community forest in 1980. All decisions about resource use are made by the village assembly, while enforcement of regulations is overseen by the *panitia tana' ulen*, or committee for restricted land, within the village administration. It is still forbidden to clear these forests for swidden fields, and outsiders are excluded from all resource use there. Hunting, once limited for commoners (but not aristocrats) to short seasons before festivals and holidays, has now been prohibited for all Long Pujungan residents in order to protect specific important wildlife food resources, such as the bearded pig (*Sus barbatus*) and the sambar deer (*Cervus unicolor*). Rattan collection is restricted to a limited open-access season (two to three weeks), when cash is in high demand, such as before Christmas. During this limited season, only one of the two protected watersheds is opened for collection. The following year, the other watershed will be opened. Although it is not required these days, all villagers still follow the tradition of giving 5 percent of their rattan harvest to the highest-ranking aristocratic family. The Penan Benalui of Long Lame, a village adjacent to Long Pujungan, were often permitted to collect rattan during these open-access seasons, but they had to give as much 20 percent of their harvest to the aristocratic leaders.

In contrast to the stratified societies such as the Kenyah and Kayan, the Punan, Penan, and other egalitarian hunter-gatherer groups rarely had a tenurial concept such as a customary restricted forest. Many of these groups were allies and trading partners of nearby longhouse groups, such as the Kenyah, Kayan, Merap, and Saben. This economic relationship was usually of a patron-client type, where the hunter-gatherers were given the use of the headwaters areas of longhouse territories in exchange for an exclusive trading relationship with the village chiefs for valuable forest products. In several cases, Kenyah aristocrats gave their customary restricted forest to their Penan clients, while excluding commoners from their own villages. With the establishment of their own officially recognized villages and lands, and increasing numbers of traders coming upstream, Penan collectors have begun to bypass their former patrons and establish direct trade relations with outside traders. In cases where these groups have been granted official village lands and administrations, leaders have begun to claim customary land rights. The Penan Benalui, who have two of their own villages on the Lurah River, have followed the Kenyah model and declared specific tributaries of the Lurah as their customary restricted lands. The Tubu' Punan claim ancestral rights to the Tubu' River watershed, even though they are now resettled far downstream on the Malinau River.

## Incorporating Customary Lands into Parks

National parks and other protected areas in Indonesia, particularly those in remote areas, are underfunded, understaffed, and under threat from competing claims and uses. There is an urgent need for a more effective role in park management by local residents (Caldecott 1992). There is also a need to integrate conservation programs and development programs, particularly at the local and provincial government levels in the so-called outer islands beyond Java and Bali, where the nation's rapid modernization over the past generation has largely bypassed and marginalized remote rural communities. Providing a supportive legal and policy framework that enables these communities to participate in the development of natural resources in their own homelands is a necessary condition for sustainable economic development and conservation of biodiversity in these areas. Such "co-management" (McCay and Acheson 1987:31) is being developed in some Indonesian protected areas—including a number of those where WWF is active—in the framework of national park zoning systems. These can now include "traditional-use zones," in which residents of local communities are permitted to harvest or otherwise use in a sustainable way certain kinds of resources according to regulations agreed on mutually by government and the communities. Community participation is also encouraged in tourism zones within national parks and in the buffer zones outside the park boundaries. Our own work and that of our colleagues and partners in and around the Kayan Mentarang National Park in East Kalimantan exemplifies this approach in that it focuses on the incorporation of community-based forest management into the national park as an integrated conservation and development project (ICDP).

In this chapter, we recount how both local residents and outsiders collect gaharu, currently one of the most valuable forest resources in the area, and explain how traditional regulation has worked—or failed to work—in the context of heavy exploitation by both local and extralocal collectors. We structure our presentation around the following questions: (1) With regard to the Kayan Mentarang area, how and why has gaharu been collected by the local people? (2) What are the economic and political conditions that contribute to the currently unsustainable rate of gaharu collecting? (3) How have local people responded to the pressures of outside collectors and the prospects of overexploitation of gaharu? (4) What measures can and should be taken to create or modify laws, policies, and institutions in order to balance economic fairness and conservation of gaharu?

Research on people–forest interactions in this area began in 1979 with the inception of the U.S.-Indonesian Man and the Biosphere (MAB) field program in East Kalimantan (Kartawinata and Vayda 1984; Mackie and Jessup 1986). Since 1990, the WWF, together with the Indonesian Directorate General for Forest Protection and Nature Conservation (PHPA) and the Indonesian Institute of Science (LIPI), have been carrying out conservation field work in and around the Kayan Mentarang National Park. This has included a substantial body of research in which the three of us have participated.[3] The data for this paper are drawn from research conducted in the Kenyah and Penan Benalui villages of the Bahau River watershed in the Kecamatan (subdistrict) Pujungan, and from brief visits to the Lun Daye and Punan villages of the Kerayan Plateau and the Malinau and Tubu' watersheds.

Although the low population densities, remote location, and customary rules of resource use in the vicinity of the Kayan Mentarang National Park should encourage sustainable resource use, we conclude that local residents have failed, in recent years, in their attempts to restrict outside collectors' access to their resources because of a combination of circumstances: first, the very high price of gaharu on international commodity markets, which makes it attractive to commercial traders even in previously marginal areas; second, the increased technological capability of outside commercial collectors to penetrate remote areas to get at the resource; and third, the neglect by government of local community-based forest tenure together with local people's inability to enforce their own claims. Our suggested solutions to the problem of overharvesting include the formulation and enforcement by government authorities of rules for forest management, based on control at the local community level, within agreed-on and demarcated extraction zones surrounding and extending into portions of the national park. We believe that this will contribute to greater incentives to exploit forest resources at economically and ecologically appropriate levels, as well as greater public acceptance of and support for the park itself.

## *AQUILARIA* SPECIES AND GAHARU

*Gaharu* is the common Malay trade-term for the resinous, fragrant heartwood of several *Aquilaria* and *Gonystylus* species in the family Thymeleaceae, although wood from the latter species is considered a poor substitute for the true gaharu of *Aquilaria* species. While some gaharu is collected and used locally in Southeast Asia for medicinal purposes, most of this product is traded to China and the Middle East, where it is used in the manufacture of incense and medicinals (Burkill 1966; Li 1979; Saletore 1975; Watt 1908).

## Biological Characteristics

*Aquilaria* are understory trees, scattered throughout the tropical lowland and hill forests up to 750 m above sea level (asl) (Whitmore 1972:386). In the Kayan Mentarang area, trees have been reported as high as 1,000 m asl. LaFrankie (1994), in a study of a population of *A. malaccensis* in a 50 ha plot at the Pasoh Forest Reserve on the Malay Peninsula, found a density of about 2.5 trees per hectare [>1 cm diameter at breast height (d.b.h.)] distributed evenly over a study area that included wet and dry ground, hill slopes, clay, and sand. "There is no indication of the strong spatial patterning sometimes found in other forest species, . . . the density . . . is a typical figure for trees and shrubs in the lowland forest of Malaysia" (LaFrankie 1994:304). The distribution of growth rates had a median value of 0.22 cm/yr, and the fastest rates were relatively high for forest-grown trees. Also, based on a median intercensus time of 2.81 years, the recruitment rate was 1.13 percent annually and the mortality rate was 1.42 percent annually. All three trees that broke during the study period coppiced.[4] From this study, LaFrankie (1994:306) concludes that the growth, coppicing capacity, recruitment, and mortality of this population of *A. malaccensis* represents roughly the median for tropical trees in Asia. The results of this study correlate well with the opinions of forest product collectors such as the Kenyah and Penan Benalui, who, with LaFrankie, recognize that the trees are evenly distributed across the landscape, which makes them costly in terms of time and energy to find and control, compared to resources more concentrated in time and space. The inability of local inhabitants to effectively control this resource, and thus to protect it from illegal exploitation, is in part a result of this characteristic of *Aquilaria* (Jessup and Peluso 1985).

Aquilaria is unusual because, for reasons that are still uncertain, the normally soft, even-grained, white wood becomes saturated with a brown resin that hardens the heartwood. The fragrant wood is said to be "pathologically diseased" (Burkill 1966:199), and "it is usually, if not always, found where some former injury has been received" (Watt 1908:78). Fungi have also been found associated with the exudation in *A. agallocha* (Jalaluddin 1977:222), so it is commonly thought that gaharu is produced by fungal infection in all *Aquilaria* (Chin 1985:125).

## Uses and Value of Gaharu

In northeast India and Burma, the resinous wood is extracted from *A. agallocha* and is known as *aguru* or *agaru* (heavy, in Sanskrit), *aloeswood* (also in Sanskrit, a lighter, floating form), *sasi* (in Assam), and *akyaw* (in Burma). In Sanskritic times the peoples of the Ganges reportedly used the wood as a perfume or as an "anodyne fumigation for surgical wounds" (Burkill 1966:199). In Cambodia, *kalambak* is extracted from *A. baillonii* and *A. crassna* and is said to be of the highest grade (Burkill 1966:202). *A. malaccensis*, *A. hirta*, and *A. rostrata* are all found on the Malay Peninsula, but most if not all the gaharu extracted comes from *A. malaccensis*. Burkill (1935:200) mentions "the Chinese, who had thrown forward a colony, in the first century, into Annam, there touched the best source of supply, and venturing forward, and finding more in the forests of Malaysia, they became busy traders in it." In Indonesian Borneo, gaharu is obtained from *A. beccariana, A. malaccensis,* and *A. microcarpa*. The most widespread in interior Borneo is, again, *A. malaccensis*; both this species and *A. beccariana* have been identified in Kayan Mentarang and both produce gaharu that is harvested by local collectors.

In the Kayan Mentarang area, both the tree and resinous heartwood are known as *sekau* among the Penan Benalui, Kenyah, and Kayan peoples, as *tah lela* among the Punan Tubu, and as *tera ala* among the Lun Daye. As an occasional trade item in the past, gaharu was used by these groups to obtain trade goods such as salt, tobacco, and cloth. In some cases, trade items were used to buy rice from other villages or downriver traders during harvest failures. Today, small-scale collecting, usually for a duration of less than a week, pays for school supplies, church expenses, and rice shortages as well as traditional trade goods. Large-scale collecting expeditions, sometimes lasting for a month or more, are mounted by local people to pay off debts incurred by purchasing motor boats, engines, radios, watches, and whole wardrobes from traders and part-time expedition sponsors.

Chinese medical journals dating back to the first century A.D. document botanical and pharmacological data of forest products that can be found in Indonesian Borneo (Hu 1990), but evidence for actual trade is circumstantial. After direct trade relationships were established between China and the northeast Bornean states in the fifteenth century, significant amounts of gaharu began to be collected and traded out of Borneo (Peluso 1983a). Beginning in the nineteenth century, the sultans of Borneo established alliances, first with English merchants, then with Dutch administrators and entrepreneurs. Chinese merchants and their families, along with a few wealthy Arab merchants, began to migrate to east Borneo in approximately 1850. Most of the gaharu was then traded through Chinese companies in Singapore. Favored by the Dutch, the Chinese took over trade in the Zulu Archipelago in the last quarter of the nineteenth century (Peluso 1983a:210–212). With Indonesia's independence, direct state control of forest resources was established and forest exploitation increased as transportation technology and infrastructure improved and expanded. Accompanying these changes has been an increase in the penetration and appropriation of local control over resource collection by external political and economic forces (Peluso 1983a:213–214).

Burkill (1966:200) writes, "The names used by the traders in the ports implied no knowledge on their part of the trees whence the wood came; nor as a rule did they have any knowledge. They developed, for trade purposes, a classification by grades, which, among the Chinese, were very minutely distinguished." Quality classes of gaharu in the coastal markets are based on criteria such as intensity of fragrance and resin content, indicated by the degree to which the wood approaches a pure black color. Quality is also determined by the integrity of the heartwood: many holes indicate a lower resin content. Finally, large pieces of heartwood, although covered with dark resin, are also classified in a lower class because of the possibility of their containing little or no resin in the interior. Although grade criteria apply as well in upriver areas, the definitions of classes and prices are less fixed and are subject to negotiation (table 10-1). Upriver traders tend to undervalue gaharu collected by local residents, both because local collectors are indebted to the traders and because without transportation to large markets there are no alternative buyers. Local collectors often frustrate traders by incurring huge debts and then delaying collecting expeditions, by hiding high-quality product in anticipation of a trip to the coast, or by taking so long to pay off the debt that the trader loses profit in the exchange.

Depending on its quality and specific fragrance properties, gaharu has different export markets. The highest qualities (1A, 1B, 2, 3) are exported via Singapore to the Arabian peninsula, mainly Saudi Arabia, Kuwait, and the United Emirates. The central market is Abu Dabi. In the Arab countries, gaharu is burned as incense before Islamic prayers. There are annual fluctuations in prices caused by increased demand just prior to the pilgrimage season to Mecca, usually beginning around March. Indonesian pilgrims buy large amounts of gaharu, which they sell in Saudi Arabia or along the pilgrimage route to finance their trip. Since 1980, before and during the pilgrimage season to Mecca, prices rise between 10 and 20 percent.

A wider quality range (1A, 1B, 2, 3, Tri A, Tri B) is exported for its medicinal properties via Singapore to Hong Kong and China. Tri A and Tri B class gaharu is also traded to Bangladesh, where it is processed to perfume oil. Lower grades of class 4 or 5, known as *kemedangan*, are traded to Taiwan for the production of fumigating candles, *hio*, and joss sticks. Smaller amounts are traded to Japan.

**Table 10-1**
Gaharu Prices (US$) Along a Kayan River Trade Route, 1994

| Class | Location | | | |
|---|---|---|---|---|
| | *Long Pujungan* | *Long Bia* | *Tarakan* | *Pa' Upan* |
| IA | 250–300 | 500 | 600 | 375 |
| IB | 150–175 | 325–425 | 450 | 325 |
| II | 90–100 | 150–300 | 300 | 140 |
| III | 15–50 | 12–15 | 175 | 100 |
| Tri Kacang | 60 | — | — | 140 |
| Tri A | 30 | 125–175 | 100 | 175 |
| Tri B | 17 | 50–62 | 50 | 37 |
| Kemedangan | 1–5 | 1.5–5 | 1–25 | 1.25 |

Long Pujungan is an upriver village, Long Bia is mid river, and Tarakan is an island near the coast. Pa' Upan is far inland and accessible from Tarakan only by small aircraft or a long overland journey. Data are from interviews with traders.

## COLLECTION AND TRADE OF GAHARU BY KAYAN MENTARANG INHABITANTS

The professional gaharu collectors operating in remote places such as the Apo Kayan have little or no interest in conserving the sources of gaharu there. Rather, they go into the forest for several months at a time and attempt to get as much gaharu as possible as quickly as possible (before the price falls or their supplies run out), and so they cut down every tree they come across that might contain gaharu.

In contrast, local collectors—when they are not under the imminent threat of heavy competition by outsiders—tend to take a more relaxed and restrained approach. Such a conservative approach to collecting on the part of local "owners" of the forest (as they see themselves) makes sense as long as there is little fear that outside collectors will come along and cut down the trees anyway, whether or not they are likely to contain gaharu. Kayan Mentarang residents undertake short-term collecting trips, lasting less than a week, at any time during the year, but usually they go between periods of peak agricultural activities such as when newly cleared swiddens are drying or just after planting before weeding begins. Men and (less often) women, even schoolchildren, also collect for short periods before holidays and festivals or before traveling, to earn money to buy provisions, pay their school fees, and so on. Longer expeditions, of more than a week, are made to more remote areas, where the risks are greater but so also are the potential fortunes to be made. Collecting parties number from a single adult to groups of six or seven adults and young people; young boys often accompany their fathers. While women are not prohibited from collecting, they usually do not participate in extended expeditions. Women are often present, however, on hunting expeditions or rattan-collecting expeditions, when *Aquilaria* trees are encountered, checked, and gaharu extracted.

### *Traditional Knowledge, Beliefs, and Practices of Collectors*

Gaharu-collecting expeditions by Kayan Mentarang residents are often very casual affairs, with collectors spending as much time hunting for pigs, fruit, and other forest products as in finding and extracting gaharu. Thus, the style of searching is fast paced and incomplete, but it certainly is not random. Collectors usually go to areas where they or others they know have previously encountered gaharu, and they search the forest along trails and ridge tops, where they can see relatively far through the forest. Experienced collectors with acute eyesight can identify *Aquilaria* trees from far away and thus do not need to walk through the whole forest. They also use botanical and ecological clues to locate areas where *Aquilaria* trees occur or are likely to contain gaharu.

Even though this "search strategy" makes use of expert local knowledge and skills, it is by no means perfect and many trees are missed, some within a few feet of well-traveled trails. The low-intensity searching and collecting without exhausting the potential resources means that even in areas where a lot of gaharu has been found there is always the possibility of finding more.[5] Furthermore, not all trees encountered by local collectors are cut down. Experienced collectors know what signs to look for and, if these are not apparent, they will leave trees behind for another day, by which time (it is hoped) they will have developed more of the valuable resin.

Among the Penan and Kenyah of the Kayan Mentarang area, there is a traditional belief in malicious ghosts and forest-dwelling spirits that like to trick hunters and collectors. Consequently, many precautions are taken before and during an expedition, such as never announcing aloud the purpose and route of the trip. It is feared that if the spirits learn of the plan, they will put dangerous obstacles in the collectors' path or hide the desired forest products. One way of counteracting the trickery of malevolent spirits is by means of sympathetic magic (i.e., "like attracts like") whereby black items, such as clothes and packs, are brought along to help find the blackest of gaharu resin, which is usually the most valuable. No ceremony is involved here; just the wearing of black clothes is said to help draw the collectors to the resin-loaded trees. Also, any item of black color found before or during a trip is taken as a sign of imminent fortune. With so much of the success of an expedition in the hands of spirits, magic, and fate, it is not surprising that many Kenyah and Penan metaphorically describe their collecting efforts as *cari nasib* (in Indonesian, to seek good fortune). The Punan in the Malinau and Tubu watersheds sometimes hold a ceremony before an expedition starts. One Punan shaman in the upper Malinau is still performing such ceremonies, reportedly serving as a medium between local collecting parties and the forest spirits. Offerings of eggs, fruits, and rice are presented to entreat the more benevolent spirits in the forest (De Beer 1993).

In addition to relying on magic, the Penan use several ecological indicators to locate areas where *Aquilaria* trees are likely to contain gaharu. Many of these indicator species are palms, such as *Arenga brevipes* and *Oncospermae horridum*, which grow on the banks of steep ravines in the mountainous headwaters area. Although many trees may be missed by using such techniques, this is not perceived as a problem by local collectors, who, on returning to a previously searched area, take delight in the accidental (they say possibly spirit mediated) encounter of trees that were missed, or hidden, earlier.

When a tree thought to contain gaharu is identified, collectors examine it for signs of rot at dead branch stubs or elsewhere, as this is believed to indicate the presence of fragrant heartwood. The collector will then cut into the bole with a machete or ax to check its content. Usually no gaharu is found and so the collector moves on. At some later time, the collector may return to reexamine the tree, since the wounding of the tree is believed to help stimulate the production of gaharu in the future. (Although we have no data on how these cuts affect mortality among the trees, the fact that previously examined trees can be found and checked again months or years later suggests the cuts are not necessarily serious wounds.)[6] Only if gaharu is believed to be distributed throughout the trunk and roots is the whole tree felled, although sometimes the only way to check upper portions of the trunk is to cut the whole tree. (Saplings are also checked, but they are rarely felled.) The softer, pale-colored wood surrounding the darker, resinous gaharu is then chipped away to extract the valuable product. Back at the collector's village, the wood is cleaned by carefully shaving and carving away the pale, less odorous or nonodorous parts. High-quality gaharu is stored in tins to prevent desiccation until the product is sold locally to an itinerant trader or taken by the collector to one of the coastal markets.

Traditional low-intensity searching and harvesting practices such as those we have described may well have contributed to sustainability of gaharu collecting in the past, before the widespread advent of the intensive, mechanized methods now employed by some traders. The traditional practices seem to have reinforced the effects of incentives both to economize on effort (and so to fell only those trees unlikely to yield much more gaharu in the future) and to leave a standing stock of *Aquilaria* trees for future collecting trips (which depends on some surety of tenure, or at least an absence of outside interference). Given the remote location and rugged topography of the Kayan Mentarang area, the small number of people inhabiting the area, and the ecological characteristics of *Aquilaria* (low density, even distribution, and coppicing ability), even a moderate level of restraint on the part of collectors and a degree of inefficiency in their searching was probably enough to ensure viable, continuously reproducing populations of *A. malaccensis* and *A. beccariana* throughout the area. Nowadays, however, in the face of mounting economic pressures and encroaching commercial interests from outside the area, more formal means of regulation have become necessary.

### The Gaharu Trade in East Kalimantan

Trade networks for gaharu and other commercial nontimber forest products in the East Kalimantan interior consist of the collectors, middlemen (e.g., police, military, local government staff, schoolteachers, or village shopkeepers), and traders. Middlemen are sometimes sponsors of small-scale expeditions and may even be local village chiefs who request that tributes or debts be paid in gaharu. In the past, local collectors rarely had the means to transport their products to the traders, so middlemen were an important link in the forest products trade. Today, that situation is changing rapidly, as will be described in more detail. Traders market large amounts of forest products in the major trading centers of Indonesia and Southeast Asia, such as Surabaya, Jakarta, and Singapore. They almost always control the local means of transportation, such as river taxis and shipping companies, and they have investment capital in the form of expedition provisions and equipment, manufactured goods for trade, and, in some cases, a labor force of professional collectors. Most traders establish and maintain alliances with influential individuals in the prevailing governing bodies at all levels, from village chiefs to provincial officials. The oldest, largest, and most profitable trader in the Kayan Mentarang area is an ethnic Chinese family that has plied the Kayan and Bahau rivers since early in the century.[7] Their trade network is organized in the form of *kongsi*, a Chinese term for a firm or partnership (see Peluso 1983a:151). Today, they transport people and products from the mouth of the

Kayan River to the upper reaches of the Bahau and have shops and trading partners all along the route. They regularly sponsor expeditions to collect rattan, gaharu, fruit, and timber, and they buy and sell rice, vegetables, and meat from domestic and wild animals. Their 1992 trade volume in gaharu was worth roughly US$500,000.

For a local or professional collector, the return may vary from US$5 to US$1,000, and when high costs or past debts are included, the typical expedition rarely breaks even. Still, in the Bahau and Tubu watersheds, the sale of gaharu now accounts for 50–70 percent of annual household income.[8] Even salaried government employees, such as school teachers, earn a significant surplus from buying and selling gaharu, serving as middlemen or even collectors.

## THE ECONOMIC AND POLITICAL CONDITIONS OF OVEREXPLOITATION

Gaharu collecting has become the most significant source of income to local and professional forest product collectors in Kayan Mentarang and elsewhere, as well as a significant threat to local control of lands and resources, not to mention the stability of the *Aquilaria* populations. Both professional and amateur collectors, many of whom do not know and use traditional means for detecting gaharu, have been scouring the forests of the Kayan Mentarang, felling nearly every adult *Aquilaria* encountered (and even saplings), regardless of its gaharu content. Faced with such competition, local collectors have increased the frequency and duration of their own collecting trips to have a chance to get their share of gaharu before its depletion. Among these local collectors now are teenagers and young men with less experience in the traditional collecting methods, who therefore have adopted many of the more destructive practices of the outside professional collectors. Inexperienced collectors, whether local or not, are also more prone to mistaking non-*Aquilaria* species for gaharu and cutting them, too. Given the stakes, it is not surprising that the recent "gaharu rush" in the Kayan Mentarang has all the feverish intensity, destructiveness, secrecy, and potential for violence of any gold rush anywhere.

In the years prior to 1975, the price of gaharu was such that it was considered just another forest product. Since that time, and for still unknown reasons, the price has continued to rise (Chin 1985:124) (table 10-2). Since 1991, there has been a tremendous increase in gaharu collecting, most likely because of a combination of increased demand for gaharu in the Middle East and China, resulting in higher prices in East Kalimantan (figure 10-4), and drastically decreasing prices for rattan since the export ban imposed in 1987 (figure 10-5).

The increasing number of collectors in the Kayan Mentarang is partially due to political and economic changes in neighboring Sarawak and Sabah. Since the 1990s, the Malaysian government has been refusing foreign workers, and with shrinking forests and closing concessions, job opportunities for migrant workers are now much rarer. Young men, and sometimes whole families, from the Kayan Mentarang would

**Table 10-2**

Gaharu Exports from Indonesia, 1989–1993

| *Gaharu resin* | 1989 | | 1990 | | 1991 | | 1992 | | 1993 | |
|---|---|---|---|---|---|---|---|---|---|---|
| | *kg* | *US$* | *kg* | *US$* | *kg* | *US$* | *kg* | *US$* | *kg* | *US$* |
| Singapore | 38,067 | 697,340 | 37,095 | 1,516,877 | 25,976 | 256,889 | 106,206 | 1,078,146 | 150,275 | 1,678,047 |
| Taiwan | — | — | — | — | 16,390 | 156,675 | 74,688 | 863,194 | 51,999 | 357,444 |
| Saudi Arabia | 913 | 45,900 | 6,896 | 364,622 | 734 | 28,121 | 20 | 3,565 | 1,315 | 24,862 |
| Japan | 20 | 2,000 | 359 | — | 4,000 | 16,000 | — | — | — | — |
| Korea | — | — | — | — | — | — | — | — | — | — |
| Belgium | — | — | — | — | — | — | — | — | — | — |
| Kuwait | 50 | 2,604 | — | — | — | — | — | — | — | — |
| Total | 39,050 | 747,844 | 44,350 | 1,881,894 | 47,100 | 457,685 | 180,914 | 1,944,905 | 203,591 | 2,060,353 |

| *Gaharu wood* | *kg* | *US$* | *kg* | *US$* | *kg* | *US$* | *kg* | *US$* | *kg* | *US$* |
|---|---|---|---|---|---|---|---|---|---|---|
| Singapore | 11,169 | 108,930 | 6,664 | 74,789 | 78,939 | 826,879 | — | — | 13,308 | 475,600 |
| Taiwan | — | — | — | — | — | — | 82,178 | 112,073 | 9,694 | 25,917 |
| Saudi Arabia | 150 | 26,240 | — | — | — | — | — | — | 137 | 24,862 |
| Japan | — | — | — | — | — | — | 640 | 2,784 | — | — |
| Korea | — | — | — | — | — | — | 14,342 | 30,761 | — | — |
| Belgium | — | — | — | — | — | — | 380 | 1,140 | — | — |
| Maylasia | — | — | — | — | 10,000 | 2,750 | — | — | — | — |
| Total | 11,319 | 135,170 | 6,664 | 74,789 | 88,939 | 829,629 | 97,540 | 1,155,408 | 23,591 | 503,225 |

Indonesian Foreign Trade Statistics, Exports, vol. 1, 1989–1993, Bureau of Trade and Service Statistics.

US$1 = Rp2,000 (Rp = rupiah).

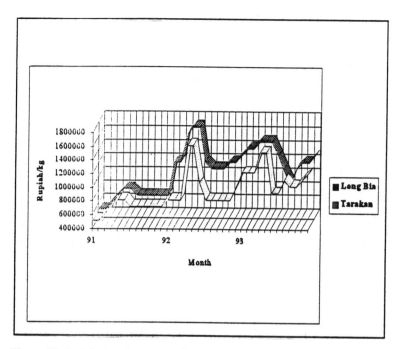

**Figure 10-4** Gaharu Prices for the Highest Grades in the Kayan Mentarang, 1991–1993. (Data from interviews by Momberg.)
US$1 = Rp2,000.

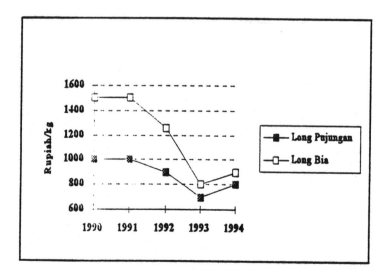

**Figure 10-5** Rattan Prices in the Kayan Mentarang Area, 1990–1994. (Data from interviews.)
US$1 = Rp2,000.

journey to these areas to work from one to five years in such enterprises as shipping, quarrying, and logging. Now, most of these young people are staying home and seeking their fortunes in gaharu collecting. Among this population of young gaharu collectors are relatives of local residents, and the children of former residents, who have migrated to the coast or other areas of Kalimantan. Being members of the same ethnic group, and having relatives in the villages, gives these collectors special privileges not enjoyed by other outsiders, such as free access to lands and resources and the ability to borrow tools and provisions as needed. This "insider" advantage has its costs as well, since the collectors are obligated to participate in village activities, such as swidden agriculture or house building.

Many of these young men and other local residents have joined professional collecting expeditions. Local Penan Benalui serve as guides for those collectors coming to the Lurah and Bahau river valleys. Often, traders use relatives of local residents to gain permission to collect on otherwise restricted lands and pay little, if any, user fee. Most of the professional collectors who come to the Kayan Mentarang are Kenyah, Punan, or Lun Daye people with at least ethnic, if not actual kinship, ties to local residents. These trips are, however, organized and provisioned by ethnic Chinese and Bugis traders from downriver towns and villages such as Long Bia and Malinau. Some of the outsiders have long experience collecting gaharu. For example, men from the predominantly Kenyah village of Long Bia, on the lower Kayan river, have been full-time collectors since the 1970s, in contrast to their upriver kinsmen who devote much of their time to rice swidden cultivation and other subsistence activities. Some traders have also secured contracts with families or whole villages, whereby the trader becomes the sole buyer of all gaharu collected by the residents on their own lands and, in return, he provisions their expeditions and advances them trade goods such as clothes, kitchenware, radios, and watches. As families and whole villages slide into debt, collecting supersedes other activities and extraction methods become more intensive and destructive.

After 1991, nonresident collectors outnumbered local collectors in the Pujungan subdistrict; by 1993 as many as 200 nonresident collectors were reportedly working in the Bahau watershed. Since 1990, four companies, based in Samarinda and Pontianak, have organized gaharu expeditions with nonresident collectors to the Apo Kayan (subdistricts Kayan Hilir and Kayan Hulu). These groups have entered the area covertly from the upper Mahakam River or by private aircraft to small landing strips. Despite official notifications and warnings prohibiting gaharu collection by nonresident collectors in the subdistricts along the international border, in 1994 the four urban-based companies continued to move nonresident collectors into the Apo Kayan (Foead and Lawei 1994; *Manuntung* 1994a). Starting in January 1994, one of these companies (CV Sumber Daya Alam) cleared helicopter landing pads at river tributaries of the Kayan River.[9] About 500 nonresident collectors, mainly Javanese transmigrants from south Kalimantan, were flown to these helicopter pads between January and June (*Manuntung* 1994f).[10] Even after CV Sumber Daya Alam lost two rented Sikorsky helicopters, one with sixteen collectors on board, when they crashed in the remote mountains east of Kayan Mentarang in May and June (*Manun-*

*tung* 1994b, 1994c, 1994d, 1994g), the company continued to operate with 100 non-resident collectors (*Manuntung* 1994e). The three other companies operated from the villages of Long Nawang and Long Ampung in the Apo Kayan with at least sixty nonresident collectors. Their expeditions were to the border region and the ridges between the Mahakam tributaries (e.g., the Boh River) and the Apo Kayan, and thus not into the park.

## LOCAL RESPONSES TO OVEREXPLOITATION

The most apparent response of people living in the Kayan Mentarang area to the "gaharu rush" has been to collect more intensively on their own lands, as just described. Greater involvement in the gaharu economy has had both positive and negative consequences for local inhabitants. On the one hand, they have indeed grown richer because of their participation in the collecting boom. On the other hand, as the rate of exploitation increases, local collectors are contributing to the exhaustion of their forest resource base, and indeed gaharu in some areas of the Bahau, Lurah, and Malinau watersheds was already said to be exhausted by 1994. Local leaders also responded to overexploitation by trying either to exclude outside collectors or else to regulate their access. Not only did they fear the loss of their forest products, but also they feared the appropriation of their traditional control over forest lands by nonlocal elites, such as traders and government officials (e.g., Dove 1993b). Their actions have included issuing proclamations closing lands, catching and reporting trespassers, as well as charging high user fees. In some instances, they have been successful in stemming the influx or exacting a high tariff, but in most cases collectors and traders manage to circumvent the regulations, usually with the help of government officials, or by just ignoring local protests and forging ahead.

In the early 1990s, when the first large groups of professional collectors arrived in the Pujungan subdistrict, village leaders were somewhat reluctant to challenge the outsiders. When confronted, the most common arguments heard from collectors were that they had permission from the subdistrict head (*camat*) or that Indonesia's lands were now held by the national government for everyone's use and so local villagers had no legal basis to restrict access. Many a confrontation ended with the collectors admitting that they had only a letter of introduction from the subdistrict officer, which carried no legal force, but warning that the subdistrict officer might be disappointed if they were not able to continue with their expedition. The most timid of village leaders would invariably let the collectors go, the more daring would demand user fees or a share of the harvest, and the most defiant would threaten the collectors with tales of headhunting, ghosts, or misfiring shotguns.

As the number of intruders increased, the village leaders became more daring. In Long Pujungan, the restricted forest committee (*panitia tana' ulen*) became the village's predominant means of stemming the tide of outside collectors on local lands. In 1992, members of the committee were able to detain outside collectors from Lumbis in their customary restricted forest. All gaharu (about 5 kg) was confiscated and the sponsoring trader, from the downriver village of Long Peso', was fined Rp250,000 (roughly US$125). In May 1994, residents of Apau Ping, the most upstream village on the Bahau river, raided the camp of outside collectors deep in the forests of their village territory. While they reportedly destroyed all the collectors' equipment, the villagers were unable to detain the intruders and bring them downstream to the village.

Not all villages have been so determined to deter outsiders. In some cases, leaders have decided that enough gaharu exists in their forest to allow extraction by outsiders, if they are willing to pay high user fees. Some villages have such huge territories and such small populations that village leaders see no threat to their long-term prospects for harvesting gaharu, and are willing to let outside collectors exploit the outer reaches of their territories while they can get such exorbitant user fees from them. In this situation, collectors try to sneak by village watches onto territories, thereby avoiding fees, which angers villagers, as described previously. Some corruption among village leaders has also been reported, with collected user fees being underreported to the community. Commoners have become incensed over the extortion by leaders of money collected for use of what they perceive to be their "community forests." Leaders, such as the Kenyah aristocrats, have been suffering economically since losing traditional clients such as the Penan and losing traditional authority and rights of appropriation over commoners in their own communities. Thus it is hardly surprising that they would take advantage of the opportunity to exploit outsiders. In Long Alango, a 1993 village assembly banned outsiders, but the aristocrat gave permission to former village residents to enter the restricted forest for an entrance fee of Rp50,000 (roughly US$25) for each group. By 1994, the pressure from common village people to exclude all outsiders from resource use was finally accepted by the aristocrats and an assembly of representatives from nine villages agreed to exclude all outsiders from forest product collecting on their lands. Unfortunately, determined collectors still manage to cross into village lands from neighboring river valleys.

In remote parts of the Iwan River valley and the Apo Kayan where there are no villages, encroachment by illegal collectors has been monitored by staff of the Kayan Mentarang Conservation Project (Foead and Lawei 1994). The integrity of core areas of the park is threatened by the opening of the forest for helicopter pads and shelters (figure 10-6). Village leaders are also concerned that these collectors will gain access to remote village lands via these core areas of the park, thereby avoiding the notice of residents and any taxes or user fees that might be imposed to regulate the extraction of forest resources.

## CONCLUSIONS AND RECOMMENDATIONS

With limited market options because of the remoteness of the Kayan Mentarang, local people will remain critically dependent on nontimber forest products. Many

**Figure 10-6** Illegal Helicopter Expedition of Urban-Financed Gaharu Collectors (from Cities), Poised for Take-off and Bound for Kayan Mentarang National Park.

forest products have gone through times of intensive exploitation and are still not depleted, such as gutta percha (from *Palaquium* species) and damar (from several species of Dipterocarpaceae) in the 1960s, rattan (predominantly *Calamus caesius*) in the late 1970s and early 1980s, and gaharu in the late 1980s and early 1990s. Because of the boom-and-bust character of nontimber forest product markets, it is more important to protect the forest as a whole, and local peoples' access to it, than to focus on one forest product only. Nevertheless, an understanding of the biology of particular forest species and their responses to harvesting is necessary for their sustainable management. Some of our recommendations focus on the necessity for further investigations of the biology and silviculture of *Aquilaria,* together with collecting practices and tenure arrangements, while others (already being implemented by WWF and its partners) deal with legal mechanisms of forest protection in the context of a national park.

## Further Studies of Gaharu

Our investigations of traditional gaharu-collecting practices indicate that a wealth of local knowledge exists, and that some of this knowledge can become the basis for further investigations in forest ecology, the botany of *Aquilaria*, and the formation of gaharu. The skills and practices used by these local collectors combined with scientific investigations of the knowledge implicit in them offer great promise for the development of management techniques that can enhance both the long-term survival of the tree and the production and economic value of the product. Unfortunately, in the face of threats to local control of resources from outsiders, traditional practices that on the surface appear to be sustainable, such as fast-paced, hit-or-miss search strategies, have been abandoned in favor of more intensive and destructive practices. It is our view, therefore, that developments in the ecology, silviculture, and conservation of *Aquilaria* must be accompanied by greater security of local forest tenure and regulation of forest resources. For this purpose, the study of past and present resource use and tenure, their biological consequences, and their manipulation by people in times of crisis offer important starting points for conservation and sustainable management of the resources.[11]

While collectors and scientists (LaFrankie 1994) agree that *Aquilaria* trees are evenly distributed across the landscape, local collectors insist that gaharu, the product, is clumped. A tree containing gaharu is likely to be surrounded by, and is probably related to, other trees containing gaharu. The possible explanatory factors for this are numerous, including the genetic relationship and physiology of the trees, the local climatic and edaphic environment, and the cause(s) of the production of gaharu in the first place. Until we understand this last factor, we will not be able to explain what causes the observed distribution of gaharu. A greater understanding of what might cause clumping could be used to increase the density of gaharu-bearing trees closer to local communities. This would certainly help local collectors to increase their ability to control and protect the gaharu portion of their forest product economy, since exclusion is more easily accomplished for a concentrated resource (e.g., birds' nests in caves) than for a dispersed one such as gaharu (Jessup and Peluso 1985:516). In fact, the ability to create forest gardens that include *Aquilaria* and gaharu might be a critical factor in considerations of tree and land tenure innovations.

## Extractive Zones In and Around the Park

We have seen principles of customary forest tenure systems revitalized by some Kayan Mentarang communities, transformed and adapted in others, and corrupted for personal gain in a few. The Penan Benalui have even borrowed or invented forest tenure rules based on Kenyah models, both in response to the threat of outside collectors and because of the opportunities to collect user fees. The longer-term consequences of these tenurial experiments for forest conservation and the maintenance of local control are still unknown, but it will be important for residents and park staff to monitor them and consider their implications when formulating a long-term plan for management of the park and its surroundings.

The WWF supported the change in status of the conservation area from a strict nature reserve to a national park, because the park designation permits legal recognition of some extractive uses of resources such as gaharu in so-called traditional-use zones inside the park (as well as other land uses in buffer zones outside the park). Conservationists were encouraged in the belief that this change in status and zoning would help mitigate overexploitation of forest resources by outsiders and build support by local communities for conservation of core protected zones because of WWF's experience in the Arfak Nature Reserve in Irian Jaya. There, instead of designating the entire reserve as a restricted no-trespass area, conservation officials agreed to the creation of traditional-use zones around the periphery of the reserve, which contributed to strong community support for the protected area and enlisted local landowners' participation in helping to exclude nonresident collectors of forest products (Craven 1990).

Traditional-use zones and buffer zones in and around Kayan Mentarang National Park will be designed through a process of participatory community resource mapping involving local residents and government officials and facilitated by WWF project staff (Momberg et al. 1996; Sirait et al. 1994). This approach has been shown to provide a structured process for learning with and from communities about patterns of people–forest interactions in space and time, as well as social, economic, and human–ecological characteristics of communities living at the fringe or within conservation areas (e.g., Momberg 1993). The maps, overlaid and integrated with other information in a Geographic Information System (GIS), clearly show the local residents' own designations of forest ownership and the distribution of important economic and cultural resources. Recent experience of WWF staff in Kayan Mentarang and other Indonesian protected areas show that the participatory mapping approach is not only enthusiastically supported by local communities but also can gain support from local governments eager to solve or avoid land use conflicts.

Both resistance to encroachment by outside collectors and WWF's mapping program have influenced the revival, reinterpretation, and adaptation of restricted forest (the tana' ulen) in Kenyah communities and its invention among the Penan. Acknowledgment of customary claims to and uses of forest in and around the national park was indicated by the strong support given to such claims by the minister of forestry during a visit to the area in 1994; by a decree issued by the governor of East Kalimantan in 1995 giving exclusive gaharu extraction rights to local residents; and by the 1996 ministerial proclamation establishing the park, which specifically refers to the interests of local communities. Government edicts such as these have reinforced local residents' access rights and pressed local government officials to enforce stricter control over encroachments by outside collectors. Implementation of a management plan for the national park (now being developed by WWF in collaboration with government and local communities) should consolidate these gains and institutionalize them through zoning and management systems that link economic incentives for conservation with effective enforcement and biological monitoring. Without strong institutions tying together the interests and values of conservation at local, national, and international scales, the economic logic of overexploitation will triumph again in the future.

## NOTES

1. Anderson (1987:341) argues that the proper role of government is "to act as a disputing ground or judicial authority" rather than as a resource-controlling agency.

2. The last great *paren* leader in the upper Bahau, father of the present so-called Kepala

Adat Besar (Great Customary Leader), was imprisoned by the Dutch in the 1930s for beheading some Bugis gutta percha collectors who strayed into his territory and refused to pay tribute.

3. Some research results from the Kayan Mentarang Project have been presented at the Second and Third Biennial Conferences of the Borneo Research Council in Kota Kinabalu, Malaysia (1992) and Pontianak, Indonesia (1994). A volume of some twenty papers, tentatively entitled "Culture and Conservation in East Kalimantan," is in preparation.

4. *A. malaccensis* in Kayan Mentarang was observed by one of us (Puri) to coppice from stumps as well as fallen trees.

5. For instance, during one expedition, four of the twelve *Aquilaria* trees encountered in one day showed no evidence of ever having been cut and examined for gaharu: apparently they had been missed by previous collectors, being hidden behind a grove of sago palms (*Eugeissona utilis*). One of the trees was felled and found to contain a small quantity of gaharu, which was considered a lucky find in an area that had already been intensively searched.

6. On the collecting trip previously referred to, the collector said the eight trees showing signs of old cuts had been checked by him five to ten years before. Some low-quality gaharu was found in the healed cuts, although our informants said that, given time, a higher quality would develop. Only two of the eight trees were felled.

7. The father of the present boss of the family enterprise emigrated from China to what was then known as Southeast Borneo in the Dutch East Indies. His son is well established, economically, socially, and politically, in the local Dayak (and the wider Indonesian) community. Such descendents of Chinese immigrants to Southeast Asia, many of whom still speak dialects of Chinese at home and maintain other aspects of Chinese culture, are known in the ethnographic literature as overseas Chinese or ethnic Chinese.

8. Household incomes of Kenyah villagers vary from US$250 to US$2,500 per year.

9. Only the helipad at the Iwan River is well inside the park boundaries, while the helipads at the Kat River, Laham River, and the Abung rapids of the Kayan River are outside the park (Foead and Lawei 1994). Nevertheless, most collecting expeditions head for the mountains between the Pujungan and the Kayan, inside the park.

10. They have to collect gaharu for at least two months, as the helicopter return flight already costs US$325 per collector, and all supplies have to be flown in and are sold at exorbitant prices, whereas the price for top grade gaharu reaches only US$325 per kilogram.

11. See Vayda (1996) for methodological approaches to these questions.

# CHAPTER 11

## The Meaning of the Manatee: An Examination of Community-Based Ecotourism Discourse and Practice in Gales Point, Belize

*Jill M. Belsky*

> The whole community ecotourism thing is very funny. It's as if Gales Point has become this make belief [make-believe] thing or image. But it's a shaky image, like a house of cards that if you blow will fall right over.
> *Dolores Godfrey, former executive director, Belize Audubon Society*

Community-based conservation (CBC) in general, and ecotourism in particular, arose to correct human injustices and social impacts wrought by a prior model of protected area management that subordinated resident peoples' welfare and rights, and local economic development, to environmental preservation (West and Brechin 1991). Proponents of community-based ecotourism defend it as a locally beneficial way to use rural landscapes and cultures, especially relative to timber or mineral extraction, that contributes to both local economic development and the conservation of threatened habitats and species (Boo 1990; Whelen 1991; Lindberg and Hawkins 1993; Horwich et al. 1993; Western and Wright 1994).

However, despite good intentions and high hopes, community-based ecotourism is not necessarily benign, nor is it always benevolent. In the case of the Gales Point Manatee Community Conservation project in central Belize, community-based ecotourism is an invention of and intervention by conservationists and their allies in government ministries and nongovernmental agencies, and rural elites. Rather than resolving contradictions between environmental protection and use, community-based ecotourism has reinforced historic political struggles within the community and with imperial national and global forces, and it has intensified human injustice in the process. I offer the case of Gales Point as a cultural and political critique of community-based ecotourism, and as an example of the need for greater attention to political struggles and outcomes.

## POLICY AND THEORETICAL CONTEXT

The thin slip of a peninsula nestled between the Caribbean Sea and the Southern Lagoon situated in central Belize has been associated with the manatee[1] since the place was first settled and named (figures 11-1, 11-2). As people in the village tell it, Gales Point was named by a Belizean man, Mr. Gale, who took refuge there during a particularly nasty storm. While riding out the storm, he witnessed a large number of manatees swimming offshore, so he called the place Gales Point Manatee after himself and these large aquatic "sea cows." "We owe a lot to the manatee," explained an elderly Gales Point resident. "The manatee give us plenty to eat, and something to focus the mind on when we sit and think and look out at the lagoon." The greatest concentration of the endangered West Indian manatee (*Trichechusmanatus*) in Central America exists in the Southern Lagoon surrounding Gales Point. They have become the focus of a major international conservation program. As a result, the importance of the manatee as a source of local sustenance—both material and cultural—has taken on new and often contradictory and contentious dimensions.

Ecotourism (otherwise described as responsible, alternative, caring and green tourism) has brought a new cachet to Belize, which was virtually unknown two decades ago as a British colonial backwater, and to the international traveler for whom the experience has become a key commodity and cultural good (Patullo 1996). Most of the literature on ecotourism in general and Belize in particular focuses on its value as an economic commodity and conservation approach, emphasizing the material constraints and contradictions yet to be worked out to implement it. Observers sug-

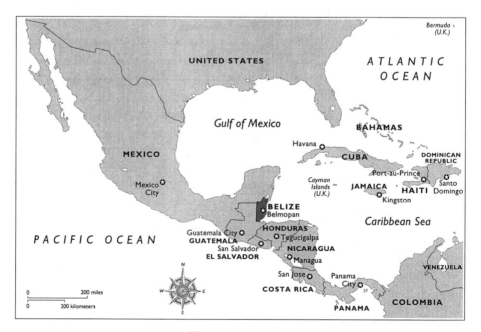

**Figure 11-1** Belize

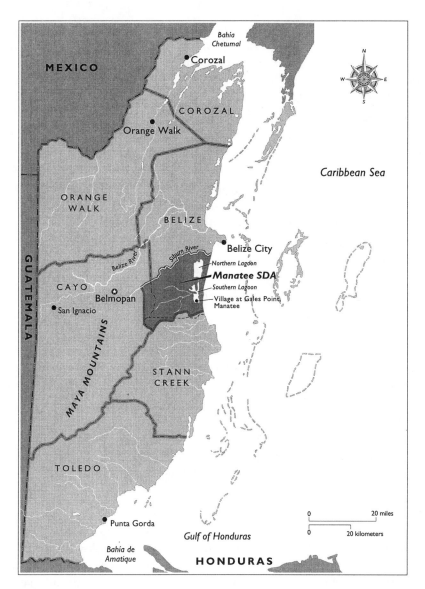

**Figure 11-2** Manatee Special Development Area and Gales Point, Belize.

gest that local peoples and physical environments continue to bear the cost of eco-tourism while benefits accrue to affluent national and foreign entrepreneurs (Boo 1990; Woods et al. 1992; Lindberg and Enriquez 1994; Place 1995). Yet for eco-tourism to act as a catalyst for conservation and development, local communities must benefit from the influx of tourists and participate in nature-based tourism. This is nec-essary to correct past human injustices and social impacts associated with centralized models of protected area management and tourism, which paid little attention to res-ident peoples' welfare and rights, and local economic development (Wells and Bran-don 1992; Wells 1994). Building on the community-based conservation movement, it purports to privilege the concerns of rural communities, indigenous culture, and nature in the business of ecotourism (Western and Wright 1994).

However, it is unlikely that the rise of community-based ecotourism rectifies the injustices associated with "old tourism." First, there have been too few empirical case studies to judge whether the model provides the material benefits to local communities and environments that it claims to make.

Second, "new" ecotourism appears to continue—indeed accentuate—the subordination and dependency of Third World "peripheral" peoples and places, not only in a material political-economic sense but culturally as well (Munt and Higinio 1993; Patullo 1996). By this I mean the way local communities—and their cultural and physical environments—are socially constructed and offered up as international destinations and experiences for affluent tourists from "metropolitan" advanced capitalist economies. If we accept that it is primarily experiences and symbols, or cultural goods (Bourdieu 1984), that are being created and consumed, we need to ask whose visions construct these cultural goods for whose benefit, and at whose expense? Given this approach, a new window of cultural and political critique can be opened into how we conceive of and examine community-based ecotourism, and the justice or injustices that result from it.

There is much to suggest that "new tourism" (as ecotourism and community-based ecotourism themselves are constructed in Belize) represents a form of "new colonialism." As a new cultural form of commodity exchange, the Belize tourism industry (itself composed of many North American expatriates and "eco-lodge" owners) restructured itself to meet the desires of international tourists, who were once satisfied to flock to the beaches and coral-rimmed coasts but now want to venture inland for a nature- and culture-based experience. In response, the Belize tourist industry has demassified, repackaged, and relabeled its holiday product to cater to these presumed desires.[2]

> There has been a market shift from the traditional mass packaged holidays, typically described as the "sun, sea, sand and sex." More flexibly packaged—individually oriented—tourisms are now of increasing significance, catering for a more "authentic" experience and characteristically, environmentally and culturally sensitive.
>
> Munt and Higinio (1993:61)

While ecotourism literature labels itself as supporting "sustainable development," "nature," and "culture," analysts rarely confront the ambiguous meanings of the words used to describe and market the experiences offered. Most critically, they do not explicitly recognize the political and economic ways in which words and images take on meanings (DuPuis and Vandergeest 1996). Their meanings are more than just points of view: they have consequences as people act on their understanding of key concepts such as "rural," "nature," and "wilderness," and orient them to meet their own self-interest.

> Stories about nature, the human community and history . . . are crucial because they shape our ideas of what were, are, and should be human relationships with the natural world. Moreover, when stories about nature and the human community are linked to power and funding, they have important ethical, political and legal consequences.
>
> Zerner (1994a:69)

With regard to community-based ecotourism, it is helpful to consider how others have analyzed the words often used to describe and promote these projects. A separation—of rural from urban, managed from wild, and human activity from natural processes—remains the basis for most development projects and suggests an important direction of inquiry for critically examining community-based ecotourism. "Biodiversity" (habitat, flora, fauna, ecological process, and genetic resources) is split from the sphere of human practices and becomes the privileged subject matter, while human groups and their rights are devalued or peripheralized (Zerner 1996). The construction of boundaries between these categories creates contradictions with the daily activities of rural people whose everyday lives do not adhere to these separations or whose understanding of these categories may differ from that of the people with power. When land use and resource managers conceptualize and plan biodiversity conservation programs on the bases of their abstract categories and imaginings, they often need to resort to coercion or violence to implement regulations based on these views (Peluso 1992a; DuPuis and Vandergeest 1996). These have entailed attempts to change dietary habits, restrict resource use to designated "zones," or prohibit people from using their lands, forests, and reefs completely (Zerner 1996).

Given that a "wilderness" experience has become a big seller in community-based ecotourism, how is it conceived? Many have written that the tendency is to approach the "wilderness experience" as pitting the human against the nonhuman, and to assume an almost religious meaning and conviction that can justify harsh consequences for local peoples.

> However much one may be attracted to such a vision, it entails problematic consequences. For one, it makes wilderness the locus for an epic struggle between malign civilization and benign nature, compared with which all other social, political, and moral concerns are trivial. . . . If we set too high a stock on wilderness, too many other corners of the earth become less than natural and too many other people become less than human, thereby giving us permission not to care much about their suffering or their fate.
>
> Cronon (1995:84)

A tendency to idealize wilderness and its inhabitants (when their physical existence is acknowledged) and to care little about social and political impacts is particularly germane to outsiders' views of tropical and neotropical rainforests—the modern-day Gardens of Eden. Yet these are also the places where ecotourism discourse claims to be appropriate and effective.

This brief look at policy and theory raises many questions. Does ecotourism, and community-based ecotourism more specifically, move beyond simple dualisms and conceptions of nature as sacred, timeless, and located in the past and the tropical periphery? Are affluent urban recreationists (including biodiversity planners) pitted against or working with rural people who actually earn their living from the natural resources? How does each group view nature, ecotourism, and themselves and one another? How is "traditional" Creole culture constructed, valued, and offered as an international cultural experience? These questions are highly relevant if we consider that "primitive" peoples in tropical areas are often idealized, even sentimentalized, until the moment they do something unprimitive, modern, and unnatural, thereby falling from environmental grace (Cronon 1995). What do analyses of concrete efforts tell us about the cultural frameworks utilized by planners, and the congruity between them, their activities, and the histories and lives of people in particular rural areas?

The case study that follows, although not explicitly about laws and rights, is nonetheless a story about justice. The justice question here revolves around the politics and power of controlling images and cultural meanings in the production of community ecotourism, and the profound impacts, however unexpected or unintended, they have had on the lives of people, especially those with little power. It is also about new possibilities for struggle and resistance to these controlling images (Escobar 1995).

## GALES POINT HISTORICALLY

The village of Gales Point is a four-hour boat ride from the nation's capital. Like the rest of the country, the area has been largely undeveloped, and exploited mostly for its natural resources—by Spanish, British, and, more recently, American interests[3]. In the early part of the century, few people lived on the peninsula. There was no road, only the river for passage, and that was not navigable during the dry season. The peninsula remained a nameless settlement for shipwrecked and escaped slaves who were able to remain hidden because of its remoteness, at least until Mr. Gale arrived.

After the world depression in the 1930s, Creoles (the descendants of European colonists and African slaves) came from Belize City to settle in Gales Point. A major draw was the availability of work created by foreigner Paul Merritt, who owned and operated a lumber mill near Soldier Creek beginning in the early 1940s. In the 1950s, Gales Point was an economically active village with income derived from logging, bush farming, hunting, fishing, and revenue from middle to upper class Belize City residents who built vacation bungalows along the shorelines.

Lumber operations closed after hurricane Hattie destroyed most of the structures in 1961. After that time, residents faced severe economic hardships. Social programs financed by the government focused on the concept of basic needs: in Gales Point they worked to promote home gardens, hygiene, and cottage industry (e.g., fruit production and preservation). However, in the 1970s, government monies for social welfare programs dried up as the country followed structural adjustment mandates set forth by the International Monetary Fund (IMF). In the absence of economic alternatives, residents of Gales Point depended largely on forests and sea resources for their subsistence. Residents (including the settlement of Soldier Creek, which relocated to Gales Point after the lumber mill closed) took advantage of the lagoon and creeks to fish, and to reach the bush and savannahs for farming root crops and hunting [especially for small rodents known as gibnut (*Agouti paca*), deer (brocket deer), tapir (white-lipped peccary), armadillo, turtle (green and hicatee), and the large manatee]. Villagers also extracted and sold chicle for the manufacturing of chewing gum, and other forest products such as "tie-tie" vines (*Desmoncus schippii*), which they wove into baskets for home use, in addition to a range of other plants collected as medicinals.

These activities then and now are carried out by a population characterized as "a skewed mix of children and their grandparents" (Manatee Advisory Committee 1992:1). Most of the Gales Point working-age population lives in Belize City or the United States, where they can find employment. In 1994, the population of Gales Point was approximately 400, composed of seventy-seven full-time households and sixteen seasonal ones. These statistics conceal the fact that there is much population movement in and out of Gales Point. Many young men leave the village each year for a few months to earn income in Belize City and the United States. They and others return to Gales Point for months at a time, or at least for holidays.

This brief history of Gales Point suggests the centrality of natural resources in the village economy, seasonal migrations to and from the village, and the long-term presence of foreign interests. Gales Point has not developed a diversified economy, in large part because of foreign domination, geographic remoteness, and vulnerability to extreme weather. By the late 1980s, the ability of natural resources to provide sustenance to Gales Point residents declined precipitously. This has been the result, not of population increase (given the high incidence of out-migration), but of competition with foreigners for land and marine resources. While many residents continue to cultivate small milpa farms near Soldier Creek some seven miles from the village, many more used to cultivate the fertile land just south of the village. During the 1970s, these lands were sold by the Belizean government to an American couple who developed a large citrus plantation known as White Ridge Farm. High debt and IMF-imposed structural adjustment policies in the 1970s had turned the Belizean economy further toward foreign trade and ownership, increasingly with the United States rather than with the ex-British colonists (Moberg 1992; Shoman 1994). The Belizean government asserted its ownership over the property to which villagers lacked legal title, claiming that it inherited the land as "eminent domain" from the British colonial regime. The Belize government offered residents other land to farm, but these lands were infertile and located far from Gales Point village. Additionally, while the plantation provides some local employment, agrochemicals also run off from agricultural fields to the stream that provides Gales Point's drinking water (Greenlee, personal correspondence, April 1996).

During the 1970s, another foreigner secured title to the tip of the peninsula, and the Manatee Fishing Lodge was created. The advent of commercial fishing in the lagoons created severe competition with local fishermen. Whereas local fishermen in the past fished with rods from small nonmotorized dories, and largely for home consumption, both commercial and local fishermen periodically used large, finely meshed "gill nets" and fiberglass motorized boats, which greatly expanded their catch and range (eight villagers owned gill nets in the village in 1994).[4] Local fishermen said they now fish for as much as they can catch to sell in the local and Belize City markets. By 1994, the Manatee Fishing Lodge had changed owners and officially dropped *fishing* from its name. The name change symbolizes the demise of world-class sport fishing in the area, which the lodge had created just a decade earlier. The lodge now markets its main attractions as sport fishing and diving off of the cays (islands), the scenic seaside views of Gales Point, and Mayan archaeological ruins.

As a result of increasing foreign investment, Gales Point villagers turned to the sea and bush for their livelihood. Unlike in indigenous Mayan villages located elsewhere in Belize, there were no common-management properties, customs, or ethics regarding resource access, allocation, control, or use in Gales Point. Natural resources were de facto open access; that is, people could extract or use nondeveloped bush and marine environments as they pleased, which they did. A 1994 random household survey that my students and I conducted in Gales Point found that 89 percent of Gales Point residents obtained their primary food through purchase (consisting of rice, red beans, and flour to make biscuits known as Johnny Cakes), rather than from home production. They purchased this food from one small store in the village, or in Belize City. Approximately a third of the residents earn income to purchase commodities through wage-work (such as working at the White Ridge citrus plantation or the Manatee Lodge, teaching, postal work, or doing boat repairs/carpentry). Another third of the households earn income from selling bush meat, especially gibnut.[5] The remaining households obtain income from relatives abroad who send remittances and, since 1992, from providing ecotourist services. A very small minority earned their income primarily from selling fish or farm crops (root-crops, bananas, vegetables, cashew, and coconuts).

Wildlife to most Gales Point residents has historically been valued as an important source of food and income, as have the sea and the lagoons; some are also dangerous predators or annoying pests. Wild peccary are killed for meat but also when they root and consume foods cultivated on small farms. Morelet crocodiles are killed by residents as they float into the lagoon during the rainy season, threatening small dogs and children. Villagers also hunt gibnut with no restraint, since they view these large rodents as aggressive and capable of rapid reproduction. Freshwater and sea turtle (hicatee, green, hawksbill) are also easily hunted and the meat highly favored in Creole cuisine; the hawksbill shell is also used to make jewelry and other ornaments.

Gales Point residents' utilitarian approach to nature, however, does not preclude their appreciation or value for nature as noncommodity. The lagoon and sea are appreciated for their view, as a place for rest and relaxation, and as a source of contemplation while watching the gentle manatees. One older resident spoke regretfully of cutting back bush and the resultant loss of bird species as the village is extended so that the wealthy from Belize City can purchase and construct vacation homes on prime beach-front lots:

> I am sad that my grandchildren will not see the scarlet macaw and curaso that I watched every day as a young child. The big birds have all been killed or disappeared with the clearing of the bush to make houses.

In the context of limited economic opportunity and mounting resource scarcity brought on by the encroachment of wealthy nonlocal vacationers and profiteers, Gales Point residents have survived by cutting the bush; killing, consuming, and selling wild game; fishing; and, to a lesser extent, farming. Nature in this rural Creole community is dynamic, resilient, and inclusive of human activity. Even in the absence of local resource management customs (offset by high rates of out-migration), nature has, until recently, provided. Environmental degradation and scarcity intensified with foreign commercial enterprises that significantly raised the scale of extraction and limited the access of local villagers to fertile land and remaining fish. In the 1990s, competition with foreign interests have further intensified, this time over the meaning and control of wildlife.

## CONSTRUCTING COMMUNITY-BASED ECOTOURISM IN THE GALES POINT MANATEE CONSERVATION PROJECT

Many articles have been published on the origin, intention, and operation of the Gales Point Manatee Community Conservation project, most authored by project founder Robert Horwich and his associates (e.g., Horwich 1995; Horwich and Lyon 1995; Horwich et al. 1993). A close examination of project proposals, plans, and brochures will be provided, with special attention to the planners' conception of the nature, culture, and community to be conserved, organized, and marketed for ecotourism, as well as the factors they claim make the project a community-based effort.[6]

Since 1968, the area surrounding Gales Point has been proposed as a protected area (Zisman 1989), and various visitors have initiated efforts to protect turtles and other wildlife. However, the current effort was begun with a proposal to the government of Belize in February 1991, drafted by wildlife biologist Robert Horwich and Jon Lyon. They proposed that the area, including the village of Gales Point, be designated a biosphere reserve. The proposed reserve encompassed approximately 170,000 acres of a variety of habitats, endangered and threatened species, and property rights (Horwich 1995). Horwich and Lyon had begun community conservation in Belize with the Community Baboon Sanctuary, a project involving 100 or more private landowners (Horwich 1990; Horwich and Lyon 1990), and they were eager to build on what they had learned (Horwich and Lyon 1998). The proposed Gales

Point project provided an opportunity to pursue community-based conservation across a mosaic of habitats and property rights regimes, together with government and private actors, and to give more attention to constructing village support and community-based management structures. According to Horwich (1995:8), the proposal evolved in the following way:

> During a trip to the area by Horwich and Chris Augusta, a long-term part-time resident of the area, a plan was proposed to the village in which villagers signed a supporting document. An initial proposal for a biosphere reserve for the area was written and submitted to the village council of Gales Point and the Ministry of Tourism and the Environment. . . . Further discussions with government staff led to the government organizing a meeting at Gales Point which included politicians and staff of the Departments of Forestry and Archeology as well as members of the tourism community and villagers. A follow-up proposal was written for a multiple land use protected area to incorporate the additional ideas from the various government meetings.

Close attention was given by project planners and government agencies to mapping, zoning, and regulating human activity across the Manatee Special Development Area (SDA). Using the biosphere philosophy, the plan provides for core areas where human disturbance will be minimal, buffer zones where specific human uses are designated, and transition zones where human activities will be restricted. Human use will be limited to low-impact ecotourism in core areas, which were selected for specific endangered species, specific ecosystems, and watershed protection (Horwich and Lyon 1998). "Proper" land use was determined through the lens of biosphere philosophy and scientific wildlife management.

The Belizean government responded by designating the Manatee Special Development Area in November 1991, and the Ministry of Tourism and the Environment headed by Minister Glen Godfrey committed funds to begin the project and to construct a hotel that was to be turned over to a village cooperative once it was formed. The Belize Audubon Society directed by Dolores Godfrey lent its support, as did the U.S. Peace Corps, which assigned two volunteers to the project. Later on, the project received grants from the United States Aid for International Development (USAID) and the United Nations Development Program (UNDP) Global Environmental Fund for biodiversity assessments and village improvement, such as loans to improve homes and build septic systems, administered through the Belizean nongovernmental organization (NGO) known as BEST (Belize Enterprises for Sustainable Technology).

Horwich et al. (1993) put forth three main objectives of the Manatee Community Reserve (also at times referred to as the sanctuary). These are (1) to develop a locally supported reserve that integrates multi-land use for private and government-owned lands and ensures sustainable use of resources; (2) to maintain and strengthen the local rural culture (based on farming, fishing, and hunting); and (3) to give the village a supplementary source of income through tourism, resulting in economic self-sufficiency and less pressure on natural resources.

> The sanctuary will concentrate on developing tourism around the community lifestyle, giving tourists an authentic experience of village life, something like the exposure to Creole culture at the Community Baboon Sanctuary. Allowing tourists to enjoy an "intercultural experience" should also relieve villagers of pressure to invent a sense of opulence for tourists.
>
> Horwich et al. (1993:162)

The planners insist that conservation in the area will encompass preservation of the rural lifestyle, and change will occur in accordance with community wishes and under community control. The project will "keep the cultural unity and integrity of the village intact" while providing opportunities for pleasing tourists.

> At Gales Point, certain areas where manatees and American crocodiles reside would please naturalists. Setting up permanent viewing sites such as an anchored raft or a viewing tower in these situations would further enhance the possibility of viewing wildlife. Often things that villagers and local guides take for granted will thrill foreign visitors.
>
> Finally, with an eye to pleasing tourists, Gales Point villagers should appraise their village. With village consensus, improvements might include alternative toilets to accommodate both village and tourist wastes. . . . Planning and constructing boat moorings and piers should be under strict local control.
>
> Horwich et al. (1993:163)

Horwich writes that it was the villagers of Gales Point who formed a cooperative to manage ecotourism activities, including community associations focused on delivering the various ecotourism services.

> In January, 1992, villagers of Gales Point formed the Gales Point Progressive Cooperative (GPPC) to promote sustainable economic development and to conserve the natural environment of the region. . . . A number of associations were created under the Gales Point Progressive Cooperative which included a bed-and-breakfast association, a tour operators association, a farmers association, and a local products association. Since the community requested help, the Manatee Advisory Team (MAT) . . . was created in mid-July, 1992.

Despite the lead role he and other foreigners played in instigating and implementing the project, Horwich represents the project efforts as home-grown, claiming that the project has had strong village support, "with over 50 percent of the adult community getting involved in at least one of the cooperative's programs" (Horwich 1995:9).

Elsewhere, he and coauthor Lyon write that a primary emphasis in their programs was the empowerment of local people. They claim to have advanced this goal in Gales Point through the operation of community-based programs that involve local people not only in managing ecotourism-income generation, but in research and conservation activities that promote local conservation awareness. Furthermore, they write that local empowerment was also enhanced and strengthened through encouraging participants of different community associations in Gales Point to interact among themselves, as well as with similar communities, for information sharing and morale boosting (Horwich and Lyon 1998).

## IN ANOTHER VOICE: THE CONTRADICTIONS AND IRONIES OF COMMUNITY ECOTOURISM

The preceding narratives suggest the large role played by American wildlife biologists in conceptualizing, organizing, funding, and implementing activities in Gales Point around the foreign concept of community ecotourism. That the project was not initiated from within the Gales Point community, nor built upon historic community-based resource management traditions, contests its representation as "community-based conservation"—at least by some definitions (e.g., Western and Wright 1994). How do planners' views of the community, nature, and ecotourism activities mesh with the cultural frameworks and daily lives of Gales Point residents generally, and by class and gender more specifically? If there are costs, what are they and how are they differentially borne across the physical and social landscapes?

### What Community? Whose "Community Lifestyle?"

Project proposals, plans, and brochures speak about "a" Gales Point "community" and "community lifestyle," which are mobilized into "community associations," and generate widespread "grassroots" support for ecotourism and conservation activities. Who is this assumed community? And whose lifestyle is the project conserving, and for whose benefit and at what cost?

The image of Gales Point community that is pictured and marketed to international and wealthy Belizean ecotourists, and which, I argue, is an invention of and intervention by project planners, is presented in a video on community conservation produced by the Belize Ministry of Tourism and Environment. The text also suggests that Belizeans themselves are implicated in this project, and in ways that reflect their own varying identities and positions. A Belizean narrator introduces Gales Point to the audience with the following words: "We drive into Gales Point village on the only street in this sleepy easy-going village." She boasts of the attraction of Gales Point to tourists viewing "natural sights," and residents "valuing the pristine state of their facility, especially the lagoon where the manatee live." Aboard a motorboat to view the manatee, she says, "Now we are going there to see if Kevin will call out any of his pets." She later concludes "that everyone feels encouraged to continue developing Gales Point. . . . Conservation of the resources which makes Gales Point and its surrounding area unique is vital to the village's very existence."

The segment on Gales Point concludes with village councilman and cooperative president Walter Goff celebrating the project and the income it generates. Chair of the tour guide association, Kevin Andrewin, concurs: "I live better. I love when people come and I take them out and around."

The words and pictures that accompany the narrative paint a picture of a quaint, clean, seaside rural village welcoming visitors to come and enjoy first-hand their natural and cultural amenities. The economy is presented as dependent on local natural resources provided from a still "pristine" physical environment, a Creole culture intact and traditional, and a community devoid of complexity, activity, conflict, and diversity (symbolized by the single dirt path). The Gales Point in this gaze is not tinged by internal conflict, poverty, violence, environmental change—let alone degradation—nor even human history.

Alternate images and a historical perspective challenge and dismantle this fiction. As noted previously, the landscape evolved from the hand of human activity: logging, farming, hunting, and fishing, as well as an unsentimental approach to animals. Wild animals have been hunted as game or eradicated as pests, not revered as symbols of wilderness or "endangered species." That the video's narrator, most likely an urban, upper-class Belizean, refers to the manatee as "pets" provides a third meaning of the manatee most familiar to Belizean city dwellers—neither as untamed wildlife nor endangered species, but as domesticated pet.

A differently framed picture of Gales Point reveals "quaint" houses as rundown and broken, beaches as garbage littered and dotted with lagoon outhouses, and an "easy-going" population preoccupied with survival and uneasy relations with each other. The feelings produced from this alternate angle are decidedly less tidy and comforting than those prompted by the promotional video.

A jolting image is the abandoned car parked a few houses down from Miss Samuel's dry goods store. The insignia and name "Bloods" scribbled across this car in bright red paint signals that drugs and an international drug culture and economy are very present in remote Gales Point, even subsidizing some of the most important elements in community ecotourism.

While men in the village talk about construction as the work they do when they seasonally migrate to the United States, a few share the truth that these trips bring considerable and fast income from marketing illegal drugs (mostly cocaine and marijuana), which they secure through growers in western Belize or by transshipment from South America.[7] Drugs enter the village through other venues as well. A few years back, a group of young Gales Point youths found a bag of cocaine that had washed up on a nearby cay. They divided the booty among themselves. Some of the youths immediately consumed the drugs, while others sold them in Belize City and were arrested. Still others arranged the marketing with greater caution. One youth carefully sold his portion of the drugs, saving the income and eventually purchasing a speedboat, which he uses today to guide ecotourists for day trips to the Manatee watch, the Caribbean seashore, or ancient Mayan caves.

The seasonal migrations of urban Belize City residents and those from the United States to Gales Point also carry with them worldly connections, material possessions, and notions that challenge the limited albeit opportunistic view of the village as traditional, rural, self-contained, and remote—the perfect place for a "getaway" and "natural" vacation.[8] Deeply involved in circuits of migration, people from Gales Point challenge the ease and simplicity of a spatially and culturally stable Gales Point community. Foreign remittances provide much needed income in Gales Point and Belize more widely. But their checks, urbanized dress, occupations, and "modern" education and ways are not incorporated into the image sold to ecotourists. Nonetheless, despite living most of their lives in Chicago, Brooklyn, or Texas, many seasonal returnees maintain a strong attachment to Gales Point, one that also buys into memories and desires for a simple, quaint home village. I had a few conversations with visitors one Easter morning, which spoke to their attachment, as well as their resentment of tourists and planners intruding with their own projections and plans for Gales Point. No doubt, crowded beaches, car-strewn streets, and lounging well-dressed urbanites—typical scenes during major holidays—are not the views highlighted and reproduced in promotional brochures.

## UNINTENDED CONSEQUENCES AND IRONIES

While promoting cultural events to attract tourists to Gales Point and provide opportunities for tourist income to replace "unsustainable" livelihood practices, some unintended and environmentally dubious outcomes have developed. These suggest not only the comedy implicit in community ecotourism, but the ironies produced by it as well.

In line with delivering an "authentic Creole" experience, residents are resurrecting an old tradition of evening "drumming." For a cost of around US$45, young men in traditional dress drum while villagers and guests dance around a blazing bonfire. Many neighborhood families come out to watch and dance, and sing songs in Creole which suggest African rhythms and antislavery lyrics (such as lampooning white masters). Whether or not the performances are authentic or a spectacle put on for tourists, they have generated renewed interest in drummings and a demand for drums made by local residents. Drums are made from various woods, with different animal skins pulled across the top, which produce different sounds. The best ones are made from wildcat skins, especially ocelot—a species on every conservationist's protection list. A long-term wildlife biology consultant has suggested that they shift to goat skins.

As the planners encourage men and women in the village to offer tourists an "authentic" experience of Creole culture, for bed-and-breakfast (B&B) operators this translates into the food they serve their guests. However, many providers are confused about what to serve, as well as about the government's intentions regarding endangered wildlife. Many of the most traditional or "authentic" Creole dishes involve bushmeat, which is no longer legally hunted in the Manatee SDA [i.e., turtle, armadillo, and iguana (known locally as bamboo chicken)]. A B&B provider I stayed with complains, "I want [to] prepare for you real Creole food. But the gov-

ernment [doesn't] want us [to] collect turtle eggs or kill turtles. How [are] we supposed to provide Creole culture when we can't serve hicatee and white rice. It's
ridiculous."

Another residents adds, "The government [is] always pushing advertisements
on TV and radio about traditional Creole dinner, telling tourists to come eat hicatee
and white rice. What is the government trying to push? Is it telling people to kill turtle or not?"

Instead, tourists are routinely served the same main meal: a stew of beans, rice,
and chicken. Guests frequently receive tinned spaghetti, sardines, or a sandwich for
lunch[9], and, although desired by tourists, few vegetables, and fresh fruit only when
seasonally available. Although gibnut is traditional food in Gales Point, "we know foreigners don't like to eat rat so we don't serve it to them though this is too bad, because
gibnut is good meat, and we can hunt if or buy it in the village when we need it."

Project planners have also encouraged the production of native crafts for sale to
tourists, although few "natives" still retain knowledge about plaiting baskets. Building on successes in other villages with selling baskets as "jungle products," an
elderly woman named Miss Iris, who spends half of each year in Los Angeles and
is interested in renewing pride in Creole culture, has been teaching a small group of
interested women and men to weave baskets. Miss Iris explains:

> So few people here know how to plait baskets anymore. Back then we all knew how
> to do it because there wasn't any plastic. But plastic is now cheap, widely available
> and doesn't leak, so most people do not want to spend the time or money buying
> the vines and weaving baskets. Many too do not want to invest the time to make
> baskets until they are ensured of a market. We do have a problem with marketing.

Baskets are woven exclusively for tourists, and there have been significant problems with poor quality, high pricing, and limited markets. Additionally, whether the
extraction of "tie-tie" vines for baskets and other handicrafts disrupts local flora and
fauna, and whether they can be sustainably harvested under higher collection pressures are questions that need to be answered before assuming the enterprise is ecologically sustainable (Belsky and Siebert 1997). Similarly, promotion of local crafts
has led to villagers buying black coral and constructing jewelry and other handicrafts
to sell to tourists. Although illegal, harvesting of coral persists in Belize, with negative repercussions for reef ecology and efforts to protect coastal environments.

Another unintended consequence of rural ecotourism is that capital accumulation from operating B&Bs and tour boats is being used to develop a new, locally initiated brand of ecotourism. In 1996, residents shared with us their dreams to construct a cabana or two in their yard, and even possibly a small restaurant. Some have
already begun to live this dream. A lesson learned from hosting B&Bs is that both
residents and foreign guests prefer a little privacy and social distance. But unlike the
community-based rural ecotourism project, which has tried to facilitate tourist development within a broader context of land-use planning and community participation
(including a concern for equity), the private-cabana approach will be available to

only those who can afford it and will not be coordinated for any equitable distribution of resources, benefits, or service delivery—and certainly not for environmental conservation or cultural preservation. In the current political economic climate of privatization and export production, it is likely that this trend will garner both national and international support.

Indeed, community-based ecotourism (and community-based conservation more generally) suggests an approach at odds with countervailing globalization forces and ideology. Globalization supports an urban-biased, export-led model of development that privileges markets and privatization over nonmarket, socialized, or common-managed approaches to development, and that entails significant environmental degradation (McMichael 1996).

## UNEVEN PARTICIPATION AND NEW VULNERABILITIES

Contrary to claims put forth in the promotional video and publications, widespread and grassroots participation in community ecotourism activities has not developed, nor has anything close to "empowerment" been achieved in Gales Point. Indeed, participation and benefits have been uneven, and new vulnerabilities associated with community ecotourism have arisen, especially for village women.

In 1994, two years after the project was up and running, ten women operated B&Bs and were members of the B&B association; seven women joined the craft association; eight men provided services with the boat operator association, and fourteen men were members of the farmers' association. Importantly, most of the men and women participating in these enterprises belonged to the same five or so households. The manager of the B&B association, Hortense Welch, is the mother of the chairman of the tour guiding association (Kevin), and wife of the most popular bush guide (Moses Andrewin). She is also a long-term member of the village council headed by Walter Goff, also president of the umbrella-management organization, the Gales Point Progressive Cooperative. That a few so-called progressive households and individuals dominate village political and economic affairs is not uncommon in rural development, but its reality shatters the fragile illusion of community-wide participation.

Additionally, in 1996 we found that two-thirds of the women operating B&Bs were employed either by the Manatee Lodge, the White Ridge citrus plantation, or the Belizean government. This pattern is noteworthy for a number of reasons. First, it suggests that a small oligarchy of community residents and households usually seizes the opportunity to access new resources. Second, it reveals that the individuals and households that do participate and receive material benefits from ecotourism are not those most dependent on the "subsidy from nature"—those who should benefit and who in turn will promote conservation, according to ecotourism discourse (Hecht et al. 1988:25). Logically, it is the better-off households who can meet the standards for offering B&B services, including bedrooms with specified furniture, and cooking and bathroom facilities with basic sanitation. And third, it suggests added burdens on female B&B operators striving to combine historic domestic responsibilities with new paid-labor and ecotourist-related services.[10]

In the first two years after tourism activities began, substantial tourism money was earned for those providing B&B and tour guiding services, especially because of the annual visits of student groups such as our own. However, income from B&B activities declined from 1994 to 1996 as a result of reductions in the number of tourists, a rise in food prices, and the establishment of new B&Bs in the village. In 1996, the price for one night of lodging and three meals increased from US$15 per person to $20 to cover added costs (including a value-added tax of 15 percent instituted in 1996), but still women complained, "Even with the raise it don't bring much profit to providing cooked meals." B&B operators also lament that nonassociation members are opening B&Bs, providing significant competition, and undercutting their prices. The nonassociation members claim that they can provide quality services without the assistance (and the obligations) of the B&B association. With regard to B&B management, women raised concerns that the B&B association manager assigned guests on the basis of favoritism rather than following a rotation schedule.

These complaints were exacerbated by new debt. In 1994, loans from the Belizean organization BEST were made available to improve B&B accommodation. In 1996, B&B members were encouraged to take loans to construct indoor plumbing and septic tanks at an interest rate of approximately 10 percent monthly. Residents not affiliated with the project expressed great resentment that they did not receive assistance for building toilets. Ironically, however, because of design choices and water shortages in the village, many of the new toilets do not function and cannot be used. The reduction in tourists combined with repayment obligations leave many B&B operators worried that they will not be able to repay loans.

Some B&B operators are concerned that they cannot combine historic domestic responsibilities with providing ecotourist services and new employment, and these worries are leading to new conflicts at home. With employment opportunities (and increased debt), B&B operators are faced with choices such as whether to transfer B&B hosting duties to other family members (e.g., keeping young girls home from school); offering lesser-quality service (e.g., serving bagged lunches), or not accepting guests entirely (and forgoing needed income). The costs of these choices are borne differently across households, and by male and female family members of different ages within these households.

## REINFORCING COMMUNITY RIVALRIES

That the project has wrought resentment between households, between local residents and project planners, and between ecotourists and the state was strongly communicated to us during our last two visits to Gales Point. In 1994, a small amount of project funds was allocated to build a community craft center in Gales Point to facilitate marketing of locally produced handicrafts. However, labor had to be hired to build the center since residents were not willing to assist in its construction. In 1995, when the building was half constructed, it was destroyed by arson by members within the community as a result of disputes over the land upon which it was built, the construction methods utilized, and the perception that the same few families were receiving the majority of employment opportunities from the project.[11]

Another example involves utilities. Government officials connected to the project were instrumental in bringing diesel-generated electricity and pumped water (from a nearby river) to the village, and in facilitating collection and payment of fees, which has become contentious and even violent. Households are charged a flat fee of US$10 per month for water and electricity, rather than an amount based on individual household usage. In addition to lacking an incentive to conserve fuel and water, some households refuse to pay their bill because they are unwilling to subsidize the B&Bs, which consume more water and energy in the form of refrigerators, lights, fans, and other appliances used by or for overnight tourists. During our stay in 1993, there were neither lights nor water available because a majority of villagers declined to pay the bill.

In addition to not wishing to subsidize B&B services, nonparticipating community members lament the rise of stratification in the village, or what they express as "not coming up together." One woman protested that a rival had commissioned the feared obea man (a specialist in black magic) to place a spell on her and her household because they generated more income from ecotourism activities and from paid employment. One resident I knew well told me, "I [am] going to let the garbage pile up on the beach. This is what we do with garbage in Gales Point. Maybe if the tourists don't like it, they won't come. And they will leave Gales Point to us."

Refusing to pick up his garbage is this man's way of resisting and speaking back to foreign-led ecotourism and conservation, the greater capital accumulation of his neighbors, a loss of personal control, and an affront to his ethics and aesthetics associated with all of these. His critique is lost on project planners, who call for more training in "hospitality" and providing international standards of tourism "service." They miss the point or choose to ignore that the failure to pick up garbage, or "impoliteness," is not just the result of different approaches to service but is related to the unequal and racist historical relationship between blacks and whites, masters and slaves, and colonizers and the colonized.

Predictably, members of households receiving income from ecotourism are more sympathetic to the project and its environmental conservation principles and strategies. "I hunt less because of tourism," acknowledged a tour guide whose wife operates a B&B. "I used to hunt more but now I don't fool with it as much because staying out at night makes me too tired to guide tourists during the day." He finds nature guiding more lucrative and less arduous than hunting for gibnut.

However, households that depend on hunting for their primary income and who do not receive income from ecotourism resent attempts to impose limitations on their hunting activities. They criticize government officials and foreign consultants who advocate new conservation regulations that require the purchase of hunting permits, that restrict hunting of particular species to particular seasons and uses, and that prohibit hunting of certain species altogether. Many protest what they perceive to be an insensitivity to their limited material resources. "We don't hunt, we don't eat," one woman reminds a student. "We [do] not get back for our loss," complained an older bush hunter who does not receive any money from ecotourism. "Maybe others do, but not us." Another man complained that government officials do not enforce the laws anyway, and after instructing you not to hunt, they will buy or confiscate precious turtles from you for their meat and shells.

The project has also exacerbated historic political rivalries. It is no secret that then-minister of tourism and environment Glen Godfrey financially and logistically supported the Gales Point community ecotourism project (and the village council through which it worked) to award and promote patronage for his political party, the Peoples United Party (PUP), which was in power from 1989 to 1993. Although not all community members in Gales Point are aligned with the PUP, most are. Those who favor the opposing party complain that they are not able to join community ecotourism associations even if they desire to do so. Since the country's other major political party, the United Democratic Party (UDP), regained power in 1993, financial support for most project activities has ceased. No funds have been forthcoming to complete the Manatee Cooperative Hotel.[12] In 1996, there was a visible void with regard to management of ecotourist activities and governance of the village as a whole. Some residents, such as Hortense Welch, have responded by realigning themselves with the other party, the UDP; others have retreated from community activities entirely.

When the new Manatee Road was constructed in 1992 linking Gales Point village to the country's two major highways, political patronage was offered to Gales Point residents to support community conservation and ecotourism with the promise of land titles to adjoining parcels. Offers to provide some forty acres to Gales Point were made by both PUP and UDP politicians. However, in neither case did residents actually receive land titles, and in one instance a Gales Point resident with established perennial crops on his farm was informed that the property had been sold to a Jamaican. Nonetheless, villagers continue to clear bush for farms. This includes clearing fertile parcels adjoining rivers and streams, despite the riparian conservation regulations of planners forbidding clearing land unless it is at least sixty-six feet away from rivers. A resident summed up the sentiment of many villagers:

> We [are] all vying to be close to the government. But it [doesn't] matter PUP or UDP. Either one [is] going to sell that land to make big money. But without land to farm we in Gales Point [will] never be able to get ahead. Food is so expensive and now the government [doesn't] want us to hunt anymore. What are our kids going to eat?

## STATE ACTIONS AND GLOBAL POLICIES

It is unlikely that ecotourism can or should be more than a supplemental source of income, and planners agree. Livelihood security is more likely the result of a diversified economy that modifies rather than replaces traditional food- and income-generating activities, and that is directed toward local rather than foreign markets. However, the political economy of development in Belize, spurred by debt and structural adjustment mandates, continues to adhere to an export-led, productivist model of development with little or no regard for environmental costs (Deere et al. 1990; Shoman 1994), and unequal concentrations of land ownership (King et al. 1993). These trends suggest further contradictions between community-based ecotourism discourse and practice, namely, that focusing attention solely on the community ignores social and environmental impacts of state actions and global policies.

Despite the designation of Gales Point and its surroundings as a Special Development Area, the Belizean state is intensifying production of commercial exports in the SDA that are environmentally suspicious, and Gales Point residents themselves observe this and question it. In 1994, the government permitted gravel and dredging operations in a major tributary to the Manatee River (Soldier Creek) with no regard for environmental impacts or damage to a farm cultivated by a local Gales Point resident. The state has also failed to respond to concerns regarding petrochemical contamination of Gales Point's drinking water (Greenlee, personal correspondence, April 1996). And in 1996, our group observed logging trucks emerging from the forested headwaters of the Sibun, which drains into the lagoon complex comprising the Manatee SDA. Mennonites were hired to cut timber in the Sibun drainage with permits from the Belizean government to clearcut parcels for a flat fee of Blz\$30,000—without any prior inventory or environmental impact assessment as required by law. Cleared land (readied for conversion to citrus) fetches the government a higher price than forested land. In addition to potential negative environmental impacts, logging reduces access to forest and bush resources utilized by Gales Point residents. Conversion of forest or bush to export-crop plantations places more pressure on existing resources (including legally protected areas) for poor residents lacking alternative means of livelihood. And it fuels resentment when scarce resources continue to flow predominantly to international, national, and local elites, and to exports.

Watching the process of "development" and reflecting on the ecotourism project in the village, a Gales Point resident asks, "Why should we respect the government and its rules for hunting when they are cutting in the reserve? We know it; we see it."

## CONTESTED LESSONS: WHOSE MEANING, WHOSE BENEFIT?

The lessons drawn from community-based ecotourism in Gales Point are not self-evident, nor are they uncontested. In the eyes of project coplanner Dale Greenlee, there were problems encountered in the Gales Point community ecotourism project that centered around the failure of instituting wildlife conservation, and the blame is largely placed on the community itself. While acknowledging encroachment and other threats created by vacation home development and the construction of the Manatee Road, he nonetheless chides the community for its inability to work "cooperatively," "logically," and pointedly toward the goal of "conservation." In an interview with our students in April 1996, he shared some of the following reflections:

> This village doesn't know how to cooperate. Gales Point villagers want to do things on their own.

> The buoy system which was set up to protect the manatee . . . was taken apart and taken down by the village immediately after [the landscape architect] left. She was the one who orchestrated that effort with the full cooperation and consent of the village. But as soon as [she] left they cut the ropes and took out the buoys. Villagers continue to overfish and use gill nets. Why don't they ride each other to stop? They all know its bad but say its not their business.

> To me, ecotourism and this whole cooperative was about manatee conservation and enhancement of wildlife habitat. The social framework of this village is interesting, for outside people to come and experience a Creole village on its own terms. It is a draw. But the real draw is and has been the manatee, the turtles and the monkeys up river—the whole environment. This is an incredibly diverse ecosystem around here where several ecosystem types are joined together in this estuary. It's a wonderful place and that whole concept has been swept aside. Nobody is talking about that anymore.

Former executive director of the Belize Audubon Society Dolores Godfrey (herself of Mayan descent) also acknowledges that a particular cultural context informs the project in Gales Point and its shortcomings. However, rather than focusing on its "uncooperativeness," she emphasizes the community's construction historically from the mixing of escaped slaves, renegades, and urban elites, along with the absence of common-managed property resource regimes and the difficulties in conserving species such as the manatee. She is also sensitive to the different optics employed by Western environmentalists and their gaze on Belizean nature and community. She makes the following observations (1994):

> Horwich and others were very well meaning, but they were not capable of communicating with villagers about their conservation plans. They tried to build them around a "community lifestyle" in Gales Point. But in fact there are many life styles in that village. Culture is not static.

> The original idea was that tourists would convey a conservation ethic to B&B providers and guides, and in that way people in Gales Point would be too self-conscious and stop doing things which hurt the environment. But there was too much pressure from the outside to stop collecting turtle eggs, as an example; it was if they were always making a religious confession. I had hoped that the project would bring money to the community, not salvation.

> Yes, people in Gales Point would talk about the value of conservation. But it was a pseudo-conservation ethic. Horwich and the ecotourists always telling the villagers not to kill or hunt, and were actually turning the residents against them and conservation

with all their urging. What was needed really was educating the tourists about wildlife in that place, that hunting is okay.

The Caribbean tourist industry also laments shortcomings of "locals," particularly their "poor attitude" and "service performance gap" and interprets the problem pragmatically as an educational exercise in which the burden and learning process is placed on the Caribbean national rather than on the tourist. But as Dolores Godfrey questions, who really needs to be educated?

I suggest a different optic. Polly Patullo (1996) in *Last Resorts* reminds us that many of the "problems" associated with the Caribbean tourist industry can be viewed as stemming from a deep-seated resentment of the industry at every level of society because of its historic sociocultural associations of race, colonialism, and slavery. The difference between "service" for a wage and coerced "servitude" is subtle, and echoes of this association may be read into the dynamics of the tourist industry today; indeed, the languid behavior of black employees evokes the passive resistance practiced in slavery. That in Gales Point resistance to ecotourism is louder and clearer may be linked to its escaped slave past and historic ability to subsist from the natural resource base. The all-white American planners ignore this history as they impose a form of cooperation and environmental ethic foreign to the place and peoples of Gales Point.

While the movement has been "based" in the community, it was not initiated there, and it involved neither historic social traditions of cooperation/common management nor local understandings of nature and wildlife. When viewed through the variable optics of women and men in the village, it can be seen that rather than resolving conflicts between environmental conservation and local development, the Gales Point Manatee Community Conservation project has reinforced historic conflicts within the community, with the state, and with global forces, and it has created new ironies and vulnerabilities. The rights and welfare of the poorest households, those not aligned with the proper political party or traditional elite families, are not advanced by this effort. Furthermore, women in the households who are benefiting are encountering new work burdens and conflicts. The community-based conservation/ecotourism discourse fails to specify class and gender-associated impacts.

No wonder nods of approval are replaced by acts of outward resistance when planners leave, even though they take with them opportunities for political patronage and development aid, which were the reasons the villagers had joined the associations and had taken up ecotourism activities in the first place. Resistance also symbolizes rejection of factional disputes, which deepened in Gales Point, between those who benefit from and thus support rural ecotourism, and those who are left out or who reject what the project symbolizes.

Thus I doubt that there ever was "full cooperation and consent of the village" for ecotourist and conservation activities. Rather, I suggest that the periodic display of acquiescence was a performance necessary for accessing new sources of resources, such as loans. I do not think that residents are unaware of the ecological consequences of using gill nets or hunting certain species aggressively. On the contrary, they continue to use the nets and not stop their neighbors from doing so because of (1) a deep appreciation of their mutual limited control over fishing grounds before the Manatee Lodge began operation (or over farmland before the White Pine citrus plantation), and (2) a steady decline in their standard of living.

There is much to suggest that rural ecotourism in Gales Point particularly, and in Belize more widely, has become a key commodity, packaged and sold largely to an international middle-class consumer. What is being sold is a Western, idealized image of tropical rurality and exotic culture devoid of the ugliness associated with real Third World poverty, inequities, and globalization. In their consumption of "natural" Belize, international tourists bypass the reality of British colonialism, slavery, racism, and extensive forest extraction, farming, and gathering. Bypassing the reality transforms what northern visitors "see," and enables them to experience a "natural" forest landscape and a "natural" Belize. Tourists flock to the celebrated Mayan ruins while ignoring the living real-world Mayans economically and politically marginalized around the edges of national parks and protected forests. Although brochures, promotional videos, and ecotourist B&Bs celebrate and manipulate images of rural Belizeans and the subsequent emergence of a "new and improved" community-based ecotourism, they represent, in contrast, a new and more subtle form of domination.

## NOTES

1. Manatees are plant-eating, aquatic mammals that can grow to be thirteen feet long and weigh up to 4,000 pounds. They have no natural predators.

2. See Belsky (1999) for a more thorough discussion of the rise of ecotourism in conservation and economic development policy in Belize.

3. Belize was granted its political independence from Britain in 1981.

4. I was told that Paul Merrick, an American owner of the lumber mill, brought the gill nets to Gales Point for his son to use (cost, $400.00). While in the States, villagers purchased nets, which they brought back for use in Gales Point.

5. A large gibnut can weigh over sixteen pounds, and one pound of gibnut can fetch Blz$4 dollars in the village or Blz$5 to $6 in Belize City.

6. See Belsky (1999) for a discussion of the rise and incorporation of community ecotourism discourse in Belizean development planning more generally.

7. In 1996, Belize was placed on the U.S. warning list of countries known to be engaged in illegal drug trafficking.

8. A popular tour book on Belize names it as *A Natural Destination* (Mahler and Wotkyns 1991).

9. Limited food self-sufficiency is a historic outcome of colonial and current government policies. Consequently, there is a large dependency on imported foods, from England and more recently from the United States.

10. See Belsky (1999) for a more detailed analysis of the B&B operations, and the contradictions between community-based ecotourism and global political-economic forces in Belize.

11. A similar incident occurred in the village of Maya Center, which was relocated to its

present site when the Cockscomb Jaguar Preserve was created. Assisted by Belize Audubon and others to develop crafts to sell to tourists who must pass by on the way to the preserve, we were told by some community members that the craft center was torched by some members of the craft association, who felt that their products were not being displayed and promoted as strongly as others'. During interviews our group held in the village, we learned that there was great resentment over the economic rise of a few families, which runs contrary to the Mayan tradition of sharing and homogeneity, and that divisions also overlaid deeper and stronger community rifts based on religious affiliations.

12. Before a brief visit by UDP officials earlier in that year, villagers weeded and cleaned the grounds, hoping to get additional funds to finish the project. However, a UDP official said he would not provide funding until its land tenure status could be clarified. The project planners expected someone from the community to obtain this information, and as of now the issue is still unresolved and the hotel is "left sitting in the weeds."

# CHAPTER 12

## Profits, Prunus, and the Prostate: International Trade in Tropical Bark

*Anthony Balfour Cunningham and Michelle Cunningham*

*Prunus africana* (Rosaceae), known as the African cherry or red stinkwood (sometimes called *Pygeum africanum*), is a wild relative of almond, apricot, cherry, peach, and plum trees. Throughout its range in the mountain forests of Africa and Madagascar (figure 12-1), *P. africana* is valued by local people. The durable timber is a favored source of wood for grinding pestles and for hoe and axe handles, and the bark is used medicinally. It is this medicinal use and the consequent international trade that are the focus of this chapter, which illustrates the "footprint" of Europe on African forests and a medicinal resource within them.

Traded internationally and harvested from the wild, *P. africana* is hardly a minor forest product. Between 3,200 and 4,900 tonnes of bark is exploited annually for export to Europe, either as dried bark or as bark extract (Cunningham et al. 1997). This is the largest volume of any African medicinal plant in international trade, and it provides a case study with practical implications for policy on harvesting and sale of forest products. Both the uncontrolled harvesting of wild *P. africana* populations and tree cultivation (one alternative to wild harvest) involve problems that are difficult to resolve. Four issues of equity and environment are raised in this assessment of *P. africana* bark trade. These have policy implications in the search for a better balance between medicinal bark harvesting, forest conservation, and local economic development.

First, who were the beneficiaries of this link between indigenous medical knowledge and commercial production of an herbal extract? This is an interesting question for a product developed in the 1960s, well before any international policy linking intellectual property rights and indigenous knowledge. In this case, the knowledge of Zulu men in South Africa, communicated to a medical doctor there and then to a French chemist, stimulated a multimillion-dollar industry based on bark trade from Cameroon, Kenya, Uganda, Zaire, Equatorial Guinea, and Madagascar,

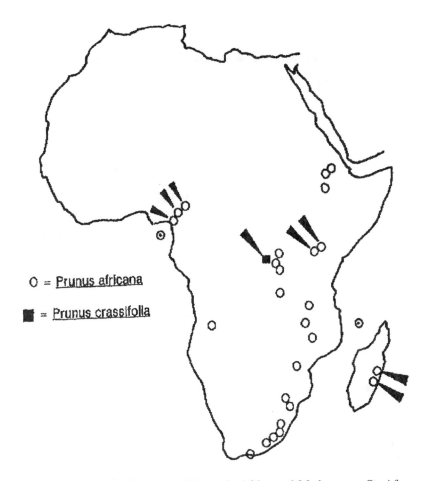

**Figure 12-1** Range of *Prunus africana* in Africa and Madagascar. On Afro-montane islands, it is often locally a common tree. Arrows indicate where commercial bark harvesting is taking place.

but not South Africa. Beneficiaries were the companies that developed the product and the trade and their employees, entrepreneurs, and national governments in bark-exporting countries. While the doctor who originally acquired the knowledge from Zulu people benefited through payments from the company, Zulu people and other South African farmers (who could cultivate *P. africana*) did not.

Second, monopolistic control by Plantecam, a French-owned company, of bark harvest in Cameroon had benefits for sustainable resource extraction, enabling a ban on felling or complete debarking of trees. This continued until 1985, when bark harvesting licenses were granted to 50 Cameroonian entrepreneurs. However, monopolistic control of bark exports led to bark prices being paid to harvesters, and this was an incentive to cultivate trees. In 1994, illegal bark exploitation, fueled by the higher price offered by an Italian company to three Cameroonian entrepreneurs, resulted in extensive damage to *P. africana* trees on Mount Cameroon. Ironically, the loss of monopolistic harvest status and the partial ban on export in fact led to a higher level of destructive harvesting and a greater volume of bark harvested than ever before.

Third, for the preceding reasons and those that will be explained later in this chapter, sustainable harvest from wild populations has unfortunately not occurred. Bark overexploitation and tree death are of concern to local people, who have lost local self-sufficiency in an important traditional medicine. The Cameroon and Madagascan Forest Departments and international conservation organizations are also worried about the effects on forest structure and biodiversity. One result, prompted by Kenya, was the listing of *P. africana* by the Convention on International Trade in Endangered Species (CITES). Government regulations in Cameroon for a partial ban on harvest and on tree cultivation by Plantecam have largely been ineffective. The same applies to Forest Department requirements in Madagascar for retention of two trees per hectare (ha) for seed production.

Finally, there is an encouraging side to the story. With no support from government, and despite the disincentives of the long period to marketable maturity of trees, at least 3,500 small-scale farmers are already planting this tree in agroforestry systems in north-west Cameroon. One result has been the experimental cultivation of *P. africana* trees by some farmers—several of whom are former bark harvesters employed by Plantecam— not just for their bark, but for axe and hoe handles. The opportunities that this offers, the longer-term outlook for the international market for the bark, and the constraints on some aspects of *P. africana* cultivation, are discussed in this chapter.

## MEDICINAL USES OF *PRUNUS AFRICANA*

### Traditional Knowledge, and How It All Began

The first step in the multimillion-dollar trade in *P. africana* extracts sold to treat prostatitis in aging European men started in the Vryheid district of KwaZulu/Natal, South Africa, when a medical doctor recorded Zulu men using *P. africana* bark to treat urinary problems, an "old man's disease." As a result, he contacted Dr. Jacques Debat, who founded the pharmaceutical company Laboratoires Debat. Tests on rats and dogs performed by Dr. Debat confirmed the results, leading to a search for reliable sources of the bark and the filing of a patent in 1966 (Br. App. 1966).

*Prunus africana* is used not only by Zulu men in South Africa, however, but widely as a traditional medicine in southern, East, and Central Africa (Watt and Breyer-Brandwijk 1962; Kokwaro 1976; Jeanrenaud 1991). In the Mount Cameroon area, 88 percent of people collect traditional medicines and 14 percent of households surveyed collected *P. africana*, which was recorded as being the fourth most important medicinal plant species to the Bakweri people of this area (Jeanrenaud 1991). Similarly, it is an important medicine in the Ijim montane forest area, where it is used for curing malaria, stomach ache, and fever (Nsom and Dick 1992). Local use rarely kills the trees, as only small patches of bark are taken. In contrast with endemic plant species localized in just one country, with a small group of local healers knowledgeable about its use, knowledge of *P. africana* uses is widespread. This greatly complicates the already complex questions about whose knowledge it is, and who should benefit from commercial use of that knowledge.

### International Medicinal Use

Today, herbal preparations from *P. africana* are manufactured in eight European countries (most importantly, France, Italy, and Spain), as well as in the United States, Venezuela, Brazil, and Argentina. In all cases, the herbal extracts, usually in capsule form, are taken to relieve the symptoms of prostatitis or benign prostatic hypertrophy (BPH). These prostate problems, common among older men, result from a swollen prostate gland that reduces the ability to urinate. *P. africana* extracts increase bladder elasticity and stimulate prostatic secretion, reducing the symptoms of BPH (Bruneton 1995).

The market for these products is large, as is the market for saw palmetto (*Serenoa repens*) fruit extracts, which are also used in the United States and Europe to treat the same medical problem. It is now anticipated that one out of every two men in Western countries will live longer than eighty years, with the result that 88 percent have a chance to develop histologic evidence of BPH. Although surgery is commonly performed and effective, it is expensive, it can cause impotence, and there is a 1–3 percent postoperative mortality (*Hospital and Specialist Medicine* 1992). Therefore, medical therapy and phytotherapy are popular alternatives.

*Prunus africana* bark extracts contain fatty acids, sterols, and pentacyclic terpenoids (Longo and Tira 1981; Catalano et al. 1984; Uberti et al. 1990) and have a sitosterol glucoside content of 11 mg/100 g bark (Longo and Tira 1981). These extracts have been shown to be effective on rats (Thieblot et al. 1971, 1977), as well as in clinical trials conducted in Austria (Barlet et al. 1990). The latter, detailed clinical trial showed that 66 percent of patients treated with Tadenan containing *P. africana* bark extract had improved urine flow, compared to 31 percent given the placebo; 1.9 percent of the patients showed gastrointestinal side effects.

## HISTORY OF *PRUNUS AFRICANA* BARK EXTRACTION

Following their experimental work, Dr. Debat and his colleagues decided to go into commercial production of herbal extracts from *P. africana* bark. They conducted searches for good wild stocks of this species in many mountain forests in Africa, including Ethiopia, Kenya, and Cameroon. By 1972, Cameroon (figure 12-2) was established as the major source of supply, followed later by Kenya and Zaire. By 1995, the Plantecam factory was reported to have an annual turnover in Communauté Financiäre Africaine (CFA) francs of 2,000 million (about US$4 million), and it employed 250 permanent workers. Synthesis of active ingredients in *P. africana* extracts did not take place, as activity was considered to depend on the whole extract rather than on a single active ingredient. Thus, for over twenty years of exports, the

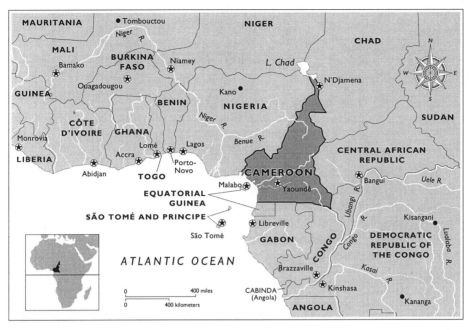

**Figure 12-2** Cameroon

emphasis has been on bark harvesting and preparing the bark extracts into tablet form. In 1992, nearly 2,000 tonnes of bark was harvested annually from Cameroon (table 12-1) (Cunningham and Mbenkum 1993), all of it from the wild. In Cameroon, virtually all bark is processed and exported by Plantecam (formerly Plantecam Medicam), a subsidiary of the French company Groupe Fournier, which bought Plantecam from Laboratoires Debat in the early 1990s. In Madagascar, bark is bought from villagers who harvest it from the wild in highland forests, and all of it is processed by the Société pour le Développement Industrielle des Plantes de Madagascar (SODIP), which then exports to the linked companies Indena Spa. and Inverni della Beffa in Italy.

Plantecam began harvesting bark from forests on Mount Cameroon before expanding into northwest Cameroon (Parrott and Parrott 1990). According to discussions with villagers during this survey, Plantecam visited Ntingue village, near Dschang in west Cameroon, in 1972, where they recruited at least 180 workers. Twenty years later, many of these bark harvesters are still working for Plantecam. Each worker was expected to collect a bundle weighing at least 40 kg, and bonuses were paid for heavier bundles. For fifteen years, Plantecam employed its own workers to harvest *P. africana* bark. The monopoly that the company held facilitated controlled harvesting, which was systematic: bark was removed from opposite sides of the trunk to avoid girdling (the local term is *ringbarking*) the tree, and felling was not allowed. Even with this system, some trees died, but the majority survived. In stark contrast is the uncontrolled bark exploitation in Madagascar, where trees are routinely felled and stripped of bark (Walter and Rakotonirina 1995).

**Table 12-1**

Sources and Destinations of *Prunus africana* Bark

| Source Country | Locality of Forests | Annual Bark Mass (tonnes/yr) | Importing Country |
|---|---|---|---|
| Cameroon[a] | Mt. Cameroon, Bamenda highlands (Mt. Oku, Mt. Kupe, Nso) | 1,116–3,900 (mean, 1,923)[b] | France |
| Kenya | Mau and other Afromontane forests in western Kenya (new tea estates, etc.) | 208[b] | France |
| Equatorial Guinea | Mountain forests on the island Bioko in the Gulf of Guinea | 200[c] | Spain |
| Uganda | Bushenyi district, including overexploitation within forest reserves | 96[d] | France |
| Zaire | Mt. Kivu, eastern Zaire[e] | 300 | France |
| Madagascar[a] | Montane forests in eastern Madagascar | 78–800 | Italy |

From *New Vision* (1993), Cunningham and Mbenkun (1993), Rasoanaivo (1990), and Cunningham et al. (1997).

[a]In Cameroon and Madagascar, bark is processed for extract, which is then exported.

[b]Over a 6-yr period (1986–1991).

[c]Estimate for 1994.

[d]Ceased in 1992.

[e]Probably a mixture with *P. crassifolia*.

In 1985, the harvesting monopoly in Cameroon broke when about fifty additional licenses were provided to Cameroonian entrepreneurs. Bark harvesting was therefore no longer controlled by Plantecam, although the company remained the sole exporter of *P. africana* bark and bark extract. Local contractors were licensed in order to increase income to Cameroonian entrepreneurs. Bark quotas were determined in the capital Yaounde, not on the basis of forest inventory and yield assessments, which merely encouraged overexploitation of wild stocks. This has been a particularly serious problem in northwest Cameroon, where bark harvesting has been in the hands of these local entrepreneurs since 1987. During the 1991/92 fiscal year, twenty-four entrepreneurs supplied Plantecam with *P. africana* bark (table 12-2). All Special Permit holders, including Plantecam, pay a regeneration tax (2 percent of the value of the raw material) and a transformation tax to the Forestry Department. These taxes are supposed to cover forest regeneration costs, but whether they are used for this purpose is doubtful.

In the late 1980s, export of dried bark from Cameroon and Madagascar changed to local processing of bark extract. This required major financial investments in equipment for bark maceration and production of bark extract in relatively remote factories in Mutengene, Cameroon, and in Fianarantsoa, Madagascar. This has led to additional local value-adding for the exporters and more local employment. Although unprocessed dried bark is still exported by ship from Zaire, Kenya, and Equa-

**Table 12-2**
Quantity of *Prunus africana* Bark Harvested by Plantecam Medicam, or Sold to Them by Special Permit Holders, 1991/1992[a]

| Company Name | Bark Mass (t) | Company Name | Bark Mass (t) |
| --- | --- | --- | --- |
| Tedongeh | 363 | Le Bien | 60 |
| Plantecam Medicam | 350 | Essama | 44 |
| Erimon | 237 | Emana | 38 |
| Mokom | 170 | Scao | 39 |
| Mbah | 113 | Effa | 37 |
| Tchiaze | 112 | Penandjo | 37 |
| Nguenang | 97 | Mama | 36 |
| Tioveh | 93 | Acome | 29 |
| Lutah | 86 | ITTC | 27 |
| Ngoko | 86 | Amougou | 22 |
| Mballa | 81 | ECIC | 18 |
| Tedongho | 71 | Jahoung | 11 |
| I K Ndi | 68 | | |

From Cameroon, Divisional Service of Forestry (1992).
[a]During the financial year.

torial Guinea, export from Cameroon and Madagascar has shifted to air-freight of the extract.

There is no doubt that at a national level, income from *P. africana* processing and exports is important. Income from bark harvesting also provides a highly significant source of cash income to villagers in Madagascar, despite price differences of up to 60 percent between buying sites (Walter and Rakotonirina 1995). The major profits are made by the exporting companies, with few benefits from local value-adding trickling down to bark gatherers (figure 12-3).

## FOREST CONSERVATION AND THE CONSEQUENCES OF HARVESTING BARK FROM THE WILD

### Theory and Practice of Sustainable Forest Harvest

*Prunus africana* is one of the few African tree species whose cambium usually survives after severe debarking, thus enabling bark regeneration. Even those trees that exhibit crown die-off after harvest (figure 12-4), or after being attacked by woodborers, show remarkable bark regrowth in moist sites. The ability of *P. africana* to withstand bark damage offers the potential for sustainable harvesting, but what is possible in theory is sometimes difficult to implement in practice.

Plantecam made a real effort to ensure that all the bark harvesters they employed were shown the procedure for removing bark. "Quarters" of bark were to be removed from opposite sides of the trunk (figure 12-5), beginning at a specified number of "cutlass lengths" above the ground but not above the level of the first branch;

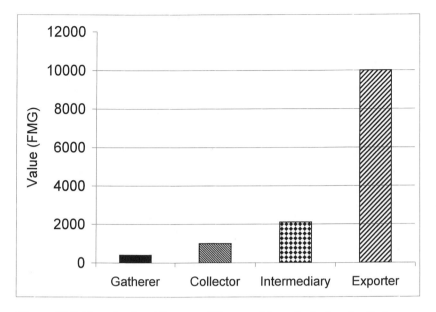

**Figure 12-3** The price for a kilogram of *Prunus africana* bark, showing the income to the gatherer, the bark collector, the intermediary, and the exporter. At the time of the study, this fluctuated between 1,500 and 2,060 FMG (Malagasy francs) per US$1. From Walter and Rakotonirina (1995).

this facilitated bark regrowth and decreased the risk of die-off. From 1972 to 1987, while this company held a monopoly, the harvesters were threatened with dismissal if they girdled or felled trees, or if they were caught removing all bark from the trees. This contrasts with the opportunistic scramble for bark since 1985 by licensed entrepreneurs, who felled or completely stripped trees, including those inside high-priority "protected" areas. Ever since Plantecam lost their monopoly, efforts at sustainable bark harvesting have been losing ground, particularly in northwest Cameroon.

In regard to species conservation, there are two effects of this uncontrolled harvesting that are of particular concern. First, a high level of bark exploitation is affecting wild populations of what Kalkman (1965) recognized as *Prunus crassifolia* in eastern Zaire. More research is required to confirm that this is a separate species from *P. africana*, but if Kalkman is correct, this endemic species, found nowhere else in the world, is under threat. Second, the removal of large, reproductively mature *Prunus* trees reduces seed dispersal and genetic flow between already isolated montane "islands," further increasing their isolation.

### Habitat Destruction and Change

A far more immediate concern is the effects of bark exploitation on mountain forest habitat. In Central Africa (Rwanda, Burundi, Uganda, western Kenya, and eastern

**Figure 12-4** *Prunus africana* tree isolated by forest clearing for small-scale farming. There is crown die-off after removal of quarter strips of bark from opposite sides of the tree trunk.

Zaire) and tropical West Africa (Cameroon), *P. africana* grows in areas with a cool, malaria-free, highland climate with good rainfall. These factors have attracted dense human populations (200 to 1,600 people/km$^2$) (figure 12-6). Clearing for agriculture is the primary cause of afromontane forest destruction. In Cameroon, Madagascar, and Zaire, this is compounded by the debarking of *P. africana* in remaining areas of forest. Recruitment of *P. africana* trees is from seed, usually in canopy gaps or disturbed areas, and large trees produce the largest quantities of seed. Bark gatherers

**Figure 12-5** Stripping of *Prunus africana* from one of two opposite
quarters of the trunk.

selectively harvest bark from the largest trees available, and when large *P. africana*
trees die off or their lifespan decreases after poor bark regrowth, there is less seed
production for the next generation. Furthermore, because of a forest population shift
to trees with smaller diameters, there are changes in forest canopy-gap dynamics:
Other forest species, including invasive introduced species, move in. This affects the
species composition in Afromontane forest where *P. africana* was formerly a dominant tree.

The decrease in seed production also means lower food availability for fruit-

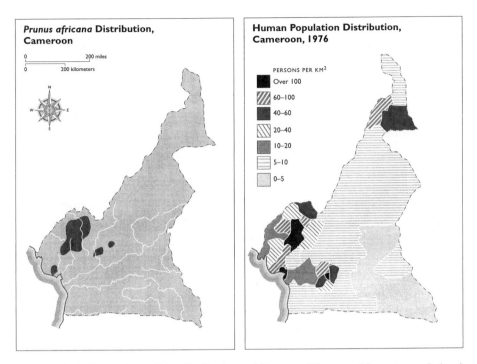

**Figure 12-6** Comparison of the distributions of *Prunus africana* and human population in Cameroon.

eating birds and mammals. Afromontane forests of Cameroon contain some of the most important habitats in Africa for bird conservation, with a high number of endemic bird species (Collar and Stuart 1988). Examples are the endangered Bannerman's turaco (*Turaco bannermani*) and the near-threatened Cameroon mountain greenbul (*Andropagus montanus*). These birds and the near-threatened primate Preuss's guenon (*Cercopithecus preussii*) are endemic to the Bamenda highlands, Cameroon. All feed on *P. africana* fruits in addition to that of other trees (Fotso and Parrott 1991; S. Tame, personal communication). At an average *P. africana* density of 5.5 trees/ha recorded for Mount Cameroon (Eben Ebai et al. 1992) and a bark yield per tree of 55 kg/tree, the commercial harvesting of 1,923 tonnes of bark per year would hypothetically affect over 6,300 ha of Afromontane forest annually. Commercial harvesting of *P. africana* bark has been taking place in the three Cameroonian forests that are most important for bird conservation (Mount Oku, Mount Cameroon, and Mount Kupe), as identified by Collar and Stuart (1988). For these reasons alone, it is important that *P. africana* cultivation be implemented as an alternative to wild stocks.

## Density and Recruitment of Prunus africana

In 1991, Plantecam funded a Cameroon Forest Department survey to determine the availability and distribution of *P. africana* on Mount Cameroon. The study highlighted the relatively low density (5.5 trees/ha) of *P. africana*; 83 percent of trees were immature, with a diameter of less than 20 cm in all sample plots (Eben Ebai et al. 1992). It should be noted that the survey benefited from the guidance of knowledgeable local people to sites where *P. africana* occurred. It can be expected that random sampling would result in an even lower population density.

The study by Samuel Eben Ebai and his colleagues also showed a low level of recruitment of *P. africana* trees in all six sample sites. Whether this was a result of episodic events, limited forest disturbance, twenty years of bark harvesting on Mount Cameroon, or a combination of these factors requires further investigation. The data emphasize the serious effects of felling or killing large *P. africana* trees in a situation where there is both a low population density and low recruitment, even though the majority of debarked trees had survived. Eben Ebai et al. (1992) recommended that Etinde Forest Reserve on Mount Cameroon, which excluded bark harvesting, be maintained as a control area for comparison with other sites where commercial bark removal is taking place. In late 1992 and in 1994, however, harvesting did take place in this protected area, as well as in other sites on Mount Cameroon. The highest mortality rates due to debarking and felling have occurred in the forests of Cameroon's northwest province.

## CONTRASTING VIEWS: COMPANIES, COMMUNITY LEADERS, AND HARVESTERS

The views of the director of the French company marketing *Prunus* extracts and those of a prominent traditional leader who was the source of information about the bark illustrate very divergent views on bark exploitation. To Dr. Jacques Debat, the Director of Laboratoires Debat, the situation in 1992 was one in which

> . . . in perfect agreement with local authorities we are seeing to it that the sustainability of the species is ensured, especially through significant reforestation measures.[1]

In Banso, northwest Cameroon, in the same year, debarking and die-off of *P. africana* trees extended even into a sacred forest that is the burial site of seventeen previous fons.[2] As a result, the present fon of Banso had a very different view, sending the following message to the owner of Plantecam:

> He [the bark buyer supplying Plantecam] came here and paid a small amount to a few people. The result was major destruction of the forest, our cultural heritage. If he comes to this region he will not be welcome.

The fon of Banso also considered that commercial harvesting of *P. africana* bark changed local perceptions of the forest, from that of a community asset (in this case referring to the forest directly under his control, where his ancestors were buried), to that of a resource to be exploited for personal gain (A. Hamilton, personal communication 1991). This was given as a major reason for the surge in forest destruction after 1985. It is difficult to evaluate whether this is correct, but there is no doubt that massive forest clearing took place after 1985, when coffee prices collapsed and farmers expanded their fields to grow alternative crops such as beans and potatoes.

Concern about the killing of *P. africana* trees was also expressed by all traditional leaders we spoke to (Chiefs Evakise, Ewome Linelo, and S. K. Liwonjo, personal communication 1992). Nevertheless, harvesting was still permitted by self-employed local harvesters and the permanently employed *Prunus* bark harvesters working for Plantecam in the areas these chiefs controlled.

Local harvesters each collected a 50 to 70 kg bundle of bark per day, working six to seven days per week. Common complaints were the difficulty of climbing trees, the poor food supplies, and the lack of social benefits and accident insurance for self-employed local harvesters. There was also widespread concern about the preferential employment of outsiders and the fact that the nonlocal, permanently employed bark harvesters were starting fires and robbing beehives in the forest. Bark harvesting is very hard work, most suited to strong young men who are capable of climbing *Prunus* trees and carrying the heavy bundles of bark through the mountain forest (figure 12-7). Consequently, no women harvest bark. All bark collectors interviewed were aware of the requirement to remove bark only from opposite quarters of the trunk and to return to the debarked tree only after four to five years. Although no bark collectors interviewed were cultivating *P. africana*, all expressed an interest, although they would require support with materials (implements, fencing) and information on planting methods.

## VOLUME AND ECONOMIC VALUE OF THE BARK TRADE

Twenty different European companies produce and sell herbal preparations containing *P. africana* extract (Cunningham et al. 1997). Several sell *P. africana* products under a licensing agreement with Groupe Fournier (France) or Indena Spa./ Inverni della Beffa (Italy). The product Tadenan, produced by Groupe Fournier, is one example: the same brand name is used by five different European companies. A wide variety of brand names in a range of dosages are available and are marketed in Europe, including Scandinavian countries, and in Australia. In the United States, *P. africana* is marketed as an extract mixed with other ingredients [including saw palmetto fruit extract and pumpkin (*Cucurbita pepo*) seeds]. The over-the-counter value of the world trade in this species is estimated to be US$220 million/year (Cunningham et al. 1997).

In 1976, only 10 tonnes of *P. africana* bark were exported from Cameroon (United Republic of Cameroon 1976). By 1986, this had increased to nearly 1,500 tonnes, with a total of 11,537 tonnes sold during the six-year period from 1986 to 1991 (Cunningham and Mbenkum 1993).

**Figure 12-7** Bark harvester with a 50 to 60 kg load of *Prunus africana* bark.

## Market Forces and Prunus Cultivation

In Cameroon, farmers are aware of the debarking of wild *P. africana* populations and the resultant scarcity of bark that has occurred in some areas. This has stimulated the farmers in northwest Cameroon to cultivate *P. africana* from seed on a much larger scale than Plantecam. With no support from government or non-government organizations (NGOs), at least 3,500 small-scale farmers are already planting this tree in agroforestry systems in northwest Cameroon (Cunningham et al., in press). Cultivation takes place in an agroforestry system, where *P. africana* trees are cultivated with food crops (yams, beans, taro), coffee, and other fruit trees. Farmers started *P. africana* cultivation in about 1977, and several did so experimentally, not just for their bark but for axe and hoe handles. This multiple-use value ensures local benefits to farmers even under the unlikely worst-case situation of a collapse in international demand for bark. Ideally, this offers the opportunity to supplement cultivation by Plantecam and Special Permit holders with production from farmers in montane Africa. As a fast-growing indigenous tree, *P. africana* also has great agroforestry and reforestation potential, especially on steep slopes where farming of annual food crops is unsustainable because of high rates of soil loss.

If small farmers are to become further involved in *P. africana* cultivation, however, there has to be a long-term market for the bark at a price that makes production profitable. In 1992, Plantecam said that they could offer a market for *P. africana* bark for the next five years, but they were unable to give any guarantee beyond that. This is unrealistic for a tree crop that is unlikely to reach harvestable maturity in less than eighteen years. Continued demand for herbal products produced from *P. africana* bark depends on competition from new treatments for BPH, and on public support for unconventional (or alternative) medicines in Europe and the United States.

There is competition from synthetic compounds such as antiandrogens and 5-alpha reductase inhibitors, and from the use of urethral probes. Studies by Merck, Sharpe & Dohme, for example, have shown that finasteride is a more effective synthetic 5-alpha reductase inhibitor than Tadenan (Rhodes et al. 1993). There is widespread support, however, for unconventional medicines in the United States (34 percent of respondents) (Eisenberg et al. 1993) and in Europe (Grunwald and Buttel 1996). The trend in saw palmetto fruit extract sales, also to treat prostate problems, indicates possible international trade in the future. So does the trend in incidence of prostatic disease. With people living longer, BPH will become more common among men in western Europe and the United States, increasing the market demand for treatment. At present, most *P. africana*-based products are sold within the European Union (EU), not in the potentially large and lucrative markets of North America and Japan. Most industrial companies producing herbal preparations from *P. africana* are located in EU countries with the most lucrative markets for herbal preparations. The United Kingdom is currently an exception to this. This will change in 1998 when European law will require European registration of herbal preparations.

A recent survey in selected health food stores showed that *Serenoa repens* was the sixth best selling herbal preparation in the United States, representing 4.4 percent of total sales (Brevoort 1995). Similarly, in Germany, *S. repens* herbal preparations were among the top ten most frequently prescribed herbal preparations (Grunwald and Buttel 1996). With the consolidation of the rapidly growing, lucrative market for herbal preparations to treat BPH, North America and Germany are considered to be the major potential markets for herbal preparations from *P. africana* in the future.

In southern Africa (South Africa, Zimbabwe), there is potential for cultivation of *P. africana* by small-holder farmers and, on a larger scale, in plantations similar to those for Australian black wattle (*Acacia mearnsii*, family Mimosaceae), which is grown in this region for its tannin-rich bark. The slower growth rate of *P. africana* trees compared to *A. mearnsii* is compensated for by the much higher value of the *P. africana* bark (US$1,000/ton versus US$77/ton). Because of the high density of rural farmers in the highlands of East and West Africa, there is little spare land for plantation-based cultivation of *P. africana*, so cultivation would need to be in the agroforestry system already implemented in Cameroon. In both cases, strong property rights over the land and trees is crucial. Weak tenure is considered to have been a key factor in the failure of the cinchona plantations in Cameroon.

## Lessons from Cinchona Bark Production: West Cameroon

In the late 1970s, in an effort to reduce dependency on imported antimalarial drugs, considerable effort and expense were put into establishing plantations of *Cinchona*, the South American tree whose bark is a source of quinine. The aim was to establish a 200 ha plantation in west Cameroon with an accompanying factory at Dschang through initial involvement of the Ministry of Agriculture, with the project to be handed over to the Union de Coopérative de Cacao et de Café de l'Ouest (UCCAO) (United Republic of Cameroon 1981). Expansion was planned from 10 ha plantings in the first year to 20 ha by 1982/83 and 30 ha/yr in the following years (United Republic of Cameroon 1976). After 1983, the plan was to cultivate 30 ha of smallholdings in west and northwest Cameroon. Bark production was projected at 300 kg/ha after five years.

A fully equipped factory was built at Dschang with supportive funding from the international community, accompanied by staff housing and imported equipment for processing of *Cinchona* bark right through to the production of quinine. In many ways, this factory symbolized an ideal often expressed in Africa: to reduce dependency on expensive imported pharmaceuticals and produce effective drugs for common and often fatal tropical diseases (in this case malaria). In addition, as *Cinchona* is an exotic species, production had to be based on cultivated stocks rather than exploitation from the wild, as is the case with *P. africana*. Despite the infrastructure and ideals, the project failed entirely after a few years. Today, the factory and its processing equipment are derelict, with little to show for the millions of dollars invested, in stark contrast to the thriving Plantecam factory in southwest Cameroon. The *Cinchona* plantation was felled for housing and bean fields: by 1992, only a single, debarked *Cinchona* tree remained.

If cultivation is to provide a viable alternative to harvesting of wild *P. africana* stocks, then it would be important to ensure that the incredible waste of energy and money on the *Cinchona* project is not repeated. This requires that bark yields from cultivated stocks not be overestimated, affirming that cultivation is an economic proposition, with fair prices paid for bark produced so that small farmers are involved.

## ENVIRONMENTAL AND CULTURAL ACCOUNTING

The lucrative market for herbal extracts from *P. africana* in Europe has generated a massive commercial demand for its bark from the highlands of Africa and Madagascar. In contrast to companies producing palm oil and rubber, which are both produced from large commercial plantations and small-holder production, *P. africana* continues to be taken from the wild. Plantecam deserves credit for attempting sustainable harvesting by removing bark from opposite sides of the trunk rather than girdling the trees. However, quotas have been awarded by the Forestry Department without adequate forest inventories and with limited resources to control exploitation. Overexploitation of *P. africana* in northwest Cameroon has worsened since bark harvesting permits were awarded to fifty entrepreneurs, and Plantecam lost its monopoly over bark harvesting. As the sole exporter of bark and bark extract, however, Plantecam must bear a large degree of responsibility for the overexploitation taking place and needs to remedy this situation. Although Plantecam has recently been supplying *P. africana* seedlings to local farmers, the practical impact of this needs to be assessed.

Despite the assurances of Laboratoires Debat that measures for conservation and sustainable use of *P. africana* have been taken by their subsidiary Plantecam, this has not applied to the entrepreneurs who are their main suppliers. Plantecam buys bark from Special Permit holders regardless of how the bark was harvested. A survey in 1992 showed that Plantecam had also not fulfilled afforestation requirements for cultivation of 5 ha/yr under their current license agreement (Cunningham and Mbenkum 1993). The Forestry Department has limited money and manpower, which limits the taking of forest inventory and control of bark exploitation. This is worsened by corruption and the poor economic situation in Cameroon, particularly since the 40 percent devaluation of the CFA franc in 1994. The situation casts serious doubt on managed, sustainable harvesting from wild populations and probably enrichment plantings. The 1991 "ban" on bark harvesting is a good example of this: it had the opposite effect of doubling the quantity of bark sold to Plantecam.

Many national governments frequently consider that their forests and wetlands are "free goods," and that degradation does not count as depreciation in calculations of gross national product (WRI 1992). *P. africana* bark harvesting is a good example of this. Quotas for bark exploitation were awarded without adequate knowledge of the quantities or consequences involved. The quantity of bark harvested has increased annually (Parrott and Parrott 1990; Besong et al. 1991), which has led to the "mining" of a natural resource rather than its managed use.

Although environmental accounting studies are in their infancy, a case study would be useful to show the economic benefits of *P. africana* bark harvesting and the real costs involved. Local people and traditional leaders are aware of many of these costs: the waste of *Prunus* timber from girdled trees that die and are left to rot; reduced accessibility to *P. africana* as a source of traditional medicine; and the degradation of country roads by heavy vehicles transporting the bark (Macleod and Parrott 1991). The 2 percent "regeneration" tax is also less than the cost of tree replacement.

Plantecam and other Special Permit holders have to take responsibility for establishing large-enough *P. africana* populations to replace harvesting of wild stocks. They should acquire land to cultivate *P. africana*, rather than expect land to be provided by the Forestry Department. If cultivation is to replace harvesting of wild stocks, then cultivation of *P. africana* needs to be an economic proposition. This largely depends on whether the price being paid for the bark reflects its *real* value in terms of the time it took to produce.

## POLICY AND PRACTICE

Detailed resource management recommendations were made by Parrott and Parrott (1990) of the Birdlife International project at Kilum Mountain as the problem worsened and affected important conservation areas in northwest Cameroon, stressing the need for prompt action (table 12-3). In January 1991, three senior members of the Forestry Department visited areas of southwest, west, and northwest Cameroon where *P. africana* harvesting has taken place. They met with harvesters, conservation NGOs, and representatives of Plantecam (Besong et al. 1991). In their report they pointed out the risks of permitting the exploitation of *P. africana* bark when the extent and the productivity of the resource were unknown. Besong et al. (1991) also record the insufficient control of bark exploitation, lack of respect for quotas and forestry regulations, and consequent resource degradation. A major reason identified for this is that people did not feel responsible for the forest resource, resulting in opportunistic overexploitation. Key recommendations made by Besong et al. (1991) are shown in table 12-4.

It was specified in the current special license that Plantecam should plant 5 ha of *P. africana* per year for the next five years. A survey in 1992 showed that at most,

**Table 12-3**
Summary of International Council for Bird Preservation (ICBP) Recommendations

---

1. There should be a complete ban, pending further study.
2. An inventory should be carried out to determine the status of *Prunus* throughout Cameroon's forests.
3. An independent study should be made to determine the best harvesting techniques and intervals for sustainable harvesting of *Prunus* bark.
4. On the basis of points (2) and (3), annual harvesting quotas should be calculated for each forest area and each Division.
5. On the basis of points (2), (3), and (4), permits should be issued against a license fee and a deposit. The deposit would be returned when the Forestry Department is satisfied that the contractor has complied with recommended harvesting procedures.

---

From Parrott and Parrott (1990).

**Table 12-4**
Summary of Department of Forests Recommendations

1. Reduce quotas and limit the number of permits given out.
2. Restrict the activities of permit holders to the forest zones allocated to them. Inventories should be done in these areas by the permit holders and their accuracy certified by the local Conservator of Forests.
3. Permit holders should be held responsible for damage caused to the trees. Bark harvesting techniques need to be respected (i.e., bark removal from opposite quarters of the trunk, up to the first branch). Killing of trees is forbidden in forestry legislation. This gives adequate power to Conservators to control this situation.
4. The minimum exploitable diameter should be increased to 40 cm diameter at breast height (DBH).
5. The total volume of bark harvested annually must not exceed 1,500 tonnes/yr.
6. Reforestation needs to be introduced.
7. Forest regeneration taxes need to be raised and linked to this exploitation, to finance a silviculture program.
8. An education campaign should be started to raise awareness of the importance of trees and their value as a source of revenue.
9. A meeting should be held with all interested parties (government, Plantecam Medicam, permit holders, etc.) to discuss the problems of production, allocation of permits, pricing, and the role of each permit holder.
10. Plantecam Medicam must create its own plantations, following the example of private oil palm and rubber companies such as HEVECAM and SOCOPALM in Cameroon.

From Besong et al. (1991).

the forestry nursery at Buea near Mount Cameroon only held 1,000 seedlings on behalf of Plantecam and this would be enough to plant only 1 ha. Only 2 ha have been planted at Buea, and there is excessively wide spacing of those trees: 5 m apart and 5 m between rows. There is little doubt that the requirement to plant 5 ha of *P. africana* per year has not been fulfilled. It is also clear that even if the goal were achieved, it would not be sufficient to replace the wild harvested trees. The number of trees that need to be planted if bark production is to be sustainable can be estimated by examining the well-documented cultivation of *Acacia mearnsii* for its tannin-producing bark and comparing it to records of total bark yields of *P. africana* trees from wild collections in the forests of northwest Cameroon.

In the northwest Cameroonian forests, mean bark yield per tree was 55 kg, with variation in yields of 38 to 73.8 kg/tree, with the exception of April 1985, when mean bark yield was very high (128.2 kg/tree) (Macleod 1987). This could have been a result of overharvesting or of undercounting the number of trees harvested. The annual quantity of *P. africana* bark harvested (1,923 metric tonnes/yr) would represent nearly 35,000 trees/yr, assuming that regulations were followed and bark was removed on opposite quarters of the trunk. From field experience and the observations of others (Besong et al. 1991; Parrott and Parrott 1990) this is clearly not the case today, when tree felling and total bark removal are common.

The mean quantity of *P. africana* bark processed annually from 1986 to 1991 by Plantecam was 1,923 metric tonnes/yr. Assuming that bark production and growth rates in *P. africana* plantations would be similar to those for *A. mearnsii*, this quantity of bark would be produced by 93,229 trees/yr (68.4 ha of 1,363 trees/ha) felled and totally stripped of bark each year (Schonau 1973, 1974). A twelve-year rotation would therefore require 820.8 ha, and even if the planting of 5 ha of *P. africana* trees/yr for five years were being carried out by Plantecam, this would be totally inadequate for the existing demand for bark. The current efforts at *P. africana* cultivation can in no way be expected to take the harvesting pressure off wild stocks.

As a relatively fast-growing indigenous tree, *P. africana* also has great potential for reforestation and agroforestry systems in deforested areas around forest remnants in Cameroon, Madagascar, Kenya, Uganda, and Zaire (Cunningham et al., in press). Managed sustainable harvesting is theoretically possible with *P. africana* because of the remarkable bark recovery of a high proportion of trees after partial bark removal. In practice, however, the high inputs required of Forestry Department money and manpower are not available and are unlikely to become so with the current economic situation in all source countries. Bark exploitation has caused serious damage to *P. africana* populations, particularly in Madagascar and Cameroon, where Afromontane forests are small and tree population densities are low. Bark exploitation to meet high export demand even takes place inside "protected" forests and national parks. For these reasons, with support from Kenya, *P. africana* was placed in appendix 2 of the CITES report in 1994, so that international monitoring of this trade could take place.

Cultivation in plantations or agroforestry systems can potentially meet future demand, and thousands of Cameroonian farmers are now growing this species in agroforestry systems. It is now necessary that international awareness be raised concerning the destructive effects of this trade on wild populations. Solutions should involve changes in bark price and a shift to cultivated sources of bark.

## ACKNOWLEDGMENTS

I would like acknowledge the assistance of C. A. Asanga, J. B. Besong, G. Bockett, B. N. Nkongo, S. Eben Ebai, S. Gartlan, T. Mbenkum, S. Tame, V. Tame, N. Ndam, M. Bovey, B. N. Ewusi, S. Ekema, C. Dig, F. C. Foncham, E. Legendre, and R. Ordot for information provided in the survey that formed the basis of this chapter. Help and advice was also provided by Chief Evakise (Bokwango), Chief Ewome Linelo (Bwassa), Chief S. K. Liwonjo (Mapanja). Particular thanks go to A. Hamilton for his great interest and encouragement in this project and M. Cunningham for editing. Financial assistance from the World Wide Fund for Nature (WWF), UNESCO, and the Tropical Forest Programme (United States Department of Agriculture Forest Service) (grant no. 93-G-001) enabled the work to be done in Cameroon as part of the People and Plants WWF/UNESCO/Kew joint program in ethnobotany with WWF-Cameroon and the Centre for the Study of Medicinal Plants, Yaounde.

## NOTES

1. "Et en parfaite entente avec les autorités locales, nous veillons à ce que la pérennité des espèces végétables concernées soit assurée, notamment par d'importantes mesures de reboisement."

2. The term *fon* (plural, fons) refers to the traditional leaders (kings) of what were small precolonial kingdoms scattered through the fertile mountainous landscapes of north and northwest Cameroon.

# CHAPTER 13

# A Tale of Two Villages: Culture, Conservation, and Ecocolonialism in Samoa

*Paul Alan Cox*

It was the best of times and the worst of times for two villages in Samoa. One village enjoyed cooperative efforts between villagers and foreign donors to establish and develop a large rain forest preserve. A second village experienced intense conflict between indigenous people and a Western conservation organization. What caused the difference in the conservation outcomes in these two villages? Why in the first case were the donors and villagers happy, while in the second case the antagonism between the villagers and a foreign conservation organization captured the attention of international media?

The stories of Falealupo and Tafua villages in Samoa suggest that issues of justice perceived by Westerners are not always the same as those prized by indigenous peoples. Where ecosystem protection, legal process, and intellectual property rights hold center stage in Western conservation discussions, respect for traditional leaders, human dignity, and observance of cultural forms loom large for many indigenous peoples. And where Western conservation ethics are based on principles of resource protection and wise use, indigenous conservation imperatives are often founded on traditional use and religious obligations. Issues of justice and conservation are often seen differently through Western and indigenous eyes, yet indigenous voices have long been ignored in the establishment of conservation areas, even though indigenous people control the majority of the world's natural areas. As a result, the presence of indigenous peoples historically has been seen (by Westerners) as a complicating factor in the creation of nature preserves rather than as a tremendous asset.

The moral of this tale of two villages is simple to state but broad in its implication: Western values of justice, equality, and equanimity are not universal and often differ from indigenous beliefs. Imposition of such Western values on indigenous peoples, even when done with the best of intentions, smacks of colonialism; if done in the name of conservation, it becomes ecocolonialism.

## BACKGROUND

Samoa is an archipelago of fifteen inhabited islands in the South Pacific, located approximately halfway between Hawaii and New Zealand (figure 13-1). The eastern and more industrialized part of the archipelago is known as American Samoa, a U.S. territory with a population of 47,000. The Western half of the archipelago comprises Samoa (which recently changed its name from Western Samoa), an independent nation that includes four inhabited islands with a total land area of 1,104 square miles and a population of 182,000 (figure 13-2). The least developed, largest, and most species-diverse of these islands is Savai'i.

In 1984, I began an extended study of Samoan herbal medicine, focusing on healing practices in one the most remote and un-Westernized villages in Savai'i, Falealupo village. I chose Falealupo as a study site primarily because of its large, 30,000-acre lowland rain forest, its remoteness, and the largely traditional lifestyles of its inhabitants. Lowland rain forests in Western Samoa have been under serious threat during the last two decades, with significant areas being logged for timber. In addition, drastic declines in populations of the keystone pollinators, flying foxes, have alarmed both villagers and foreign scientists.

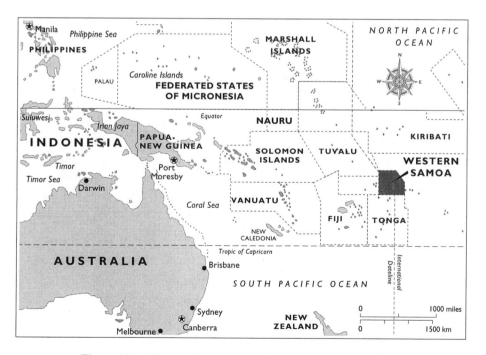

**Figure 13-1** Western Samoa, east of Indonesia and west of Hawaii.

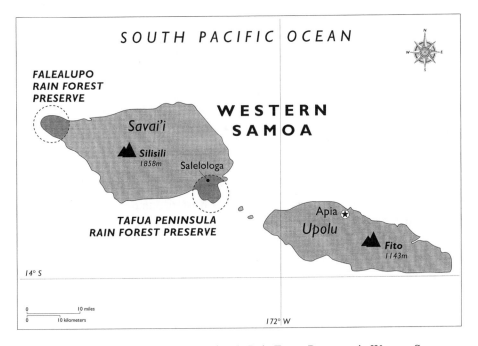

**Figure 13-2** Falealupo and Tafua Peninsula Rain Forest Preserves in Western Samoa.

At the initiation of my ethnobotanical studies, few protected areas existed in Western Samoa, and those that had been established, such as 'O Le Pupu'pue National Park on the island of Upolu, were nascent in the development of their management philosophy and operations. None of the Savai'i forests enjoyed any formal protection, and only a few areas of coastal and lowland rain forest on the island remained undisturbed.

Yet traditional land tenure systems remain strong in Samoa, with most of the undisturbed forests and other natural areas occurring on communally held land. Such land, owned by extended family groups or villages, is under the stewardship of village chiefs chosen in part for skill in making wise land allocations. Although increased Westernization has begun to erode this system, under Samoan laws such communal lands can be neither bought nor sold. As my ethnobotanical studies progressed over the next decade, it became increasingly apparent that conservation efforts in Samoa were unlikely to be successful unless they were based on the traditional land tenure system under the control of village leaders.

## TRADITIONAL SAMOAN HEALING SYSTEMS

The value of communally owned forests as sources of novel pharmaceuticals also became increasingly clear during the research. A description of Samoan ethnopharmacology appears elsewhere (Cox 1990b, 1991, 1994a), but to summarize, except for a group of basic remedies known to nearly all Samoans, the practice of Samoan herbalism is carried on by healers called *taulasea*, women who learned their craft from their mothers or other female relatives. However, the few remaining Samoan herbalists are old and seldom have apprentices.

Samoan ethnomedicine differs significantly from Western medicine in its concepts of disease etiology (Cox 1990b). As a result, many Samoan diseases are not directly translatable into Western terminology. If a healer diagnoses a disease requiring an herbal treatment, she collects plants from the forest or village to treat the disease. Only fresh plants, usually angiosperms, are collected and then formulated as water or oil infusions. Some are even ignited and inhaled. Many remedies, including those for internal ailments, are applied externally. Most treatments are prepared as multispecies combinations.

In my field work, I have recorded eighty-one ethnotaxa comprising seventy-five plant species that are used by Samoan healers to treat the ill, a figure that represents only a subset of the total Samoan ethnopharmacopoeia (Cox 1995). Analyses of these eighty-one ethnotaxa reveal some interesting patterns (Cox 1994b). Although cultivated plants are used in Samoan herbal medicine, most Samoan medicinal plants are gathered from the wild, a practice common throughout Polynesia (Zepernick 1972). Many of these plants are feral, growing near the beach or village areas, but some are collected from primary and secondary forests, highlighting the importance of rain forest conservation for Samoan culture: logging primary rain forest in Samoa is the equivalent of destroying traditional pharmacies.

Yet the importance of the Samoan forest as a source of indigenous pharmaceuticals has not been appreciated by Western visitors, who question the efficacy and authenticity of Samoan herbal remedies. Accounts of early missionaries have led some to conclude that Samoan herbal medicine developed after European contact (Macpherson 1985), but since Samoa was colonized by the Polynesians as early as 940 B.C. (Davidson 1979), it is unlikely that Samoan ethnomedicine is an introduced tradition. In general, there is an absence of ethnobotanical information in missionary accounts, including plant uses that would capture the attention of any botanically adept observer. For example, the widespread use of plants as fish poisons (Cox 1979) and in food-preservation technologies in Samoa (Cox 1980a, 1980b) received little mention by the early missionaries, who denigrated the medical acumen of Samoans (Williams 1837; Turner 1861, 1884; Stair 1897). "The Samoans in their heathenism," British missionary George Turner (1884:139–140) wrote,

> seldom had recourse to any internal remedy except an emetic, which they used after eating a poisonous fish. Sometimes juices from the bush were tried; at other times the patient drank on at water until it was rejected; and, on some occasions, mud, and even the most unmentionable filth, was mixed up and taken as an emetic draught.

Eventually, the confidence of the Samoan people in their traditional medicine was reduced by such disparagement. "The trust of the Samoans in their medical science continuously disappears," German botanist Franz Reinecke wrote in 1895 (p. 22), "in favor of the viewpoint of the [Western] medical representatives." Yet despite this denigration, Samoan herbalism continued. "A majority of the Samoans believe that their own crude drugs and harsh medical treatment are more efficacious than the purer manufactured drugs supplied to them through the Medical Department," the Lieutenant Commander Hunt of the U.S. Navy Medical Corps in Pago Pago wrote in 1923. Rather than ignore Samoan herbalism, some Western physicians and investigators began to record the tradition (Stephenson 1934; Christopherson 1935, 1938; Crosby and Brown 1937; Williams 1952; McCuddin 1974). Recent botanists have also recorded many medicinal plants (Uhe 1974; Whistler 1984), yet several anthropological studies (Forsyth 1983; Gerding 1986; Macpherson 1985; Macpherson and Macpherson 1990) relegate plants to a minor role in Samoan healing.

The first studies of possible pharmacological activity of Samoan medicinal plants were made in 1973 (Norton et al. 1973); they showed that a majority of the plants produced significant hypotensive effects in rats; 38 percent showed antibacterial activity, and 15 percent showed antiviral activity. But no significant attempt to isolate or develop pharmacologically active compounds from Samoan medicinal plants was made prior to the 1980s (Cox 1990a).

## CONSERVATION AND TRADITIONAL MEDICINE: DESTRUCTION OF THE SAMOAN FOREST

During the destruction of much of the Savai'i forest, Falealupo village resisted all approaches by companies to log their forests. But in 1988, the village desperately needed money to construct a school required by the Western Samoan government. Since per capita annual income in Falealupo is less than US$100, the logging company's offer represented the only potential source of funding. If Falealupo did not build the school, village children would not be allowed access into the Western Samoan educational system. Reluctantly, the village decided to allow logging, but on a license basis rather than on the leasehold basis requested by the company. Under this scheme, the company would pay the village for each tree removed from the forest and would be under strict instructions to immediately cease logging after the US$65,000 required to build the school was accumulated.

Logging of the Falealupo forest commenced in June 1988 to the considerable dismay of the villagers who openly wept as they saw the bulldozers push over the rain forest trees and scoop the soil aside. I and other members of my research team shared the villagers' distress. It was clear that the destruction of the Falealupo forest would result not only in a significant loss of biodiversity, but also in a loss of potentially important pharmacological compounds. Collaborative efforts with the University of Uppsala had indicated significant pharmacological activity in a variety of Samoan medicinal plants (Cox et al. 1989) and, subsequent to logging, several important anti-inflammatory compounds were isolated from Falealupo plants by a group headed by V. D. Hegde at the Schering-Plough Research Institute in Bloom-field, New Jersey. Of particular interest was the discovery, through collaboration with the National Cancer Institute (NCI), of a potent and novel antiviral compound from a Samoan plant: prostratin.

Samoan healers use water infusions of the macerated wood of *Homalanthus nutans* (Euphorbiaceae) as a tonic and to treat hepatitis. Extracts of *H. nutans* screened for anti-human immunodeficiency virus (HIV) activity by a team led by Michael Boyd at NCI exhibited potent activity. Bioassay-guided fractionation resulted in the isolation of prostratin (12-deoxyphorbol 13-acetate). At noncytotoxic concentrations, prostratin prevented HIV-1 reproduction and fully protected human cells from the lytic effects of HIV-1. Since phorbols are known to be tumor promoters (Evans 1986), identification of the active component as a phorbol raised questions concerning its therapeutic potential. However, in contrast to many other phorbol derivatives, prostratin is not a tumor promoter (Zayed et al. 1984; Gustafson et al. 1992). The NCI team concluded that prostratin "represents a non-promoting activator of protein kinase C which strongly inhibits the killing of human host cells in vitro by HIV. By these criteria, prostratin is unique" (Gustafson et al. 1992). Currently, NCI is accepting bids from pharmaceutical companies to license prostratin as a part of a combination therapy for AIDS.

Although *Homlanathus* occurred elsewhere in Samoa, we could not discount the possibility that populations of the plant differed in chemical composition and concentration, in the same way that *Cinchona* populations differ dramatically in quinine concentrations (Balick and Cox 1996). A great deal was potentially at risk as the first bulldozer blades bit into the Falealupo forest.

## CONSERVATION CRISIS AND THE FALEALUPO COVENANT

After the loggers arrived, I offered to raise money to build the school if the village could stop the logging. The village chief's council accepted the offer, so I signed the mortgage for the Falealupo primary school as a personal debt. In the following months, approximately $85,000 was raised from private donors for this cause, as well as three large contributions from the business community: Nature's Way, a manufacturer of herbal medicine in Springville, Utah; Ever Living Products, a marketer of aloe vera products in Phoenix, Arizona; and James MacEwan and Associates, a dealer in rare botanical art based in London. Subsequently, Scandinature Films in Sweden donated proceeds from video sales of a documentary film on Falealupo to the cause.

In November 1988, a covenant was negotiated in the Samoan language between the donors and the villagers to protect the Falealupo forest. In January 1989, a representative group of donors met with Falealupo villagers and signed the Falealupo covenant. In this covenant, the donors explicitly renounced any rights in village lands and forests, but they pledged to build the school in exchange for the village's promise to continue to protect the rain forest for a fifty-year period. The covenant allows village use of the forest for medicinal plants, kava bowls, canoe construction, and other cultural purposes, but it forbids logging or any other activities that could significantly damage the forest. The Falealupo covenant also addressed the disposition of royalty income for any discovery of new drug compounds from the Falealupo forest.

The advantages of the Falealupo covenant as a conservation instrument soon became apparent. Relatively small sums of money were required, far less than would be required to purchase the land, which in any case was not for sale. Donated funds were not distributed to individuals but instead used to construct a needed public work. The covenant relationship entailed few enforcement or survey problems; the villagers, who by covenant are responsible for protection of the forest, use traditional enforcement procedures, which are very effective, and they are required to protect only the lands that they control by tradition. And finally, by placing administration of the preserve completely in the hands of the villagers, conflicts between reserve management and indigenous cultural and spiritual values were minimized.

## INDIGENOUS KNOWLEDGE AND INTELLECTUAL PROPERTY RIGHTS

Before beginning studies of traditional medicine in Falealupo village, I sought permission of the Falealupo chief's council and healers. I explained that my intent was to discover new medicines and, while the chances for success were low, I would protect their financial interests should any novel compounds be found.

This understanding was codified in the Falealupo covenant, which calls for a return to the village of 30 percent of patent income generated by any pharmaceutical discovered in the Falealupo forest. As this book goes to press, there has yet to be any commercial development of a pharmaceutical from the Falealupo forest. But in the case of the antiviral compound prostratin, the NCI has guaranteed that a significant portion of any royalty income will be returned to the Samoan people, a guarantee that appeared in the Federal Register: the NCI requires any drug company wishing to license prostratin to negotiate directly with the Western Samoa government about the appropriate shares of the proceeds of gross sales. Furthermore, in recognition of the terms of the Falealupo covenant, Brigham Young University has guaranteed to return to the village of Falealupo one-third of its royalty shares from the development of Prostratin. Finally, should Prostratin be developed as a commercial drug, it is my hope that sourcing of plant materials will focus on plantations in Falealupo and other Samoan villages that have courageously protected their rain forest. Yet, as in so many drug discovery programs, royalty income may never be generated, or it may occur only in the distant future. Hence it is important to ensure that the village enjoy some immediate benefits. To that end, Seacology, a 501(c)3 nonprofit foundation (www.seacology.org), raised funds to build the Falealupo primary school, to rebuild it after it was destroyed by cyclone Ofa, and to rebuild an entirely new school after hurricane Val. We have also donated personal funds to build a small clinic, water tanks, and houses for several villagers. In excess of $235,000 has been transferred to village accounts for these projects, which have been completely designed, controlled, and administered by the village chief's council.

Trail construction and a rain forest information center have been supported by $16,000 forwarded to Seacology from the Australian Conservation Foundation, and a $10,000 grant was recently made by Seacology for village water catchments. Most recently, on May 11, 1997, a $75,000 aerial walkway, funded by a generous donation from Nu Skin International, was dedicated by government officials, village leaders, and Nu Skin representatives. In addition to providing a unique research platform, the rain forest walkway has already generated significant ecotourism income for the village. Finally, my half of a $75,000 Goldman Environmental Prize shared with Falealupo chief Fuiono Senio has been donated to establish a permanent endowment for the Falealupo Rain Forest Preserve. Generous matches of the prize from Nature's Way and Nu Skin International have allowed $100,000 to be set aside for the endowment, with an additional $12,500 being allocated to a similar endowment for Tafua village. Annual checks will be issued in perpetuity to Falealupo as long as the rain forest stands. In total, over one-half million U.S. dollars in cash and hundreds of hours in donated labor have been returned to the Falealupo well in advance of the commercialization of any pharmaceutical compound. Both donors and villagers are delighted with the way the covenant relationship has evolved. They have enjoyed reciprocal visits, and have each independently pledged to extend their covenant obligations forever. A detailed narrative of events leading to the Falealupo covenant can be found in Cox (1997).

## CULTURAL CONFLICT AND THE TAFUA COVENANT

News of the Falealupo covenant and reserve spread to Tafua village, which faced a similar problem: the need to raise funds to build a school. Because of the proximity of the Tafua peninsula to the Salelologa wharf, logging companies had made numerous attempts to harvest this forest. The Tafua peninsula, like the Falealupo, contains a largely undisturbed lowland rain forest. In the Tafua peninsula, the rain forests are controlled by three villages: Fa'aala, Salelologa, and Tafua; of these, Tafua village has by far the largest percentage. Thomas Elmqvist and I sought to use the Falealupo covenant as a model for covenants with the three villages on the Tafua peninsula.

Dr. Elmqvist and I traveled to Stockholm and asked the Swedish Society for Nature Conservation (SSNC) to serve as a repository for funds we intended to raise in Europe for the school. SSNC agreed and launched an appeal to its own membership. Soon we negotiated a covenant between SSNC and Tafua village for creation of the Tafua Rain Forest Preserve, and similar covenants with the other two villages followed. All three covenants were patterned after the Falealupo covenant. The SSNC renounced all rights to the land, pledged to respect village sovereignty, and promised to build the school. The villagers guaranteed to preserve the forest for fifty years, and, unlike the Falealupo covenant, also promised to protect contiguous marine resources, including a ban on hunting of turtles. The Tafua covenant was signed during a kava ceremony in January 1990. Two other covenants were signed with Fa'aala village in October 1990, and with Salelologa village in January 1991, completing the efforts to make the entire peninsula a preserve.

Unfortunately, the covenants were soon threatened by a long and complicated sequence of deep conflicts between Tafua village and the SSNC (Cox and Elmqvist 1993). Initially, relations between SSNC and Tafua village were very good. The chiefs of Tafua village assisted SSNC in negotiating similar covenants for contiguous parcels of rain forest in Salelologa and Fa'aala villages. Based on this success, an application for supplemental funding from the Swedish International Development Authority (SIDA) was prepared in consultation with the village chiefs. It was proposed that SIDA funds be used to assist Tafua, Fa'aala, and Salelologa villages (which by covenant had been denied logging revenues) in a variety of projects, including rain forest information centers, solar power installations, trails, signs, and rain forest information material. However, although over $800,000 was awarded for these projects, few were ever initiated. Instead of forwarding the funds to the village, SSNC used most of the monies to support environmental bureaucracy. In May 1997, an hour-long documentary film (*Snakes in Paradise*) on the disposition of these aid funds aired on prime-time television throughout Sweden.

Of greater importance to the villagers than the derailing of the aid funds, however, was the profound cultural conflict during the course of the project. On 23 January 1993, three years after initially signing the rain forest covenant, Tafua village renounced any continued association with SSNC and refused any future financial assistance from them. The village announcement received considerable television and newspaper coverage in Sweden. Since that time, Salelologa has announced its intention to develop part of its covenant preserve as a township expansion. Recently, a road was carved through the preserve by the government as the first part of the proposed development.

Why did these profound changes in covenant agreements occur in the Tafua peninsula? The advent of the SIDA funds resulted in the injection of new personnel and attitudes into the Samoan project. Uncomfortable about working directly with village leaders, the new personnel chose to channel support through a newly created local nongovernmental organization (NGO), formed primarily for that purpose, called O Le Si'osi'omaga. Until 1991, SSNC had direct liaison with the traditional leaders of Tafua village. In 1991, however, SSNC staff sought to place the Western-educated staff of the new NGO in control of the project. The board and membership of the small NGO were composed primarily of Westerners and Western-educated elites who lived in the Samoan capital of Apia. In a contract signed on 22 April 1991, SSNC engaged the services of the NGO with a salaried "project coordinator" to be appointed on the basis of agreement solely between SSNC and the NGO, without any input from the covenant villages. For these services, SSNC agreed to pay the new NGO a fee of US$150,000. The local NGO sought and obtained funding from SSNC for an Apia office, staff salaries, computers, a fax machine, a business vehicle, and overseas air tickets, including trips to Stockholm, New York City, and the environmental summit in Rio de Janeiro. Perhaps because of the rigor of international travel and the press of their concurrent consulting obligations, the NGO staff found it difficult to find time to visit the covenant villages. Yet they publicly decried what they termed a "handout mentality" on the part of village leaders, who were puzzled by the repeated delays in the initiation of promised development projects.

Although they were residents of the same country, cultural differences between staff of the local NGO and the villagers rapidly emerged. For example, important meetings in Tafua village are always preceded by ceremonial administration of kava, a beverage made from water infusions of the roots and rhizomes of *Piper methysticum*. During kava ceremonies, ancient rhetorical forms are observed. However, NGO staffers, uncomfortable with traditional forms of village etiquette, requested that no kava ceremonies be held and increasingly requested village leaders to travel to their Apia office for meetings.

As further difficulties emerged, village leaders made repeated attempts to reconcile the differences. Chief Ulu Taufa'asisina, who neither requested nor received any payment for protecting Tafua's rain forest during his lifetime, was astonished by the insertion of highly paid local NGO staffers into the covenant agreement. "Then the Si'osi'omaga entered between us in our relationship," Ulu wrote in a letter to SSNC. "This became a continuing source of confusion and continuously generated innuendo and rumors about this and that. But in our covenant there is no provision for such people. I solemnly testify there were included in the covenant only representatives of your country. This is absolutely clear—the correct truth of all these things" (Taufa'asisina 1993a).

The original idea of village control of the project was rapidly eroded as more and more power and authority was transferred by SSNC staff from the village councils to the NGO for allocation of funds and decision-making authority.

In his study of the Tafua situation, Hardie-Boys (1994) wrote,

> The establishment and incorporation of O Le Si'osi'omaga into the project was alien to the local culture. Operating as it does outside the established social structure, it is a foreign concept. . . . The technical language of the largely Western and academic Si'osi'omaga then went on to clash with the cultural language of the village.

The villagers began to believe that SSNC and the local NGO did not respect their culture and their traditional power structure, and they wrote a series of letters in protest. Eventually, the village banned NGO representatives from the village and the forest preserve and urgently requested a meeting with the SSNC president. SSNC staff, in turn, continued to direct village leaders to the local NGO for assistance.

In Samoan culture, it is believed that disputes can be best resolved through face-to-face discussions. So, with funding from overseas friends, Chief Ulu Taufa'asisina traveled to Stockholm where he made a dramatic appearance at an SSNC board meeting. But even this effort at direct negotiation failed to convince the SSNC board to remove the local NGO from project oversight. In frustration, Tafua village finally decided to renounce its association with SSNC. In his letter composed immediately after the Stockholm meeting, Chief Ulu Taufa'asisina wrote,

> To the SSNC: I have the conclusion of our discussion as we met at your meeting. I tell you that I will return completely unsatisfied with your decision. Enough of that. You have struggled in crooked tricks and ways that are filled with Satanic opinions to corrupt the righteousness from God.
>
> You have reached your decision. I will tell you the decision of our village. At this time I tell you that our relationship is severed and our covenant ended. Even with my poverty I do not want much money that comes through filthy paths. The truth is that my trust, which I thought was true compassion, has been betrayed, but now it appears that it was the kiss of Judas Iscariot.
>
> I am Ulu Taufa'asisina Tausaga, the representative of the chiefs and orators and myself. I confirm this day that our friendship is broken. You continue to support the Si'osi'omaga but my village will resign from it.
>
> Taufa'asisina (1993b)

The ongoing conflicts were muted when the president of SSNC, Ulf von Sydow, personally came down to the village and apologized. The president reassured the villagers that SSNC from now on would respect the will of the village people and listen to them. The president also reassured Tafua that the covenant was signed by two equal parties, and the local organization in Apia was being removed from further interference in the relationship between Tafua and SSNC. Yet although the apology satisfied dictates of Samoan custom, the SIDA funds still have not been forwarded to the village but continue to flow to the NGO.

Ersson, a Swedish journalist who studied the Tafua peninsula projects, believes that the difficulties between SSNC and Tafua village revolved around "mutual respect and understanding" (Ersson 1994), a conclusion shared by Hardie-Boys (1994). Ersson believes that had SSNC staff shown more trust in the Samoan village leaders, and had they complied with requests for direct dealings without an intermediary urban NGO, most of the difficulties could have been avoided.

## LESSONS FOR THE FUTURE

A review of the writings of nineteenth-century colonial administrators reveals a startling fact: most of the colonialists believed their political and cultural subjugation of indigenous peoples were acts of altruism. Today, well-intended conservation efforts based on the best of intentions can go seriously awry if care is not paid to listen and learn from village leaders. In Samoa, the SSNC staff acted in a manner it believed to be in Tafua's best interest, but they were unable to bridge the immense gap between Western and Samoan culture. This prevented them from seeing that their counterparts in the village chief's council shared, and in many areas exceeded, their own competency and commitment to conservation. In turn, village leaders were deeply disturbed by the SSNC's belief that the Tafua chief's council was unprepared to correctly determine conservation and funding priorities for the village, was unequipped to deal with foreign aid agencies and environmental organizations, was lacking sufficient integrity to ensure that aid funds would be used for their intended purpose, and was bereft of a culture that espoused Western standards of democracy and gender equality. Rather than resolving these cultural concerns directly with the village, SSNC placed local English-speaking elites of a newly formed NGO in an oversight position. For a culture based on respect of traditional leaders and consensus in decision making, unilateral insertion of a newly formed NGO as an overseer seemed to the villagers at best paternalistic and at worst imperialistic. The villagers were deeply offended by the NGO's allegations that they lacked integrity and were motivated by a "handout" mentality. These injustices, and particularly SSNC's disrespect for village leaders, so violated the indigenous sense of etiquette that the villagers severed their written relationship with SSNC and refused all further aid from the organization—a remarkable stance for a village with an average annual per-family income of less than $100.

Although in most respects, the SSNC had fulfilled the letter of their covenant obligations—the school, after all, had been built—in the villagers' minds, the SSNC staffers had violated the spirit of the covenant, which was negotiated as an agreement between equal partners. However, despite these conflicts, Tafua village continued to protect its forest and conserve its marine resources, and the village began to view the SSNC as an impediment to their conservation initiatives.

The belief that non-English-speaking chiefs require oversight by local elites to effect significant progress in conservation was shattered when the Tafua chiefs independently created an island-wide indigenous conservation organization called Fa'asao Savai'i ("Savai'i Conservation"). They chose as the president of the new organization one of the few female chiefs in Samoa, Va'asilifiti Moelagi Jackson. Without any paid staff and little funding, Fa'asao Savai'i rapidly emerged as the most effective conservation force in the country. They demonstrated against loggers, boycotted the sawmill, encouraged village leaders to resist logging of their rain forests, initiated a campaign to plant indigenous tree species, published a nature guide to Savai'i preserves, and orchestrated a widely publicized campaign against French nuclear testing in the South Pacific. Fa'asao Savai'i is also pursuing an effective program of conservation education in village schools and churches. Rather than signing up individual members, Fa'asao Savai'i enrolls only entire villages that have satisfied a unique entrance requirement: creation of a village-based preserve.

Fa'asao meetings are conducted entirely in the Samoan language according to dictates of Samoan custom, and participation in village kava ceremonies is a highlight of Fa'asao activities.

## LESSONS FROM THE TALE OF TWO VILLAGES

In many respects, the events that recently transpired in Falealupo and Tafua villages represent a controlled, if unanticipated, experiment in social ecology and cross-cultural conservation initiatives. Two similar villages with similar forests and identical cultures entered into similar covenants with foreign conservationists to build similar schools. In Falealupo, management and administration of the preserve was left in the hands of village leaders, a strategy termed "indigenous control" (Cox and Elmqvist 1993, 1997). In Tafua, administration and management of the preserve was left in the hands of a local NGO created by the donors. Although this was a subtle difference in preserve establishment, the results could not have differed more dramatically.

The lesson of trusting and empowering local institutions, such as village councils that have been in existence for centuries, is applicable to conservation projects elsewhere (Edwards 1997). "A key point is that local responsibility should follow local institutional patterns," writes Jeff McNeely of the International Union for the Conservation of Nature and Natural Resources (IUCN) (1993). "It is usually better to strengthen local institutions than to create new ones." McNeely suggests several principles for using indigenous peoples as full-fledged partners in conservation initiatives: (1) Build on the foundations of the local culture, (2) give responsibility to local people, and (3) consider returning ownership of at least some protected areas to indigenous people.

In this tale of two villages, it appears that issues of justice concerning plants and people must go far beyond equitable returns from drug development—indeed equitable returns are a necessary but insufficient component of Western–indigenous relationships. A cynic might ask what returns indigenous peoples have realized from drug discovery programs. Many drugs discovered from plants have dramatically improved the health of nearly all peoples throughout the world; aspirin and quinine, for example, are used throughout the tropics. However, most of these drugs were discovered during the colonial phase of world history, when the rights and aspirations of indigenous peoples were routinely overlooked (Cox 1993; Cox and Elmqvist 1993). Modern ethnobotanists consider indigenous peoples to be colleagues rather than informants, and they have almost uniformly rejected exploitation as an inevitable consequence of joint scientific investigation with indigenous peoples (Balick and Cox 1996). Yet promising vast financial gain to indigenous peoples, many of whom live in poverty, from such an uncertain enterprise as drug discovery is as cruel as promising unproven cures to Western peoples with terminal illness. Monetization may also conflict with some indigenous views about the sacredness of the creation and the selflessness of healing, views that could inform much current Western debate on both conservation and health care (Cox and Elmqvist 1993).

Yet there seems to be no paucity of ethnocentric Westerners who wish to force indigenous peoples to accept Western terms of both equity and equitability, as well as to accept burdensome Western infrastructures to administer such "equity." Unless Western NGOs are willing to subject themselves and their own budgets to complete control by indigenous peoples (Cox and Elmqvist 1991, 1993), such imposition of Western concepts of equitability can lead to ecocolonialism, merely substituting ethnocentricity and cultural imperialism for the economic and political colonialism of the nineteenth century. And, unfortunately, ecocolonialism, like its political antecedents, is not carried out entirely by Westerners. In many developing countries there are profound cultural differences between the older, traditional village leaders and the younger, Western-educated elite. Conservation efforts may flounder when the Western-educated elite perceive themselves as the sole legitimate interface between village leaders and the outside world.

In April 1997, Goldman Prize winners Chief Fuiono Senio from Falealupo village and Chief Loir Botor Dingit from East Kalimantan met with the undersecretary general of the United Nations and argued that indigenous leaders have the right to speak directly to Western decision makers, unfiltered by NGOs, Western advocates, or other self-appointed representatives. Both were alarmed to learn that on that very same day a meeting of "indigenous peoples" as part of the U.N. environmental summit was in progress, and that their respective countries were represented by NGOs rather than by traditional village leaders. Fuiono, who had saved his 30,000-acre rain forest and was honored with an award that has been called an alternative Nobel Prize, was particularly nonplused to bump into a member of the SSNC's local NGO in the United Nations lobby who had flown from Apia for the meeting and "a family vacation" at NGO expense. Solia Papu Va'ai, the minister of justice for Western Samoa who accompanied Fuiono to the United States, shared Fuiono's displeasure. "Who are these guys?" Fuiono asked.

The Kalimantan chief concurred in his negative assessment of the NGO. "We control the forests," Dingit said to the undersecretary. "Why were we not invited to this meeting?"

We in the Western scientific and conservation communities must learn to listen directly to such traditional leaders who control and protect their forests. In many cultures, the time and diligence required for training as an indigenous leader equals or exceeds that required by Westerners to win a law degree or a Ph.D. Western conservationists must be prepared to deal with these indigenous leaders as intellectual, moral, and spiritual equals. We must seek to solve issues of justice on terms acceptable to both Western *and* the particular indigenous culture involved. As a result, each conservation solution involving indigenous peoples, while incorporating some features common to others, will be unique. *Consent* of the indigenous people, *respect* for their culture, and *submission* to indigenous political control are features that should characterize all responsible agreements. Indigenous peoples have, in fact, been stewards over a majority of the world's biodiversity for thousands of years. Partnership with indigenous peoples is crucial for the future of conservation, particularly in the tropics. But true partnership demands full and equal footing for both parties. Since indigenous peoples protect not only natural areas of extraordinary significance but also generations of priceless knowledge on how to use that biodiversity, it behooves us to listen carefully to their voices. We must learn to use our skills and our funds in a supportive role acceptable to our indigenous partners. Only in that way can a repeat of this tale of two villages be avoided in the future. If we are truly to protect the biodiversity of this planet, we must seek to become one village, one village in which indigenous as well as Western perspectives are accorded equal dignity.

## ACKNOWLEDGMENTS

Some of the material in this case study comes from articles previously published in *Pacific Conservation, the Journal of Ethnopharmacology, Ambio,* and CIBA foundation symposium #185.

This paper is dedicated to the memory of Fuiono Senio, colleague, village leader, and internationally renowned conservationist.

# CHAPTER 14

## One in Ten Thousand? The Cameroon Case of *Ancistrocladus korupensis*

*Sarah A. Laird, A. B. Cunningham, and Estherine Lisinge*

*Ancistrocladus korupensis* is a woody climber found in the tropical forests of Cameroon and Nigeria. The epithet *korupensis* refers to Korup, the people, and the national park that bears their name in the Southwest Province of Cameroon (figure 14-1). It was in the Korup National Park that *A. korupensis* was first collected, a forest vine with no reported local use, or name. *A. korupensis* was originally collected by staff of the Missouri Botanical Garden under contract from the Natural Products Branch of the National Cancer Institute (NCI). Since that time, it has yielded the anti-human immunodeficiency virus (HIV) naphthyl-isoquinoline alkaloid michellamine B, generating a complex debate on access and benefit-sharing (ABS) issues associated with the commercialization of biodiversity[1] (see, for example, Adams 1993; Gustafson 1993; *Le Messager* 1993; Katz-Miller 1993; *African Wildlife Update* 1993; *La Nouvelle Expression* 1995).

The issues addressed in this debate were expressed in, and now grow in part from, the documents signed at the 1992 Rio de Janeiro United Nations Conference on Environment and Development (UNCED, or Earth Summit), in particular the Convention on Biological Diversity (CBD).[2] The policy process leading up to and following the UNCED, and environmental organizations' international public information campaigns highlighting the medicinal riches of the rainforest conducted around this time, were manifested in interesting ways in the case of *A. korupensis* in Cameroon. In part, this was a constructive influence, and one that helped to steer government, nongovernmental organizations (NGOs), universities, and other stakeholders through the myriad issues raised by the case. This included the implications of sovereignty over genetic resources, and the sharing of ABS strategies with other high-biodiversity countries grappling with these issues, such as Australia and Costa Rica.

*Ancistrocladus korupensis* was collected in 1987, early on in the biodiversity

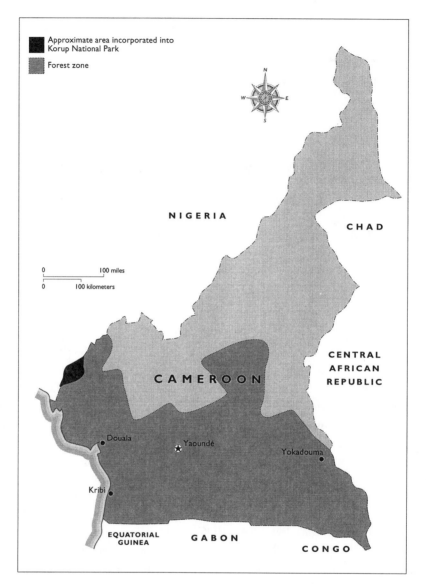

**Figure 14-1**  Korup National Park, Cameroon.

prospecting policy discussions (and, indeed, years before the term *biodiversity prospecting* was coined), but it was not until 1990 that the NCI found compounds of interest in *A. korupensis*, a year after NCI staff, in conjunction with their contracted collectors (the New York Botanical Garden, the Missouri Botanical Garden, and the University of Illinois), had developed a draft letter of intent (LOI). This LOI—for all its subsequently perceived faults—was at that time a progressive step forward for a government collection program, and a significant advance in addressing the ABS issues involved in biodiversity prospecting. In 1993, the LOI—by then renamed a letter of collection (LOC)—was signed by parties representing the NCI and

Cameroon. By that time, the National Institute of Biodiversity (INBio) in Costa Rica had signed and staff were actively discussing their $2 million deal with Merck and Company for the supply of more than 1,000 samples over two years (Reid et al. 1993b; Sittenfeld 1996; Joyce 1994; Balick et al. 1996).

The development of thinking on biodiversity prospecting issues in Cameroon grew from, and had the benefit of, a rapidly expanding international policy discussion, and the specific experiences of a number of groups actively working in tropical countries. This not only included INBio, which is one type of biodiversity prospecting "model," but the groups working with Shaman Pharmaceuticals, such as the Aguarana Federation in Peru, and other countries and institutions negotiating with the NCI regarding promising compounds, such as the Department of Conservation and Land Management (CALM) in Western Australia, which provided assistance and advice during the early stages of the *A. korupensis* case (Katz-Miller and Dayton 1993; Armstrong and Hooper 1994). The Manila Declaration (1992) was also consulted by both governmental and nongovernmental parties within Cameroon actively searching for general guidelines for a relationship of the type that had been established with the NCI.

While contributing constructively to the development of ABS measures within Cameroon, the international policy process and public information campaigns also brought with them some problems. This was in part because the issues had been distorted and oversimplified when international policy discussions lacked an adequate grounding in practical experience. Oversimplification can facilitate agreement and create order in a policy equation at the international level, but, afterward, implementation on the ground can be fraught with problems.

The underlying distortions and gaps with relevance to the case of *A. korupensis* include the ill-defined scope and nature of industry demand for genetic resources; the inherent and perhaps misplaced optimism that nationalizing genetic resources will likely lead to conservation and sustainable development, or the sharing of benefits beyond the national level; and the enormously difficult task of "equitably sharing" benefits with NGOs, research institutions, indigenous peoples and local communities, and others where "true" owners of forest resources are not easily identified, in countries where political and economic power are centralized, and where established patterns of resource extraction are exploitative.

All too frequently, governments investing in ABS policies to control access to something for which the market is uncertain—for example, genetic resources—also tolerate or promote the clearing of high-biodiversity forests for commercial agriculture or unsustainable logging, often by foreign-owned companies. In Cameroon, more than half of timber exploitation is undertaken by foreign-controlled companies, and foreign companies severely overharvest medicinal barks for shipment to Europe, with little return to local communities or serious investment in the sustainability of these practices (Sikod 1996; Ekoko 1997; Cunningham and Mbenkum 1993; Sunderland et al. 1997; Laird and Lisinge 1998; also see chapter 12). Logging companies exported more than 1.8 million tons of timber products in 1996, making

Cameroon the third largest exporter of timber in Africa (Sikod 1996). Natural resource extraction is a major part of the national economy, yet a great deal less attention is paid to the conservation of biodiversity, sustainable use and development, and the fair and equitable sharing of benefits than is given to genetic resources, which, by and large, are part of a poorly defined and poorly understood, and certainly to date within Cameroon a largely unrealized, market.

Governments and stakeholders involved in the process of drafting access and benefit-sharing measures must do so with a firm understanding of industry demand for genetic resources. There is great variation among the industries targeted by these measures, including the ways in which they seek access and generate benefits (ten Kate and Laird 1999). Countries must also develop a strategy that balances the need to control exploitative practices on the part of industry, with the desire to promote new and varied commercial uses of biodiversity, in order to generate a range of benefits and create incentives for conservation and sustainable development. The lack of organized industry participation and/or opposition within countries developing access and benefit-sharing regimes is pronounced. Beyond concerns relating to intellectual property rights and excessive bureaucracy, the industry reaction is small and silent, or derisory. In many countries, the ABS policy process does not proceed on the basis of a sound understanding of the industries involved, nor does it grow from a well-articulated national strategy. As a result, many measures have missed the target, which—we must assume—is the coming together of the three objectives of the Convention on Biological Diversity.[3]

In Cameroon, the absence of sufficient information and a national strategy was reflected in early attempts at access control and requirements for benefit sharing. Within a forestry law (Government of Cameroon 1994) then under consideration, the government included provisions regulating genetic resources, including establishing national sovereignty over genetic resources, requiring prior informed consent from the Government of Cameroon (GoC), and benefit sharing, mainly in the form of royalties. The will to implement these ABS measures, and general interest in these issues has dwindled, however, alongside commercial prospects for michellamine B. Perceptions that outsiders are rapaciously exploiting Cameroon's genetic resources, growing originally from international policy and public education programs, and finding fertile ground within the country, persist at all levels, however. Outsiders in forest villages are frequently asked if they are in search of a million-dollar drug; or the town a foreigner inhabits might teem with rumors of their undercover explorations in the forest for pharmaceutical medicines. At the government and NGO level, suspicion of outside researchers has become routine.

An analysis of the relationship between justice and biodiversity prospecting is most compelling when it uses the language and framework of resource extraction industries such as timber and mining; when a long history of "common heritage" is placed under today's very different ethical and legal microscope; or when the bad guys and the good guys are clearly motivated and easy to identify. There are many examples of unethical collections, companies motivated by the need to re-collect a promising species and doing so destructively, an absence of prior informed consent from local people or governments, and so on. However, it is also common for the motivations of the collectors to vary, or to be noncommercial (e.g., a publication being the desired object rather than cash), and for the company's objectives to be more complex than commonly thought (a reliable partner to provide consistent services, rather than an undocumented, untraceable load of plant material). The "good" developing country government might exploit and draw benefits away from local communities, and the "bad" company might tend to business as usual, while adhering to national ABS measures. Biodiversity prospecting does not always fit easily within the framework of exploitation most commonly cited at an international level today. While research and commercial collections should be carefully monitored and regulated, our analysis of justice and equity must incorporate the complexity of relationships, and the importance of histories and patterns of exploitation within high biodiversity countries.

As affirmed under the Convention on Biological Diversity, national governments have sovereignty over genetic resources. Within biologically rich countries, many governments have also concentrated power over natural resources to a great extent. Resource management and use, as administered by these centralized bodies, have historically proven anything but "equitable," or "fair," nor has there been the broad and creative sharing of benefits that is required if biodiversity prospecting is to contribute to the conservation of biodiversity or sustainable development. A minority of high-biodiversity countries have the technological and scientific muscle in place to capitalize on commercial collaborations and to make full use of "capacity building" and "technology transfer," and other benefits called for under the Convention. Most of these also have the business and legal acumen to develop frameworks for collaboration that control access and ensure benefit sharing. But as yet, few even in this minority have the political and social will to establish strong links between the commercialization of biodiversity, conservation, and sustainable development, and to draw a range of stakeholders into the process.

As this case study indicates, biodiversity prospecting easily falls into established, inequitable patterns of resource extraction. There is a good chance that a benefit-sharing process, including royalties and up-front benefits such as schools, roads, and health clinics, will follow much the same pattern and will be tied to long-term conservation and development objectives in much the same way as those previously or currently supplied by timber, mining, and oil companies. The timber industry is often required to provide something like a "fair and equitable sharing of benefits"; oil companies in West Africa support the building of schools for local communities. What has this meant in practice? Access and benefit-sharing policies will be effective only if their design takes into account these historical patterns of resource use and the economic, social, and cultural context within which they occur. Taking into account local histories and cultures, as onerous as this may seem to some, and as rote and empty as the concept has become through cynical overuse, is the only sure way any effective ABS policy measure will be developed and implemented.

## THE CASE[4]

### Ancistrocladus korupensis

Ancistrocladaceae is an unusual family of about twenty species of forest climbers from tropical Africa and Asia, in the single genus Ancistrocladus. There is growing scientific interest in this genus, in part because of the uncertainty that remains with regard to the relationship between the Ancistrocladaceae and other plant families.[5] However, interest has been piqued mainly because of the presence in many species of a unique group of chemical compounds, the naphthyl isoquinoline alkaloids (Gereau 1997; Bringmann 1986; Bringmann et al. in press; Manfredi et al. 1991; Hallock et al. 1994). Although the recently described *A. korupensis* (Thomas and Gereau 1993) appears to have no local use in its native range in Cameroon, there are several records of local peoples' use of other *Ancistrocladus* species in traditional medicine. For example, M. M. Iwu reports on the use of aerial parts of *A. abbreviatus*, a species very similar to *A. korupensis*, in the treatment of measles and fever in Ghana (Iwu 1993; Irvine 1961). The boiled roots of *A. extensus* are used to treat dysentery in Malaysia (Burkill 1966). *A. robertsoniorum*, which is a restricted endemic of coastal forest in Kenya and was described only in 1984, exudes the insecticidal compound droserone (Leonard 1984).

*Ancistrocladus korupensis* is a tall (up to 25 m) canopy liana[6] with stems sometimes exceeding 10 cm in diameter. The sparingly branched main stems climb by means of numerous short, hooked, lateral branches. Leaves of the lateral branches are borne in dense evergreen rosettes, and each leaf apparently lives for over one year. Mature leaves contain the highest concentrations of the alkaloid michellamine B. Little is known of the phenology, although flowers have been collected in November, and dense crops of fallen fruit were seen in February and March of 1993 (Jato and Thomas 1993b; Thomas et al. 1994). The density of stems is estimated at one to two mature climbers per hectare. The area in which *A. korupensis* is known to grow lies at 50 to 160 m above sea level with highly acidic (pH range, 3.9 to 4.5), leached, and infertile soils with a high sand content (60–91 percent) and little clay (Thomas and Gereau 1993; Thomas et al. 1994; Gereau, personal communication, 1995).

### Collections

*Ancistrocladus korupensis* was first collected by botanists in the early part of the twentieth century near Oban in the Cross River State of Nigeria (Talbot 1726, BM), but it was not identified to the species level. The second collection (Thomas 6889, MO, YA) was made in 1987 in the Korup National Park,[7] about 50 km from Talbot's locality. The Thomas collection was a voucher for a 0.5 kg sample of dried stems and leaves, collected under a Missouri Botanical Garden (MBG)–National Cancer Institute contract. These collections were conducted in conjunction with the Center for the Study of Medicinal Plants, Yaounde (Thomas and Gereau 1993; Jato and Thomas 1993a).

A number of years later, in 1990, researchers at NCI discovered that extracts from *A. korupensis* inhibited the ability of HIV to kill human cells; the HIV-inhibiting alkaloids michellamine A and B were subsequently isolated (Manfredi et al. 1991). Following on this research, NCI sought out additional supplies of what it thought was *A. abbreviatus* Airy Shaw, a species widespread in west and central Africa (Thomas et al. 1994). MBG collectors in central Africa (including Gabon and the Central African Republic) conducted subsequent collections of *A. abbreviatus*, *A. ealaensis*, and *A. letestui*; however, these samples showed no activity against HIV. The original voucher specimen was reexamined, and it was found that the species in question was in fact new to science. Cameroonian scientists from the National Herbarium, the Center for the Study of Medicinal Plants, and other institutions, were hired to locate the species in Cameroon. In 1991, the original collector, Duncan Thomas, was able to re-collect *A. korupensis* in Korup National Park. In 1992, the inflorescence axis was found, confirming that the species was not *A. abbreviatus*. In June 1992, the Ancistrocladus Project technician at Korup, Emmanuel Jato, found *A. korupensis* fruit and, in early 1993, both fruit and flowers, which led to the description of *A. korupensis* as a new species (Thomas, personal communication, 1995; Jato, personal communication, 1995; Thomas and Gereau 1993).

## Development of a Sustainable Supply: Botanical and Horticultural Research

In 1992, after michellamine B was approved for preclinical development at the NCI, the Missouri Botanical Garden, in conjunction with the University of Yaounde and funded by the NCI, expanded its research program in Cameroon to assess the density and distribution of the population of *A. korupensis*. Researchers found that *A. korupensis* is very localized in its distribution but within this area is fairly common (Thomas et al. 1994), with vines existing in localized patches.

Over the next few years, while research continued on the distribution and taxonomy of *A. korupensis*, large collections were made of the active species, and smaller collections of other *Ancistrocladus* species, for testing by the NCI. During this time, as part of work funded by the NCI, researchers in Cameroon also undertook preliminary propagation trials and evaluated possible methods of production from wild and cultivated sources. Seedlings were collected in the forest and planted in a forest nursery and in the Korup Project Nursery in Mundemba, at the Limbe Botanic Garden, and at the compound of a University of Yaounde researcher. By July 1993, 5,000 seedlings had been raised in the Korup Project Nursery, and many others were planted out in preliminary trials to study the effects of various cropping systems (D. Thomas 1992; A. Thomas 1993; Symonds, personal communication, 1994.)

Leaf harvesting trials began in April 1992 in forests under threat from shifting cultivation on the edge of the Ndian oil palm estate, outside the national park. Sourcing of raw leaf material presented a number of problems because the harvest of live plant material from a national park is not legal, and leaf harvesting trials demonstrated that an interval of at least two years between harvests was required. Researchers turned to leaf litter in an attempt to develop a sustainable supply. This approach proved successful since all samples of leaf litter showed high levels of michellamine B (Thomas et al. 1994).

In November 1993, Purdue University received the NCI contract for work on cultivation of *A. korupensis* at Korup. This was a three-year program, designed to determine the feasibility of cultivating *A. korupensis*.[8] The budget for this research program was subsequently scaled back from original estimates because of severe budgetary cuts in the NCI Developmental Therapeutics Program, but it was still by far the largest investment made by NCI in sourcing to date (Cragg, personal communication, 1994). By investing in this research, despite uncertainty about the future of michellamine B, the NCI hoped to insure itself against sudden supply shortages such as those experienced when taxol (from the bark of *Taxus brevifolia*) passed into clinical trials. Now that michellamine B appears too toxic to pursue, the Purdue University cultivation program at Korup is winding to a close. Findings will be published and made available to the Prime Minister's Committee, which will then decide whether to make it publicly available (Cragg, personal communication, 1997).

Michellamine B has been synthesized, but synthesis is still not economical and a licensee would likely have to work with the GoC and the Korup Project to source raw materials affordably in the future. This might prove the point at which an equitable deal can be struck, but it is questionable whether hard bargaining could take place, and a company might be tempted to throw itself into research on more affordable synthesis, or it might try other sources, rather than risk dependence on future supplies of raw material from one source.

## The NCI Letter of Collection[9] and the Government of Cameroon

> Under one such arrangement the National Cancer Institute is studying a vine in Cameroon that contains a potentially promising anti-HIV agent; should this particular substance fulfill its initial promise, Cameroon would realize significant benefits from development of this resource
>
> Timothy Wirth, Undersecretary of State for Global Affairs, in April 1994 testimony before the U.S. Senate Foreign Relations Committee

In August 1992, under the auspices of their extended work for NCI in Cameroon, MBG staff met with staff at the University Center for Health Sciences of the University of Yaounde to discuss the NCI Letter of Intent, which was then signed in early 1993 by the Dean, Pierre Cateret. This LOI was subsequently revoked by the Government of Cameroon because it considered the University an inappropriate body to represent the country's interests, its view being that such a document should be signed by a minister in the GoC. As of July 1997, the LOC had not been signed by the GoC.

The process by which the government became actively involved in the case of

*A. korupensis* began in 1993. During this time, concerns relating to access and benefit-sharing issues, as they related to the Korup Project and Cameroon as a whole, were increasingly raised. Korup Project, World Wide Fund for Nature (WWF), and government officials entered into direct dialogue with the NCI, one result of which was the revoking of the University Center for Health Sciences and NCI LOI. However, although this document was determined to be no longer legitimate, there existed no substitute agreement or framework that outlined the terms of the NCI-GoC relationship, including the supply of raw materials for testing, as well as requirements for "fair and equitable sharing of benefits" with Cameroon, the Korup Project, and local communities in the Korup Project area.

Meanwhile, the variety and number of parties involved in the process expanded, causing confusion over roles and responsibilities; parties included the University of Yaounde, the many GoC ministries, the Korup Project, the MBG, and Purdue University. It remained unclear who had final responsibility for negotiating and dealing with the practical realities of the NCI research and development effort, as well as for brokering the various national and local interests involved. Confusion and, as a result, ill-defined suspicion resulted, with no single party appearing to have all the necessary information on hand at one time.

Without a framework agreement, for example, the Korup Project staff were reluctant to send further plant material to the United States. The NCI, meanwhile, was extremely eager to acquire additional raw materials for animal toxicology studies and was actively working to explain the benefits of the LOC. In August 1993, an interministerial committee was established within Cameroon, and a meeting was held to address the issues raised by *Ancistrocladus*. The results of the meeting included declaring *A. korupensis* a "national treasure"; restricting the number of multiplication plots of *A. korupensis*; prohibiting the export of live plant material and seeds; conducting research into capabilities within Cameroon for the establishment of research partnerships with NCI; and the establishment of three committees with the following well-articulated agendas: (1) production/exploitation, (2) laboratory research, and (3) legal aspects.[10] Unfortunately, the interministerial committees did not follow up effectively on this initial meeting, largely because they failed to clarify their respective ministerial responsibilities in the case, which resulted in confusion in the design of negotiating strategies with the NCI.

The Government of Cameroon ministries most directly involved in biodiversity prospecting–related issues are the Ministry of Environment and Forestry (MINEF) and the Ministry of Scientific and Technical Research (MINREST), although a number of other government ministries such as the Ministries of Industrial and Commercial Development, Health, Higher Education, Justice, and Finance, as well as the Prime Minister's office have become involved in the *A. korupensis* case to varying degrees. Today, the bulk of responsibility for *A. korupensis* and other "medicinal plants" lies within the Prime Minister's Follow-Up Commission for the Exploitation and Conservation of *A. korupensis* and MINEF, but there remains a great deal of confusion as to respective responsibilities, and no real movement toward the development of a competent national authority to oversee and implement permitting procedures for ABS.

Following the August 1993 meeting, MINEF sent a letter to NCI stating that further raw material would not be sent to NCI without a full agreement between the NCI and the Cameroon government. The government also requested information on any live material of *A. korupensis* outside of Cameroon with which the NCI was working, and it asked that all cultivation work be done in Cameroon (restricted to Korup, or under the direction of the Korup Project). It also requested that immediate benefits be returned to Cameroon, including research on propagation and cultivation at Korup; provision of a field herbarium at Korup; provision of training courses in plant taxonomy in Cameroon; assistance with the development of appropriate capacity within Cameroon for the evaluation of new natural products and authentication of traditional medicines; provision of a full list of all biological samples obtained for the NCI in Cameroon (where and how collected, and all lab results); and a moratorium on collection of further samples in Cameroon until general terms for collection of such samples had been determined.

The NCI responded that they would be willing to replace their agreement with the University Center for Health Sciences for one with the GOC, and that their staff were preparing a new draft, later received by MINEF. The NCI said that it could not provide herbaria, but it saw this as the type of program that the United States Aid for International Development (USAID) would fund. The NCI offered to train researchers at their laboratories and agreed to send a summary of all biological samples collected in Cameroon and submitted to them, together with an assessment of their biological testing.

The NCI is no longer pursuing research and development on michellamine B because of its toxicity. Within the NCI research and development program, it is effectively shelved; however, the NCI would like to find a company to conduct further research on it.[11] Because Cameroon has not signed the NCI LOC, the NCI cannot *require* that a licensee do more than negotiate in good faith.[12] A signed LOC would have the added benefit of covering all of the materials collected in Cameroon in 1987 for NCI, many of which might prove of value in the future.[13]

## Intermediaries

The case of *A. korupensis* highlights some of the complexities and potential problems associated with the NCI-contracted collector model. Because NCI depends on independent subcontractors to carry out collections, these subcontractors and their in-country collaborators determine the nature of benefits associated with the collection phase, and they identify in-country beneficiaries. The NCI can constrain and motivate collectors in particular directions through its contracts and funding, but it is ignorant of conditions within countries in which collections take place.

Collectors, in turn, are selected because of their abilities in plant collection and identification techniques, not because of their capacity to mediate the various national and local interests with regard to the commercialization of biochemical and genetic resources. Beyond the obvious need for contracted collectors to follow high professional ethical standards themselves, they must also often provide advice, information, brokering, and negotiation assistance in the early stages of research and development to local partners. If they do not, the type of confusion and rumor that has typified the *A. korupensis* case is likely to result.

At the stage of collection, the NCI does not become directly involved in compensation and benefit sharing, but it has supplied additional funds to their contracted collectors to allow them to implement short-term infrastructure- and expertise-building measures in countries of collection. These types of benefits can be written into research agreements and, had the Missouri Botanical Garden and the Korup Project established a better-defined working relationship early on, and preferably some form of research agreement, a package of "process" benefits (see following discussion) might have been supplied to Korup as part of the plant collection process.

In a departure from past practices, the NCI is now ceasing to use intermediary collectors in some regions, instead entering into direct Memoranda of Understanding (MOU) with source country collaborators, particularly in South and Central America, but also in South Africa, China, and Zimbabwe. The collaborations defined in these MOUs are far more involved, and they place an emphasis on discovery taking place in the source countries. However, this approach requires a significant level of research and development capacity within source countries, and so it is limited to around a dozen or so high-biodiversity, relatively high-capacity countries (Cragg, personal communication, 1997).[14]

## Forestry Law No. 94/01 and Implementing Decrees

In December 1993, the Cameroon National Assembly passed a new forestry law (Government of Cameroon 1994; concerning forests, wildlife, and fisheries) and implementing decrees (Government of Cameroon 1995a, 1995b; relating to forestry and wildlife, respectively). The law and its implementing decrees are the main legal instruments for implementing the Forest Policy. They outline the administrative procedures and norms relevant to the attribution and management of the forest. Included are provisions relating to the collection and use of genetic resources.

Article 12 of the forestry law establishes national sovereignty over all genetic and biological resources and requires prior informed consent from the GoC prior to any scientific, commercial, or cultural exploitation. A permitting process for exploitation of forest products, guidelines for the collection of genetic resources, and the equitable sharing of benefits are also detailed. Article 12 also channels all benefits in the form of royalties to the GoC. Other articles relating to benefit-sharing with local communities include 68, 51, and 85, but they address timber exploitation and do not mention genetic resources.

More recently, a framework law relating to environmental management (Government of Cameroon 1996) was adopted. Articles 64(1)(c) and 65(1) and (2) recognize the need for a system of access control for genetic resources. This framework law further states that scientific exploration of genetic resources should benefit Cameroon and should be carried out under conditions of transparency and in close collaboration with national research institutions and local communities, as stipulated in relevant international conventions signed by the GoC, in particular the Convention on Biological Diversity. The law further calls for an enabling decree to define the contractual relationship that should exist between foreign and Cameroonian research institutions and local communities. The provisions cited establish a sufficient legal base on which subsequent access-control agreements and benefit-sharing mechanisms can be developed.

Although the forestry law and implementing decrees are silent on article 8j of the Convention on Biodiversity, which includes language to "respect, preserve, and maintain knowledge, innovations, and practices of indigenous and local communities embodying traditional lifestyles . . . encourage the equitable sharing of the benefits arising from the utilization of such knowledge, innovations, and practices," the framework law stipulates in article 65(1) that the exploration and exploitation of genetic resources should be in accordance with the provisions of the CBD. This article, therefore, implicitly incorporates the relevant provisions of article 8j.

A significant institutional obstacle to implementing the genetic resources provisions of the forestry law, and to addressing issues raised by the case of *A. korupensis*, has been the lack of a clearly defined national authority to oversee access and benefit-sharing issues. As a result, rather than a strategic approach to negotiations with the NCI, and ABS issues in general, the GoC largely pieced together policies in response to events. Even the Prime Minister's Committee set up in 1993 appears defunct, not having met since 1997.

### Community Forests

With the 1994 forestry law innovation of "Community Forests" (article 37), community control over forest resources, including their role in access and benefit-sharing arrangements for genetic resources, is in flux (Besong 1995; Government of Cameroon 1997). The 1994 forestry law classifies the national forest estate into two categories of forest: permanent and nonpermanent forest. The nonpermanent forest includes communal forest, community forest, and forest belonging to individuals.[15] There is some question as to whether the community forests will differ substantively from existing forest in the public domain to which communities have usufruct rights (1974 Land Tenure Act, ordinance 74–1, July 6, 1974), and whether communities will feel any greater guarantee of their long-term control of the resource base.

The application of the provision relating to community forests is complicated by ambiguity in the definition of *community*. Both the law and its decree of implementation see *community* as an entity provided for under existing Cameroonian legislation [article 28(2)]. Applications for community forests have been delayed by the GoC pending clarification of the legal status of the groups involved, and a redefinition of procedures for inventories and management plans (Sharpe 1997). A draft

MINEF manual on procedures and norms for the attribution and management of community forests defines a community as a legal entity duly registered under the existing legal text as either an association, a cooperative, a common initiative group, or an economic interest group (Government of Cameroon 1997).

## THE CULTURAL CONTEXT: INDIGENOUS PEOPLES AND LOCAL COMMUNITIES

The people living in the Korup area, like those in all of southwest Cameroon, are a combination of indigenous villagers, settlers from Nigeria and the Bamenda Highlands, and migrant laborers. In the precolonial period, the forest of the Southwest was inhabited by a large number of small linguistic and cultural groups known in the ethnographic literature as Bakweri, Bambuko, Bafaw, Balong, Bakundu, Balue, Bai, Mbonge, Ngolo, Batabga, Korup, Batoke, Mbo, Bakossi, Basossi, Elung, Ninong, and so on (Sharpe 1994). Within the Korup Project area, the main ethnic groups are the Bantoid Ekoi, including the Ejagham tribes, and Ibibio, including the Korup people; the Cameroon-Congo Bantu in the area include the Oroko tribes, and Mbo tribes to the east (Thomas et al. 1989; Tchounkoue and Jenkin 1989).

The 100 or so villages within the Korup Project area have largely mixed subsistence and cash crop economies. The primary cash crops are cocoa and coffee, with other cash and subsistence crops including cassava (*Manihot esculenta*), plantains (*Musa* species), bananas (*Musa* species), cocoyams (*Colocasia esculenta* and *Xanthosoma sagittifolium*), maize (*Zea mays*) and yams (*Dioscorea* species). The typical holding is between five and ten hectares (ha) per household, with 2 ha or less under full cultivation at any one time. Fishing and hunting (often within Park boundaries) are important subsistence and economic activities throughout the Korup Project area, and to a lesser extent the harvest of various forest products, such as cane, foods, spices, medicinal plants, and dyeing and carving materials for both subsistence and sale in local markets (see, for example, Malleson 1987, 1993; Okafor 1992; Thomas et al. 1989; Wood 1993; Devitt 1988; Carter 1992; Jeanrenaud 1991; Laird and Sunderland 1996). Timber extraction from concessions surrounding the National Park is also underway. However, the Ndian Division economy is dominated by the production of palm oil and kernels, largely through the Plantations Pamol du Cameroun (PAMOL), which was previously a subsidiary of Unilever, but also through oil palm smallholders (Tchounkoue and Jenkin 1989; Wicks et al. 1986).

Over the past century, the indigenous groups of southwest Cameroon have experienced forced labor under German colonial rule, the establishment of plantations, and in-migration by large numbers of plantation workers and settlers from Nigeria and the Bamenda Highlands, the latter of which continues today (Sharpe 1994; Kofele-Kale 1981; Watts 1994). The Southwest Province continues to be characterized by plantations, mainly those under the control of the parastatal Cameroon Development Corporation, which produces palm oil, rubber, bananas, and tea, as well as those of

PAMOL (Tchounkoue and Jenkin 1989). As a result of these factors, there is no "over-arching" ethnic identity in this area (Sharpe 1994). Although there is a clear division between indigenous villagers and the "strangers" to whom they sell their land,[16] and who settle in a client relationship to the village (often in former slave towns), within and between communities there is considerable heterogeneity, and complex relationships exist that belie a simplistic distinction between the indigenous and the nonindigenous.

Richards (1993) suggests that this complex mix of old and new (colonial and postcolonial) migrant populations characterizes the forest margins throughout West Africa, and that narrow definitions of the category *indigenous peoples* should be avoided. The term *indigenous peoples* should, he argues, be used to cover all groups in West Africa with effective local knowledge of the forest.[17] Sharpe (1997) similarly criticizes a perspective on indigenousness that led a recent World Bank report on southern Cameroon to class only the 40–60,000 "pygmies" as indigenes, out of a population of 4–6 million. They argue that the Amazonian concept of undisturbed autochthonous groups makes little sense in West Africa, where most forest areas have a long and complex history of settlement and resettlement.

This long-running dynamic of migration into and out of the forest, however, does not mean that a concept of indigenousness, albeit a very different one from that found in the Amazon, is not important to local communities. In particular, the delineating of complex bundles of rights based on relative indigenousness, the recognition of these rights by others, and the mediation of disputes arising therefrom, absorb a great deal of cultural energy in West Africa (Richards 1993; Sharpe 1997). Being a "son (or daughter) of the soil" is the single most important political identity in South West Province, and it is crucial to village politics, legal systems, and land holdings, although degrees and types of indigenousness are recognized (Sharpe and Malleson, personal communication, 1996).

Today, in response to European-led conservation and development programs that now emphasize "indigenous or local communities," some have observed a tendency, as there was in colonial times, to generate "ethnic federations." Sharpe (1994) sees the emphasis on ethnicity and "nativeness" on the part of these programs as not only misplaced but potentially divisive, and likely to create interethnic conflict.[18] Richards (1993) also describes the conservation agencies' confused, but prevailing, belief that forests and forest margins in West Africa have a single " `true' owner with whom a once-and-for-all resource management deal might be struck," and that all other local interest groups are in some sense "imposters." Burnham (1993) refers to government and NGO planning documents, drafts of new laws, and publicity statements that are "shot through" with references to participatory forest management by "traditional communities," while little attention is paid to how these communities are constituted and defined.

Ironically, the forested areas that have of late become reserves or national parks, such as the Korup National Park, are typically land that either was reserved by colonial forestry departments, was depopulated by local conflicts in the centuries before colonial rule, or once served as boundary wildernesses between neighboring precolonial societies (Richards 1993; 1996a). The Korup National Park includes previously abandoned farm sites, settlements, and forest managed for valuable species such as the oil seed tree *Baillonella toxisperma*, and during colonial and postcolonial times it served as a border traversed by traders and smugglers. These forested areas survive because they are old contested domains, no-man's-lands, or boundary wildernesses over which no single authority has been able to assert undisputed control. Local groups that settle in these areas are thus more "fluid" and "labile" than those elsewhere and are engaged in "competitive redefinition over time" (Richards 1993, 1996b; Burnham 1993; Sharpe 1997). As a result, in many of these forested areas it is especially difficult to identify a "true" owner or stakeholder to whom one could assign the right to negotiate access to local resources and subsequent benefit sharing.

Korup was demarcated as a forest reserve in 1937. From the beginning there was strong local opposition to the reserve in a manner that is telling for the case of *A. korupensis*. Forest reserves, purportedly for conservation, were in fact a form of timber concession. The local people throughout the southwest of Cameroon knew that this often meant little benefit for them and potential harm in the form of migrant workers in the logging camps. In some cases, chiefs might gain personally at the expense of their community. Within and between villages and native authorities there existed conflicts over the control of forest resources, and a great deal of suspicion. When approached regarding the demarcation of the forest reserve in 1936, the villagers of Korup were reportedly "universally suspicious" (Sharpe 1994). Numerous well-documented cases of dispute shortly after World War I were linked to the actions of the Forestry Service in creating forest reserves, as well as to the actions of timber companies within their concessions; in the mid 1950s, continual protests in the newly created legislative assembly led the French colonial government to suspend the creation of new forest reserves for fear of negative publicity reaching the United Nations Mandate Commission (Burnham 1993). Conservation projects and national parks, relatively new arrivals on the West African scene, are seen by many as only the latest manifestation of this historical intervention in local resource use and management. In response to these concerns, conservation policies, projects, and parks must be designed with a very real understanding of local culture and the history of resource extraction in the area.

There are three central themes with regard to the *A. korupensis* case that emerge from the history of the peoples in this area. First of all, the nature of "indigenousness" in the area is in no way clearly defined. To attempt to concentrate efforts on assigning ethnic provenance in the case of *A. korupensis* would likely prove disastrous and divisive. Although *A. korupensis* appears to have no local use (nor even a name),[19] an argument can generally be made for recognizing the contribution made by indigenous stewardship of forest resources (Laird 1994; Posey 1994; Posey and Balee 1989; Posey and Dutfield 1996; Posey 1996; Brush and Stabinsky 1996; Greaves 1994). However, in this case, stewardship cannot be assigned to any particular group, and efforts to do so would likely result in conflict, playing the indigenous peoples off against each other, and against strangers.[20] In addition, specialist, as well as common, knowledge of plant uses is often shared among ethnic groups in the Korup area.

Second, the people living in the Korup area have long experience with outside agencies that mine resources and renege on promised benefits. There is, therefore, an understandable cynicism and suspicion remaining in the area, coupled with a hard-earned ability to manipulate the situation to address local and individual needs, and this had run over into the activities of the Korup Project prior to the arrival of *A. korupensis* on the scene. This suggests that the reality behind *A. korupensis* research and development activities could never be as controversial as the suspicions it raises, and that relative transparency and active communication in dealing with the variety of local peoples, rather than unnecessarily raising expectations, would help clarify their role in the process. Clearly, local communities do not believe that historical patterns of resource exploitation have changed, that government institutions that facilitate forest exploitation have undergone a transformation, or that foreign interests (including conservation projects) have become other than economic, or primarily self-serving.

The Korup Project has, to date, attempted to include a variety of community members, largely through employment, in the sustainable sourcing of *A. korupensis*.[21] Should royalties result from a commercial product developed from *A. korupensis*, it is likely that the Korup Project would be a beneficiary of a portion of the funds, which might then be applied to the needs of the community; these include roads, health clinics, schools, water, electricity, and training and support for alternative income-generation activities. The mechanism by which this would occur, and the role of communities in detailing the exact nature and distribution of benefits, remains undefined to date. Under the present circumstances, however, it is possible to conclude that the benefits that have accrued to local communities take the form primarily of temporary employment, some training, and limited equipment and technical support.

And, finally, the divisions between anglophone and francophone Cameroon, and between urban elites and rural communities, are likely to be further aggravated should any commercial product come from *A. korupensis*.[22] Already, control over the case of *A. korupensis* has migrated closer to the center as awareness of the implications and potential value of the plant has grown. It has finally settled in the prime minister's office. This follows the pattern of access and benefit sharing experienced with other natural resources (Ekoko 1997; see discussion in Dove 1993b).[23] Timber revenues that once found their way to local councils for improving infrastructure in the Korup area are now put in a central fund in Yaounde and do not find their way back to Ndian Division. *Prunus africana* bark, used in a medicine to treat prostate hyperplasia in Europe (see chapter 12), is harvested in large quantities in other parts of South West Province, but similarly it yields minimal benefits for local communities, and there is little serious investment in sustainability that would ensure future supplies and reliable income for local collectors.[24] The tendency to nationalize species or products of great economic value should be factored into a realistic assessment of the potential impact of benefits on local communities, sustainable development, and the conservation of biodiversity.

## BENEFIT SHARING

### *Endemism and Point of Collection*

As *A. korupensis* was new to science in 1987, its distribution was not known at the time of collection, nor during the following few years. At one time it was suggested that *A. korupensis* was a common species; however, this was then revised and some considered it locally endemic to the Korup area. Although narrowly endemic, it has been found in the forest on both sides of the border between Cameroon and Nigeria. Had *A. korupensis* proved to exist only in the area around Korup, where the plant was originally collected, few questions would be raised regarding sovereignty and benefit sharing from commercialization of the compound michellamine B. However, *A. korupensis* is found in forest type shared by Cameroon and Nigeria, spanning a border that, similarly, separates people of similar ethnic heritage.

Indeed, as we have discussed, forest settlements in West Africa are often characterized by migration in and out of the forest. Richards (1993) argues that rarely, if ever, is it realistic to think of forest in West Africa as "empty" land not yet passed into human ownership and use, but it is equally distorting to treat the local groups currently found in possession of the forest edge as sole custodians. Sharpe (1997) describes West African forest settlements as occupying "social rather than geographical boundary zones," with almost all forest societies claiming to have met with previous inhabitants, or evidence of other settlements.

There are, therefore, no defining ethnic or geographic limits for *A. korupensis* that fall within national political boundaries. This raises a number of questions with regard to who should benefit from the commercialization of *A. korupensis*, which can be addressed by a range of approaches, including (1) *point of collection*—benefits should be negotiated by and returned to communities, institutions, or governments (or all of these) in areas where a species or knowledge of that species is collected; (2) *bioregional approach*—benefits should be returned to a bioregion, that is, the area to which a species is native; and (3) *global funds*—benefits should be fed into a global fund that will return benefits to communities and institutions throughout the developing world.

The *point of collection* model, based in practice on a variety of institutional mechanisms, is currently the most common. It can be seen, for example, in the prior informed consent process called for by the Convention on Biological Diversity, in the access and benefit-sharing measures under development in a number of countries and regions (such as the Philippines, the Andean Pact countries, South Africa, Fiji, Brazil, and Australia), and in many examples of two-party agreements, such as the National Institute of Biodiversity, Costa Rica–Merck & Co., the NCI Letter of Collection, as well as in the Shaman Pharmaceuticals approach.[25]

The point of collection approach avoids the need to trace ethnic or geographic provenance of samples or knowledge supplied, and it assumes that collaborators, or the state and national governments through their policies, will ensure an equitable and effective application of benefits to conservation and development needs. These relationships are based on the services and expertise applied to biodiversity prospecting—collection, taxonomic identification, extraction, screening, and so on—which are prerequisites for the subsequent collection and organization of the sample or information itself. Point of collection arrangements will yield the most benefits, in terms of conservation and development objectives, if regulated and monitored by a range of external measures, including community research agreements, professional codes of conduct, institutional policies, national access and benefit-sharing legislation, and international policy and law such as the Convention on Biological Diversity.

The point of collection approach might reduce problems deriving from bureaucracy, potential corruption, and concentration of the decision-making process away from local communities and institutions, which is common to large administering agencies, whether bioregional or global. Employing a combination of *research agreements* (dictating the terms of collection), *contracts* (detailing and formalizing commercial agreements), *national measures* (ensuring that agreements distribute a portion of the benefits to national conservation and development priorities, and ensuring a framework for equitable partnerships), and *international policy and law* (outlining the principles behind best practices), point of collection makes use of the market, and a variety of relationships developed along the way, to return and distribute benefits throughout all stages of biodiversity prospecting research and development. However, the point of collection approach might also result in national institutions competing with each other, the concentration of power in the hands of collectors, and a few individuals enriching themselves or their institutions at the expense of wider conservation and development for the country or region in which collections take place.

The *bioregional* and *global fund* approaches, while they avoid some of the dangers inherent in the point of collection system, have their own characteristic set of potential problems. Both recognize the importance of ethnic and geographic provenance as a factor in the determination and distribution of benefits (which point of collection does not explicitly do), but both acknowledge that in practice it would be virtually impossible to implement a system based strictly on these boundaries. As a result, each attempts to "scale up" the model to include a very broad suggestion of boundaries.

The bioregional approach is based on the assumption that the countries or ethnic groups within whose territories a genetic resource is found all gain a share of the benefits, either because of some property-like right or because this creates an incentive to conserve. It attempts to respond to the fact that neither species nor cultures conform to political boundaries, but, rather than scaling up to a global level, it suggests that benefits should be based on the geographical distribution of plant species. The distribution of plant species is considered more stable and more easily traced than knowledge, and this approach would employ taxonomists, biogeographers, and natural-products chemists to set biogeographic limits and identify areas with highest levels of endemism (Cunningham 1994).

Bioregional funds, as they have been proposed, would act as brokering bodies and would be funded from biodiversity prospecting payments made by large companies, or equities from smaller companies. The same bioregional organization could receive and disperse benefits from royalties to member countries. Countries would pool their resources for screening and capacity building. As a group, there is more chance of a "hit" than as separate competing organizations within countries or among countries in a region. In some cases (e.g., Madagascar and Australia), nation-states also represent a bioregion because of the high level of endemism within a discrete unit. In other cases (e.g., Southeast Asia, southern and western Africa), political boundaries do not conform to biogeographic ones, hence the need for collective bargaining, just as is happening in the world economy with the North American Free Trade Agreement, the European Community, the Association of South East Asian Nations, and the Southern African Development Community (SADC) (Reid et al. 1995; Eisner and Beiring 1994).

The bioregional approach would clearly require strong regional collaboration and legislation such as that detailed in the July 1996 Common System on Access to Genetic Resources developed by the Commission of the Cartagena Accord (the Andean Pact), which will regulate access to the genetic resources of the member countries and their derivatives according to agreed-upon principles and objectives (see Andean Pact 1996; Secretariat of the Convention on Biological Diversity 1996a; Gollin and Laird 1996; ten Kate 1997).

A global fund for natural products, like the one for Farmers' Rights, would not "depend on a detailed accounting of genetic contributions of peoples, communities, or nations" and, fed by commercial funds, would be dispersed widely to local communities throughout the developing world (see also, for example, discussions by Kloppenburg and Balick 1996; Richards 1993; Brush 1992). Some have suggested that the fund would apply only to those species with no biogeographic endemism, and others that a global fund for *all* species would best assist local communities.

In many ways, the concept of a global fund takes the wind out of the sails of biodiversity prospecting as it might be understood in the Convention on Biological Diversity and other international fora—that is, as an engine for the creation of incentives for conservation and sustainable development. Biodiversity prospecting at its most effective for conservation and development makes wide and varied use of local biological and intellectual resources, and it spreads benefits throughout the research, as well as the commercial, phase to a wide variety of parties including communities, research institutions, and national governments. A global fund severs the direct link between the generation and the distribution of benefits, but it could still create economic incentives through wise grant making. However, grant making through intergovernmental funds does not have a track record that suggests this is the most effective mechanism to create incentives for conservation and sustainable development.[26]

Currently, there is no bioregional fund, nor a global fund, that would negotiate terms for the relevant countries, and into which benefits would be channeled and from there dispersed. The NCI Letter of Collection states that the terms requiring compensation do not apply to "organisms which are freely available from different countries (i.e., common weeds, agricultural crops, ornamental plants, fouling organisms) unless information indicating a particular use of the organism (e.g., medicinal, pesticidal) was provided by local residents to guide the collection of such an organism from their country, or unless other justification acceptable to both the *country organization* and DTP/NCI is provided. In the case where an organism is freely available from different countries, but a genotype producing an active agent[27] is found only in *name of country*, sections detailing the nature of benefits due country organizations shall apply."

Suppose, however, that *A. korupensis* was in fact a common weedy species. Should the application of sections detailing benefits in the NCI Letter of Collection depend on the arguments of taxonomists as to the plant's original habitat, or should it depend on the negotiations and understandings reached by scientists, communities, and government officials actively involved in the collection and subsequent supplies of materials for NCI testing? The issue is further complicated by the fact that biochemical research often changes ideas about the systematic positions of species. Taxonomists have been called as expert witnesses in court before, and this would likely increase under a system of compensation based on taxonomy (A. Hamilton, WWF, personal communication, 1995).

The rosy periwinkle, *Catharanthus roseus,* is one of the best examples of the endemism versus point of collection dilemma. Indigenous to Madagascar, it is a weedy ornamental that had spread throughout the tropics by the time Eli Lilly, which developed vincristine and vinblastine, undertook any research. It first sparked the interest of researchers at Lilly, however, not because of ethnobotanical collections done on the part of Lilly staff in Madagascar, but because Gordon Svoboda and other researchers conducted a literature search for plants from the Australasian region with "folkloric usage of believable quality and the reported presence of certain types of plant ingredients" (Svoboda 1992:1). Subsequent research led them to vincristine (Velban) and vinblastine (Oncovin) for the treatment of childhood leukemia.

It is often claimed that Madagascar should receive benefits from the commercialization of compounds from the rosy periwinkle. In fact, the argument seems stronger for the Philippines or Jamaica, where local use of the plant for medicine first attracted the interest of scientists. The rosy periwinkle case would seem to argue for a global fund approach, which can cover the large number of useful species that have crossed numerous international borders. The rarer and more localized a species is, the stronger the case for national sovereignty, such as the Kenyan case for the rare endemic *Ancistrocladus robertsoniorum*. The case of Shaman Pharmaceutical's main species of interest, blood of the dragon, *Croton lechleri*, from South America, or *Garcinia livingstonei* in Africa, on the other hand, would appear to argue for a bioregional approach, in that both species are common and are widely used in their respective regions. Similarly, bioregional sovereignty might apply in the case of a regional endemic such as *A. korupensis*, which occurs across the national boundary between two countries.

Today, support for a global fund has significantly diminished. The approach most in favor, and appearing to take advantage of the best range of benefits that can derive from these partnerships, is one that employs a point of collection model, including two-party agreements, guiding institutional and professional codes and policies, strong national ABS legislation, and regional cooperation.

## "Process" Benefits

The development of *A. korupensis* into a commercial drug could yield a range of benefits for Cameroon, the conservation of local biodiversity, and communities in the Korup area. These benefits might include advance, milestone, and royalty payments that could contribute to technology transfer, training, conservation "overhead," and local community development programs, a license to manufacture a commercial product for in-country or regional consumption, the development of supply industries for raw materials or extracts, commercial drugs at cost, assistance with the development of screening capabilities for tropical diseases, and so on (for a list and discussion of benefits, see, for example, Laird and Wynberg 1996; Laird 1995a; Secretariat of the Convention on Biological Diversity 1996b; 1998).

To date, and possibly in total, however, the actual benefits resulting from *A. korupensis* stem not from commercialization—since no commercial product is yet developed—but from the research and development (R&D) process. In fact, these R&D benefits are often the most significant, since even compounds of great interest, such as michellamine B, might never make it into commercial product development. For biodiversity prospecting to maximize its contributions to both conservation and development, a wide spectrum of individuals and groups must benefit, often in distinctly different ways, and this must occur in the short, medium, and long term. Royalty payments into a global fund ten years down the road, no matter what the magnitude, will never have as great an impact as benefits scattered both spatially and temporally. It is in the wide and creative dispersal of benefits throughout the R&D, as well as the commercialization, phase that biodiversity prospecting will have the most lasting effect. One must look at the *process* by which samples are collected, chemicals extracted, R&D conducted, and sources of raw materials developed, in order for the many spin-off benefits for biodiversity science, medicinal plant research, conservation, and overall development to become fully apparent.

## Setting the Stage: Research Agreements and National Legislation

Much of the confusion surrounding negotiations and the assignment of responsibility within Cameroon for the *A. korupensis* case could have been avoided had research agreements with original collectors (based on the prior informed consent of the Korup project, local communities, the University of Yaounde, and the GoC), and national legislation guiding these collections, been in place. Instead, there were no terms set for the potentially commercial NCI collections conducted in 1987 and, once interesting compounds had been identified, no framework to guide the activities of local organizations, research institutions, and the government. Additionally, it was unclear to which body in government the responsibility for administering the case fell. At that time, this was not unusual, but much has changed since then, and awareness has been raised on the importance of having good access and benefit-sharing frameworks in place.

The drafting and development of national ABS legislation is receiving a great deal of attention, as countries work to implement the Convention on Biological Diversity (see, for example, Glowka 1998; Secretariat of the Convention on Biological Diversity 1996a; Government of the Philippines, 1995, 1996; Andean Pact 1996; Barber and LaVina 1997; ten Kate 1997; Laird 1995a; Mugabe et al. 1997; Review of European Community and International Environmental Law 1997). Research agreements have received less attention, but they are extremely important as a complement to national legislation, and more so in its absence; it is here that the details of relationships are hammered out.

### Research Agreements

The technologies and expertise resulting from the R&D phase of biodiversity prospecting will often be far more important than the commercial revenues. Thus, a research relationship must be established early on that reflects the best possible terms for local communities and tropical countries. Tropical country research institutes and conservation projects have long collaborated with outside scientists, relying heavily on their expertise to develop a research and knowledge base, and management plans for conserved areas. Tropical countries will continue to depend on outside expertise, particularly in light of the responsibilities of country parties to the Convention on Biological Diversity to inventory and monitor their biodiversity. But, increasingly, this will be done on terms set by biodiversity-rich governments, institutions, and local communities.

To set these terms, the biodiversity-rich institutions must negotiate research agreements that clearly outline the responsibilities and expectations of each party and ensure that all research, whether academic or commercial, contributes in some way to conservation and development activities in the areas in which it takes place (Laird 1995b; Cunningham 1993). This must be done carefully, however, so as not to create numerous bureaucratic obstacles to important academic research. Two elements form the core of research agreements: *control* by local projects and communities over the nature of research projects, and the use of resulting information, whether commercial or academic (including the choice not to commercialize); and the *contribution* of research programs to the management costs or needs of conserved areas, and the equitable return of benefits to local projects and communities from any commercial activity.

It is important for conservation projects, research institutions, universities, and local communities to tackle the implementation of research agreements as a precursor to any biodiversity prospecting–related program.[28] The manner in which research is conducted creates a framework of community, project, and institution or government control over resources and knowledge, which is often carried over into commercialization. If groups do not provide prior informed consent, and if they do not assert control over the manner in which academic research is conducted, they are unlikely to have a say in the extension of this research into commercial areas, or any control over the dissemination of material and knowledge through academic publications, databases, and other forms of distribution common to pure research.

## CONCLUSION

The case of *A. korupensis* is one of only a handful in recent years in which a rain-forest species yielded compounds of great interest for drug development. As such, it can provide valuable lessons in the practical realities of the relationship between biodiversity prospecting, conservation, and sustainable development.

It would be an exaggeration to say that this case created incentives for conservation and sustainable development. On a local level, many forest communities now know that the forest might contain a million-dollar drug, but this does not appear to have changed local peoples' relationship with the forest, nor has it spurred local-level conservation efforts. At the national level, we find a government that, unlike most, has come across a potential source of "green gold." Since this discovery, however, government-sanctioned forest clearance for agriculture, and dramatically increased timber exploitation, have caused significant damage to biodiversity and ecosystems within Cameroon; option values do not appear to be a heavily favored decision-making tool within the government.

Access and benefit-sharing language in the 1994 forestry law has not been implemented, and the Prime Minister's Committee has ceased to meet. NGOs, conservation projects, universities, and other groups within the country, which might undertake the difficult job of brokering various interest groups involved in ABS issues, and of assisting the government in operationalizing the concept, are not, for the most part, much better suited to this task than they were ten years ago.

The case of *A. korupensis* in Cameroon is a very particular one, but it serves to illustrate the difficulty in predicting the form international concepts and policies will take when laid on top of existing social, economic, and cultural systems. In some countries, practical implementation will follow with surprising consistency the international framework. In others, it will not. ABS, as articulated in the Convention on Biological Diversity, is a particularly complex policy to implement, involving a wide range of commercial sectors, scientific disciplines, cultures, and governments. ABS, like the Convention itself, reflects the grouping of extremely heterogeneous, and potentially inconsistent, spheres of activity and thought: the conservation of biodiversity, sustainable development, and the fair and equitable sharing of benefits resulting from the commercial use of genetic resources. Given this complexity, a dogged commitment to practical realities and an on-going redefinition of the ABS concept based on experiences such as those described in this case study are required.

## ACKNOWLEDGMENTS

We would like to thank the Korup Project and WWF Cameroon for their valuable assistance with this research, in particular Steve Gartlan. Support and insight have also been provided by Alan Hamilton, WWF-International; Maurice Iwu, Bioresources Development and Conservation Program; Terry Sunderland and Nouhou Ndam of the Limbe Botanic Garden; David Downes of the Center for International Environmental Law; and Dr. B. Hemarplagh. Michael Gollin contributed substantively to discussions on the NCI Letter of Collection early in this process. Thanks are also due to Barry Sharpe, University College, London, for generously sharing with us his field notes, and for commenting on drafts of this document. We would also like to thank the following for their comments on earlier drafts of this manuscript: Johnson G. Jato, University of Yaounde and the Bioresources Development and Conservation Program; Thomas Tata-Fofung, BDCP and formerly MINEF; Ruth Malleson, University College, London; Gordon Cragg, the National Cancer Institute; Duncan Thomas; Paul Symonds, Willy De Greef and Catherine Butcher of the Korup Project; Jim Miller, Roy Gereau, and Dan Harder of the Missouri Botanical Garden; Walter Reid of the World Resources Institute; Lyle Glowka of IUCN-ELC; and Michael Dorsey of Johns Hopkins University.

All errors and omissions are, of course, the sole responsibility of the authors, who have endeavored to "get the story straight," but know it is a complex one to tell.

## NOTES

1. *Biodiversity prospecting* has become a widely accepted term used to describe the collection, screening, and development of new commercial uses of biochemical and genetic resources (see Reid et al. 1993b). Although the scope of the Convention on Biological Diversity is limited to "genetic resources," the scope of most national ABS measures drafted to implement the Convention reflect an expanded understanding of biodiversity prospecting that includes biochemicals, derivatives, by-products, and traditional knowledge (see, e.g., Government of the Philippines 1995, 1996; Andean Pact 1996). This reflects a broadening of the term *biodiversity prospecting*, which is now applied not only to the pharmaceutical, biotech, and agriculture industries, but also to phytomedical, personal care and cosmetics, horticulture, and other industries that explore, or prospect, for new leads in biological diversity.

2. The objectives of the CBD are the conservation of biodiversity; the sustainable use of its components; and the fair and equitable sharing of the benefits arising out of the utilization of genetic resources, including appropriate access to genetic resources and appropriate transfer of relevant technologies (article 1) (see Glowka et al. 1994).

3. In drafting national ABS measures, governments, NGOs, researchers, and others involved in the process need information on the industries they intend to regulate. National measures drafted to date, such as the Philippines' Executive Order and Implementing Rules and Regulations (Government of the Philippines 1995, 1996), are built on an assumption of demand by industry that overlooks the unique manner in which genetic resources and biochemicals are used (see chapter 15 on the similarities between natural products and "infor-

mation"-based industries). These measures bluntly go after "benefits" more effectively extracted with a finer tool than those employed. In the case of the Philippines, for example, companies are loathe to sign on to anything that looks like compulsory licensing, and they are publicly stating their reluctance to work there; in other cases, the procedures for prior informed consent are so elaborate that they strike some as a "nightmare" (see ten Kate and Laird 1997, 1999).

4. Aspects of this case have been documented and analyzed in light of the benefit-sharing provisions of the Convention on Biological Diversity, as part of case studies drafted for the United Nations Environment Program (UNEP) for the fourth meeting of the Conference of the Parties in Bratislava, May 1998 (Laird and Lisinge 1998).

5. Recent investigations based on molecular evidence place the Ancistrocladaceae near the Droseraceae (D. Harder, personal communication, 1995).

6. In providing an example of ecologically driven collections, Gentry (1993) said the following: "Climbing plants (lianas), by their very nature, are mostly diffusely branched, fast-growing and light-demanding. Thus, their leaves are likely to be more scattered throughout the rainforest canopy, shorter-lived, and generally less apparent to herbivores than those of many other mature forest plants. Therefore ecological theory might predict that vines should be expected to concentrate more of their resources in specific highly active 'qualitative' (i.e., toxic) defensive compounds rather than energetically expensive broad-spectrum 'quantitative' (i.e., mechanical) defenses like tannins and lignins. It is precisely these low-molecular weight toxic compounds that tend to be biodynamic and medicinally effective."

7. The Korup National Park was established by the Government of Cameroon in 1986. The National Park covers 1,259 square kilometers and is rich in biodiversity, with over 3,000 species of plants and animals recorded (Tchounkoue and Jenkin 1989). In February 1988, a Project under World Wide Fund for Nature (WWF) management was formed with the overall aim of conserving biodiversity within Korup National Park. In April 1993, the European Union signed a four-year funding agreement with the GoC to support the Korup Project. Funds, personnel, and resources are provided by a number of other donors and the GoC, and the Project is managed by WWF. Aspects of the Korup Project include parks management, conservation education, research coordination within the park, and a rural development component, which is involved primarily in the 300,000 ha "Support Zone" surrounding the National Park. One hundred seventy-two villages lie within this area, twenty-seven of these lying 3 km or less from the park boundary; a further seven villages are located inside the National Park (C. Butcher, personal communication, 1995).

8. The directive (MAO/RFP) received by the NCI asked Purdue University to address the following objectives: (1) to study the feasibility of cultivating the plant to develop a reliable biomass source from which to obtain sufficient quantities of michellamine B for clinical evaluation; (2) to investigate the selection and propagation of high-yielding phenotypes; (3) to develop production systems to optimize the yields of michellamine B; (4) to examine the biology of the plant so that its growth and development and the accumulation of the secondary product can be predicted and understood.

9. For general background and more information on the NCI LOC, and on the NCI approach to ABS issues, see, for example, Cragg et al. 1994; Mays et al. 1993; Baker et al. 1995.

10. The Production and Exploitation Committee was established to oversee the inventory of A. korupensis, to produce a monograph, and to research and establish the methods of harvesting and cultivation. It was agreed that all work on A. korupensis and other medicinal plants within the Korup area would be coordinated and supervised by the Korup Project to

"safeguard the interests of the indigenous people of the area and the interests of Cameroon as a whole." The Research Committee was intended to study and develop a plan of action for the medicinal, chemical, and processing aspects of *A. korupensis*. The committee was to look into local capabilities with regard to exporting extracts rather than raw materials, the requirements of doing so, the location for such a committee, the timeline for such program, and the potential for developing screening capabilities within Cameroon for local medicinal plants. The Legal Aspects Committee was to devise a first instrument of negotiation with NCI and subsequently draft a long-term agreement. Initially, the committee would draft a document based on the Manila Declaration, the Department of Conservation and Land Management, the Australia draft agreement, and the NCI Letter of Collection. The committee would reconvene to discuss this document. This committee was also intended to determine the assignment of responsibilities within the government and the need for a permanent commission to deal with biodiversity prospecting issues.

11. Dr. Bringmann in Germany, with a longstanding interest in this genus, continues to conduct research on *A. korupensis*. This species has also yielded the korupensamines, antimalarial alkaloids that are under investigation by local scientists within Cameroon and Nigeria (Bringmann et al. 1994). The University of Yaounde continues to acquire small amounts of material from the Korup Project for antimalarial testing.

12. The NCI LOC requires direct negotiations between a successful licensing pharmaceutical company and appropriate source country organizations. Kaufman (1993) argued that the NCI has means at its disposal to insist that companies provide compensation to source countries. For example, the NCI license to a company could be made void if no mutual agreement between the company and the country of collection were reached. By holding the patent, the NCI has a certain amount of control over the ultimate marketing of beneficial drugs, therefore helping to guarantee the source country a fair return. Overall, although the NCI cannot commit to explicit royalty and licensing provisions, staff seem to feel that the support of the NCI and the National Institutes of Health (NIH), as well as members of the U.S. government (see, for example, the statement by Timothy Wirth quoted earlier), for the policies formulated in the LOC will guarantee source countries the types of compensation outlined in the LOC. For example, according to Dwight Kaufman (Deputy Director, Division of Cancer Treatment, NCI, 1993), "The LOC can't be more specific in guaranteeing recompensation to the host country, since by U.S. law the NCI, as a U.S. government agency, is not authorized to promise or encumber future intellectual property or patent rights. . . . The LOC is, nevertheless, a firm commitment to ensure that royalties and other forms of compensation shall be provided to the host country. We assure the world community that this commitment shall be honored."

13. Down the road, Camptothecin (isolated from the Chinese tree *Camptotheca acuminata*) was dropped from NCI clinical trials in 1972 because of severe bladder toxicity, and for ten years research was put on hold, until the mechanism of action of the antitumor activity of camptothecin was understood. In the 1980s, SmithKline Beecham developed a derivative of camptothecin—topotecan—that has lower toxicity and better selectivity (Carte and Johnson 1996). This research was funded by the National Cooperative Drug Discovery grants for the NCI, with industry and academic collaboration. The NCI sponsored clinical trials. Also, a Japanese pharmaceutical company, Yakult Honsha, developed a soluble camptothecin derivative, CPT-11-irinotecan.

SmithKline Beecham, the NCI, and Yakult Honsha currently obtain natural camptothecin from Chinese and Indian pharmaceutical concerns. Although total syntheses for camptothecin exist, and yields for synthetic materials are constantly improving, they are still not competitive with semisynthetic production from the natural product.

The delay between plant collection of *Camptotheca acuminata* (in the 1950s) and product development (in the 1980s) clearly argues the case for strong agreements between companies (or the NCI) and source countries. The commercial potential of a species long outlives the professional relationships on which collections are based, and agreements are needed to protect the interests of source countries and local communities. Whether or not research into *Ancistrocladus* comes to anything in the near future, the government of Cameroon should ensure that an effective agreement is in place to guarantee returns from any future work on this species, as well as on any others collected as part of the NCI program in the 1980s.

14. In addition, some intermediary collectors have changed their practices. As James Miller, Head of the Applied Research Department at the Missouri Botanical Garden, put it, botanic gardens in 1987 still operated under a series of unwritten principles and practices, while today the MBG (along with other botanic gardens such as the Royal Botanic Gardens, Kew, and The New York Botanical Garden) has a written Natural Products Research Policy. "While Missouri Botanical Garden would no longer even think of exporting samples for commercial development without a series of signed accords in place, in the mid-1980s all of our programs operated on spoken agreements, These were generally informal agreements with the national herbaria and other botanical organizations with which we collaborated. We generally relied very heavily on their help and logistical support and in turn provided training, some institutional support and the opportunity for their scientists to participate in our research" (J. Miller, personal communication, 1995).

15. The British Colonial Forestry Ordinance of 1917 did vest forest reserves in designated local councils—so-called Native Authorities—although these laws were superseded by the 1974 forestry law (Sharpe 1997). Under the British colonial law, "vacant" lands were considered to belong to the local communities, in the form of the Native Authorities. Under the French colonial law, however, all lands "vacant and without a master" belonged to the state. The criteria for recognition of personal ownership were strict, and swidden cultivation fallow were not included unless perennial cash crops such as coffee and cocoa had been planted. When the two federated territories were unified in 1972, the British concept of "communal land" was scrapped in favor of the French system (Burnham 1993).

16. Watts (1994) defines *strangers* as "nonindigenous people who have lived in the area for up to three generations." Some dispute the concept that land can be sold and transferred away from a community's control, and that this is what communities intended when they "leased" rights to the land to Europeans and "strangers." Ejedepang-Koge (1975), for example, describes the communal system of land ownership employed by indigenous peoples in the South West Province: "Every individual native can use as much of it as possible but has no right whatever to dispose of it. He has the crops he plants, the house he builds, the animals he keeps on it, but not the land itself, for it belongs to the community, to the ancestors. As such he has only the traditional user's or occupancy right to the land. It can be said that an individual has only holding rights on the land he occupies, under strict traditional conditions. . . . Following this concept, being dispossessed of one's land is a very grave thing for its means, among many of these people, that ancestors can no longer live on such land, that the dispossessed person can no longer enjoy the support of his ancestors who henceforth will have no permanent place of contact with him, and then his labours on any other land cannot therefore be adequately regarded; he can't be satisfied."

17. Human populations in the forested "boundary wildernesses" are not indigenous in the sense of being undisturbed autochthonous groups, Richards (1993) maintains, "but rather they are divided into factions that have negotiated a place for themselves in a complex and labile sociological landscape where the prime criterion for success is the local knowledge that

allows for survival in a harsh and isolated environment beyond the effective scrutiny or assistance of central authority."

18. See, too, Oates's (1995) account of the confusion in objectives and activities that results when conservation programs try to incorporate a "development" component for local communities without a complete understanding of what comprises the "community," and how development activities might influence a long history of migration into and out of forest areas. In the Korup area, for example, "development" to local communities includes the return of migrants who left before the establishment of the Park. This is an objective of many community development associations, and it could have significant repercussions for conservation programs in the area (Sharpe and Malleson, personal communication, 1996).

19. Because *A. korupensis* is a canopy liana (growing to 30 m) its leaves are very difficult to spot in the forest. It is too brittle for cordage and construction purposes, and it does not appear to have been used by local communities. Thomas (personal communication, 1993) maintains that most old-growth species are not used by the local people, most of whom are migrants recently settled in the area, who tend to utilize the more common, widespread secondary species. Mbenkum (1993) reports that local people tend to use plants from lower altitudes (where *A. korupensis* is found) in far greater proportion than higher-altitude species because of historical migration patterns. Although there are no widely known or reliable reports of local uses of *A. korupensis,* M. M. Iwu (personal communication, 1996) makes the point that uses for highly toxic species such as *A. korupensis* are often not revealed to lay people, and that specialist uses might exist.

20. A centralized administrative or organizational structure for the communities in the Korup area, which might assist in the determination and distribution of benefits in this case, does not exist. Land tenure and forest management are determined on a village basis, administered by chiefs and elders (in village councils) according to traditional laws, and each village has a "territory" recognized by others. All land in the Project area is or was claimed by villages, which will often have clear boundaries, particularly between different ethnic groups (Thomas et al. 1989; Tchounkoue and Jenkin 1989; Wood 1993). Should rights then be assigned to villages that control, or once controlled, land on which *A. korupensis* collections take place? This would likely create more cause for conflict than any benefits would warrant.

21. For the most part, expertise brought in as part of the *A. korupensis* research phase has not been applied directly to what might be considered the priority needs of local communities. However, during the research phase, local communities have benefited from the employment of approximately ten staff members in leaf collection, cultivation, and research. It is also thought that supplying *A. korupensis* to industry might make it an alternative agricultural crop (although as a crop, *A. korupensis* would be even less reliable and less under the control of local farmers than cocoa, their main cash crop). A variety of methods for its cultivation were researched to promote local community participation in industrial supply: interplanting with oil palms, both mature and juvenile; cultivation in traditional fields and fallows; and planting out in primary and secondary forest. Some individuals and community groups have supplied land in exchange for oil palms or cash, and should *A. korupensis* prove a valuable crop, they will have ownership over the plants on their land. Should it fail, crops such as the oil palm supplied by the Project will continue to produce income (Symonds 1994, personal communication; Thomas 1993).

22. See also, for example, discussions by Wamulwange (1993) on the conflicts between traditional and political governments in Zambia, and Agyare's (1993) discussion of a case from Ghana. In many areas of the Southwest Province, for a number of generations at least,

the village was the highest level of political organization. Emphasis on centralized authority in village chiefs and Paramount Chiefs was a product of colonial administrative needs to organize and govern local communities (Sharpe 1994; Kofele-Kale 1981; Geschiere 1993). During the colonial and postcolonial period, power to control the use of forest resources has steadily moved farther away from the villages and now rests largely in the hands of the central government in Yaounde. It is from here that new laws and policies regarding forest and resource use emanate, although in practice, "the legal framework governing forest resources is inextricably entwined with local level informal accommodations between staff of different ministries, the political authorities, and wealthy and powerful villagers" (Sharpe 1994).

23. In Ndongo, in South East Cameroon, a French logging company brought money, housing, education, and social services to a remote area, but when they had finished logging, according to local chief Comada Marcel, "They just left us with nothing but the road they built and that is just a path today" (*Cameroon Post* 1997).

24. Dorsey (in press) makes the point that "equitable benefit-sharing schemes," as promoted under the Convention on Biological Diversity, become questionable and perpetuate injustice when local communities benefit the least and incur the greatest costs from biodiversity conservation.

25. These examples of two-party arrangements based on point of collection are distinctly different from each other, however. For example, in the case of INBio-Merck, the collecting institution—INBio—receives all of the benefits, but it makes a commitment to share them with the National Park System and to apply them to the objectives of INBio, which include serving the national good. Shaman Pharmaceuticals, on the other hand, does not usually sign agreements with collaborators for the return of benefits, but promises to distribute benefits to all of the company's collaborators throughout the world. This is, in effect, an extended version of the point-of-collection model, although it will operate more like a fund.

26. For example, after years of discussion and planning, the Commission on Plant Genetic Resources agreed at its fifth session that a number of issues must be resolved with regard to farmers' rights, including the nature of funding (voluntary or mandatory); the question of linkage between financial responsibilities and the benefits derived from the use of plant genetic resource; the question of who should bear financial responsibilities (countries, users, or consumers); how the relative needs and entitlements of beneficiaries, especially developing countries, were to be estimated; and how farmers and local communities would benefit from funding (United Nations Environment Program 1994).

27. For example, in Sarawak, it appears to have been a single individual of *Calophyllum lanigerum* that produced the anti-HIV compound calanolide A (Soejarto 1993). Following collection, this particular tree was felled by loggers, prompting an extensive effort to locate other individuals with the active compound. In subsequent years, calanolide A was synthesized, but, based on the requirement included in the NCI LOC, the company currently holding the license to calanolide A is undertaking negotiations with the government of Sarawak (G. M. Cragg, personal communication, 1997).

28. In 1994, in South West Province Cameroon, for example, staff of the Limbe Botanic Garden drafted a research agreement and initiated a process of discussion with researchers, government officials, and local communities to better define and articulate the relationships involved in research collaborations, and general terms for access and benefit sharing.

# CHAPTER 15

## The Fate of the Collections: Social Justice and the Annexation of Plant Genetic Resources

*Bronwyn Parry*

The ethical and political implications of contemporary bioprospecting have been the subject of an extraordinarily heated debate over the past decade. Given that the term conjures up an image of a mad gene rush, a rabble of bounty hunters rummaging through biodiversity for that elusive "big find," it is unsurprising that it has ignited some serious concerns about the commercialization and potential monopolization of that most fundamental constituent of life—genetic material. On reflection, it is apparent that this debate has taken an interesting shape. Arguments about the perceived opportunity or threat posed by these "explorations of nature" have tended to coalesce around two opposing schools of thought and one spatial scale. While critics of bioprospecting have drawn a parallel with earlier colonial projects of appropriation and exploitation, suggesting that it constitutes a new and even more pervasive form of biocolonialism or bioimperialism, advocates have alternatively argued that "done right, . . . bioprospecting can bolster both economic and conservation goals while underpinning medical and agricultural advances" (Reid et al. 1993b).

What is clear is that, to date, most of the evidence to support either set of claims has come from investigations at either a local or a regional scale. Well-publicized accounts of bioprospecting agreements between corporations and suppliers in developing countries such as that between Merck and Costa Rica's National Institute of Biodiversity (INBio) have served to keep attention closely focused on what might be termed the point of supply—the locations from which the material is sourced—and the terms of trade—who bought what, from whom, and for how much. Within this context, a resolution of the question of whether contemporary collecting programs are just or unjust awaits nothing more than a simple audit of exchange, a determination of how much material was supplied and how much compensation was forthcoming in return.

What has been missing from this debate so far has been any discussion of, or investigation into, the question of what happens to these samples of genetic and biological materials *after* they leave the localities in which they were collected. Considering that the current wave of bioprospecting has been underway for at least a decade, it is remarkable that there has been so little research into the fate of the collections of genetic and biochemical material generated by these programs. There is, in fact, very little public knowledge that such collections even exist. Perhaps this is not entirely surprising. A striking feature of much of the literature on contemporary bioprospecting is the emphasis it places on the notion of searching. Bioprospecting was defined in 1993 as "the *exploration* of biodiversity for commercially valuable genetic and biochemical resources" (italics added) (Reid et al. 1993a:1). Many recent contributions make explicit references to "examining the plant kingdom," "exploring biological wealth," "making systematic searches," and the like (see, e.g., Cox 1990a:41; Sittenfeld and Lovejoy 1994:20; Farnsworth 1988:83; Newman 1994:479). In the rush to characterize, perhaps even valorize, bioprospecting as an activity dedicated to exploration and discovery, it has been easy to overlook the fact that bioprospecting is, fundamentally, about the practice of *collecting*, that is, removing material from one location and concentrating it in another.

Collecting—which has been variously described as "the appropriation of exotic objects" (Clifford 1989:218), or more broadly as "the gathering together of chosen objects for purposes regarded as special" (Pearce 1995:vii), has long been central to the practice of natural history. Within this context, collecting has traditionally been understood as either a benign aesthetic activity or alternatively as a dispassionate scientific activity, but rarely as an inherently political activity (Brockway 1979; Kloppenburg 1988). Yet collecting can be understood in these terms, as it is a process that enables individuals or groups to alienate (both territorially and epistemologically) particular bodies of material for their exclusive use. In light of this, greater attention must be given to exploring the dynamics of collecting and the profound ethical, political, economic, and cultural implications of collecting practices.

To achieve this end, I would argue that it is necessary to examine both the social and the spatial dynamics of collecting, processes that must be understood as mutually constitutive. Consequently, I begin this chapter by setting out a threefold typology of what I refer to as the social and spatial dynamics of collecting. My intention here is to illustrate the central role that spatial relations play in the actual mechanics of collecting and thus in the inherently political process of annexing and monopolizing specific collections of material.

Collecting is often perceived as being a one-off appropriation and transference of individual objects. However, these are perhaps best thought of as simply the first acts in a complex *process* of collection that entails not only the acquisition but also the concentration, disciplining, circulation, and regulation of flows of material. To gain a more nuanced understanding of the political and ethical implications of contemporary collecting practices, it is necessary to look beyond initial acts of appropriation and transference to begin to consider how, and in what ways, the value of collections can escalate with their movement over time and space, and most importantly to question who will ultimately enjoy the benefits of these progressive revaluations.

Part of the danger in focusing myopically on the point of collection and fixating on the specifics of particular compensatory agreements is that in so doing it is possible to lose sight of much larger and far more significant questions surrounding the ownership, potential value, and future usage of the collections. It is impossible to accurately assess the adequacy or ethics of existing compensatory agreements until something more is known about what sort of a resource these collections constitute, where they are concentrated, how they might be used now and in the future, and how returns from their use might be disbursed.

Having drawn out the neglected role of spatial relations in collecting practice through a brief history of collecting within natural history, I want to turn attention to the fate of contemporary collections of genetic and biological material that are currently used as sources of raw material by the pharmaceutical industry. In so doing, I will illustrate how the construction, uses, and regulation of biological materials are being transformed by processes of technological, economic, and regulatory change; I will then draw attention to the serious political and ethical implications of the profound shift in the nature of the social and spatial dynamics of collecting induced by these changes. However, before it is possible to assess the future implications of contemporary collecting programs, it is essential to begin by situating these practices within a broader theoretical framework on the nature of collecting within natural history—to begin with some understanding of what it is to collect, of why collecting programs are instituted, of how collections acquire their value, and of the power relations that are inherent in many collecting practices.

## REVEALING THE SOCIAL AND SPATIAL DYNAMICS OF COLLECTING

### Collecting as Simple Acquisition: Mobilization, Decontextualization, and Exoticization

Spatial dynamics have always been central to the practice of collecting, although this has not always been explicitly recognized. Space is implicated in collecting practice in various distinct but interconnected ways. Collecting is, by definition, a bringing together,[1] a process that necessarily involves transfer across space. At its simplest, the value of a collection can derive from the fact that this process involves removing material from one place and relocating it in another. This process of decontextualization *alone* can serve to exoticize and valorize the objects in question and confer on them a particular value. The collection of rare natural materials as a project for the "acquisition of the exotic" has of course a long and well-established history. As Findlen (1996:57) reminds us, late medieval princes from Frederick II to the Duke of Burgundy included rare and exotic natural objects among their most coveted (and private) possessions.

Objects such as those displayed within the "cabinets of curiosities" so popular in European society in the early eighteenth century were valued not only for their rarity or peculiarity, but also for the way in which they came to stand "metonymically, for a whole region or population" (Clifford 1989:227). The collected objects thus enjoyed a certain sovereignty, with each valued independently as a self-contained microcosm of a wider universe. They were not, however, at this time, principally collected or valued for their relationships to each other. It was not until the Enlightenment that assemblages of natural material came to be elaborated into sets of formally related categories—to acquire a value that was not determined self-referentially. It was not until the Enlightenment that the appropriation of the exotic, the very culture of collecting, began to acquire the status of a science.

It is often assumed that the process of gaining power or profit from decontextualizing and exoticizing natural or cultural objects necessarily demands that such objects be transferred between autonomous and geographically distant localities. It is worth remembering that processes of decontextualization and exoticization *are* usually dependent on movement but *not* at any predetermined scale. Objects can be revalued both economically and culturally by virtue of their relocation from one country and cultural context to another, but they may as easily be revalued, for example, by virtue of their relocation from one institutional setting to another—say from a museum or a factory to an art gallery not more than a few city blocks away. It is not the length of the movement but the degree of decontextualization involved in the movement and the relative novelty of that process that, as we shall see later, are of greatest significance.

## Collecting as Concentration and Control

### Cycles of Accumulation

The Enlightenment project of understanding and studying nature was underwritten by a conviction that "Nature, in being known, may be mastered, managed and used in the services of human life" (Thomas 1979:27). Within this epistemological framework, the utility of a collection, in both a scientific and mercantilist sense, was of the essence. The potential utility of a collection was increasingly determined by how systematically it had been ordered. Within this new context, as Gasgoine has suggested, "the happy randomness of many of the earlier virtuosi, with their bowerbird collections of curious objects which had no apparent relation, were of little use" (Gasgoine 1996:108). The value of a collection no longer derived simply from accumulating exotic but unrelated objects. An additional value could now be derived from determining and explicating the *relationships* or *correlations* between collected materials. It is interesting to take a moment to reflect on *what* was being valued here. For it is here that we see the genesis of a process in which materials came to be valued not just for their corporeality, but also for the information about the relationships existent between them, and, much later, for the information embodied within them.

As the historian of science Bruno Latour (1987:215–257)[2] suggests, the great European voyages of discovery of the seventeenth and eighteenth centuries were undertakings dedicated not simply to investigation but to accumulation. But what was being amassed with each successive voyage was not just an undisciplined array of trophies or souvenirs but a systematically organized body of information about the coastlines, flora, fauna, language, and cultures of distant peoples. This information could be employed to re-create, within particular centers in Europe, a scaled-down universe that could be surveyed panoptically. It is here that the interdependent relationship between the social and spatial dynamics of collecting (and thus of annexation and monopolization) becomes more apparent. The development of a hierarchical system of classification was based on both observation (the objective gaze of the naturalist) and categorization. Explicating the relationships or correlations between particular specimens was ideally achieved through *direct comparison*—an activity that demanded a concentration *within a fixed locality* of a number of like (probably related) objects. Comparison is not rendered impossible without spatial juxtaposition, but it is certainly facilitated by it. As Foucault (1970) noted, although the establishment of botanical and zoological gardens in the eighteenth century was thought to reflect a new curiosity in exotic plants and animals, this interest had been in existence for some while. What had changed was the institutions, which now constituted new dedicated spaces within which collected material could be concentrated, surveyed, and disciplined. By centralizing information and specimens in one location, it became possible for the collectors to gain an overview of the relationships between the specimens that would be all but impossible for a single person in a single location to obtain.

By these "extraordinary means" (Foucault 1970:224), it became possible to "see" places, events, and experiences that were distant, without actually experiencing them firsthand. As Latour (1987) suggests, it was at this point that something akin to a Copernican revolution took place: "Those who were the weakest because they remained at the center and saw nothing start becoming the strongest, familiar with more places not only than any native, but than any travelling captain as well" (Foucault 1970). To use Latour's term, a *cycle of accumulation* was thus set up. Armed with "foresight," a familiarity with events, places, and materials that could be acquired prior to contact, future emissaries were able to refine their searches, to search more effectively, to combine collected materials and information about them in order to further the process/cycle of accumulation.

We would, however, be succumbing to an attack of spatial fetishism if we were to imagine that the built spaces *alone* had the power to invest materials collected therein with a particular value. Of course they did not. Rather, they acted as focal points for the cycles of accumulation that became concentrated upon them. Individual institutions such as the Royal Botanical Gardens at Kew became constituted not only as master repositories but also as centers of expertise—what Latour refers to as "Centers of Calculation." Scientists within these institutions were able to draw on this concentration of knowledge to obtain a mastery of it. Minions in the field were no longer entrusted to undertake classifications of material in situ, and materials began to be drawn, as if in a vortex, toward the center. As Browne has suggested, when Joseph Hooker undertook his major taxonomic studies of the New Zealand and Indian floras in the late nineteenth century, he chastised foreign collectors for their inability to understand their own plants. Without access to the specialized knowledge and range of specimens now available in centralized institutional collections, the field collectors' purview was deemed to be limited and they were directed in no uncertain terms to send plants to Kew—"the hub of the colonial scientific enterprise" (Browne 1996:313). Natural materials, it seemed, could acquire value only when extracted from their chaotic surroundings and subsumed within a particular system for codifying knowledge, a process that now demanded their export to those centers within which the classificatory systems were both constructed and maintained.

### Technologies of Exploration and Exploitation

This project of "stabilizing" and "mobilizing" nature (Latour 1987:224) could not have been effected without the introduction of various new technologies of exploration and exploitation such as the herbaria and the greenhouse. Greenhouses were a new technology in the 1700s and they played a pivotal role in enabling both the mobilization and the further fragmentation of nature and in the further fragmentation of nature. With the introduction of the greenhouse it became possible to mobilize *environments*, which until that point had been an integral and inextricable part of the constitution of particular *places* (Hix 1996; Woods and Warren 1988). This technology enabled the "specificities of place," so necessary for the propagation and sustainment of particular plants, to be unprolematically reproduced in any location. However, what was being reproduced was *simply environmental conditions, not entire ecosystems*. Although not as radically stabilized as their counterparts now glued to the herbarium sheets, greenhouse plants were nonetheless still decontextualized— extricated from the tangle of their complex natural environments, *concentrated, controlled*, and able to be mass produced on demand. While theoretically it had become possible to reproduce environments and plants in any location, it was in reality only possible to reproduce them in locations where greenhouses were to be found. Not surprisingly, they were concentrated only within "the centers of calculation" or at their satellite out-stations located at strategic points across the globe. As we shall see, greenhouses were to play a vital role in a much wider imperial project of redeploying plant resources to economic and political advantage.

### *Collecting as Recirculation and Regulation*

This brings us to the third and most significant of the ways in which spatial relations are implicated in collecting practice. Analysis of the dynamics of collecting practice reveals how value is compounded with each stage of the collecting process. A certain amount of power and value accrue when objects are first moved about, decontextualized, and exoticized. A further value accrues when objects begin to be concentrated in certain places where they can be ordered, controlled, and disciplined within particular systems for codifying knowledge—where information about the relationships between the objects can be identified and explicated. But it is not until the third stage that these two factors are squared and an even greater value obtained. By applying the knowledge, technologies, and expertise concentrated within the centers of calculation to the collected materials, it becomes possible to combine either the living materials or information about them, or both, to produce entirely new materials or bodies of information. In this instance, these might be entirely new varieties of plants suited to particular climates or localities, or information or data about which related varieties might substitute for those proving difficult to obtain. The collector can then derive further power or value from *recirculating* the materials or information, *by redeploying them to advantage* through strategic utilization, exchange, or trade.[3]

Within this context, it becomes apparent that the centers of calculation—botanical gardens in particular—played a vital role not only in the collection and disciplining of material, but perhaps more importantly in regulating the flow of collected material as it was recirculated and redeployed on a global scale. Using the technologies of exploration and exploitation available in the centers of calculation, it became possible to hold these materials in a suspended state, keeping the specimens alive in a transitory life-world (the greenhouse) until a time and a place could be identified in which they could be successfully redeployed. This usually involved a further recirculation of collected materials to colonial dependencies via a series of linked collecting institutes. Botanical gardens located in colonial outposts had a dual role to play, acting not only as a focal point for collecting operations, but also as holding and redistribution stations for recirculated materials (Brockway 1979; Kloppenburg 1988). Together, the colonial network of botanical gardens formed a series of nodes in an imperial circuit of exchange, with each acting as a capacitor regulating the recirculation of botanical materials and information.

The collection of botanical material by the European powers did not end with the collapse of the empire, and has continued throughout the twentieth century. Interestingly, though, the rationale for undertaking such activities changed somewhat. Although the symbiotic relationship between scientific study and economic advancement still flourished, attempts were made to create a distinction between the two. *Scientific* collection and investigation of plant material was to be undertaken for purely scientific purposes: classification, analysis, and the advancement of "the common good." This rationale underpinned the establishment of a series of new International Agricultural Research Centers (IARCs) in the 1950s and 1960s, which were to play a central role in a global program to improve agricultural productivity. As a consequence of their involvement in the development of new plant, crop, and seed varieties, these centers became repositories for collections of plant genetic resources donated by developing countries to aid agricultural research. It was assumed that material collected for these scientific purposes would not be directly accessed or utilized by corporate interests. Revelations that this material had in fact been accessed by multinational corporations who used it to develop proprietary and patentable varieties prompted commentators such as Kloppenburg to conclude that IARCs and other ostensibly scientific collecting institutes were at risk of becoming "the modern successor to the eighteenth- and nineteenth-century botanical gardens that served as conduits for the transmission of plant genetic information from the colonies to the imperial powers" (Kloppenburg 1988).

At this point in our fragmentary history and geography of collecting practice within natural history, the trail goes cold. This review has established that the power that accrues from collecting derives not only from possessing decontextualized and exoticized materials, but from concentrating, controlling, and redeploying those materials to strategic advantage. Given this, it is surprising that we know so little about how plant material is currently collected, where it is being concentrated, how it is being disciplined, where and how it is being redeployed, and, most important of all, to whose advantage. Is plant material still the same resource that it formerly was? Where are the new centers of calculation—or are they the old ones? What new technologies have been devised for the disciplining of plant material? Who is controlling the utilization, circulation, exchange, and trade in such material these days; and what are the justice and equity issues implicated in these processes?

## "BIOIMPERIALISM"—EVERYTHING OLD IS NEW AGAIN?

Having devoted the first half of this chapter to outlining the long history of collecting natural specimens, it would be foolhardy to begin this half by suggesting that contemporary bioprospecting is in any way a new phenomenon. Although *bioprospecting* and *the bioprospecting industry* were terms that were bandied about with increasing frequency from the early 1990s onward, there is little immediate evidence to suggest that any such "new industry" was actually constituted at this particular juncture. In fact, most of the contemporary collecting carried out under the broad rubric of bioprospecting is still undertaken by the same scientific, academic, and commercial collecting agencies that have a long history of involvement in the collection of natural materials.

However, there can be no question but that the last decade has witnessed an extraordinary resurgence of interest in the collection of biological materials. This interest has been fueled by a demand for genetic and biochemical raw materials, which soared throughout the 1980s as the agriculture, pharmaceutical, and cosmetics industries became increasingly biotechnology dependent. This demand translated directly into an exponential increase in bioprospecting. My own research into bioprospecting within the pharmaceutical industry[4] indicates that a series of new collecting programs were progressively instituted by, or on behalf of, the industry throughout the 1980s. These collecting programs have been underwritten by government agencies such as the U.S. Department of Agriculture or the National Institutes of Health (NIH), by corporations such as Merck, Bristol-Myers Squibb, or Glaxo Wellcome, as well as a host of smaller companies with an interest in genetic and biological material, the majority of which are headquartered in the developed world, principally in the United States, Europe, and Japan.

The fact that materials collected under these programs are being transferred to the developed world and there commodified and patented has prompted commentators such as Shiva (1993a, 1993b) and Mooney (1983, 1993) to argue that contemporary bioprospecting programs mirror earlier colonial projects of appropriation, thereby constituting a new form of biocolonialism or bioimperialism.[5] Enticing as it is, I would argue that it is in fact difficult to oversimplify parallels between past and present collecting practices. Part of the risk associated with conflating understandings of historical and contemporary collection programs is that of reducing the complexity of the current situation by creating a comparison that is insensitive to changes that have occurred in the social and spatial dynamics of collecting over time.

It would, for example, be foolish to argue that there have not been significant changes in the organization, the administration, and, perhaps, even the rationale behind collecting programs in recent years. While previously plants were appropriated without any form of recompense, considerable attempts have been made over the last decade to ensure that they are now collected and utilized in a "just and equitable" fashion. Indeed, with the ratification of the Biodiversity Convention[6] in 1992, this became a formal requirement. It could be argued that the Convention interprets the notion of justice and equity rather narrowly, so that it is possible to secure a just and equitable outcome simply by rejigging the distributive paradigm in such a way that the allocation of *financial* compensation is more evenly apportioned. Nevertheless, measures such as article 15[7] were introduced with the clear aim of ensuring that suppliers of genetic resources (many in developing countries) benefit from the uses that are made of collected materials, both now and in the future. It is hoped that these returns will also illustrate the commercial value of endangered ecosystems and thus provide an important incentive for their conservation.

These changes have been widely discussed, even celebrated as important breakthroughs. But other changes have also occurred over time that are equally significant, although less well celebrated. First, there have been fundamental changes in the way in which plant material is both constituted and valued as a resource. Second, it is apparent that this change is linked to and driven by wider changes that have been occurring in the organization of the global economy. These changes bear closer examination because they have the capacity to *speed up* the social and spatial dynamics of the collecting process from whence power and value derive. If, as the historical review suggests, power and value derive from an ability to acquire, concentrate and control, and recirculate and regulate collected materials, then it is possible to theorize that the amount of power and value that derives from being in possession of a collection of materials may well be *increasing*, (1) as processes of technological innovation fundamentally alter the nature of biological materials so that they become infinitely more amenable to collection, concentration, and control; and (2) as processes of global economic and regulatory change improve collectors' ability to recirculate and regulate the flow of collected materials more strategically and thus to further advantage.

Understanding the nature of these changes is particularly important as they have the capacity not only to undermine measures introduced to ensure the just and equitable utilization of this material, but also to generate an extraordinarily complex range of new justice and equity issues. What are these technological, economic, and regulatory changes, then, and what impact are they having on the social and spatial organization of this new global trade in samples of plant genetic material? We now turn to an examination of these questions.

## Collections of What? Retheorizing Plants as Material and Informational Resources

As the historical review established, there have been progressive changes in the way biological material has been valued over time. Whereas during the Renaissance natural objects were valued largely for their exoticism, they were subject to a revaluation during the Enlightenment when they became additionally prized for the unique position they occupied within a wider set of related objects and a wider system of knowledge about natural history. At the turn of this century, the study of biology underwent another profound transformation as new sciences such as genetics and experimental embryology allowed living material to be opened up, both literally and figuratively, to more detailed investigation. Since their introduction, but particularly since the 1970s when the first experiments in genetic engineering were conducted, biological material has been undergoing another progressive reevaluation. The definition of bioprospecting by Reid et al. (1993b) as "the exploration of biodiversity for commercially valuable genetic and biochemical resources" alerts us to the nature of this change. Natural materials are no longer valued just for their corporeality or for the information that they provide about relationships between particular species. They are now principally valued for the genetic and biochemical information embodied within them.

Extraordinary advances in DNA extraction, identification, and sequencing; biochemical assays; and roboticized screening have fundamentally changed the way in which plant bodies are both constituted and valued as a resource. These new technologies enable biological material to be disaggregated and reduced to a series of constituent parts able to be utilized independently of the whole. These constituent parts cannot, however, be understood simply as biological material at a finer scale of resolution. Mooney (1983), Krimsky (1982), and Doyle (1985) were among the first to recognize that what had once been understood as simply a *material resource*—plants, animals, some bark or seeds—had become also, with biotechnology's capacity to access the genetic or biochemical components within it, what could be referred to as *an informational resource*. It is now possible to extract genetic or biochemical information from living organisms, to *process* it by replicating, modifying, or transforming it, and to *produce from it* minor modifications of this information that are themselves able to be utilized as raw materials, commodified as resources. The genetic information embodied within material resources has become, in effect, the *instrument of production, not only for that resource, but also for a range of other potential resources that could be produced by recombining the information in an almost limitless number of ways.*

This new valuation is linked to and indicative of wider changes in the organization of the global economy. In a capitalist economy, continued accumulation relies on reducing turnover times and speeding up the flow of capital through circuits of exchange. Any ability to reduce "the friction of distance" by either improving technology or accelerating the flow of information provides a means to attain these objectives. As Harvey suggests (1990:159), in an informational society, "in a world of quick changing tastes and needs and flexible production systems . . . access to the latest technique, the latest product, the latest scientific discovery implies the possibility of seizing an important competitive advantage . . . *knowledge itself becomes a key commodity.*" Access to, and control of, information—particularly privileged information—not only provides a competitive advantage within particular production systems, it now also forms the basis of an entirely new economy within which *information itself* is the commodity being exchanged. An array of specialized consultancies and business service providers speculate, deal, and trade in raw information, as well as in organizing the marketing and exchange of informational products such as videos and computer software. The speed with which the information or informational products can be recirculated determines the rate of financial return.

Within an informational economy, then, improved productivity and competitive advantage rely not so much on having command of fixed infrastructure and stock resources as on the ability to control networks through which information may be traded, rented, circulated, recirculated, or, even, where strategically necessary, withheld or monopolized. Like other informational resources, genetic and biological materials are fluid, transitional, and dynamic. They are able to be readily conveyed, preserved, modified, transformed, circulated, and even replicated over both space and time. However, to create an informational resource from collections of plant genetic material and to control the flow of that resource requires access to specific technologies. The question of who, then, has access to these technologies and as a consequence the ability to create, concentrate, and control these resources is of the utmost significance given that access to such facilities is by no means equitably distributed within society.

## *Decontextualization and Exoticization*

### The New Technologies of Exploration and Exploitation

Creating and controlling informational resources from biological material cannot be achieved without access to new, sophisticated technologies of exploration and exploitation. A popular argument offered in defense of collecting projects is that expressed recently by a director of one of America's leading botanical gardens: that there is nothing inequitable about collecting botanical material as "the resource is still there for others to capitalize on," that it is not in any way monopolized or alienated. While it is true that the material resources are not being monopolized (they continue to exist in many locations), the far more valuable informational resources contained within them certainly are. For although it is possible to capitalize on the material resource without access to technology—for example by selling plants at a market—it is impossible to capitalize on the *informational resource* housed within them without access to appropriate technologies. The most basic of these, technologies for screening and extracting active chemical compounds, are available in most pharmaceutical laboratories in developing countries. However, the most sophisticated are concentrated principally within the new centers of calculation, the corporate laboratories of the first world. Even in this high-tech age, our first principle of spatial relations in regard to collecting practice still holds true: material becomes more exoticized and hence more valuable as it becomes more decontextualized. In this era, the rarity and thus the value of genetic or biochemical information derives from the fact that comparatively few groups or institutions within society have the technologies necessary to reveal or access it. It is the ability to acquire (that is, to secure access to), concentrate, control, and rapidly recirculate genetic and biochemical information that undoubtedly forms the basis of competitive advantage within today's global biotechnology industry. Technological, economic, and regulatory changes are, however, fundamentally altering the ease with which particular groups can control these processes. Let us take processes of concentration and control to begin with.

## *Old and New Centers of Calculation: Concentration and Control in the 1990s*

Technological developments such as roboticized screening and extraction, biosynthesis, plant cell culture, and combinatorial chemistry have led to a valorization not of cumbersome whole plants or animals, but rather of the fragments of genetic and biochemical information embodied within them, and this is evidenced by the changing way in which biological materials are now collected, stored, and used. Whereas plants were once collected whole—for transportation and reproduction in colonial outposts using technologies such as the greenhouse—or collected in fragments and archived in herbaria for classification purposes, they are now collected and stored for future use, not just as reference materials but as reproducible industrial commodities, in the form of tissue samples or biochemical extracts. Public institutions such as the NIH, as well as various corporate concerns, are currently amassing large collections of material in this form, primarily for use in the production of pharmaceuticals. The NIH collection consists of some 50,000 bulk samples of plant, fungal, microbial, and marine material from which over 114,000 different extracts have been refined. This collection, which constitutes a living archive, was sourced from over thirty tropical countries and is stored cryogenically in forty-one walk-in freezers in their Natural Products Repository in Maryland. The corporation Merck Sharpe & Dohme holds a collection of a similar size. It is unusual for organizations such as these (which are principally interested in screening plant materials for active chemical compounds) to house collections of live higher plants. These are more likely to be stored for them by the botanical garden or intermediary institution contracted to collect such materials on their behalf. These institutions may keep plants as live, fertile or infertile specimens, or alternatively as herbarium voucher specimens.[8] Academic or scientific research institutes may hold collections of live plants, but they may also find it expedient to store collections of plant material as a "library of extracts." These libraries consist of active chemical compounds extracted from plant samples by a process of fractionation. Other materials, including live culture collections and cell lines, are also are archived in public and private repositories for future use.

A serious debate surrounds the swirling and constantly evolving question of how much and what types of biological material will be able to be reproduced from living samples using technologies such as plant cell culture and tissue cloning. While it will always be necessary to maintain large living populations in the interest of safeguarding diversity, it must also be acknowledged that repositories of cryogenically stored biological materials represent an increasingly valuable source of raw material for the biotechnology industry. As studies of the history of the archiving and retrieval of other informational resources attest, processes of miniaturization greatly facilitate the storage, retrieval, and circulation of information. This is now as true for biological informational resources as it is for any other type of informational resource. It is clearly more efficient to move, store, and concentrate genetic or biochemical resources in a miniaturized form as living cell lines or libraries of extracts than it is to conserve or maintain large living collections of material in situ. The first change in the social and spatial dynamics of collection induced by technological advances is now evident. The informational resources embodied within biological material have become, like other informational resources, infinitely more amenable to collection and to concentration.

Biological materials are also more readily controlled when in fragmented form. Lefebvre said of space that its domination is most readily effected through a process of "pulverization" or fragmentation in which it is reduced to freely alienable parcels of private property able to be bought and traded at will on the market (1991:385). Perhaps the same could now be said of nature. Reflecting the importance of informational commodities in the global economy are new global regulations such as GATT TRIPs,[9] which allow for the commodification and monopolization of embodied alterations made to the genetic and biochemical information. All the modified compounds or genetic structures produced from natural material will almost certainly be patented by corporations, or private or public research institutes, as new products in their own right, or stored as matter that may be used as the basis of a new patentable product at some later date. Each of these combinations and recombinations, manipulations and remanipulations, has the potential to add value to the material. The patent holders within the new centers of calculation continue to play an important role as gatekeepers, although they now act to regulate the flow of biochemical information rather than the movement of whole plants. Whereas, in the colonial era, power and profit could accrue from an ability to rapidly circulate plant material to various strategic locations, power and profit now derive from the ability to actively restrict the circulation of information through the application of the patent. As one pharmaceutical company executive explained to me,

> A patent is a legal document that doesn't give you the right to do anything—it gives you the right to exclude others. So if I hold a patent and there are no licenses to it, I can exclude everybody else from using this unique compound.

## TAMING THE SLIPPERY BEAST: TRACKING AND CONTROLLING THE USE OF PLANT GENETIC RESOURCES

While it could be argued that processes of fragmentation have made it easier to control biological materials, it could equally be argued that these processes are making the task of tracking and controlling the *use* of such materials infinitely more difficult. Genetic and biochemical resources could now be said to share particular attributes with other types of informational resources such as computer software, electronic data, videos, recorded music, and the like. These attributes include the following:

- The ability to be used successively without diminution of the original source material
- The ability to be replicated away from the original source
- The ability to be readily conveyed and circulated and disseminated
- The ability to be readily modified or transformed to produce new versions
- The ability to be archived for future use without decomposing

Some of these attributes have always inured in biological materials. For instance, it has always been possible to replicate or modify biological material through traditional breeding practices, to circulate it through plant or animal exchange, to store it as seed. It is not the fact of this transmissibility, but rather the nature and scale of it that has undergone such a recent and profound change. Technological and regulatory changes and the development of new forms of commodity exchange are dramatically improving the transmissibility of genetic and biochemical information, and with it the collectors' ability to profit from the strategic recirculation of this informational resource.

Although it has always been possible to store, replicate, modify, and circulate biological materials, these transactions have traditionally taken place within certain biological, geographical, and temporal parameters. Different types of biological materials could be combined, but only between related species. Moreover, the transmission of biological information between like species could be effected only through the interbreeding of whole plants or animals, processes that were constrained by the need for proximity and certain fixed timescales of gestation. Plant or animal materials could be collected, stored, or exchanged, but replication of these or related species could be effected only if the collected materials were capable of self-reproduction—that is, if they existed as living beings with reproductive organs or as seeds capable of self-regeneration. The utility of the collected specimens as a source of reproductive material lasted only as long as these capabilities remained intact, and, as the earlier review demonstrated, such materials historically have often been either irrevocably corrupted with their transference/transmission over both time and space, or so radically stabilized to withstand transportation as to be rendered infertile.

Today, these parameters have all but collapsed, and with them the traditional paradigm that governed the nature and scale of transmission of genetic and biochemical information. It is now possible to remove specific genes from one organism and insert them into other completely unrelated organisms, or to extract particular molecular compounds and combine them with an almost endless number of others extracted from a variety of organisms. Hybrids have been overtaken by transgenic organisms and recombinately engineered molecular compounds. There are now few "natural" limits to what Latour calls the "combinability" of these materials. Whereas the reproduction or transmission of genetic or biochemical information once demanded an "up-close" sexual interaction between two living beings, it can now be achieved either by cloning isolated cells or by extracting—perhaps I should say excerpting—information from one organism and inserting it into another. Consequently, systematic reproduction of biological materials or transmission of certain characteristics no longer requires the amassing in one location of a viable fertile population. It no longer even relies on having possession of a single whole organism, but can be achieved simply by conveying the requisite raw materials (informational resources in the form of milligrams of genetic or biochemical material) from one location to another.

The fact that the information necessary to reproduce entire organisms, or the desirable characteristics of particular organisms, is now available in a compressed version/form is ushering in a new era in the history of the supply of industrial raw materials, which might be termed microsourcing. With access to advanced screening technology, it is, for instance, currently possible to extract lead compounds from as little as four milligrams of living material. But these resources are of little commercial value unless they can be reproduced in quantity. Some stored materials such as cultures or germplasm can be reproduced the old-fashioned way by simply regrowing them. This usually entails "farming" them, either in the ground or in large vats. Frequently, though, it is more efficient or desirable to secure industrial quantities of particular resources by simply *replicating* specific biochemical or molecular components in isolation from the clumsy baggage of the whole plant structure, an objective that is now being realized through technological advances in the fields of synthesis, semisynthesis, and plant cell culture. The advantages of replicating natural material in this way were described to me in a recent interview with a senior corporate research chemist:

> The fact that we can identify the compound, and we can now do that routinely on a milligram or less of material, basically opens the door for the synthetic chemist to come in and—if the activity that we are seeing is something that is sufficiently interesting—build it for us. And as soon as he [sic] can build it for us we are no longer talking of having a couple of milligrams to work with but having grams or kilos or whatever from a well-documented, repeatable procedure—which from nature is just not quite the same . . .

By employing these technologies, it has become possible to reproduce viable quantities of some active compounds from very small amounts of source material. With the aid of these technologies, small initial samples do indeed have the potential to become a "means of production for that resource" as another pharmaceutical company executive recently confirmed:

> There are no guarantees, but we've made a lot of progress. One can envision perhaps doing an initial collection and storing some plant material and then when you did have your hit—it's an arduous process as well—establishing a cell line and culture and trying to tease out the compounds of your choice. But once this art is developed—great! Now it is a fairly crude science, but it will become an art, then a science, and I think . . . that this will be the way of natural products in the future.

Compounding the effects that processes of compression or miniaturization have had on collecting practice are those effects induced by the introduction of technologies such as cryogenic storage, which have the capacity to *extend* the temporal limits to the viability of these new miniaturized collections. Cryogenically stored materials such as live cultures and tissue samples can exist in a state of suspended animation for an almost indefinite period of time. This technology allows materials that are collected now to be conveyed intact over both space and time for future reutilization. Of course, not all material will survive the journey and not all material will remain viable for replication over time. Material that can be regrown from germplasm or as a live fungal or bacterial culture most easily retains its authentic form and is most easily reproduced without distortion. Although some chemical compounds break down or become volatile over time, others remain remarkably stable. While the amount of biological material required to constitute a viable industrial resource grows ever smaller, the same material remains viable as an industrial resource for ever longer periods of time. "Micro-sourcing" gives collectors the capacity to reduce not only the amount of material required for contemporaneous reproduction, but also the amount that they need to store to ensure viable future reproduction.

Although collected materials may be used for the purpose of "self-replication," this is not the only way in which they may prove to be of value to collectors. Following Castells' argument (1989:13), it is clear that the biochemical or genetic information embodied within samples of material can now be used not only as a means to replicate that resource, but can, in addition, be modified or combined to create entirely new collections of valuable and commodifiable genetic or biochemical resources. The development of a new science, combinatorial chemistry, enables entirely new compounds to be created by randomly combining various existing natural and synthetic compounds of proven efficacy. This process has the potential to dramatically increase the value of existing collections by increasing the number of ways and times specific compounds may be used or reused. The introduction of new techniques such as these increases both the demand and the market for natural products research specimens, which, within the pharmaceutical industry alone, was recently estimated to be worth some US$30–60 million per annum.[10] Reviewing all of these developments in the light of our typology of the social and spatial dynamics of collecting, it is apparent that processes of technological change are enabling collectors to not only concentrate and control genetic and biochemical information more readily, but also to recombine and recirculate it more readily. Opportunities to profit from the recirculation of these informational resources are also being enhanced by recent changes in the organization of the global economy.

### New Links Between the Centers of Calculation: Redeploying and Recirculating Plant Genetic Resources

Given that genetic and biochemical materials are now being utilized principally as informational rather than material resources, it should come as no surprise to discover that new forms of commodity exchange are evolving that mirror those that currently govern the exchange of other informational resources. As we know, once ownership of a computer program, song, or moving image is secured, a continuing return can be generated from selling access to that information. The same is now true for samples of genetic and biochemical material. Evidence suggests that a range of organizations have been quick to capitalize on this realization. They have declared their intention to specialize in "renting out" samples of genetic and biochemical materials from their own libraries of living plant tissue and biochemical extracts. The notion of a rental firm perhaps conjures up an image of a small commercial enterprise, something akin to the local video store. However, the organizations that intend to specialize in the rental of material are surprisingly diverse, ranging from small commercial enterprises, such as the Knowledge Recovery Foundation, to the corporate subsidiaries of publicly funded universities, such as the Strathclyde Drug Research Institute. Each operates on the premise that libraries of material can be utilized and reutilized by a number of interested parties. With each transaction, the material resource is screened for different types of biochemical or genetic information embodied within it. As each client is searching for different types of information, the same material may be recirculated and rescreened a number of times before becoming exhausted. Brokerages and individual brokers similarly intend to act as clearing houses for the sale and resale of collections of specimens and extracts jettisoned by academic laboratories or "asset-stripped" from failing biotech companies.

This raises several important political and ethical questions. First, both rental firms and brokerages secure a financial return on each transaction involving the material, but is there any effective way of tracking these transactions or of ensuring that some form of compensation is returned to the supplier for each rental of the informational resource? Second, is there any effective way of monitoring how genetic or biochemical materials may be manipulated or modified by successive recipients? If such material were to form the basis of some new compound, or if it were to be used as a starter for processes of replication or synthesization, would the supplier have any way of securing compensation for contributing the necessary raw materials? None of these rental agencies or brokerages appear to be owned or run by suppliers from developing countries. Is there any reason to assume that these suppliers will be assured of any compensation for the use of material circulated or exchanged in this way?

Changes in the organization of the wider economy—notably a trend toward cutting public funding, corporatization, and downsizing, and a move toward subcontracting out rather than in-house production—are also beginning to have their effect even in the apparently rarefied atmosphere of botanical gardens and museums of natural history. Funding for scientific or academic collecting programs is dwindling and public institutions are, as a consequence, being pressed to secure income from other sources. The need to generate income has led many scientific and academic institutions to offer their services as commercial collecting agents. Since the mid 1980s, institutions such as botanical gardens, museums of natural history, and academic laboratories have become increasingly involved in collecting specific raw materials for corporations on a contractual basis or, alternatively, allowing corporations to underwrite the costs of scientific collecting programs in exchange for the provision of specific samples. There is of course nothing unethical in the establishment of a formal subcontractual relationship between collecting institutions and corporate interests, as long as it is subject to established protocols and public scrutiny. The potential difficulty lies in the possibility that the establishment of such a relationship might encourage commercial interests to exert undue influence over the direction of research studies, or to attempt to access material in the larger existing collection that is not covered by the contractual agreement.

Materials accumulated prior to 1992 and held in ex situ collections are currently exempted from the protocols introduced in the Biodiversity Convention. Commercial interests may consequently find it more expedient to bioprospect living collections held by scientific or academic institutions than to have to collect in situ in developing countries. The ethical implications of remining existing scientific or academic collections in this way are particularly disturbing. For example, material collected for scientific purposes is now being used as a commercial resource without the consent of the supplying parties (which may now be impossible to obtain) and in many cases without public knowledge or scrutiny. Some senior executives in scientific collecting institutions are beginning to express considerable unease about the ethical implications of these activities:

> I shouldn't even really mention the name of the botanical garden, because I hope they're not doing it, but I had a long talk with them about exactly that. A pharmaceutical company approached them and said, "Can we bioprospect your living collection?" and I don't know whether they said yes or no, but I personally have a real problem with them saying yes, because those things were collected for a different purpose. Maybe some of them were collected a way long time ago under no purpose at all—you know, just colonialism. You go and find some cool plants and bring them back and you don't have to ask the natives—that was acceptable then, that was what people did, but the fact is that to then take those collections and to reap a profit from them and to have *no way* to return a benefit to them . . . you know, I have a real problem with that.

Recent findings from my research suggest that this practice is not an isolated occurrence but rather is becoming an increasingly common phenomenon as scientific and academic collecting agencies search for alternative sources of funding, and as rapid technological change creates lucrative new markets for existing collections of biochemical and genetic information.

Conservation agencies are, similarly, surviving harsh economic times by becoming increasingly corporatized. Organizations such as Conservation International (CI), among others, are setting up corporate enterprises and partnerships to fund their operations. In 1995, CI entered into a commercial partnership with the cosmetics corporation Croda International. Under the terms of this agreement, CI is being paid to facilitate the collection of ethnobotanical information and biological specimens on behalf of Croda. CI is already involved in a wider bioprospecting program in collaboration with other government and corporate collecting agencies such as the NIH and Bristol-Myers Squibb. Although CI would argue that such agreements afford an opportunity to develop biological resources in an equitable and sustainable fashion, there is no question that the creation of new linkages and partnerships between these actors also facilitates the circulation, exchange, and trade of such material. This by itself need not necessarily be cause for concern, as long as there is some certainty that each subsequent usage of the material can be effectively monitored and will benefit the source country in some way. But is there? And how might this be done? This then brings us to the crux of the issue.

These new institutional linkages and the new technologies of exploration have together created a new resource (genetic and biochemical information) that is slippery and, because of its inherently dynamic and mutable constitution, extremely difficult to monitor or track. Plant material has been fragmented to such a degree that valuable chemical and genetic components can be extracted from as little as a few milligrams of living material. These component parts are infinitely mutable and subject to myriad processes of transfiguration—they can be modified, synthesized, cultured, multiplied, and duplicated. The timescale and the geographic scale over which this material may be deployed and redeployed is also undergoing a profound process of "extantiation." The introduction of sophisticated processes of fractionation, screening, and cryogenic storage has made these resources eminently portable—able to be utilized and replicated over space and time. A resource has been created that can also be readily circulated through the conduits created by new institutional links and partnerships. Although commentators such as Kloppenburg and Balick (1996:174–181) still insist on characterizing the institutional, governmental, and corporate actors involved in the exchange and trade of genetic and biochemical information as separate actors with separate interests, it is important to challenge the presumption that these actors are somehow distinct from one another, either organizationally or in terms of their interest in the collected material. On the contrary, it appears that changes in the organization of the global economy have promoted the creation of new links between these groups, which share increasingly similar interests in trading this material. Institutional linkages and partnerships facilitate the flow or circulation of the material; it is through these partnerships and contractual arrangements that samples of genetic material are being marketed, traded, bought and sold, and, most important of all, *deployed and redeployed.*

As we have already established, value accrues not simply from owning collected materials, but from being able to repeatedly redeploy or recirculate them to advantage. We are now witnessing the development of entirely new forms of commodity exchange in relation to plant genetic material and perhaps the genesis of an entirely new political economy or trade in samples of genetic and biochemical material. This new economy is based on the rapid recirculation and reutilization of samples of genetic and biological material. How will it be possible to regulate this trade, and to do so in a just and equitable fashion, to ensure that suppliers ultimately benefit from the successive uses that are made of the resources they have contributed?

## REGULATING THE SPACE OF FLOWS: GEOGRAPHY, JUSTICE, AND THE CIRCULATION OF COLLECTED MATERIAL

Manuel Castells (1989, 1996) introduced the concept of a "space of flows" to illustrate that even though they remain largely invisible to us, the exchange and circulation of informational resources creates an information flow that has taken on a particular form and occupies a particular space within the global economy. The new trade in circulating, exchanging, buying, and selling samples of genetic material could be understood as being one of these flows of information, a particular space of flows. The question is, is it possible to regulate this space of flows in a just and equitable manner?

It has been argued that speeding up the rate of exchange of these informational resources can only benefit supplying countries, as they will gain directly from this trade through the direct receipt of monetary compensation. For this to be the case, however, it would be necessary for the protocols that govern the exchange of such material to be adequate to the task of tracking multiple transactions involving a resource that is inherently fluid, mutable, and dynamic. It is no longer sufficient to simply pay a sum of money for the sample, to compensate only for the corporeal value of the biological material. To draw an analogy from cutting-edge work that is currently being undertaken in the area of copyright law and information technology, what is of greatest value and requires the most protection is *no longer the form of the work but the transmissible content of that work*.[11]

A brief review of the contractual agreements designed to compensate suppliers of genetic materials reveals that only the most sophisticated of these agreements have attempted to address this reality. Both prior to and following the Biodiversity Convention, groups of senior ethnobotanists, scientists, and managers within the bioprospecting industry began to devise contractual agreements to govern the collection and use of genetic resources. These agreements included specific measures to ensure the just and equitable sharing of benefits. The details of these agreements vary, but in principle they rely on a number of compensatory mechanisms. The most basic compensatory agreements fail to take account of the informational nature of these resources and continue to treat biological material in very much the same way as other stock resources such as coal or tin. Compensation comes in the form of a fee per sample of material collected. The amount for which samples are sold varies according to demand and to the rarity of the sample, but it is rarely in excess of US$250. The closest these agreements come to engaging with the fact that the collected material may be recirculated or reused over time and space is in their commitment to ensuring that suppliers receive up to 50 percent of the income derived from any resale of the material, for example through a brokerage. Although these terms may appear to be generous, the actual return to the supplier is minimal. In an agreement in which 1,000 samples are supplied for, say, $50 each, and 50 percent are resold, the return on this arrangement would amount to, at most, US$62,500. This return seems rather paltry when compared to the profit potential for a major product that, to use the example of Merck's Methaclor, is currently in excess of US$1 billion per year.

It is important to remember that not every sample will prove useful, and this is part of the explanation for why initial fee-per-sample payments are so low. There is a presumption that further benefits will be forthcoming, when and if collected samples prove to have some utility. It is assumed, for instance, that companies will return to the source country for bulk supplies of material once they move into commercial production, and that this will provide ongoing income and an incentive for conservation. In addition, it is also assumed that supplying countries will benefit substantially from the forms of infrastructural investment made by companies in particular regions.

There are several difficulties with these assumptions. First, technological changes are releasing large corporations from their dependence on in situ collecting. Corporate executives have made it clear in a series of interviews, moreover, that the growing cost of this type of production (inflated expectations, demands for infrastructural investment, and political instabilities in source countries) are making in situ collecting an increasingly unattractive proposition. It is increasingly evident that these executives do not have, and may never have had, any intention of securing commercial quantities of genetic or biochemical materials from either national parks or extractive reserves in source countries. They at least are quite clear about how such material will be produced now and in the future:

> We set up on the assumption that anything that we find will eventually be made by total synthesis. We're not really looking towards producing material from wild specimen collection.

and from another interviewee:

> As far as sourcing goes, you are absolutely right. . . . I don't think the answer, the best answer, is to go back and even cultivate it, let alone collect it in the country of origin. My particular feeling is that the right answer is plant cell culture.

Some specimens may continue to be collected in the wild, but only as sources of lead compounds. Only compounds that prove too complex to replicate via existing technology will be resupplied from wild collection. As technology advances further, these will probably fall in number. Even if it proves necessary to source material through in situ collecting, there is no guarantee that the country that originally supplied the material will be the one resupplying it in quantity for commercial production. Most contractual agreements contain an escape clause that allows the buyer to opt out of the resupply agreement if the source country "cannot provide adequate amounts of raw materials at a mutually agreed upon fair price." As plants do not respect national boundaries, most source countries do not enjoy a monopoly on these resources. These suppliers are as vulnerable to the pressures of global sourcing as producers of any other raw materials. They may well be consulted, but they are unlikely to be able to exercise any leverage in negotiations to improve what the buyer deems to be a fair price.

These developments have been anticipated, to a certain degree. Buyers who do not seek resupply from the source country are—where they are party to such an agreement (which is by no means always the case)—liable to pay to that source country, by way of compensation, a "negotiated fee," which might be used for conservation purposes. However, the negotiated fee will almost certainly be based on the market value of the material or on a percentage royalty that may be paid on net income from products derived from the supplied material. There is no question that the collection of biological materials has created a market, but it is for the informational resources rather than the material ones. As the ability to reproduce these resources in laboratories continues to improve, the incentive to collect them and conserve them in situ declines. Companies are unlikely to continue to invest in the preservation of tracts of pristine wilderness for wild specimen collection, and there will be even less demand to create labor-intensive extractive reserves. As one executive from a major pharmaceutical company recently suggested, they may even cease collecting altogether:

> Personally, I can see the current heyday of collecting ending by the turn of the century. The politics involved in this will get severe enough that we may wind up collecting in two or three countries. It's unfortunate, but it's an economic situation. . . . We may just retract to North America and spend our dollars here.

In this event, the only way bioprospecting initiatives can possibly make any long-term contribution to the conservation of the environment in source countries is in circumstances where a negotiated royalty or fee for successive use of the collected material is actually paid, and then only when a percentage of this is committed to and used explicitly for conservation purposes.

The second difficulty with assuming that supplying countries will benefit, although there is no doubt that companies have shown a willingness to invest in "capacity building" and infrastructural development, is that much of what has been supplied to date (the provision of airstrips, the training of parataxonomists, the building of screening laboratories, etc.) is designed principally to improve the speed and efficacy of the extraction process. Without ensuring access to more sophisticated technology (which is at best unlikely), these collecting programs will only reinforce existing spatial divisions of labor by relegating developing countries to the role of supplying raw materials.[12] Meanwhile, the recipient countries will be able to employ the latest technologies to isolate, manipulate, combine, and even replicate the genetic and biochemical information within these materials to produce new combinations of material that can be employed to advantage over and over again. It is only through this ongoing processes of exploration and exploitation that the full informational value of these resources will be realized.

The possible curtailment of collecting in situ should serve to concentrate attention even more sharply on the question of the fate of material that has already been collected. Clearly, this material is being accessed by a number of interested parties. It is circulated relatively freely, it is used repeatedly, and it is being stored for future use. If this new phase of resource extraction is not to be unfavorably paralleled with earlier colonialist exercises, then effective measures for securing informed consent and effective compensation for the use of these informational resources must be devised. Only the most sophisticated compensatory agreements brokered over the last ten years or so have attempted to address the question of how to compensate for the successive use of the informational resources embodied in biological material. Perhaps not surprisingly, these agreements rely on a mechanism, the royalty, that is frequently used to compensate for the use of other types of informational resources. At present, attempts are being made to pay a royalty to the suppliers of material whenever that material is used as the basis of a commercial product, even if the product is ultimately produced by replication, modification, or synthesization. However, it is impossible to pay a royalty unless it is possible to determine to whom that royalty should be paid. This requires that each successive user be able to establish what could be referred to as the provenance of the material. Given that such material is likely to be subject to processes of modification, synthesization, or replication, that it may well have been circulated to, and utilized by, a variety of different recipients, or, in addition, cryogenically stored and subject to similar processes of manipulation in the future, it is clearly apparent that tracing the provenance of the material is going to prove extremely difficult.

It will certainly be impossible without the existence of protocols such as Material Transfer Agreements (MTA), and may well prove impossible even if these protocols are in place. These agreements require successive users of the material (1) to establish from where the material has been acquired and (2) to agree to return compensation to those suppliers either directly or through the intermediary collecting institute. Although the very largest collecting agencies, which are subject to heightened public scrutiny, have voluntarily introduced MTAs, there is at present no legal requirement that bioprospecting agencies make themselves a party to MTAs or otherwise monitor the use, circulation, or exchange of the material they collect. Assurances of commitment to this principle are made by many small collecting agencies, brokerages, and rental firms, but surprisingly few have formal MTA agreements in place.

Even when companies are well intentioned and introduce such agreements, the extantiation of production over time and space may present insurmountable barriers to the effective execution of these agreements. The lead time for drug development is currently ten to fifteen years. With the introduction of cryogenic storage, this lead time could in effect be a hundred years longer than that. Can these agreements possibly remain effective over such a long period of time? Many working within the industry already harbor serious reservations about whether these agreements can withstand even the minor corporate restructurings that may occur in the next decade. As one senior researcher with a major pharmaceutical company suggests,

> You can envisage a situation wherein small company x goes out and collects materials with all the appropriate material transfer agreements, royalty agreements, etc., and then sells samples to major company y and then basically consumes all of their capital and goes out of business. Big company y would then have a situation wherein the company that they bought the materials from no longer exists, so the contracts that they signed really are not enforceable any more, so that could make things a little bit on the sticky side.

There is also a quiet acknowledgment within the industry that although intermediary agencies may collect material under particular contracts, they have no real power to control how that material is utilized once they hand it over to the recipient. The very existence of an MTA confirms that the first recipient, which may well be a government agency or a scientific institution, is trading or exchanging those resources with other interests. These exchanges are taking place, hopefully with, but often without, the direct approval or involvement of either the intermediary collecting institution or the source country, which clearly gives the director of one such collecting institute great cause for concern:

> Those material transfer agreements mean that they are taking some of those samples that were collected and they're providing them to companies outside of the original agreements, and that is a very serious issue for us . . . but, you know, we've given up control of them by giving them to the United States government who'll presumably follow a series of rules—so we have no control anymore, we have no control in the same way that the village that they were collected in has no control anymore. We've worked with them [the recipient] to put a series of mechanisms in place to make sure that *maybe* they won't be abused . . . we hope, but you'd have to talk to them about that.

By the frank admission of one senior corporate research chemist, there simply is no way of establishing or confirming the provenance of collected materials, of determining exactly where material has been sourced from, after it has been collected and circulated:

> If you are talking about physically proving, I don't see any way in the world that they can. About the best that you can hope for is a relatively consistent paper trail, and beyond that, you basically have to have a small chunk of faith.

Technological, economic, and regulatory changes are in the process of fundamentally altering the ease with which genetic and biochemical information can be exchanged and traded. These developments have irrevocably changed the ways in which biological materials are now constituted, used, and traded as resources, creating a host of complex and as yet unanswered questions about how, or if, it will be possible to maintain control of these resources as they flow around the new circuits of exchange. Every actor involved in this new trade in samples of genetic and biochemical material has an obligation to remain cognizant of the unique nature of the commodity in which they trade, and of the special responsibility they bear in drafting agreements that purport to regulate this trade. For, as the director of the Institute of Economic Botany (IEB) reminded me recently,

> It's very different now because all you need is a bit of the plant. You only have to go there one time and then it's over as far as control of that resource and access goes for the people who own it, so the deal has to be brokered with them in mind from day one. So absolutely, it's not like you have to go back to the village to keep filling up your truck with coal, you have the plant and that's it.

## Future Fates

One last issue remains to be addressed: What is the future of these collections of biological materials? Much of the preceding discussion hinges on the supposition that while these collections will become increasingly valuable as both an informational and a material resource, the probability that any benefit from their exploitation will be returned to suppliers is inexorably diminishing. It is impossible to speculate with any accuracy about the potential value these collections may have in the future or the potential uses to which they may be put. There is every reason to assume, though, that these collections will become progressively more valuable as biotechnology improves and, sadly, as environmental destruction continues.

However, the utilitarian value of these collections must await revelation by new technologies of exploration and exploitation. The collections may prove to be worthless (although if this is the case, it is worth questioning why they are being preserved at such enormous expense). Alternatively, they may prove to be almost infinitely utilizable. However, as discussion turns to techniques such as "isolating native DNA from plant materials and cloning it in *Escherichia coli*," or "taking the cassette of genes out of various slow-growing pacific yew trees, putting it into a rapidly growing *E. coli*, and fermenting it," or even extracting viable quantities of DNA from preserved materials, three things become increasingly clear. The difficulties of tracking or controlling successive uses of, and modifications to, collected materials are unlikely to diminish in the future and will in fact become increasingly acute. As a consequence, the difficulties of designing socially just protocols to compensate for the use of these materials will increase as the ability to establish provenance declines. Last, the collections of materials concentrated within the new centers of calculation now constitute a stockpile or warehouse of rare and increasingly valuable biological and genetic resources. There is a recognition by those working within such institutions that if experiments with DNA extraction prove successful, these repositories may well become so valuable that, as one researcher working within the Smithsonian put it,

> It really won't matter what the hell Ecuador, or wherever, has, because we will have the most biodiverse piece of real estate in the world right here at the Natural History Museum of the Smithsonian Institution.

While it is difficult to predict how these institutions might use their collected materials in this brave new world, it is clear that decisions about these issues will continue to be made from within the centers of calculation. As the material sits in cryogenic storage awaiting its second coming, the suppliers and the agreements will recede further and further into the past. By the time the material is eventually used, what record will there be of where it was collected or to whom compensation should be paid, and who will take responsibility for ensuring that any compensation finds its way back, by whatever circuitous route it has traveled, to those who are entitled to a "just and equitable" share of the benefits that will ultimately accrue from the exploitation of these materials?

## CONCLUSION

This chapter began by looking at the social and spatial dynamics of collecting. This investigation revealed that value accrues progressively with each stage of the collecting process. It derives not simply from acquiring particular materials but principally from having an ability to concentrate them in particular localities where they can be ordered, controlled, and disciplined, and ultimately redeployed to advantage through strategic utilization, exchange, or trade. The more efficiently the collected materials can be recirculated within the economy, the greater will be the rate of financial return. This is certainly true for informational resources such as computer software, music, and videos. Once ownership of the program, song, or moving image is secured, a continuing return can be generated from selling access to that information. Genetic material can also be understood as an informational resource and the same principles of exchange thus apply—i.e., most value will derive not simply from owning samples of biological material, but from having the capacity to repeatedly utilize the genetic and biochemical information stored within that material, to rent that information out within the market, and, ultimately, to use that material as the means of production of that resource as well as a range of other products.

Despite the well-intentioned efforts of those within the bioprospecting industry, the compensatory mechanisms now in place may well prove to be inadequate to the task of regulating the use of this informational resource. The very nature of the resource (fluid/mutable/replicable) and the very nature of the new institutional linkages (also flexible and mutable these days) make it easy to circulate, store, and reproduce genetic and biochemical material over space and time, factors that militate against the effectiveness of existing compensatory measures. There is a greater need to acknowledge that biological materials are now principally valued as informational resources rather than simply as material resources. Contractual agreements that continue to compensate for the use of the material resource only through the payment of a fee per sample are unjust and inequitable, as they give compensation only for the body of the plant, not for the successive use of the information within it. Agreements that do recognize the informational quality of this resource and attempt to compensate for successive uses of genetic and biochemical information through the payment of a royalty are likely to be ineffective as there is not now, and may never be, any effective means of either establishing the provenance of, or tracking successive uses of, this most mutable and dynamic of resources.

By only partially acknowledging that biological material has been reconstituted as an informational resource, by continuing to pay out small sums of money for genetic samples as if they were stock resources, by continuing to rely on "a paper trail and a chunk of faith" to track successive uses of this material, the bioprospecting industry risks inviting accusations that these contemporary collecting projects not only mirror earlier colonialist phases of resource extraction, but actually constitute a more sophisticated form of bioimperialism than any previously envisaged. Important distinctions exist between historical and contemporary collecting projects, and important measures have been introduced with the goal of ensuring that the latter are conducted in a just and equitable fashion. It is unlikely that this goal will be met, however, unless existing protocols are reviewed. In undertaking this review, it may prove helpful to reflect further on what this work has revealed about the role of the social and spatial dynamics of collecting, on the changing way in which biological materials are constituted and traded as resources in the global economy, and on the new institutional linkages that are shaping relations of exchange, as we consider the fate of the collections and the ongoing annexation of genetic resources into the twenty-first century.

## ACKNOWLEDGMENTS

I would like to thank my friends and colleagues in the Geography Department of the University of Cambridge, Drs. MacDowell, Howell, Kearns, and Spencer, for their invaluable contributions and editorial support. I would particularly like to thank Flora Gathorne-Hardy for her unwavering support, intellectual input, and faith in this project. The empirical research could not have been undertaken without the financial assistance of the British Council, the support of the Conant family, and other friends in the United States who so generously accommodated me during my fieldwork, and the forthrightness of all those who kindly agreed to be interviewed. I would like to thank Eldred Herrington and Andrew Taylor for their editorial help.

Finally, I would like to thank Catharine Hancock of Johns Hopkins University for her constancy, and for introducing me to the pleasures of the history and philosophy of science.

## NOTES

1. Based on *The Concise Oxford Dictionary of Current English.*

2. I am indebted to Latour for providing the underlying structure of this section of the argument.

3. The pecuniary value of collections, or of the pieces within them, are determined only when they are placed as saleable commodities within a circuit of exchange. This is as true of collections of cultural artifacts as it is of collections of natural materials. However, natural materials differ from artifactual materials in at least three ways, and these have the power to fundamentally alter the conventional parameters of this process of circulation and exchange. First, unlike cultural artifacts or other inanimate objects, collections of living biological materials can be reproduced over space and time. Second, if reproduction increases exponentially, the collected materials may in fact *proliferate* over space and time, increasing the size of the collection. Last, as they are living and mutable, they can in the process of reproduction be transformed to produce entirely new materials, which can, in turn, become the subject of an entirely new collection process.

4. This research, which is the basis of my doctoral dissertation, consists of a series of interviews with key actors in the pharmaceutical industry and the nascent bioprospecting industry, a survey of approximately ninety academic and scientific collecting institutes, and an extensive review of relevant literature and archival materials. This chapter is based on an analysis of the findings of this research.

5. See also the communiqués of the Rural Advancement Foundation International, Ottawa, Canada, for a further discussion of biopiracy. These are available on the internet at www.rafi.org.

6. The United Nations Convention on Biological Diversity, 1992.

7. Which obliges signatories "to ensure that benefits arising out of the utilization of genetic resources are shared fairly and equitably with the contracting party supplying the resource."

8. A voucher specimen is a sample of plant material that is dried, pressed onto pages, taxonomically classified, and catalogued.

9. The General Agreement on Tariff and Trades' global Trade Related Intellectual Property rights regulations.

10. As estimated by Mr. Neil Belson, president of Pharmacogenetics Inc., a company that specializes in biological collections and natural products research (Rural Advancement Foundation International 1995).

11. Copyright law has traditionally afforded protection against unlicensed reproduction of the information embodied within particular works by prohibiting the unlicensed reproduction of the form of the work, that is to say by prohibiting reproduction in a literal sense, the reproduction of the pages of a book for instance. The development of new information technologies such as the internet are increasingly allowing information to be conveyed/transmitted in immaterial ways, which is creating potentially insurmountable problems for a regulatory

device that relies entirely on prohibiting the unlicensed reproduction of the form, rather than the transmissible content, of the work. I am indebted to Dr. Susan Marks, Department of Law, University of Cambridge, and Dr. Dan Hunter, Research Fellow in Law, Emmanuel College, Cambridge, for alerting me to these developments.

12. In this case, genetic or biochemical information in the form of catalogued plant extracts.

# REFERENCES

Adams, J. 1993. AIDS: Garden botanists discover a plant that may lead to treatment. *Missouri Botanical Garden Bulletin*, July/August.

*African Wildlife Update.* 1993. Rare vine from Cameroon's Korup National Park continues to offer hope in fight against AIDS. September-October.

Agarwal, A. 1997. *Community in Conservation: Beyond Enchantment and Disenchantment.* Gainesville: Conservation and Development Forum.

Agyare, A. K. 1993. *Partnership with Indigenous People in Sustainable Natural Resource Use: A Case in Ghana.* Presented at the Symposium for Indigenous People, The Pueblo of Zuni, New Mexico, July.

Andaya, Leonard Y. 1993. *The World of Maluku: Eastern Indonesia in the Early Modern Period.* Honolulu: University of Hawaii Press.

Andean Pact. 1996. *The Common System on Access to Genetic Resources.* The Commission of the Cartagena Accord.

Anderson, Benedict. 1990. Cartoons and monuments: The evolution of political communication under the New Order. In *Language and Power: Exploring Political Cultures in Indonesia.* Ithaca, NY: Cornell University Press.

Anderson, D. and R. Grove. 1987. Introduction: The scramble for Eden: Past, present and future in African conservation. In D. Anderson and R. Grove, eds., *Conservation in Africa: People, Policies and Practices*, pp. 1–12. Cambridge: Cambridge University Press.

Anderson, E. N. 1987. A Malaysian tragedy of the commons. In B. M. McCay and J. M. Acheson, eds., *The Question of the Commons: The Culture and Ecology of Communal Resources*, pp. 327–343. Tucson: University of Arizona Press.

Appadurai, Arjun. 1995a (original, 1986). Introduction: Commodities and the politics of value. In A. Appadurai, ed., *The Social Life of Things: Commodities in Cultural Perspective.* Cambridge: Cambridge University Press.

———. 1996. *Modernity at Large: Cultural Dimensions of Globalization.* Minneapolis: University of Minnesota Press.

Appadurai, Arjun, ed. 1995b. *The Social Life of Things: Commodities in Cultural Perspective.* Cambridge: Cambridge University Press.

Appell, G. N. 1993. *Ownership and the Analysis of Property Relations: Observational Procedures for Land Tenure and Tree Ownership in the Societies of Borneo.* Williamsburg, VA: The Borneo Research Council.

Arana, J. C. 1913. *Las Cuestiones del Putumayo.* Declaraciones prestadas ante el Comité de Investigación de la Cámara de los Comunes, y debidamente anotadas. Folleto no. 3. Barcelona: Imprenta Viuda de Luis Tasso.

Arhem, K. 1984. Two sides of development: Maasai pastoralism and wildlife conservation in Ngorongoro, Tanzania. *Ethnos* 49(3/4):186–210.

———. 1985. *Pastoral Man in the Garden of Eden: The Maasai of the Ngorongoro Conservation Area, Tanzania*, Uppsala, Sweden: University of Uppsala.

Armstrong, J. and K. Hooper. 1994. Nature's medicine. *Landscope.*

Arnold, David. 1996. *The Problem of Nature: Environment, Culture and European Expansion*, pp. 141–168. Oxford: Blackwell.

Bailey, C. and C. Zerner. 1992. Community-based fisheries management institutions in Indonesia. *Maritime Anthropology Studies* 5(1):1–17.

Bailey, J. 1998. Green Corporatism. Unpublished manuscript. Berkeley: University of California.

Baker, J. T., R. P. Borris, B. Carte, et al. 1995. Natural product discovery and development: New perspectives on international collaboration. *Journal of Natural Products* 58(9): 1325–1357.

Balick, M. J. and P. A. Cox. 1996. *Plants, People, and Culture.* New York: Freeman.

Balick, M., E. Elisabetsky, and S. Laird, eds. 1996. *Medicinal Resources of the Tropical Forests: Biodiversity and Its Importance to Human Health.* New York: Columbia University Press.

Banerjee, Ajit, Gabriel Campbell, Chona Cruz, et al. 1994. Participatory forestry. Presented at the *World Bank Workshop on Participatory Development.* Washington, DC, May 17–20.

Barber, C. V. and A. La Vina. 1997. Regulating access to genetic resources: The Philippine experience. In J. Mugabe, C. Barber, G. Henne, et al., eds., *Managing Access to Genetic Resources: Towards Sustainable Strategies for Benefit-Sharing.* Washington, DC: African Centre for Technology Studies, Nairobi and World Resources Institute.

Barlet, A., J. Albrecht, A. Aubert, et al. 1990. Wirkamskeit eines extractes aus *Pygeum africanum* in der medikamentosen therapie von miktionsstorungen infolge einer beningen prostathyperplasie: Bewertung objectiver und subjectiver parameter. *Wiener Klinisch Wochenschrift* 102(22):667–673.

Barr, Christopher. 1997. Discipline and Accumulate: State Practice and Elite Consolidation in the Indonesian Timber Industry. Master's thesis, unpublished, Cornell University, Ithaca, NY.

Barriga López, F. 1988. *Cofanes, Etnologia Ecuatoriana*, vol. IX. Quito, Ecuador: Instituto Ecuatoriano de Credito, Educativo y Becas (IECE).

Baskin, Y. 1994. There's a new wildlife policy in Kenya: Use it or lose it. *Science* 265:733–734.

Bates, Henry Walter. 1988. *The Naturalist on the River Amazons.* New York: Penguin.

Bellier, I. 1994. Los Mai Juna. In F. Santos and F. Barclay, eds., *Guia Etnografica de la Alta Amazonia*, vol. 1, pp. 1–181. Quito, Ecuador: Flacso-Sede Ecuador.

Belsky, Jill M. 1999. Misrepresenting Communities: The Politics of Community-Based Rural Ecotourism in Gales Point Manatee, Belize. Unpublished manuscript under review at *Rural Sociology.*

Belsky, Jill M. and Stephen F. Siebert. 1997. Non-timber forest products in conservation and community development: The palm *Desmoncus* sp. in Gales' Point, Belize. In Richard Primack and David Bray, eds., *Timber, Tourists and Temples: Conservation and Development in the Maya Forests of Belize, Guatemala and Mexico*, pp. 141–154. Washington, DC: Island Press.

Benavides, M. 1992. Asháninka self-defense in the central forest region. *IWGIA Newsletter*, April-June, pp. 36–45. Copenhagen: International Work Group of Indigenous Affairs.

———. 1993. Los Asháninka, víctimas de la violencia y la guerra. *Ideele 59–50*, December, pp. 116–118.

Benton, L. 1996. The greening of free trade? *Environment and Planning A* 28:2155–2177.

Bergeret, Anne and Jesse C. Ribot. 1990. *L'Arbre Nourricier en Pays Sahélien.* Paris: Editions de la Maison des Sciences de l'Homme.

Bergin, P. 1995. Conservation and Development: The institutionalization of community conservation in Tanzania National Parks. Doctoral dissertation, University of East Anglia, Norwich, England.

Berkes, Fikret, ed. 1989. *Common Property Resources*. London: Bellhaven Press.

Berlant, L. 1991. *The Anatomy of National Fantasy*. Chicago: University of Chicago Press.

Berry, S. 1994. *No Condition Is Permanent*. Madison: University of Wisconsin Press.

Besong, J. 1995. *Implications of the New Forestry Law on Forest Communities*. Yaounde, Cameroon: Ministry of Environment and Forests.

Besong, J. B., P. Abeng Abe Meka, and S. Ebamane-Nkoumba. 1991. Étude sur l'exploitation du Pygeum: Rapport de mission effectuée dans les provinces du Sud-Ouest de l'Ouest et du Nord-Ouest. Direction des Forêts, Ministère de l'Agriculture, 25 Janvier.

Bobbio, Norberto. *Left and Right*. Cambridge, UK: Polity.

Bock, Carl. 1985. *The Head-Hunters of Borneo: A Narrative of Travel Up the Mahakam and Down the Barito*. Singapore: Oxford University Press.

Boltanski, L. and L. Thevenot. 1991. *De la Justification*. Paris: Gallimard.

Bonifaz, C. 1996. Compendium of litigation documentation and affidavits: In the United States Courts for the Southern District of New York, Maria Aguinda, et al. (Plaintiffs) v. Texaco Inc. (Defendant). 93 CIV 7527 (BDP) (LMS). Selected exhibits from plaintiffs' response to Texaco's motion to dismiss filed with the court on February 9, 1996. Law offices of Cristobal Bonifaz, Amherst, MA.

———. 1997. Aguinda v. Texaco. *Progress Report to Clients, Amherst, Massachusetts, September, 1996.*

Boo, E. 1990. Ecotourism: The Potentials and Pitfalls, vols. 1, 2. Washington, DC: World Wildlife Fund.

Botkin, Daniel B. 1990. *Discordant Harmonies: A New Ecology for the Twenty-first Century*. New York: Oxford University Press.

Bouamrane, M. 1996. A season of gold: Putting a value on harvests from Indonesian agroforests. *Agroforestry Today* 8(1):8–10.

Bourdieu, Pierre. 1984. *Distinction: A Social Critique of the Judgement of Taste*. Translated by Richard Nice. Cambridge, MA: Harvard University Press.

———. 1990. *In Other Words*. London: Polity.

———. 1993. *Outline of a Theory of Practice* (trans. Richard Nice). Cambridge, UK: Cambridge University Press, 171–183.

Bourgeois, R. 1984. Production et Commercialisation de la Résine "Damar" à Sumatra Lampung. Master's thesis, École Nationale Supérieure d'Agronomie de Montpellier, Montpellier, France.

Boyce, James K. 1996. Ecological distribution, agricultural trade liberalization, and in situ genetic diversity. *Journal of Income Distribution* 6(2):265–286.

———. 1997. NAFTA, the environment, and security: The maize connection. *Political Environments* 5(fall):S22–S25.

Brevoort, P. 1995. The US Botanical market—An overview. *Herbalgram* 36:49–57.

Bringmann, G. 1986. Naphthylisoquinoline alkaloids. In A. Brossi, ed., *The Alkaloids*, pp. 141–184. New York: Academic Press.

Bringmann, G., R. Gotz, P. A. Keller, et al. 1994. First total synthesis of korupensamines A and B. *Heterocycles* 39(2).

Bringmann, G., C. Kehr, U. Dauer, et al. In press. *Ancistrocladus robertsoniorum* "produces" pure crystalline droserone when wounded. *Planta Medica*.

Brockway, L. 1979. *Science and Colonial Expansion: The Role of the British Royal Botanical Gardens*. New York: Academic Press.

Brosius, J. Peter, Anna L. Tsing, and Charles Zerner. 1998. Representing communities: Histories and politices of community-based natural resource management. *Society and Natural Resources* 11(57):157–168.

Brosius, P. 1997. Prior transcripts, divergent paths. *Comparative Studies in Society and History* 39(3):468–510.

Browder, John O. and Brian J. Godfrey. 1997. *Rainforest Cities: Urbanization, Development, and Globalization of the Brazilian Amazon*. New York: Columbia University Press.

Browne, J. Biogeography and Empire. 1996. In N. Jardine, J. Secord, and E. Sparry, eds., *Cultures of Natural History*, pp. 305–322. Cambridge: Cambridge University Press.

Bruneton, J. 1995. *Pharmacognosy, phytochemistry and medicinal plants*. Andover, England: Intercept Limited.

Brush, S. B. 1999. Bioprospecting the Public Domain. *Cultural Anthropology* 14(4): 535–555.

———. 1992. Farmers' rights and genetic conservation in traditional farming systems. *World Development* 20:1617–1630.

Brush, Stephen and Doreen Stabinsky, eds. 1996. *Valuing Local Knowledge: Indigenous People and Intellectual Property Rights*. Washington, DC: Island Press.

Bryant, Richard. 1992. Political ecology: An emerging research agenda in third-world studies. *Political Geography* 11(1):12–36.

Brysk, Alison. 1994a. Acting globally: Indian rights and international politics in Latin America. In D. Cott, ed., *Indigenous Peoples and Democracy in Latin America*. Washington, DC: St. Martin's Press and Inter-American Dialogue.

———. 1994b. Turning weakness into strength: The internationalization of Indian rights. *Latin American Perspectives* 23(2):38–57.

———. In press. *From Tribal Village to Global Village: Indian Rights and International Relations in Latin America*. Stanford, CA: Stanford University Press.

Bullard, Robert. 1994. *Dumping in Dixie: Race, Class, and Environmental Quality*. Boulder, CO: Westview Press.

Burkill, I. H. 1966 (original, 1935). *A Dictionary of Economic Products of the Malay Peninsula*, 2 vols. Reprinted: Kuala Lumpur, Malaysia: Ministry of Agriculture and Cooperatives. (Originally, London: Crown Agents for the Colonies.)

Burnham, P. 1993. *The Cultural Context of Rainforest Conservation in Cameroon*. Presented at the 36th annual meeting of the African Studies Association, Boston, December.

Buttel, F. 1992. Environmentalization. *Rural Sociology* 57(1):1–27.

Buttel, F. and P. Taylor. 1992. How do we know we have global environmental problems? *Geoforum* 23:405–416.

Cabodevilla, M. A. 1989. *Memorias de Frontera, Misiones en el Rio Aguarico (1954–1984)*. Ediciones CICAME, Vicariato Apostólico de Aguarico, Quito, Ecuador.

Caldecott, J. O. 1992. Biodiversity management in Indonesia. *Tropical Biodiversity* 1(1):57–62.

*Cameroon Post*. 1997. Logging companies leave trail of wreckage in Cameroon. Yaounde, April 8.

Carte, B. and R. K. Johnson. 1996. Topotecan development: An example of the evolution of natural product drug discovery research. In M. Balick, E. Elisabetsky, and S. Laird. eds., *Medicinal Resources of the Tropical Forests: Biodiversity and Its Importance to Human Health*. New York: Columbia University Press.

Carter, E. J. 1992. Socio-economic and institutional study. *Final Report, Limbe Botanic Garden and Rainforest Genetic Conservation Project,* Cameroon.

Caruthers, J. 1989. Creating a national park, 1910 to 1926. *Journal of Southern African Studies* 15(2):188–216.

Casement, R. 1913. *British Bluebook. Correspondence Respecting the Treatment of British Colonial Subjects and Native Indians Employed in the Collection of Rubber in the Putumayo District*. Presented to both Houses of Parliament by command of His Majesty, July 1912.

Castells, M. 1989. *The Informational City: Information Technology, Economic Restructuring and the Urban-Regional Process*. London: Blackwell.

———. 1996. *The Rise of the Network Society*. London: Blackwell.

Catalano, S., M. Ferretti, A. Marsili, and I. Morelli. 1984. New constituents of *Prunus africana* bark extract. *Journal of Natural Products* 47(5):910.

Chaloupka, William and Jane Bennett. 1993. *In the Nature of Things: Language, Politics and the Environment*. Minneapolis: University of Minnesota Press.

Chibnik, M. 1994. *Risky Rivers. The Economics and Politics of Floodplain Farming in Amazonia*. Tucson: The University of Arizona Press.

Chin, S. C. 1985. Agriculture and resource utilization in a lowland rainforest Kenyah community. *Sarawak Museum Journal*, special monograph no. 4, vol. XXXV(56) (new series). Kuching, Sarawak, Malaysia.

Christopherson, C. 1935. Flowering plants of Samoa. *Bernice P. Bishop Museum Bulletin* 128:1–221.

———. 1938. Flowering plants of Samoa, II. *Bernice P. Bishop Museum Bulletin* 154:1–77.

CILSS and LTC. 1993. Atelier sur les Codes Forestiers au Sahel: Synthèse régionale. *Proceedings of CILSS and the Land Tenure Center Conference*. Bobo-Dioulasso, Burkina Faso, January 18–20.

Clay, Jason W. 1990. A rainforest emporium. *Garden,* Jan/Feb:1–6.

Cleaver, Kevin and GÖtz Schreiber. 1992. *The Population, Agriculture and Environment Nexus in Sub-Saharan Africa*. Agriculture and Rural Development Series, no. 1. Technical Department, Africa Region, The World Bank.

Clifford, J. 1989. *The Predicament of Culture: Twentieth-Century Ethnography, Literature and Art*. New York: Harvard Press.

Cohen, J. and A. Arato. 1992. *Civil Society and Political Theory*. Cambridge, MA: MIT Press.

Cohen, John M. and Norman Uphoff. 1977. Rural Development Participation: Concepts and Measures for Design, Implementation and Evaluation. Rural Development Monograph no. 2. Ithaca, NY: Cornell University, International Studies.

Colfer, Carol J. P. 1980. Change in an Indigenous Agroforestry System: The Kenyah of Long Segar. Preliminary report, Man and the Biosphere (MAB) Project: Interactions between People and Forests in East Kalimantan.

Collar, N. J. and S. N. Stuart. 1988. Key forests for threatened birds in Africa. International Union for the Conservation of Nature and Natural Resources (IUCN)/ICBP. *International Council for Bird Preservation Monograph no. 3*. Cambridge.

Collet, O. J. A. 1925. *Terres et Peuples de Sumatra*. Amsterdam: Elsevier.

Collier, J. 1968. *The River That God Forgot*. New York: Dutton.

*Concise Oxford Dictionary of Current English*. 1990. Oxford: University of Oxford Press.

Conklin, B. A. and L. R. Graham. 1995. The shifting middle ground: Amazonian Indians and eco-politics. *American Anthropologists* 97(4):695–710.

Costales, Piedad P. de and Alfredo C. Costales. 1983. *Amazonia-Ecuador-Peru-Bolivia. Mundo Shuar*. Sucúa: Centro de Documentación e Investigación Cultural Shuar.

Cowen, M. and R. Shenton. 1996. *Doctrines of Development*. London: Routledge.

Cox, P. A. 1979. Use of indigenous plants as fish poisons in Samoa. *Economic Botany* 33:397–399.

———. 1980a. Two Samoan technologies for breadfruit and banana preservation. *Economic Botany* 34:181–185.

———. 1980b. Masi and tanu 'eli: Two Polynesian technologies for breadfruit and banana preservation. *Bulletin of the Pacific Tropical Botanical Garden* 4:81–93.

———. 1990a. Ethnopharmacology and the search for new drugs. In A. Battersby and J. Marsh, eds., *Bioactive Molecules from Plants* (Ciba Symposium 154), pp. 40–55. Chichester, England: Wiley.

———. 1990b. Samoan ethnopharmacology. In H. Wagner and N. R. Farnsworth, eds., *Economic and Medicinal Plant Research, vol 4. Plants and Traditional Medicine*, pp. 123–139. London: Academic Press.

———. 1991. Polynesian herbal medicine. In P. A. Cox and S. A. Banack, eds., *Islands, Plants, and Polynesians*, pp. 147–169. Portland, OR: Dioscorides Press.

———. 1993. Saving the ethnopharmacological heritage of Samoa. *Journal of Ethnopharmacology* 38:181–188.

———. 1994a. The ethnobotanical approach to drug discovery: Strengths and limitations. In G. Prance and J. Marsh, eds., *Ethnobotany and the Search for New Drugs*. Ciba Foundation Symposium 185:25–41. London: Academic Press.

———. 1994b. Wild plants as food and medicine in Polynesia. In N. Etkin, ed., *Eating on the Wild Side*, pp. 102–113. Tucson: University of Arizona Press.

———. 1995. Shaman as scientist: Indigenous knowledge systems in pharmacological research and conservation biology. In K. Hostettmann, A. Marston, M. Maillard, and M. Hamburger, eds., *Phytochemistry of Plants Used in Traditional Medicine*, pp. 1–15. Oxford: Clarendon Press.

———. 1997. *Nafanua: Saving the Samoan Rain Forest*. New York: Freeman.

Cox, P. A. and T. Elmqvist. 1991. Indigenous control: An alternative strategy for the establishment of rainforest preserves. *Ambio* 20:317–321.

———. 1993. Ecocolonialism and indigenous knowledge systems: Village-controlled rain forest preserves in Samoa. *Pacific Conservation* 1:11–25.

———. 1997. Ecocolonialism and indigenous-controlled rainforest preserve in Samoa. *Ambio* 26(2):84–89.

Cox, P. A., L. R. Sperry, M. Tuominen, and L. Bohlin. 1989. Pharmacological activity of the Samoan ethnopharmacopoeia. *Economic Botany* 43:487–497.

Cragg, G. M., M. R. Boyd, M. R. Grever, and S. A. Schepartz. 1994. Policies for international collaboration and compensation in drug discovery and development at the United States National Cancer Institute: The NCI Letter of Collection. In T. Greaves, ed., *Intellectual Property Rights for Indigenous Peoples: A Source Book*. Oklahoma City: Society for Applied Anthropology.

Craven, I. 1990. *A Management Prescription for the Arfak Mountain Strict Nature Reserve 1990–1992, Irian Jaya, Indonesia*. Jakarta, Indonesia: World Wide Fund for Nature Indonesia Programme.

Cronon, William. 1995. *Uncommon Ground: Toward Reinventing Nature*. New York: Norton.

Crosby, P. T. and G. G. Brown. 1937. *A Book of Health for Samoans*. Pago Pago: U.S. Navy.

Cunningham, A. B. 1993. Ethics, Ethnobiological Research and Biodiversity. Research paper, World Wide Fund for Nature International.

———. 1994. Conservation, knowledge and new natural products development: Partnership or piracy? *Intellectual Property Rights and Indigenous Knowledge Conference Proceedings*. Lake Tahoe, June, 1993.

Cunningham, A. B., E. Ayuk, S. Franzel, et al. In press. An economic evaluation of medicinal tree cultivation: *Prunus africana* in Cameroon. *People and Plants Initiative Working Paper no. 6*. Paris: UNESCO.

Cunningham, M., A. B. Cunningham, and U. Schippmann. 1997. Trade in *Prunus africana* and the implementation of CITES. Bonn, Germany: *Bundesamt fur Naturschutz*.

Cunningham, A. B. and F. T. Mbenkum. 1993. Sustainability of harvesting *Prunus africana* bark in Cameroon: A medicinal plant in International trade. *People and Plants Initiative Working Paper no. 2*. Paris: United Nations Educational, Scientific, and Cultural Organization (UNESCO).

Curry, M. R. 1995. Rethinking rights and responsibilities in geographic information systems: Beyond the power of imagery. *Cartographic and Geographic Information systems* 22(1):58–60.

Curtin, P. 1964. *The Image of Africa: British Ideas and Action, 1780–1850*. Madison: University of Wisconsin Press.

Davidson, J. M. 1979. Samoa and Tonga. In J. D. Jennings, ed., *The Prehistory of Polynesia*, pp. 82–109. Cambridge, MA: Harvard University Press.

Debat, J. 1966. Br. App., 25.893/66, June 10.

De Beer, J. H. 1993. Rattan for rice, eaglewood for salt. Some notes on Punan resource utilization in Bulungan, East Kalimantan. Unpublished manuscript.

De Beer, J. H. and M. J. McDermott. 1989. *Economic Value of Non-timber Forest Products in Southeast Asia*. The Hague: Council for the International Union of Conservation of Nature.

Deere, C. D. (coordinator), P. Antrobus, L. Bolles, et al. 1990. *In the Shadows of the Sun*. Boulder, CO: Westview Press.

de Foresta, H. and G. Michon. 1992. Complex agroforestry systems and conservation of biological diversity. 2. For a larger use of traditional agroforestry trees as timber in Indonesia: A link between environmental conservation and economic development. In Y. S. Kheong and L. S. Win, eds., In Harmony with Nature. An International Conference on the Conservation of Tropical Biodiversity, Kuala Lumpur, Malaysia. *Malayan Nature Journal* (Golden Jubilee issue), pp. 488–500.

———. 1993. Creation and management of rural agroforests in Indonesia: Potential applications in Africa. In C. M. Hladik, H. Pagezy, O. F. Linaret, et al., eds., *Tropical Forests, People and Food: Biocultural Interactions and Applications to Development*, 709–724. Paris: UNESCO and Parthenon.

———. 1994a. Agroforests in Indonesia: Where ecology and economy meet. *Agroforestry Today* 6(4):12–13.

———. 1994b. From shifting cultivation to forest management through agroforestry: Smallholder damar agroforests in West Lampung (Sumatra). *APAN News* 6/7:12–13.

———. 1997. The agroforest alternative to Imperata grasslands: When smallholder agriculture and forestry reach sustainability. *Agroforestry Systems* 36(1–3):1–16.

Deharveng, L. 1992. Soil Mesofauna in Agroforests and Primary Forests of Sumatra. Field report.

de Holanda, Sérgio Buarque. 1985. *Visao do Paraíso. Os Motivos Edênicos no Descobrimento e Colonizaáao do Brasil*, 4th ed. Sao Paulo: Editora Nacional.

de Jong, W. 1994. Recreating the forest: Successful examples of ethno-conservation among land-dayaks in central West Kalimantan. In O. Sandbunkt, ed., *Management of Tropical Forests: Towards an Integrated Perspective*, 295–304. Centre for Development and the Environment, University of Oslo, Norway.

de Onis, Juan. 1992. *The Green Cathedral: Sustainable Development of Amazonia*. Oxford: Oxford University Press.

Departemen Kehutanan Republik Indonesia. 1988. Rapat Kerja Kehutanan Th. Konsep Repelita Kelima Kehutanan. Jakarta.

Devitt, P. 1988. The People of the Korup Project Area. Report 1 of the Socio-economic Survey. World Wide Fund for Nature and the Government of Cameroon, no. 3206/A9.7.

Dia, Ibrahima. 1985. Des hommes et leurs forêts: Le cas de Sare Lamine en Moyenne Casamance. Mémoire présenté pour l'obtention du Diplìme d'Études Approfondies. Dakar, Senegal: Institut des Sciences de l'Environnement, University of Dakar.

Diehl, C. 1985. Wildlife and the Maasai: The story of East African parks. *Cultural Survival Quarterly* 9(1):37–40.

Dingit et al. 1994. Keterangan tentang Tuntutan Masyrakat Jelmu Sibak terhadap P. T. Timber Dana, P. T. Kalhold Utama, P. T. Hutan Mahligai, P. T. Subur Kaltim, C. V. Adil Makmur. 16 July.

Djohani, Rili Hawari. 1996. The Bajau: Future park managers in Indonesia. In M. Parnwell and R. Bryant, eds., *Environmental Change in Southeast Asia: People, Politics, and Sustainable Development.* London: Routledge.

Dominguez, C. and A. Gómez. 1990. *La Economía Extractiva en la Amazonía Colombiana 1850–1930.* Bogotá: Tropenbos-Corporación Araracuara.

Dorsey, Michael. In press. Towards an idea of international environmental justice. *World Resources Report 1998–1999.* Washington, DC: World Resources Institute.

Dove, Michael. 1983. Theories of swidden agriculture and the political economy of ignorance. *Agroforestry Systems* 1:85–99.

———. 1985a. *The Agroecological Mythology of the Javanese and the Political Economy of Indonesia.* Honolulu: East-West Environment and Policy Institute.

———. 1985b. *Swidden Agriculture in Indonesia: The Subsistence Strategies of the Kalimantan Kantu.* Berlin: Mouton.

———. 1986. The ideology of agricultural development in Indonesia. In C. Mac Andrew, ed., *Central Government and Local Development.* Singapore: Oxford University Press.

———. 1993a. Smallholder rubber and swidden agriculture in Borneo: A sustainable adaptation to the ecology and economy of the tropical forest. *Economic Botany* 47(2):136–147.

———. 1993b. A revisionist view of tropical deforestation and development. *Environmental Conservation* 20(1):17–24,56.

———. 1994. Marketing the rainforest: "Green" panacea or red herring? *Asia Pacific Issues* 13:1–7.

Downs, R. E. and S. P. Reyna, eds. 1988. *Land and Society in Contemporary Africa.* Hanover: University of New Hampshire.

Doyle, Arthur Conan. 1995 (original, 1912). *The Lost World.* Oxford: Oxford University Press.

Doyle, J. 1985. *Altered Harvest: Agriculture, Genetics and the Fate of the World's Food Supply.* New York: Viking.

Dransfield, J. 1988. Prospects for rattan cultivation. *Advances in Economic Botany* 6:190–200.

Dreiling, M. 1997. Remapping North American environmentalism. *Capitalism, Nature, Socialism* 8(4):65–99.

Dunn, F. L. 1975. *Rain-Forest Collectors and Traders: A Study of Resource Utilization in Modern and Ancient Malaya.* Kuala Lumpur, Malaysia: Maylaysian Branch of the Royal Asiatic Society, monograph vol. 5.

Dupain, D. 1994. Une Région Traditionnellement Agroforestière en Mutation: Le Pesisir ("A Traditionally Agroforestry Area in Mutation: Pesisir"). Master's thesis, Centre National d'Études Agronomiques pour les Régions Chaudes, Montpellier, France.

DuPuis, E. M. and P. Vandergeest. 1996. *Creating the Countryside: The Politics of Rural and Environmental Discourse*. Philadelphia: Temple University Press.

Eben Ebai S., B. N. Ewusi, C. A. Asanga, and J. B. N. Nkongo. 1992. *An Evaluation of the Quantity and Distribution of Pygeum africanum on the Slopes of Mount Cameroon*. Divisional Service of Forestry, Limbe, Cameroon.

*The Economist*. 1989. Rainforest products' growing profits. September 9.

EcoVitality. 1998. "Welcome to the EcoVitality Website." http://www.ecovitalitv.com, 6/24/98.

Edwards, R. 1997. Beware green imperialists. *New Scientist*, 31 May, pp. 14–15.

Eisenberg, D. M., R. C. Kessler, C. Foster, et al. 1993. Unconventional medicine in the United States. *New England Journal of Medicine* 328:246–252.

Eisner, T. and E. A. Beiring. 1994. Biotic exploration fund—Protecting biodiversity through chemical prospecting. *Bioscience* 44:95–98.

Ejedepang-Koge, S. N. 1975. *Tradition and Change in Peasant Activities: A Study of the Indigenous People's Search for Cash in the South-West Province of Cameroon*. Yaounde: Ministry of National Education.

Ekoko, Franáois. 1997. The Political Economy of the 1994 Cameroon Forestry Law. CIFOR. Presented at the African Regional Hearing of the World Commission on Forests and Sustainable Development, May, Yaounde.

Ersson, B. 1994. *A Letter Came from Samoa*. Lulea, Sweden: Private printing.

Escobar, Arturo. l992a. Imagining a post-development era? Critical thought, development and social movements. *Social Text* 31/32:20–56.

———. l992b. Culture, economics, and politics in Latin American social movements theory and research. In A. Escobar and S. E. Alvarez, eds., *The Making of Social Movements in Latin America*, pp. 62–85. Boulder, CO: Westview Press.

———. 1995. *Encountering Development: The Making and Unmaking of the Third World*. Princeton, NJ: Princeton University Press.

———. 1996. Constructing nature: Elements for a poststructuralist political ecology. In R. Peet and M. Watts, eds., *Liberation Ecologies*, pp. 46–68. London: Routledge.

———. 1997. Cultural politics and biological diversity: State, capital and social movements in the Pacific coast of Colombia. In O. Starn and R. Fox, eds., *Between Resistance and Revolution: Culture and Social Protest*, pp. 40–64. New Brunswick, NJ: Rutgers University Press.

Esteva, G. l992. Development. In W. Sachs, ed., *The Development Dictionary: A Guide to Knowledge as Power*, pp. 6–25. London: Zed.

Evans, F. J., ed. 1986. *Naturally Occurring Phorbol Esters*. Boca Raton, FL: CRC Press.

Evans, P. 1996. Government action, social capital and development. *World Development* 24(6):1119–1132.

Fairhead, James and Melissa Leach. 1994. Contested forests: Modern conservation and historical land use in Guinea's Ziama Reserve. *African Affairs* 93:481–512.

———. 1996. *Misreading the African Landscape: Society and Ecology in a Forest-Savannah Mosaic*. Cambridge: Cambridge University Press.

Farnsworth, N. 1988. Screening plants for new medicines. In E. O. Wilson, ed., *Biodiversity*, pp. 83–97. Washington, DC: National Academy Press.

Fearnside, P. 1989. Extractive reserves in Brazilian Amazonia. *BioScience* 39:387–393.

Findlen, P. 1996. Courting nature. In N. Jardine, J. Secord, and E. Sparry, eds., *Cultures of Natural History*, pp. 57–75. Cambridge: Cambridge University Press.

Fisiy, Cyprian F. 1992. *Power and Privilege in the Administration of Law: Land Law Reforms and Social Differentiation in Cameroon*. Research Reports 1992/48. Leiden: African Studies Center.

Fletcher, S. 1990. Parks, protected areas and local populations: New international issues and imperatives. *Landscape and Urban Planning* 19:197–201.

Foead, Nazir and Liman Lawei. 1994. Project Kayan Mentarang Research Report on Gaharu Collecting in the Apo Kayan. Jakarta: World Wide Fund Indonesia Program.

Forrest, T. 1995. *Politics and Economic Development in Nigeria*. Boulder, CO: Westview.

Forsyth, C. 1983. Samoan Art of Healing: A Description and Classification of the Current Practice of the Taulasea and Fofo. Ph.D. dissertation, United States International University, San Diego, CA.

Fosbrooke, H. 1990. Pastoralism and land tenure. Presented at the *Workshop on Pastoralism and the Environment*. Arusha, Tanzania, April.

Fotso, R. C. and J. R. Parrott. 1991. Ecology and breeding biology of Bannerman's Turaco (*Turaco bannermani*). *Bird Conservation International,* vol. 1.

Foucault, M. 1970. *The Order of Things: An Archaeology of the Human Sciences*. New York: Random House.

Foucault, Michel. 1995 (original, 1975). *Discipline and Punish: The Birth of the Prison*. A. Sheridan, trans. New York: Vintage Books.

Fox, J. 1994. The difficult transition from clientelism to citizenship. *World Politics* 46:151–184.

Foxworthy, F. W. 1922. Minor forest products of the Malay Peninsula. *Malayan Forest Records* 2:151–217.

Frake, Charles. 1980. The genesis of kinds of people in the Sulu archipelago. In *Language and Cultural Description*. Stanford: Stanford University Press.

Freudenberger, Mark Schoonmaker. 1993. Regenerating the gum arabic economy: Local-level resource management in northern Senegal. In J. Friedmann and H. Rangun, eds., *In Defense of Livelihood*. West Hartford, CT: Kumerian Press.

Fried, Stephanie. 1992. Field notes, March.

———. 1995. Writing for Their Lives: Bentian Dayak Authors and Indonesian Development Discourse. Ph.D. thesis, Cornell University, Ithaca, NY.

Friede, J. 1952. Los Kofán. Una tribu de la alta Amazonía Colombiana. *Actas del XXX Congreso Internacional de Americanistas*. Cambridge.

Furro, T. 1992. Federalism and the Politics of Revenue Allocation in Nigeria. Ph.D. dissertation, Clark College, Atlanta, GA.

García, P. 1996. Atalaya, una historia en dos tiempos. *Asuntos Indígenas*, no. 1, Enero/Febrero/Marzo. Copenhagen: International Work Group of Indigenous Affairs.

García, Pedro, Søren Hvalkof, and Andrew Gray. 1998. Liberation through land rights in the Peruvian Amazon. In A. Parellada and S. Hvalkof, IWGIA Doc No. 90. IWGIA, Copenhagen.

Gasgoine, J. 1996. The ordering of nature and the ordering of empire: A commentary. In D. Miller, and P. Reill, eds., *Visions of Empire: Voyages, Botany and Representations of Empire*, pp. 108–114. Cambridge: Cambridge University Press.

Gentry, A. 1993. Tropical forest biodiversity and the potential for new medicinal plants. In A. D. Kinghorn and M. F. Balandrin, eds., *Human Medicinal Agents from Plants*. Washington, DC: American Chemical Society.

George, Kenneth. 1996. *Showing Signs of Violence: The Cultural Politics of a Twentieth-Century Headhunting Ritual*. Berkeley: University of California Press.

Gerding, H. 1986. Medizin in Samoa seit Beginn der Kolonialisierung. *Arbeiten de Forschungsstelle des Instituts für Geschichte der Medizin de Universität zu Köln* 42:1–118.

Gereau, R. E. 1997. Typification of names in *Ancistrocladus wallich* (Ancistrocladaceae). *Novon* 7:242–245.

Geschiere, P. 1993. Chiefs and colonial rule in Cameroon: Inventing chieftaincy, French and British style. *Africa* 63(3):151–175.

Gianno, R. 1981. The Exploitation of Resinous Products in a Lowland Malayan Forest. Research report, Museum Support Center, Smithsonian Institution, Washington, DC.

Giddens, Anthony. 1991. *Modernity and Self-Identity: Self and Society in the Late Modern Age.* Stanford, CA: Stanford University Press.

Giles-Vernick, Tamara. 1999. We wander like birds: Migration, indigeneity, and the fabrication of frontiers in the Sangha basin of equatorial Africa. *Environmental History* 4(2).

Gillis, M. 1988. *Public Policies and the Misuse of Forest Resources.* Cambridge: Cambridge University Press.

Glowka, L. 1998. *A Guide to Designing Legal Frameworks to Determine Access to Genetic Resources.* Environmental Policy and Law paper no. 34. Bonn: IUCN Environmental Law Centre.

Glowka, L., F. Burhenne-Guilmin, and H. Synge in collaboration with J. A. McNeely and L. Gundling. 1994. *A Guide to the Convention on Biological Diversity.* Switzerland: International Union for the Conservation of Nature and Natural Resources.

Godfrey, D. 1992, 1994. Personal interviews. Missoula, Montana.

Godoy, Ricardo A. and K. S. Bawa. 1993. The economic value and sustainable harvest of plants and animals from the tropical forest: Assumptions, hypotheses and methods. *Economic Botany* 47(3):215–219.

Godoy, Ricardo and T. C. Feaw. 1988. *Smallholder Rattan Cultivation in Southern Borneo, Indonesia.* Harvard Institute for International Development. December.

Godoy, Ricardo A. and R. Lubowski. 1992. Guidelines for the economic valuation of non-timber tropical forest products. *Current Anthropology* 33:423–433.

Godoy, Ricardo A., R. Lubowski, and A. Markandaya. 1993. A method for the economic valuation of non-timber tropical forest products. *Economic Botany* 47:220–233.

Gollin, M. and S. A. Laird. 1996. Global policies, local actions: The role of national legislation in sustainable biodiversity prospecting. *Boston University Journal of Science and Technology Law* 2:16.

Gómez, A., A. C. Lesmes, and C. Rocha. 1995. *Caucherías y Conflicto Colombo-Peruano. Testimonios 1904–1934.* Bogotá: Disloque.

Gouyon, A., H. de Foresta, and P. Levang. 1993. Does "jungle rubber" deserve its name? An analysis of rubber agroforestry systems in southeast Sumatra. *Agroforestry Systems* 22:181–206.

Government of Cameroon. 1992. Divisional Service of Forestry, Fako Division. *Annual Report 1991/92.* Limbe, Cameroon: Divisional Delegation of Agriculture, Ministry of Agriculture.

———. 1994. Law No. 94/01 of 20 January 1994 Concerning Forests, Wildlife and Fisheries.

———. 1995a. Decree No. 95/466PM of July 1995 Relating to Wildlife.

———. 1995b. Decree No. 95/531PM of 23 August 1995 Relating to Forestry.

———. 1996. Law No. 96/12 of 5 August 1996 Relating to Environmental Management.

———. 1997. Ministry of Environment and Forestry. Manual of the Procedures and Norms for Attribution and Management of Community Forests, June 1997.

Government of Indonesia. 1985. Law No. 9. About Living Natural Resources and Their Ecosystems.

———. 1990. Law No. 5. About the Conservation of Natural Resources and Ecosystems.

————. 1995a. Decree of the Agrarian Minister of the Republic of Indonesia. No. 375/Kpts/JK.250/5/95. Regarding the Ban on Napoleon Wrasse Fish Haul.

————. 1995b. Decree of the Director General of Fisheries. No. SK330/DJ.8259/95. Regarding Size, Location, and Manners of Hauling Napoleon Wrasse.

————. 1995c. Decree of the Trade Minister of the Republic of Indonesia. No. 94/Kp/V/95 Regarding the Ban on Export of Napoleon Wrasse Fish.

Government of the Philippines. 1995. Executive Order No. 247. *Prescribing Guidelines and Establishing a Regulatory Framework for the Prospecting of Biological and Genetic Resources, Their By-products and Derivatives, for Scientific and Commercial Purposes,* May.

Government of the Philippines. 1996. The *Implementing Rules and Regulations on the Prospecting of Biological and Genetic Resources,* June.

Gray, A. 1986. *And After the Gold Rush: Human Rights and Self-Development in Southeastern Peru.* IWGIA Document no. 55. Copenhagen: International Work Group of Indigenous Affairs.

————. 1990. The Putumayo Atrocities Revisited. Paper presented at Oxford University seminar on State, Boundaries and Indians. Manuscript.

————. 1991. *Between the Spice of Life and the Melting Pot: Biodiversity Conservation and Its Impact on Indigenous Peoples.* IWGIA Document 70. Copenhagen: International Work Group of Indigenous Affairs.

————. 1997. Freedom and territory: Slavery in the Peruvian Amazon. In *Enslaved People in the 90s,* IWGIA Document 83, pp. 183–215. London: Anti-Slavery International; and Copenhagen: International Work Group of Indigenous Affairs.

Gray, A. and S. Hvalkof. 1990. Indigenous land titling in the Peruvian Amazon. *IWGIA Yearbook 1989,* pp. 230–243. Copenhagen: International Work Group of Indigenous Affairs.

Gray, J. 1993. *Beyond the New Right.* London: Routledge.

Greaves, T., ed. 1994. *Intellectual Property Rights: A Sourcebook.* Oklahoma City, OK: Society for Applied Anthropology.

Greenlee, Dale. 1996. Personal interview, April. Gales Point, Belize.

Greenpeace. 1994. *Shell Shocked.* Amsterdam: Greenpeace International.

Grove, R. 1995. *Green Imperialism.* Cambridge: Cambridge University Press.

Grunwald, J. and K. Buttel. 1996. The European phytotherapeutics market: Figures, trends and analyses. *Drugs Made in Germany* 39:6–11.

Guha, R. and M. Gadgill. 1995. *Ecology and Equity.* New Delhi: Oxford University Press.

Gupta, A. 1998. *Indian Agro-Ecology.* Durham, NC: Duke University Press.

Gustafson, A. 1993. Botanical hunt finds a vine that fights AIDS. *The Statesman Journal,* December.

Gustafson, K. R., J. H. Carellina, J. B. McMahon, et al. 1992. A non-promoting phorbol from the Samoan medicinal plant *Homalanthusnutans* inhibits cell killing by HIV-1. *Journal of Medicinal Chemistry* 35:1978–1986.

Guthman, J. 1997. Representing crisis. *Development and Change* 28:45–69.

Greenough, Paul and Anna Tsing. 1994. Environmental discourses and human rights in South and Southeast Asia. *Items* 48(4):95–99.

Haas, P. 1990. *Saving the Mediterranean.* New York: Columbia University Press.

Habermas, J. 1994. *Between Facts and Norms.* Cambridge, MA: MIT Press.

————. 1987 *Theory of Communicative Action (II).* Boston: Beacon Press.

Hall, Kenneth R. 1985. *Maritime Trade and State Development in Early Southeast Asia.* Honolulu: University of Hawaii Press.

Hallock, Y. F., K. P. Manfredi, J. W. Blunt, et al. 1994. Korpusensamines A-D, novel anti-malarial alkaloids from *Ancistrocladus korupensis*. *The Journal of Organic Chemistry* 59:6349–6355.

Hammer, J. 1996. Nigerian crude. *Harpers Magazine*, June, pp. 58–68.

Haraway, Donna. 1991. *Simians, Cyborgs and Women: The Reinvention of Nature*. New York: Routledge.

———. 1992. The promises of monsters: A regenerative politics for inappropriate(d) others. In L. Grossberg, C. Nelson, and P. A. Treichler, eds. *Cultural Studies*, pp. 295–337. New York: Routledge.

Hardenburg, W. E. 1913. *The Putumayo. The Devil's Paradise. Travels in the Peruvian Amazon Region and an Account of the Atrocities Committed upon the Indians Therein*. London: T. Fisher Unwin.

Hardie-Boys, N. 1994. The Rhetoric and Reality of Conservation Aid in Western Samoa. M.S. thesis. Department of Geography, University of Canterbury, Christchurch, New Zealand.

Hardjasoemantri, Koesnadi. 1995. *Hukum Perlindungan Lingkungan: Konservasi umber Daya Alam Hayati dan Ekosystemnya*. Yogyakarta: Gajah Mada University Press.

Harvey, David. 1990. *The Condition of Postmodernity: An Enquiry into the Origins of Cultural Change*. Oxford: Blackwell.

Harwell, Emily. 2000. Remote sensibilities: Discoveries of technology and the making of Indonesia's natural disaster, 1997–98. *Development and Change* 31:307–340.

Hecht, Susanna B., A. B. Anderson, and P. May. 1988. The subsidy from nature: Shifting cultivation, successful palm forests, and rural development. *Human Organization* 47:25–35.

Hecht, Susanna and Alexander Cockburn. 1990. *Fate of the Forest: Developers, Destroyers and Defenders of the Amazon*. New York: Penguin.

Hefner, Robert. 1990. *The Political Economy of Mountain Java: An Interpretive History*. Berkeley: University of California Press.

———. 1998a. Introduction: Society and Morality in the New Asian Capitalisms. In R. Hefner, ed., *Market Cultures: Society and Morality in the New Asian Capitalisms*. Boulder, CO: Westview Press.

Hefner, Robert, ed. 1998b. *Market Cultures: Society amid Morality in the New Asian Cultures*. Boulder, CO: Westview Press.

Hemming, John. 1978. *The Search for El Dorado*. New York: Dutton.

Hesseling, Gerti (in collaboration with M. Sypkens Smit). n.d. (circa 1984). Le Droit Foncier au Senegal: L'Impact de la Réforme Foncière en Basse Casamance. Mimeo.

Hix, J. 1996. *The Glasshouse*. New York: Phaidon.

Hochschild, Adam. 1998. *King Leopold's Ghost: A Story of Greed, Terror, and Heroism in Colonial Africa*. Boston: Houghton Mifflin.

Horwich, Robert H. 1990. How to develop a community sanctuary: An experimental approach to the conservation of private lands. *Oryx* 24:95–102.

———. 1995. Community development as a conservation strategy: The Community Baboon Sanctuary and Gales Point, Manatee Projects Compared. Presented at *Conservation and Community Development in the Selva Maya of Belize, Guatemala, and Mexico*. Chetumal, Mexico, November 8–10.

Horwich, Robert H. and Jonathan Lyon. 1990. *Belizean Rainforest—The Community Baboon Sanctuary*. Gays Mills, WI: Orangutan Press.

———. 1995. Multi-level conservation and education at the Community Baboon Sanctuary, Belize. In S. Jacobson, ed., *Wildlife Conservation: International Case Studies of Education and Communication Programs*. New York: Columbia University Press.

———. 1998. Community-based development as a conservation tool: The Community Baboon Sanctuary and the Gales Point Manatee project. In Richard Primack, David Bray, H. A. Galletti, and I. Ponciano, eds., *Timber, Tourists and Temples: Conservation and Development in the Maya Forests of Belize, Guatemala and Mexico*, pp. 343–363. Washington, DC: Island Press.

Horwich, Robert H., Dail Murray, Ernesto Saqui, et al. 1993. Ecotourism and community development: A view from Belize. In K. Lindberg and D. E. Hawkins, eds., *Ecotourism: A Guide for Planners and Managers*, pp. 152–168. North Bennington, VT: Ecotourism Society.

*Hospital and Specialist Medicine.* 1992. Urology: Risk of benign prostatic hyperplasia (BPH) underestimated. October, p. 58.

Hu, S. Y. 1990. History of the introduction of exotic elements into traditional Chinese medicine. *Journal of the Arnold Arboretum* 71:487–526.

Human Rights Watch. 1995. *The Ogoni Crisis.* Report #7/5. New York: Human Rights Watch.

Hunt, D. 1923. Samoan medicines and practices. *U. S. Naval Medical Bulletin* 19:145–152.

Hutabarat, Christoverius. 1995. Potassium used to catch fishes and its impact on the local community in the Togean Islands, Central Sulawesi. *Tangkasi* 2(1):6–7. Depok: Yayasan Bina Sains Hayati Indonesia.

Hutabarat, Christoverius, Hadi Promono, and Susi Yuliati. 1995. Preliminary study on marine resource use in the Togean Islands, Sulawesi: With special reference to the Bajau people. *Tangkasi* 2(1):1–6. Depok: Yayasan Bina Sains Hayati Indonesia.

Hvalkof, S. 1986a. *Urgent Report on the Situation of the Ashéninka (Campa) Population of Gran Pajonal, Central Peruvian Amazon.* Research Report. Copenhagen: Danish International Development Agency and Danish Social Science Research Council.

———. 1986b. El drama actual del Gran Pajonal. Primer parte: Recursos, historia, población y producción Ashéninka. *Amazonía Indigena, Boletin de Analisis* 6(12). Lima: Copal.

———. 1987. El drama actual del Gran Pajonal. Segunda parte: Colonización y violencia. *Amazonia Indigena, Boletin de Analisis* 7(13). Lima: Copal.

———. 1989. The nature of development: Native and settlers' views in Gran Pajonal, Peruvian Amazon. *Folk*, vol. 31. Copenhagen: Danish Ethnographic Society.

———. 1990. Roller i udviklingens spil: Aktører og attituder i Peruansk Amazonas. *Den Ny Verden*, no. 1: Udviklingstemaer II. Copenhagen: Center for Development Research. (Spanish translation: Roles en el juego del desarrollo: Actores y actitudes en la Amazonía Peruana.) Unpublished manuscript.

———. 1994a. The Asháninka disaster and struggle—The forgotten war in the Peruvian Amazon. *Indigenous Affairs*, no. 2. Copenhagen: International Work Group of Indigenous Affairs.

———. 1994b. Territorial organization and democracy in the Peruvian Amazon. The current Asháninka struggle for land, autonomy and recognition. Presented at symposium, ant. 14, *Sacred Land, Threatened Territories—Contested Landscapes in Native South America.* 48th International Congress of Americanists (ICA), Stockholm/Uppsala, July 4–9.

Hvalkof, S. and P. Aaby, eds. 1981. *Is God An American? An Anthropological Perspective on the Missionary Work of the Summer Institute of Linguistics.* Document 42. Copenhagen: International Work Group of Indigenous Affairs; London: Survival International.

Hvalkof, S. and A. Escobar. 1997. Nature, political ecology and social practice: Towards an academic and political agenda. In A. Goodman and T. Leatherman, eds., *Building a New Biocultural Synthesis.* Ann Arbor: University of Michigan Press.

Hvalkof, S. and H. Veber. 1997. Los Ashéninka de Gran Pajonal. In F. Santos and F. Barclay, eds., *Guía Etnografica de la Alta Amazonia*, vol. III. Panama: Smithsonian Tropical Research Institute.

Ikein, A. 1990. *The Impact of Oil on a Developing Country.* New York: Praeger.

Ikporukpo, C. 1993. Oil companies and village development in Nigeria. *OPEC Review*, pp. 83–97.

———. 1996. Federalism, political power and the economic power game: Control over access to petroleum resources in Nigeria. *Environment and Planning C* 14:159–177.

Inyuat e-Maa. n.d. Draft Constitution of Inyuat e-Maa. Unpublished typescript.

Irvine, F. R. 1961. *Woody Plants of Ghana.* London: Oxford University Press.

Isaacman, A. 1990. Peasants and rural social protest in Africa. *African Studies Review* 33(2):1–120.

Iskandar. 1993. Studi Tentang Sistem Budidaya Rotan Secara Tradisional pada Masyarakat Dayak di Sungai Lawa Kabupaten Kutai. Unpublished manuscript (skripsi). Fakultas Kehutanan, Universitas Mulawarman, Samarinda, East Borneo.

Iwu, M. M. 1993. *Handbook of African Medicinal Plants.* London: CRC Press.

Izaguirre (OFM), B. 1922–1927. *Historia de las Misiones Franciscanas y narración de los progresos de la geografía en el Oriente del Perú. Relatos originales y producciones en lenguas indigenas de varios misioneros.* Lima: Talleres Tipográficos de la Penitenciaría.

Jalaluddin, M. 1977. A useful pathological condition of wood. *Economic Botany* 31:222–224.

James, R. W. 1971. *Land Tenure and Policy in Tanzania.* Toronto: University of Toronto Press.

Jato, J. and D. Thomas. 1993a. *Ancistrocladus korupensis* and michellamine-B in Cameroon. March.

———. 1993b. Distribution and Cultivation of *A. abbreviatus* in Cameroon. Report to the National Cancer Institute.

Jeanrenaud, S. 1991. The conservation–development interface: A study of forest use, agricultural practices and perceptions of the rainforest, Etinde Rainforest, South West Cameroon. London: Overseas Development Administration.

Jessup, Timothy C. and Nancy L. Peluso. 1986. Minor forest products as common property resources in East Kalimantan, Indonesia. Proceedings of the *Conference on Common Property Resource Management,* pp. 21–26. April, 1985. Washington, DC: National Academy Press.

Johannes, Robert E. and Michael Riepen. 1995. *Environmental, Economic, and Social Implications of the Live Fish Trade in Asia and the Western Pacific.* Consultants' Report to The Nature Conservancy and The South Pacific Forum Fisheries Agency.

Joyce, C. 1994. *Earthly Goods: Medicine-Hunting in the Rainforest.* Boston: Little Brown.

Kalkman, C. 1965. The Old World species of *Prunus* sub-genus *Laurocerasus. Blumea* 13(1):33–35.

Kartawinata, K. and A. P. Vayda. 1984. Forest conversion in East Kalimantan: The activities and impact of timber companies, shifting cultivators, migrant pepper farmers and others. In F. Di Castri, F. W. G. Baker, and M. Hadley, eds., *Ecology in Practice*, vol. 1, pp. 98–126. Paris: United Nations Educational, Scientific, and Cultural Organization (UNESCO).

ten Kate, K. 1997. The common regime on access to genetic resources in the Andean Pact. *Biopolicy*, vol. 2, paper 6.

ten Kate, K. and S. A. Laird. 1997. *Placing Access and Benefit Sharing in the Commercial Context: Private Sector Practices and Perspectives.* World Resources Institute and Royal Botanic Gardens, Kew.

———. 1999. *The Commercial Use of Biodiversity: Access to Genetic Resources and Benefit Sharing.* London: Earthscan.

Katz-Miller, S. 1993. High hopes hanging on a "useless" vine. *New Scientist*, January 16.

Katz-Miller, S. and L. Dayton. 1993. Australia takes tough line on "HIV plant." *New Scientist*, July 3.

Kaufman, D. 1993. Published letter. *Botany 2000—Asia Newsletter*. October.

Keane, Webb. 1997. *Signs of Recognition: Powers and Hazards of Representation in an Indonesian Society.* Berkeley: University of California Press.

Keck, Margaret. 1995. Social equity and environmental politics in Brazil: Lessons from the rubber tappers of acre. *Comparative Politics* 27(4):409–424.

Keck, Margaret and Kathryn Sikkink. 1998. *Activists Beyond Borders: Advocacy Networks in International Politics*. Ithaca, NY: Cornell University Press.

Khan, S. A. 1994. *Nigeria: The Political Economy of Oil.* London: Oxford University Press.

King, R. B., J. H. Pratt, M. P. Warner, and S. A. Zisman. 1993. *Agricultural Development Prospects in Belize.* Chatham, UK: Natural Resources Institute Bulletin 48.

King, Steven R., T. J. Carlson, and K. Moran. 1994. Biological diversity, indigenous knowledge, drug discovery and intellectual property rights. In S. Brush, ed., *Indigenous Peoples and Intellectual Property Rights*, pp. 161–185. Washington DC: Island Press.

Kingsbury, Benedict. 1994. Environment and trade. In A. Boyle, ed., *Environmental Regulation and Economic Growth.* Oxford: Clarendon Press.

———. 1998. Indigenous peoples in international law: A constructivist approach to the Asian controversy. *The American Journal of International Law* 92:414–457.

———. In press. The international concept of indigenous peoples in Asia. In D. Bell and J. Bauer, eds., *Human Rights and Economic Development in East Asia.*

Kiondo, A. 1992. The nature of economic reforms in Tanzania. In H. Campbell and H. Stein, eds., *Tanzania and the IMF: The Dynamics of Liberalization*, pp. 21–42. Boulder, CO: Westview.

KIPOC. n.d. Constitution of the Korongoro Integrated Peoples Oriented to Conservation. Unpublished typescript.

———. 1992. *The Foundation Program: Program Profile and Rationale.* Loliondo, Tanzania: Korongoro Integrated Peoples Oriented to Conservation. Principal document no. 4. Typescript.

Kjekshus, H. 1977. *Ecology Control and Economic Development in East African History: The Case of Tanganyika 1850–1950.* Berkeley: University of California Press.

Kloppenburg, J. R. 1988. *First the Seed: The Political Economy of Plant Biotechnology 1492–2000.* Cambridge: Cambridge University Press.

Kloppenburg, J. and M. Balick. 1996. Property rights and genetic resources: A framework for analysis. In M. Balick, E. Elisabetsky, and S. Laird, eds., *Medicinal Resources of the Tropical Forests: Biodiversity and Its Importance to Human Health*, pp. 174—181. New York: Columbia University Press.

Kofele-Kale, N. 1981. *Tribesmen and Patriots: Political Culture in a Polyethnic African State*. University Press of America.

Kokwaro, J. O. 1976. Medicinal plants of East Africa. Nairobi: General Printers.

Koon, C. B. 1995. *Environmental Assessment of the Oriente District of Ecuador.* (Annex of affidavit for court case.) Houston, TX.

Kopytoff, Igor. 1995 (original, 1986). The cultural biography of things: Commoditization as process. In Arjun Appadurai, ed., *The Social Life of Things: Commodities in Cultural Perspective*. Cambridge: Cambridge University Press.

Krimsky, S. 1982. *Genetic Alchemy: The Social History of the Recombinant DNA Debate*. Cambridge, MA: MIT Press.

Kuletz, Valerie L. 1998. *The Tainted Desert: Environmental and Social Ruin in the American West*. New York: Routledge.

Kusworo. 1997. Government Policies That Affect the Damar Agroforests in Pesisir Krui, West Lampung, Sumatra. Research report, International Center for Research in Agroforestry, Southeast Asia, Bogor.

LaFrankie, J. V. 1994. Population dynamics of some tropical trees that yield non-timber forest products. *Economic Botany* 48(3):301–309.

Laird, Sarah A. 1994. Natural products and the commercialization of traditional knowledge. In T. Greaves, ed., *Intellectual Property Rights for Indigenous Peoples: A Source Book*. Oklahoma City, OK: Society for Applied Anthropology.

———. 1995a. *Access Controls for Genetic Resources*. A WWF-International Discussion Paper, prepared for the World Wide Fund for Nature Support for the Implementation of the Biodiversity Convention Project, Switzerland.

———. 1995b. *Fair Deals in the Search for New Natural Products*. World Wide Fund for Nature International.

Laird, Sarah A. and Estherine Lisinge. 1998. Benefit-sharing case studies from Cameroon: *Ancistrocladus korupensis* and *Prunus africana*. In *Case Studies on Benefit-Sharing Arrangements, Conference of the Parties to the Convention on Biological Diversity*, 4th meeting, Bratislava, May.

Laird, Sarah A. and T. C. H. Sunderland. 1996. The Over-lapping Uses of "Medicinal" Species in South West Province, Cameroon: Implications for Forest Management. Presented at the annual meeting of the Society of Economic Botany, June, London.

Laird, Sarah A. and R. Wynberg. 1996. Biodiversity prospecting in South Africa: Towards the development of equitable partnerships. In J. Mugabe, C. Barber, G. Henne, et al., eds., *Managaing Access to Genetic Resources: Towards Sustainable Strategies for Benefit-Sharing*. Nairobi: African Centre for Technology Studies, and Washington, DC: World Resources Instititue.

Laraburra y Correa, C. 1905–1909. La opinión nacional. *Colección de Leyes, Decretos, Resoluciones y Otros Documentos Oficiales Referentes al Departemento de Loreto*, vol. 1–18. Lima.

Latour, B. 1987. *Science in Action: How to Follow Scientists and Engineers through Society*, pp. 215–257. Cambridge, MA: Harvard University Press.

Laumonier, Y. 1981. Ecological and Structural Classification of Southern Sumatra Forest Types. Unpublished report. Bogor, Indonesia: BIOTROP.

Lawrence, D. C., M. Leighton, and D. R. Peart. 1995. Availability and extraction of forest products in managed and primary forest around a Dayak village in West Kalimantan, Indonesia. *Conservation Biology* 9(1):76–88.

Leach, Melissa and Robin Mearns. 1997. *The Lie of the Land: Challenging Received Wisdom in Africa*. London: Currey.

Leach, Melissa, Robin Mearns, and Ian Scoones. 1997. *Environmental Entitlements: A Framework for Understanding Institutional Dynamics of Environmental Change*. IDS Discussion Paper 359. Sussex: Institute for Development Studies.

Lefebvre, H. 1991. *The Production of Space*. Oxford: Blackwell.

Lélé, S. N. 1991. Sustainable development: A critical review. *World Development* 19:607–621.

Lembaga Alam Tropika Indonesia (LATIN). 1995. Strengthening Community-Based Damar Agroforest Management as Natural Buffer-Zone of Bukit Barisan Selatan National Park, Lampung, Indonesia. Research report.

Leonard, J. 1984. *Bulletin des Jardin Botanique Nationale Belgique* 54:465–470.

Lesser, W. H. and A. F. Krattiger. 1994. Marketing "genetic technologies" in South-North and South-South exchanges: The proposed role of a new facilitating organization. In A. F. Krattiger, J. A. McNeely, W. H. Lesser, et al., eds., *Widening Perspectives on Biodiversity*. Geneva, Switzerland: International Union for the Conservation of Nature and Natural Resources, International Academy of the Environment.

Lev, Daniel. 1998. Religion and Politics in Indonesia and Malaysia. Talk presented at the Council on Southeast Asian Studies, Berkeley, CA, February 13.

Levang, P. 1989. Systèmes de production et revenus familiaux ("Farming systems and household incomes"). *Transmigration et Migration Spontanées en Indonésie ("Transmigration and Spontaneous Migrations in Indonesia")*. Départemen Transmigrasi—the French Research Institute for Development through Cooperation (ORSTOM), pp. 193–283.

Levang, P. and Wiyono. 1993. Pahmungan, Penengahan, Balai Kencana. Enquête Agro-économique Dans la Région de Krui (Lampung). Research report, ORSTOM/BIOTROP.

Lévi-Strauss, Claude. 1992. *Tristes Tropiques* (trans. John and Doreen Weightman). New York: Penguin.

Lewis, P. 1996. From prebendalism to predation: The political economy of decline in Nigeria. *Journal of Modern African Studies* 24(1):79–104.

Li, H. 1979. *Nan-fang Ts'ao-mu Chuang.* "A fourth century flora of Southeast Asia." Hong Kong: The Chinese University Press.

Li, Tanya Murray. 1996. Images of community: Discourse and strategy in property relations. *Development and Change* 27(3):501–528.

———. 1997a. Constituting Tribal Space. Paper presented to the University of California, Berkeley, Environmental Politics Seminar.

———. 1997b. Boundary work: A response to *Community in Conservation: Beyond Enchantment and Disenchantment* by Arun Agarwal. In A. Agarwal, *Community in Conservation: Beyond Enchantment and Disenchantment*, pp. 69–82. Gainesville, FL: Conservation and Development Forum.

Lindberg, K. and J. Enriquez. 1994. An Analysis of Ecotourism's Economic Contribution to Conservation and Development in Belize, vols. 1, 2. Washington, DC: World Wildlife Fund.

Lindberg, K. and Donald E. Hawkins, eds. 1993. *Ecotourism: A Guide for Planners and Managers*, pp. 152–168. North Bennington, VT: Ecotourism Society.

Linkenbach, A. 1994. Ecological movements and the critique of development. *Thesis Eleven* 39:63–85.

Longo, R. and S. Tira. 1981. Constituents of *Pygeum africanum* bark. *Planta Medica* 42:195–203.

Loolo, G. 1981. *A History of the Ogoni.* Port Harcourt, Nigeria: Saros.

Lowe, Lisa and David Lloyd. 1997. Introduction. In L. Lowe and D. Lloyd, eds. *The Politics of Culture in the Shadow of Capital*. Durham, NC: Duke University Press.

Lubis, Z. 1996. Repong Damar: Kajian Tentang Pengambilan Keputusan Dalam Pengelolaan Lahan Hutan Pada dua Komunitas desa di Daerah Krui, Lampung Barat. Research report, P3AE-UI and CIFOR.

Lucena Salmoral, M. 1977. *Las Ultimas Creencias de los Indios Kofan. Magia, Selva, Petróleo en el Alto Putumayo*. Colombia: Universidad de Murcia, Departemento de Historia de Amerika.

Luke, Timothy W. 1997. *Ecocritique: Contesting the Politics of Nature, Economy, and Culture.* Minneapolis: University of Minnesota Press.

Luxemburg, R. 1968. *The Accumulation of Capital.* New York: Monthly Review Press.

Lynch, Owen and Kirk Talbott. 1995. *Balancing Acts: Community-Based Forest Management and National Law in Asia and the Pacific.* Washington, DC: World Resources Institute. Calton, Australia: Melbourne University Press.

Mackie, C. and Jessup, T. C. 1986. Shifting cultivation and patch dynamics in an upland forest in East Kalimantan, Indonesia. In Y. Hadi, K. Awang, N. M. Majid, and S. Mohamed, eds., *Proceedings of Regional Workshop on Impact of Man's Activities on Tropical Forest Ecosystems*, pp. 468–518. Selangor, Malaysia.

Macleod, H. L. 1987. The conservation of Oku Mountain Forest, Cameroon. *International Council for Bird Preservation Study Report* no. 15. Cambridge: ICBP.

Macleod, H. L. and J. Parrott. 1991. Exploitation of *Pygeum* Bark in the Kilum Mountain Forest Reserve. Unpublished correspondence, Kilum Mountain Forest Project, International Council for Bird Preservation.

Macpherson, C. 1985. Samoan medicine. In C. D. F. Parsons, ed., *Healing Practices in the South Pacific*, pp. 1–15. Laie, Hawaii: The Institute for Polynesian Studies.

Macpherson, C. and L. Macpherson. 1990. *Samoan Medical Belief and Practice.* Auckland: Auckland University Press.

Mahar, Dennis. 1979. *Frontier Development Policy in Brazil: A Study of Amazìnia*, New York: Praeger.

———. 1988. *Government Policies and Deforestation in Brazil's Amazon Region.* Washington, DC: World Bank.

Mahler, Richard and S. Wotkyns. 1991. *Belize: A Natural Destination.* New Mexico: John Muir Publications.

Malkki, Lisa. 1992. National geographic: The rooting of peoples and the territorialization of national identity. *Cultural Anthropology* 7(1):24–44.

Malleson, R. 1987. Food Survey of Mundemba Town and Ndian Estate. World Wide Fund for Nature Paper no. 1 of the Korup Project socio-economic survey, no. 3206/A9.

———. 1993. Harmony and Conflict Between NTFP Use and Conservation in Korup National Park. Rural Development Forestry Network Paper 15. London: ODI.

Mallon, F. 1995. *Peasant and Nation.* Berkeley: University of California Press.

Mamdani, M. 1995. *Citizen and Subject.* Princeton, NJ: Princeton University Press.

Manatee Advisory Committee. 1992. Gales Point Progressive Management Plan, Gales Point, Belize.

Manfredi, K. P., J. W. Blunt, J. H. Cardellina II, et al. 1991. Novel alkaloids from the tropical plant *Ancistrocladus abbreviatus* inhibit cell killing by HIV-1 and HIV-2. *Journal of Medicinal Chemistry*, December, pp. 3402–3405.

Manila Declaration. 1992. Asian Symposium on Medicinal Plants, Spices and Other Natural Products (ASOMPS VII), February. Manila, Philippines.

*Manuntung.* 1994a. Pencari Gaharu Liar Tetap Beroperasi. May 4.

———. 1994b. Heli Airfast Hilang di Kutai. July 29.

———. 1994c. Heli yang Hilang Belum Ditemukan. July 30.

———. 1994d. Paranormal Ikut Diterjunkan. SDA Sendiri Belum Tahu Persis. July 31.

———. 1994e. 100 Orang Masih Cari Gaharu. August 5.

———. 1994f. SAR Temukan Reruntuhan Airfast. August 6.

———. 1994g. Hari Ini Kemungkinan Evakuasi. August 9.

Marcos, Sub-Commandant. 1997. Why we are fighting: The Fourth World War has begun. *Le Monde Diplomatique* [English edition]. www.monde-diplomatique.fr/md/en/1997/08-O9/marcos, 3/11/98.

Marsden, W. 1783. *The History of Sumatra.* London.

Martinez-Alier, J. 1990. Poverty as a Cause of Environmental Degradation. Report prepared for the World Bank, Washington, DC.

————. 1997. Environmental Justice as a Force for Sustainability. Paper presented to the Conference on Global Futures, Institute for Social Studies, The Hague.

Mary, F. 1987. *Agroforêts et Sociétés*. Analyse Socio-économique de Systèmes Agroforestiers Indonésiens. Document École Nationale Supérieure d'Agronomie de Montpellier—INRA.

Mary, F., and G. Michon. 1987. When agroforests drive back natural forests: A socio-economic analysis of a rice/agroforest system in South Sumatra. *Agroforestry Systems* 5:27–55.

Matthiessen, Peter. 1965. *At Play in the Fields of the Lord*. New York: Random House.

Maxwell, Kenneth. 1991. The mystery of Chico Mendes. *New York Review of Books*, 28 March, 38(6):38–41.

Mayer, Judith. 1989. Rattan cultivation, family economy and land use: A case from Pasir, East Kalimantan. In *Forestry and Forest Products*. Samarinda, Indonesia: German Forestry Group (GFG) report no. 13, June.

Mays, T. D., E. J. Asebey, M. R. Boyd, and G. Cragg. 1993. Quid pro quo—A reexamination of the equities in the search for new medicines from biodiversity. Presented at Intellectual Property and Indigenous Rights: Alternatives for Equity and Conservation, Lake Tahoe.

Mbano, B. N., R. C. Malpas, M. K. Maige, et al. 1995. The Serengeti regional conservation strategy. In A. R. Sinclair and P. Arcese, eds., *Serengeti II: Dynamics, Management, and Conservation of an Ecosystem*, pp. 605–616. Chicago: University of Chicago Press.

Mbenkum, F. T. 1993. Ethnobotany Survey of the Kilim Mountain Forest and Surrounding Areas in Bui Division. World Wide Fund for Nature, Cameroon.

McAfee, Kathleen. 1999. Selling nature to save it? Biodiversity and green developmentalism. *Environment and Planning D: Society and Space* 17(2):133–154.

McCay, B. M. and J. M. Acheson. 1987. Human ecology of the commons. In B. M. McCay and J. M. Acheson, eds., *The Question of the Commons: The Culture and Ecology of Communal Resources*, pp. 1–34. Tucson: University of Arizona Press.

McCormick, John. 1989. *Reclaiming Paradise: The Global Environmental Movement*. Bloomington, IN: Indiana University Press.

McCuddin, Charles R. 1974. *Samoan Medicinal Plants and Their Usage*. Pago Pago, Office of Comprehensive Health Planning, Department of Medical Services, Government of American Samoa.

McMichael, P. 1996. *Social Change and Development*. London: Pine Forge Press.

McNeely, J. A. 1993. People and protected areas: Partners in prosperity. In E. Kemf, ed., *Indigenous Peoples and Protected Areas*, pp. 249–257. San Francisco: Sierra Club.

McNeely, J. and K. Miller, eds. 1984. *National Parks, Conservation and Development*. Washington, DC: Smithsonian Institution Press.

Menon, K. P. 1989. *The Rattan Industry: Prospects for Development*. Food and Agriculture Organization of the United Nations report, Jakarta, Java.

*Le Messager*. 1993. Korup forest gives fresh hopes for cure. Tuesday, March 16, Cameroon.

Metcalfe, Simon. 1994. The Zimbabwe Communal Areas Management Programme for Indigenous Resources (CAMPFIRE). In David Western, R. Michael Wright, and Shirley Strum, eds., *Natural Connections: Perspectives in Community-Based Conservation*, 161–192. Washington, DC: Island Press.

Michon, G. 1985. De l'Homme de la Forêt au Paysan de l'Arbre: Agroforesteries Indonésiennes. Ph.D. thesis, Université des Sciences et des Techniques du Languedoc, Montpellier, France.

Michon, G. and J. M. Bompard. 1987a. Agroforesteries indonésiennes: contributions paysannes à la conservation des forêts naturelles et de leurs ressources. *Revue d'Écologie (Terre Vie)* 42:3–37.

————. 1987b. The Damar gardens (*Shorea javanica*) in Sumatra. In A. G. J. H. Kostermans, ed., *Proceedings of the Third Round-Table Conference on Dipterocarps*, pp. 3–17. Samarinda: United Nations Educational, Scientific, and Cultural Organization (UNESCO).

Michon, G. and H. de Foresta. 1992. Complex agroforestry systems and conservation of biological diversity. 1. Agroforestry in Indonesia: A link between two worlds. In Y. S. Kheong and L. S. Win, eds., In Harmony with Nature. An International Conference on the Conservation of Tropical Biodiversity, Kuala Lumpur, Malaysia. *Malayan Nature Journal* (Golden Jubilee issue), pp. 457–473.

————. 1995. The Indonesian agro-forest model. In P. Halladay and D. A. Gilmour, eds., *Conserving Biodiversity Outside Protected Areas. The Role of Traditional Ecosystems.* Cambridge, UK: International Union for the Conservation of Nature and Natural Resources (IUCN).

————. 1996. Agroforests as an alternative to pure plantations for the domestication and commercialization of NTFP's. In R. R. B. Leakey, et al., eds., *Domestication and Commercialization of Non-timber Forest Products in Agroforestry Systems.* Food and Agricultural Organizations of the United Nations (FAO). Rome: Non-Wood Forest Products, 9:160–175.

Michon, G., H. de Foresta, and A. Aliadi. 1996. Damar resins, from extraction to cultivation: An "agroforest strategy" for forest resource appropriation in Indonesia. In S. K. Jain, ed., *Ethnobiology in Human Welfare*, pp. 454–459. New Delhi: Deep.

Michon, G., H. de Foresta, and P. Levang. 1995a. Stratégies agroforestières paysannes et développement durable: Les agroforêts à damar de Sumatra. *Natures-Sciences-Sociétés* 3(3):207–221.

————. 1995b. New Face for Ancient Commons in Tropical Forest Areas? The "Agroforest Strategy" of Indonesian Farmers. Communication to the 4th annual meeting of the International Association for the Study of Common Property Resources. Bodo, Norway, May.

Michon, G. and D. Jafarsidik. 1989. *Shorea javanica* cultivation in Sumatra: An original example of peasant forest management strategy. In E. F. Bruenig and J. Poker, eds., *Management of Tropical Rainforests: Utopia or Chance of Survival*, pp. 59–71. Baden-Baden: Nomos Verlagsgesellschaft.

Miller, K. R. 1984. Regional planning for rural development. In F. R. Thibodeau and H. H. Field, eds., *Sustaining Tomorrow: A Strategy for World Conservation and Development*, pp. 37–50. Hanover: University Press of New England.

Mitchell, Timothy. 1991. The object of development: Egypt in the discourse of the World Bank and USAID. *Middle East Report*, March.

Moberg, M. 1992. Structural adjustment and rural development: Inferences from a Belizean village. *Journal of Developing Areas* 27:1–20.

Momberg, Frank. 1993. Indigenous Knowledge Systems—Potentials for Social Forestry Development: Resource Management of Land Dayaks in West Kalimantan. *Berliner Beitrage zu Umwelt unt Entwicklung* Bd. 3. Technische Universitat Berlin.

Momberg, Frank, Christianus Atok, and Martua Serait. 1996. *Drawing on Local Knowledge: Participatory Community Resource Mapping in Indonesia.* Ford Foundation, Pancur Kasih, and World Wide Fund for Nature, Jakarta.

Mooney, P. R. 1983. The Law of the seed: Another development in plant genetic resources. *Development Dialogue* 1,2:1–172.

————. 1993. Exploiting local knowledge: International policy implications. In W. De Boef, K. Amanor, and K. Wellard, eds., *Cultivating Knowledge: Genetic Diversity, Farmer Experimentation and Crop Research.* London: Intermediate Technology Publications.

Moore, Donald. 1994. Contesting terrain in Zimbabwe's eastern highlands: Political ecology, peasant resource struggles, and the ethnography of the state. *Economic Geography* 69:380–401.

Moore, Sally Falk. 1986. *Social Facts and Fabrications: 'Customary' Law on Kilamanjaro, 1880–1980*. Cambridge: Cambridge University Press.

Mugabe, J., C. Barber, G. Henne, et al., eds. 1997. *Managing Access to Genetic Resources: Towards Sustainable Strategies for Benefit-Sharing*. Nairobi: African Centre for Technology Studies, and Washington, DC: World Resources Institute.

Munt, I. and E. Higinio. 1993. Eco-tourism waves in Belize. *Spear Reports* (Belize) 9:61–72.

Murphree, M. W. 1993. Decentralizing the proprietorship of wildlife resources in Zimbabwe's communal lands. In D. Lewis and N. Carter, eds., *Voices from Africa: Local Perspectives on Conservation*, pp. 133–145. Washington, DC: World Wildlife Fund.

———. In press. Congruent objectives, competing interests, and strategic compromise: Concept and process in the evolution of Zimbabwe's CAMPFIRE programme. In J. P. Brosius, A. L. Tsing, and C. Zerner, eds., *Representing Communities: Histories and Politics of Community-Based Natural Resource Management*.

Mustafa, K. 1993. Eviction of Pastoralists from the Mkomazi Game Reserve in Tanzania: A Statement. International Institute for Environment and Development. Unpublished typescript.

Naanen, B. 1995. Oil-producing minorities and the restructuring of Nigerian federalism. *Journal of Commonwealth and Comparative Politics* 33(1):46–58.

Nadapdap, A., I. Tjitradjaja, and Mundardjito. 1995. Pengelolaan hutan berkelanjutan: Kasus hutan damar rakyat di Krui, Lampung Barat. *Ekonesia* 2:80–112.

Nasir, Josia. 1991. Tradisi Pemilikan dan Pengolahan Tanah Pada Masyarakat Dayak Bentian. In *Gaharu*. Samarinda, East Borneo: Lembaga Pengembangan Lingkungan dan Sumberdaya Manusia (PLASMA).

Neumann, Roderick P. 1995. Ways of seeing Africa: Colonial recasting of African society and landscape in Serengeti National Park. *Ecumene* 2:149–169.

———. 1996. Dukes, earls and ersatz edens: Aristocratic nature preservationists in colonial Africa. *Society and Space* 14:79–98.

———. 1997. Forest rights, privileges, and prohibitions: Contextualizing state forestry policy in colonial Tanganyika. *Environment and History* 3(1):45–68.

———. 1998. *Imposing Wilderness: Struggles over Livelihood and Nature Prservation in Africa*. Berkeley: University of California Press.

Newman, E. 1994. Earth's vanishing medicine cabinet: Rain forest destruction and its impact on the pharmaceutical industry. *American Journal of Law and Medicine* XX(4):479–501.

*New Vision*. 1993. Bushenyi trees ravaged. Friday, 5 February.

Niang, Seydou. 1985. Régénération naturelle après exploitation forestière pour le charbon de bois et le bois de chauffe dans la Zone de Dialinkine (Moyenne Casamance). Mémoire de Diplìme d'Études Approfondies. Dakar, Senegal: Institut des Sciences de l'Environnement, University of Dakar.

Norton, T. R., M. L. Bristol, G. W. Read, et al. 1973. Pharmacological evaluation of medicinal plants from Western Samoa. *Journal of Pharmaceutical Sciences* 62:1077–1082.

*La Nouvelle Expression*. 1995. Guerre de contrìle entre Camerounais et Américains. No. 240 du 17 au 20 mars, Cameroon.

Nsom, C. L. and J. Dick. 1992. An Ethnobotanical Tree Survey of the Kom Area. Unpublished document, Ijim Mountain Forest Project.

Oates, J. F. 1995. The dangers of conservation by rural development—A case-study from the forests of Nigeria. *Oryx* 29(2):115–122.

O'Brien, D. Cruise. 1975. *Saints and Politicians: Essays in the Organization of a Senegalese Peasant Society*. London: Cambridge University Press.

Offe, C. 1985. *Disorganized Capitalism*. London: Polity.

Oitesoi ole-Ngulay, S. 1993. Inyuat e-Maa/Maa pastoralists development organization: Aims and possibilities. Presented at the IWGIA-CDR Conference on the Question of Indigenous Peoples in Africa. Greve, Denmark, June 1–3.

Okafor, J. C. 1992. The Korup Project Agroforestry Programme. A Report on Consultancy Work Carried Out for the GTZ. 6 January–1 February.

Okilo, M. 1980. *Derivation: A Criterion of Revenue Allocation*. Port Harcourt, Nigeria: Rivers State Newspaper Corporation.

Okpu, U. 1977. *Ethnic Minority Problems in Nigerian Politics*. Stockholm: Wiksell.

Olarte Camacho, V. 1911. *Las Crueldades en el Putumayo y en el Caqueta*, 2nd ed. Bogotá: Imprenta Eléctrica.

ole-Saitoti, T. 1994. Local perspective of Ngorongoro Conservation Area. Presented at the Second Maa Conference on Culture and Development. Arusha, Tanzania, May 30–June 3.

Osaghae, E. 1991. Ethnic minorities and federalism in Nigeria. *African Affairs* 90:237–258.

———. 1995. The Ogoni uprising. *African Affairs* 94:325–344.

Ostrom, Elinor. 1990. *Governing the Commons: The Evolution of Institutions for Collective Action*. Cambridge: Cambridge University Press.

Padoch, C. and C. M. Peters. 1993. Managed forest gardens in West Kalimantan, Indonesia. In C. S. Potter, J. I. Cohen, and D. Janczewski, eds., *Perspectives on Biodiversity: Case Studies of Genetic Resource Conservation and Development*, pp. 167–176. Washington, DC: American Association for the Advancement of Science Press.

Padoch, Christine and Miguel Pinedo-Vasquez. 1996. Managing forest remnants and forest gardens in Peru and Indonesia. In J. Schelhas and R. Greenberg, eds., *Forest Patches in Tropical Landscapes*, pp. 327–342. Washington, DC: Island Press.

Pantir, Titus. 1990. Mengenal Masyarakat Bentian: Mengenal dan Menghormati Adat Istuadat Antara Masyarakat, Sikap Berwawasan Menjadi Mantap. Manuscript prepared for Seminar Adat Masyarakat Dayak se Kabupaten Kutai, Tenggarong.

Paoli, G. D., M. Leighton, D. R. Peart, and I. Samsoedin. 1994. Economic Ecology of Gaharu (*Aquilaria malaccensis*) in Gunung Palung National Park: Valuation of Extraction and Ecology of the Residual Population. Unpublished manuscript.

PARCE. 1983. Sénégal étude des prix des combustibles ligneux (version provisoire). Dakar, Senegal: Projet d'Aménagement et de Reboisement des Forêts du Centre-Est, MPN. Mimeo.

Parrott, J. and H. Parrott. 1990. The Overexploitation of *Pygaeum africanum* (*Prunus africana*) in Cameroon. Unpublished Report. Kilum Mountain Forest Project, International Council for Bird Preservation.

Patullo, P. 1996. *Last Resorts: The Cost of Tourism in the Caribbean*. New York: Monthly Review Press.

Payaguaje, Fernando. 1990. Presentación. La nostalgia del sobreviviente. In M. A. Cabodevilla, ed., *El Bebedor de Yaje*, pp. 5–11. Ediciones CICAME, Vicariato Apostólico de Aguarico, Quito, Ecuador.

Pearce, S. 1995. *On Collecting: An Investigation into Collecting in the European Tradition*. London: Routledge.

Peet, Richard and Michael Watts. 1996. Liberation ecology: Development, sustainability, and environment in an age of market triumphalism. In R. Peet and M. Watts, eds., *Liberation Ecology: Environment, Development, Social Movements*. London: Routledge.

Peluso, Nancy L. 1983a. Markets and Merchants: The Forest Products Trade of East

Kalimantan in Historical Perspective. Unpublished Ph.D. thesis, Department of Rural Sociology, Cornell University, Ithaca, NY.

———. 1983b. Networking in the commons: A tragedy for rattan? *Indonesia* 35(1):95–108.

———. 1992a. Coercing conservation: The politics of state resource control. *Global Environmental Change* 3(2):199–217.

———. 1992b. *Rich Forests, Poor People: Resource Control and Resistance in Java.* Berkeley: University of California Press.

———. 1992c. The ironwood problem: (Mis)management and development of an extractive rainforest product. *Conservation Biology* 6(2):210–219.

———. 1996. Fruit trees and family trees in an anthropogenic forest. *Comparative Studies in Society and History* 38(3):510–548.

———. 1998. Violence and the Environment. Unpublished manuscript. University of California, Berkeley.

Peluso, Nancy L. and Timothy C. Jessup. 1985. Ecological Patterns and the Property Status of Minor Forest Products in East Kalimantan, Indonesia. Paper presented at BOSTID-NRC Conference on Common Property Resource Management, Annapolis, MD.

Pelzer, K. J. 1978. Swidden cultivation in Southeast Asia: Historical, ecological, and economic perspectives. In P. Kundstadter, E. C. Chapman, and S. Sabhasri, eds., *Farmers in the Forest*, 271–286. Honolulu: The University Press of Hawaii.

Peters, C. M., A. H. Gentry, and R. O. Mendelsohn. 1989. Valuation of an Amazonian rainforest. *Nature* 339:656–657.

Peters, Pauline. In press. Afterward. In A. L. Tsing, P. J. Brosius, and C. Zerner, eds., *Representing Communities: Politics and Histories of Community-Based Natural Resource Management.*

Petit, S. and H. de Foresta. 1996. Precious woods from the agroforests of Sumatra, where timber provides a solid source of income. *Agroforestry Today* 9(4):18–20.

Pickett, S. T., V. T. Parker, and P. Fiedler. 1992. The new paradigm in ecology: Implications for conservation biology above the species level. In D. Fiedler and K. Jain, eds., *Conservation Biology: The Theory and Practice of Nature Conservation, Preservation, and Management*, pp. 66–88. New York: Chapman and Hall.

Pickles, John. 1991. Geography, GIS, and the surveillant society. *Papers and Proceedings of the Applied Geography Conferences.* 14:80–91.

———. 1995. *Ground Truth: The Social Implications of Geographic Information Systems.* New York: Guilford Press.

Pieterse, J. 1996. Your paradigm or mine? Working paper, Institute of Social Studies, The Hague.

Place, Susan. 1995. Ecotourism for sustainable development: Oxymoron or plausible strategy? *GeoJournal* 35(2):161–173.

Poffenberger, M. 1990. *Keepers of the Forest.* West Hartford, CT: Kumerian Press.

Polanyi, K. 1947. *The Great Transformation.* Boston: Beacon Press.

Posey, D. A. 1994. A covenant on intellectual, cultural and scientific property: A basic code of ethics and conduct for equitable partnerships between responsible corporations, scientists or institutions, and indigenous groups. In T. Greaves, ed., *Intellectual Property Rights for Indigenous Peoples: A Source Book.* Oklahoma City, OK: Society for Applied Anthropology.

———. 1996. *Traditional Resource Rights: International Instruments for Protection and Compensation for Indigenous Peoples and Local Communities.* Geneva, Switzerland: International Union for the Conservation of Nature and Natural Resources.

Posey, D. A. and W. Balee, eds. 1989. Resource management in Amazonia: Indigenous and folk strategies. *Advances in Economic Botany*, vol. 7.

Posey, D. P. and G. Dutfield. 1996. *Beyond Intellectual Property Rights*. Ottawa: International Development Research Center.

Pousset, J. L. 1992. *Plantes Medicinales Africaines 2*. Paris: Agencie de Cooperation Culturelle et Technique (ACCT).

Pratasik, Silvester Benny. 1983. *Pengaruh NaCN Terhadap Kehidupan dan Kualitas Daging Ikan Mas dan Ikan Mujair* ("The Influence of NaCN on the Life and Tissue Quality of Carp Fish and Mujair Fish"). Manado, Indonesia: Universitas Sam Ratulangi, Faculty of Fisheries.

Pred, Allan and Michael J. Watts. 1992. *Reworking Modernity: Capitalisms and Symbolic Discontent*. New Brunswick, NJ: Rutgers University Press.

Priasukmana, S. 1988. The economic study of rattan cultivation in East Kalimantan. In T. Silitonga, ed., *Final Report: Rattan Project Indonesia*. Samarinda: Research and Development, Ministry of Forestry.

Purseglove, J. W. 1972. *Tropical Crops: Monocotyledons, Dicotyledons*. Essex: Longman.

Pulido, Laura. 1996. *Environmentalism and Economic Justice: Two Chicano Struggles in the Southwest*. Tucson: University of Arizona Press.

Radin, Margaret Jane. 1996. *Contested Commodities*. Cambridge: Harvard University Press.

Rahnema, M. and V. Bawtree, eds. 1997. *The Post Development Reader*. London: Zed.

Rainforest Action Network. 1997. *Human Rights and Environmental Operations Information of the Royal Dutch/Shell Group of Companies*. San Francisco: Rainforest Action Network.

Ramberg, L. 1992. Wildlife conservation or utilization: New approaches in Africa. *Ambio* 21:438–439.

Ramos, Alcida R. 1988. Indian voices: Contact experienced and expressed. In J. D. Hill, ed. *Rethinking History and Myth: Indigenous South American Perspectives on the Past*, pp. 214–234. Chicago: University of Illinois Press.

———. 1994. The hyperreal Indian. *Critique of Anthropology* 14(2):153–171.

Rangan, Haripriya. 1995. Contesting boundaries. Local challenges to global agendas: Conservation, economic liberalization and the Pastoralists' Rights movement in Tanzania. *Antipode* 27:(4):343–362.

———. 1996. From Chipko to Uttaranchal: Development, environment and social protest in the Garhwal Himalayas, India. In R. Peet and M. Watts, eds., *Liberation Ecologies: Environment, Development, and Social Movements*, pp. 205–226. London: Routledge.

Rangel, Alberto. 1914 (original, 1908). *Inferno Verde: Scenas e Scenários do Amazonas*, 2nd ed. Famalicao: Minerva.

Rappaport, R. A. 1984. *Pigs for the Ancestors*. New Haven, CT: Yale University Press.

Rappard, F. W. 1937. Oorspronkelijke bijdragen: De damar van Bengkoelen ("The damar of Bengkulu"). *Tectona* D1(30):897–915.

Rasoanaivo, P. 1990. Rainforests of Madagascar: Sources of industrial and medicinal plants. *Ambio* 19:421–424.

Raustialla, K. 1997. Domestic institutions and international regulatory cooperation. *World Politics* 49:482–509.

RdS (République du Sénégal). 1964. Loi no. 64.46 du 17 juin, relative au domaine national. *Journal Officiel de la République du Sénégal*.

———. 1972. Décret no. 72–636 du 29 mai, relatif aux attributions des chefs de circonscriptions administratives et chefs de village. *Journal Officiel de la République du Sénégal*, June 17.

————. 1992. Code Electoral, Lois no. 92–15 et 92–16 du 7 Février, portant code électoral (partie législative), Décret no. 92–267 du (partie réglementaire). Dakar, Senegal: Ministère de l'Intérieur.

————. 1994. Projet de Décret Portant Code Forestier (Partie Reglementaire). Dakar, Senegal: Ministère du Développement Rural et de l'Hydraulique.

Reid, Anthony. 1988. *Southeast Asia in the Age of Commerce, 1450–1680.* New Haven, CT: Yale University Press.

Reid, Walter V., C. V. Barber, and A. La Vina. 1995. Translating genetic resource rights into sustainable development: Gene cooperatives, the biotrade and lessons from The Philippines. *IPGRI Plant Genetic Resources Newsletter* 102:1–17.

Reid, Walter V., S. Laird, R. Gamez. 1993a. A new lease on life. In Reid, W., S. A. Laird, C. A. Meyer, eds. *Biodiversity Prospecting: Using Genetic Resources for Sustainable Development*, pp. 1–52. Washington, DC: World Resources Institute.

Reid, Walter V., S. A. Laird, C. A. Meyer, et al., eds. 1993b. *Biodiversity Prospecting: Using Genetic Resources for Sustainable Development.* Washington, DC: World Resources Institute.

Reid, Walter and Kenton Miller. 1989. *Keeping Options Alive: The Scientific Basis for Conserving Biodiversity.* Washington, DC: World Resources Institute.

Reinecke, F. 1895. Uber die Nutzpflanzen Samoas und ihre Verwendung. *Ber. der Schles. Ges. fur vaterl. Kult. Breslau*, series II. 73:1–24.

Review of European Community and International Environmental Law (RECIEL). 1997. *Focus On: Africa.* 6(1). London: Foundation for International Environmental Law.

Rey de Castro, Carlos. 1913. *Los Escandalos del Putumayo.* Carta Abierta dirigida a Mr. Geo B. Michell, Consul de S. M. B. Acompañada de diversos documentos, datos estadísticos y reproducciones fotográficas. Barcelona: Imprenta Viuda de Luis Tasso.

————. 1914. *Los Pobladores del Putumayo. Origen-Nacionalidad.* Con un apéndice que contiene documentos de importancia. Barcelona: Imprenta Viuda de Luis Tasso.

Rhodes, L., R. L. Primka, C. Berman, et al. 1993. Comparison of finasteride (Proscar), a 5-alpha reductase inhibitor, and various commercial plant extracts in *in vitro* and *in vivo* 5-alpha reductase inhibition. *Prostate* 22:43–51.

Ribot, Jesse C. 1990. Markets, States and Environmental Policy: The Political Economy of Charcoal in Senegal. Ph.D. dissertation, University of California.

————. 1995. From exclusion to participation: Turning Senegal's forest policy around? *World Development* 23(9):1587–1599.

————. 1996. Participation without representation: Chiefs, councils and forestry law in the West African Sahel. *Cultural Survival Quarterly*, Fall.

————. 1998a. Decentralization and Participation in Sahelian Forestry. Presented to the Environmental Politics Seminar, University of California, Berkeley.

————. 1998b. Theorizing access: Forest profits along Senegal's charcoal commodity chain. *Development and Change* 29:307–341.

Ribot, Jesse C. and Reginald Cline-Cole. In press. A history of West African forestry policy. In John Middletown, ed., *Encyclopedia of Sub-Saharan Africa.* New York: Simon and Schuster.

Ribot, Jesse C. and Nancy L. Peluso. In press. *A Theory of Access: Putting Property and Tenure in Place.*

Richards, P. 1985. *Indigenous Agricultural Revolution.* London: Hutchinson.

————. 1993. Indigenous peoples. In *Thematic Issue of the Proceedings of the Royal Society of Edinburgh on the Lowland Rain Forest of the Guinea-Congo Domain.*

————. 1996a. *Fighting for the Rain Forest: War, Youth, and Resources in Sierra Leone*. London: International African Institute.

————. 1996b. Culture and community values in the selection and maintenance of African rice. In S. Brush and D. Stabinsky, eds., *Valuing Local Knowledge: Indigenous People and Intellectual Property Rights*. Washington, DC: Island Press.

Rivera, José Eustacio. 1962. *La Voragine ("The Vortex")*. Buenos Aires: Losada.

Robeson, Richard. 1986. *Indonesia: The Rise of Capital*. Canberra: Asian Studies Association of Australia.

Roe, E. 1991. Development narratives. *World Development* 19:287–300.

Roosevelt, Anna Curtenius. 1980. *Parmana: Prehistoric Maize and Manioc Subsistence Along the Amazon and Orinoco*. New York: Academic Press.

————. 1997. The excavations at Corozal, Venezuela: Stratigraphy and ceramic seriation. *Yale University Publications in Anthropology 83*. New Haven, CT: Department of Anthropology, and the Peabody Museum, Yale University.

————. In press. The lower Amazon: A dynamic human habitat. In M. Heckenberger, ed., *Pre-Columbian New World Ecosystems*. New York: Columbia University Press.

Roosevelt, Anna, ed. 1994a. *Amazonian Indians from Prehistory to the Present: Anthropological Perspectives*. Tucson, AZ: University of Arizona Press.

Roosevelt, Theodore. 1994b. *Through the Brazilian Wilderness*. Mechanicsburg, PA: Stackpole.

Rothman, David J. 1998. The international organ traffic. *New York Review of Books* XLV(5):14–17.

Rousseau, J. 1990. *Central Borneo*. New York: Oxford University Press.

Routledge, P. 1994. *Resisting and Shaping the Modern*. London: Routledge.

Rural Advancement Foundation International. 1995. *Bioprospecting and Indigenous Peoples: An Overview Prepared for the South Pacific Consultation on Indigenous People's Knowledge and Intellectual Property Rights*. Fiji: Suva, April 24–27.

Sachs, W. 1993. Global ecology and the shadow of development. In W. Sachs, ed., *Global Ecology: A New Arena of Political Conflict*. London: Zed.

Sachs, W., ed. 1992. *The Development Dictionary*. London: Zed.

Said, E. W. 1994. *Orientalism*. New York: Vintage.

Salafsky, Nick B., Barbara Dugelby, and John Terborgh. 1993. Can extractive reserves save the rainforest? An ecological and socioeconomic comparison of nontimber forest product extraction systems in Peten, Guatemala, and West Kalimantan, Indonesia. *Conservation Biology* 7(1):39–52.

Saletore, R. N. 1975. *Early Indian Economic History*. London: Curzon Press.

Salles, Vicente. 1971. *O Negro no Pará sob o Regime da Escravidao*. Rio de Janeiro: Fundaáao Getúlio Vargas.

Salm, Rodney, Yulheri Abas, and Rolex Lameanda. 1982. *Marine Conservation Potential: Togean Islands, Central Sulawesi*. Bogor: Food and Agriculture Organization of the United Nations.

San Roman, J. 1975. *Perfiles Históricos de la Amazonía Peruana*. Lima: Ediciones Paulinas/CETA.

Sardjono, M. A. 1992. Lembo Culture in East Kalimantan: A Model for the Development of Agroforestry Land-use in the Humid. *GFG-Report* 21:45–62.

Saro-Wiwa, K. 1989. *On a Darkling Plain*. Port Harcourt, Nigeria: Saros.

————. 1992. *Genocide in Nigeria*. Port Harcourt, Nigeria: Saros International.

————. 1995. *A Month and a Day*. London: Penguin.

Sather, Clifford. 1997. *The Bajau Laut: Adaptation, History, and Fate in a Maritime Fishing Society of South-eastern Sabah*. Oxford: Oxford University Press.

Satish, Chandra and Mark Poffenberger. 1989. Community forestry management in West Bengal: FPC case studies. In K. C. Malhotra and M. Poffenberger, eds., *Forest Regeneration Through Community Protection: The West Bengal Experience*. Proceedings of the Working Group on Forest Protection Committees, Calcutta, June 21–22, pp. 22–47; appendix 1, pp. 1–6.

Schama, Simon. 1996. *Landscape and Memory*. New York: Knopf.

Scheper-Hughes, Nancy. 1998. The new cannibalism. *Fellow Observer* 1(1):8–9.

Scholz, U. 1982. *Decrease and Revival of Shifting Cultivation in the Tropics of South-east Asia: The Examples of Sumatra and Thailand*. Ministry of Agriculture, Republic of Indonesia, Agency for Agrocultural Research and Development, Central Research Institute for Food Crops (CRIFC).

Schonau, A. P. G. 1973. Metric bark mass tables for black wattle, *Acacia mearnsii*. *Wattle Research Institute Report, October, 1972–1973*. Pietermaritzburg: Commercial Forestry Research Institute.

———. 1974. Metric site index curves for black wattle. *Wattle Research Institute Report, October, 1973–1974*. Pietermaritzburg: Commercial Forestry Research Institute.

Schroeder, Richard. 1995. Contradictions along the commodity road to environmental stabilization: Foresting Gambian gardens. *Antipode* 27(4):325–342.

———. In press. Community, forestry and conditionality in the Gambia. In A. L. Tsing, J. P. Brosius, and C. Zerner, eds., *Representing Communities: Politics and History of Community-Based Natural Resource Management*.

Schurz, William L., Hargis, O. D., Marbut, C. F., and Manifold, C. B. 1925. *Rubber Production in the Amazon Valley*. Crude rubber survey. Trade promotion series, no. 23. Department of Commerce, Bureau of Foreign and Domestic Commerce. Washington, DC: Government Printing Office.

Scott, J. C. 1985. *Weapons of the Weak: Everyday Forms of Peasant Resistance*. New Haven, CT: Yale University Press.

———. 1987. Resistance without protest and without organization: Peasant opposition to the Islamic Zakat and the Christian Tithe. *Comparative Studies in Society and History* 29(3):417–452.

Searle, J. 1995. *The Construction of Social Reality*. New York: Free Press.

Secretariat of the Convention on Biological Diversity. 1996a. UNEP/CBD/COP/3/20. *Access to Genetic Resources*. Paper for the 3rd meeting of the Conference of the Parties to the Convention on Biological Diversity, Buenos Aires, Argentina, November.

———. 1996b. UNEP/CBD/COP/3/Inf.53. *Fair and Equitable Sharing of Benefits Arising from the Use of Genetic Resources*. Paper for the 3rd meeting of the Conference of the Parties to the Convention on Biological Diversity, Buenos Aires, Argentina, November.

———. 1998. *Case Studies on Benefit-Sharing Arrangements, Conference of the Parties to the Convention on Biological Diversity*, 4th meeting, Bratislava, May.

Seibert, Sam. 1989. "Hug a tree," kiss an herb: Saving Brazil's forests. *Newsweek*, May 1.

Sen, A. 1980. *Poverty and Famines*. Oxford: Clarendon.

———. 1990. Food, economics and entitlements. In J. Dreze and A. Sen, eds., *The Political Economy of Hunger*, vol. 1, pp. 34–50. London: Clarendon.

Sevin, O. 1989. Histoire et peuplement ("History and population"). *Transmigration et Migration Spontanées en Indonésie ("Transmigration and Spontaneous Migrations in Indonesia")*, Departemen Transmigrasi—ORSTOM, pp. 13–123.

Sharpe, B. 1994. Study of cultural perceptions of conservation in Southwest province, Cameroon. Unpublished field notes, Department of Anthropology, University College, London.

————. 1997. Forest people and conservation initiatives: The cultural context of rainforest conservation in West Africa. In B. Goldsmith, ed., *Rainforests: A Wider Perspective*.

Shiva, V. 1993a. *Monocultures of the Mind: Perspectives on Biodiversity and Biotechnology*. Penang: Third World Network.

————. 1993b. Why we should say 'No' to GATT TRIPs. *Third World Resurgence*, no. 39, November.

————. 1993c. The greening of global reach. In W. Sachs, ed., *Global Ecology*. London: Zed.

————. 1989. *Staying Alive*. London: Zed.

Shivji, I. 1992. The politics of liberalization in Tanzania: The crisis of ideological hegemony. In H. Campbell and H. Stein, eds., *Tanzania and the IMF: The Dynamics of Liberalization*, pp. 43–58. Boulder, CO: Westview.

Shoman, Assad. 1994. *Thirteen Chapters of Belizean History*. Belize: Angeles Press.

Shoumatoff, Alex. 1986. *The Rivers Amazon*. San Francisco: Sierra Club.

————. 1990. *The World Is Burning: Murder in the Rain Forest*. New York: Avon.

Sibuea, T., T. Herdimansyah, and D. Herdimansyah. 1993. The Variety of Mammal Species in the Agroforest Areas of Krui (Lampung), Muara Bungo (Jambi) and Maninjau (West Sumatra). Final research report, ORSTOM and HIMBIO.

Siebert, S. F. 1989. The dilemma of dwindling resources: Rattan in Kerinci, Sumatra. *Principes* 32(2):79–97.

————. 1993. Rattan and extractive reserves. *Conservation Biology* 7:749–750.

Sikod, Fondo. 1996. *The Structure of the Logging Industry and Sustainable Forest Management in Cameroon*. Yaounde: World Wide Fund for Nature, Cameroon.

Sinha, S., S. Gururani, B. Greenberg, et al. 1997. The new traditionalist discourse of Indian environmentalism. *Journal of Peasant Studies* 24(3):65–99.

Sirait, M. T., S. Prasodja, N. Podger, et al. 1994. Mapping customary land in East Kalimantan, Indonesia: A tool for forest management. *Ambio* 23(7):411–417.

Sittenfeld, A. 1996. Tropical medicinal plant conservation and development projects: The case of the Costa Rica National Institute of Biodiversity (INBio). In M. Balick, E. Elisabetsky, and S. Laird, eds., *Medicinal Resources of the Tropical Forests: Biodiversity and Its Importance to Human Health*. New York: Columbia University Press.

Sittenfield, A. and A. Lovejoy. 1994. Biodiversity prospecting. *Our Planet* 6(4):20–21.

Slater, Candace. 1994a. 'All that glitters': Contemporary gold miners' narratives. *Comparative Studies in Society and History* 36(4):720–742.

————. 1994b. *Dance of the Dolphin: Transformation and Disenchantment in the Amazonian Imagination*, pp. 233–256. Chicago: University of Chicago Press.

————. 1995. Amazonia as edenic narrative. In W. Cronon, ed., *Uncommon Ground: Toward Reinventing Nature*. New York: Norton.

Smith, Nigel. 1982. *Rainforest Corridors: The Transamazon Colonization Scheme*. Berkeley: University of California Press.

Soejarto, D. D. 1993. Tropical Rain Forest Exploration and Drug Development. Presented at Bioresources Development and Conservation Program conference, Nigeria, February.

Spierenburg, Marja. 1995. The Role of Mhondoro Cult in the Struggle for Control over Land in Dande (Northern Zimbabwe): Social Commentaries and the Influence of Adherents. Amsterdam School for Social Science Research, Amsterdam.

Stair, J. B. 1897. *Old Samoa*. London: Religious Tract Society.

Steedly, Mary. 1993. *Hanging Without a Rope: Narrative Experience in Colonial and Post-Colonial Karoland*. Princeton, NJ: Princeton University Press.

Stephenson, C. S. 1934. *Report of the Department of Public Health*. Pago Pago: Government of American Samoa.

Steward, J. H. 1949. *Handbook of South American Indians*, vol. 5, pp 669–772.

Stoler, Ann. 1985. *Capitalism and Confrontation in Sumatra's Plantation Belt: 1870–1979.* New Haven, CT: Yale University Press.

Sunderland, T. C. H. et al. 1997. The Ethnobotany, Ecology, and Natural Distribution of Yohimbe [Pausinystalia johimbe (K. Schum.) Pierre ex Bielle], an Evaluation of the Sustainability of Current Bark Harvesting Practices, and Recommendations for Domestication and Management. Report prepared for the International Center for Research in Agroforestry, Cameroon.

*Supermodels in the Rainforest* (video). 1994. Big Rock Productions.

Svoboda, G. 1992. The Discovery of *Catharanthus roseus* Alkaloids and Their Role in Cancer Chemotherapy. Presented at the Rainforest Alliance symposium, Tropical Forest Medical Resources and the Conservation of Biodiversity, January.

Tall, Siriff. 1974. L'Economie du Charbon de Bois à Dakar. Mémoire du Diplìme d'Études Approfondies. Dakar, Senegal: Département de Géographie, University of Dakar, Dakar.

Tardjo, S. 1982. *Pengolahan Rotan dan Beberapa Aspek Sebagai Komodity Ekspor Non-Migas.* Kalimantan Selatan: Dinas Kehutanan Propinsi, Daerah Tingkat I.

Taufa'asisina, Ulu. 1993a. Letter of August 11, 1993, to the Swedish Society for the Conservation of Nature.

———. 1993b. Letter of January 29, 1993, to the Swedish Society for the Conservation of Nature.

Taussig, M. 1984. Culture of terror—Space of death. Roger Casement's Putumayo report and the explanation of torture. *Comparative Studies in Society and History*, vol. 26, no. 3. Chicago: University of Chicago.

———. 1985. *Shamanism, Colonialism and the Wild Man: A study of Terror and Healing.* Chicago: University of Chicago Press.

Tennant, Chris. 1994. Indigenous peoples, international institutions, and the international legal literature from 1945–1993. *Human Rights Quarterly* 16:1–57.

Thieblot, L., S. Berthelay, and J. Berthelay. 1971. Action préventive et curative d'un extrait d'écorce de plante africaine *Pygeum africanum* sur l'adénome prostatique expérimental chez le rat. *Thérapie* 26:575–580 (cited by Pousset, 1992).

Thieblot, L., G. Grizard, and D. Boucher. 1977. Étude du V 1326: Principe actif d'un extrait d'écorce de plante africaine *Pygéum africanum* sur l'axe hypophospho-génito sur énalien du rat. *Thérapie* 32:99–100 (cited by Pousset, 1992).

Thiollay, J. M. 1995. The role of traditional agroforests in the conservation of rain forest bird diversity in Sumatra. *Conservation Biology* 9(2):335–353.

Thomas, K. 1979. *Man and the Natural World: Changing Attitudes in England 1500–1800.* London: Penguin.

Torquebiau, E. 1984. Man-made dipterocarp forest in Sumatra. *Agroforestry Systems* 2(2):103–128.

Triwahyudi, M. A., Mushi, and H. A. Farchad (Tim Studi WALHI). 1992. Peran HPH Dalam Pembangunan Ekonomi Regional Kalimantan Timur. Jakarta: Wahana Lingkungan Hidup Indonesia (WALHI).

Tsing, Anna Lowenhaupt. 1993. *In the Realm of the Diamond Queen: Marginality in an Out-of-the-Way Place.* Princeton, NJ: Princeton University Press.

———. 1999. Becoming a tribal elder, and other green development fantasies. In T. M. Li, ed., *Transforming the Indonesian Uplands: Marginality, Power and Production.* Amsterdam: Harwood Academic.

———. 1997. Finding Cultural Difference Is the Beginning, Not the End of Our Work. Presented at Forum 97: New Linkages in Conservation and Development, November 16–21. Istanbul, Turkey.

Tsing, Anna Lowenhaupt, J. Peter Brosius, and Charles Zerner. 1998. Representing communities: Politics and histories of the community-based resource management: A conference report. *Society and Natural Resources* 11(57):157–168.

Tsing, Anna Lowenhaupt, J. Peter Brosius, and Charles Zerner, eds. In progress. *Representing Communities: Politics and History of Community-Based Natural Resource Management.*

Turner, G. 1861. *Nineteen years in Polynesia.* London: John Snow.

———. 1884. *Samoa a Hundred Years Ago and Long Before.* London: Macmillan.

Uberti, E., E. M. Martinelli, G. Pfifferi, and L. Gagliardi. 1990. HPLC analysis of *N*-docosyl ferulate in *Pygeum africanum* extracts and pharmaceutical formulations. *Fitoteralpia* 61(4):342–347.

Uhe, G. 1974. Medicinal plants of Samoa. *Economic Botany* 28:1–30.

United Nations. 1992. Commission on Development and Environment for Amazonia. *Amazonia Without Myths.* New York: United Nations.

United Nations Environment Program (UNEP). 1994. Ownership and Access to ex situ Genetic Resources—Farmer's Rights and Rights of Similar Groups. Report of the Intergovernmental Committee on the Convention on Biological Diversity.

United Republic of Cameroon. 1976. *Fourth Five-year Economic, Social and Cultural Development Plan (1976–1981).* Ministry of Economic Affairs and Planning, Yaounde.

———. 1981. *The Fifth Five-year Economic, Social and Cultural Development Plan (1981–1986).* Ministry of Economic Affairs and Planning, Yaounde.

UNPO (Unrepresented Nations and People Organization). 1995. *Ogoni: Report of the UNPO Mission to Investigate the Situation of the Ogoni.* The Hague: Unrepresented Nations and Peoples Organization.

URT (United Republic of Tanzania). 1964, 1965, 1966, 1967. *Annual Report of the Ngorongoro Conservation Unit.* Dar Es Saalam: Ministry of Agriculture, Forests and Wildlife.

U.S. Department of State. 1913. *Slavery in Peru.* Message from the President of the United States. Transmitting report of the Secretary of State, with accompanying papers, concerning the alleged existence of slavery in Peru. February 7, 1913. 62nd, 3rd session, House of Representatives. Document no. 1366, Washington, DC.

Valcárcel, C. A. 1915. *El proceso del Putumayo y Sus Secretos Inauditos.* Imprenta Comercial, Lima: de Horacio la Rosa.

van der Koppel, C. 1932. *De Economische Beteekenis der Ned. Indische Harsen ("The Economic Significance of Dutch East Indies Resins").* Batavia: Kolff.

Vayda, A. P. 1996. *Methods and Explanations in the Study of Human Actions and Their Environmental Effects.* CIFOR/World Wide Fund for Nature Special Publication, Centre for International Forestry Research, Bogor, Indonesia.

Veber, H. M. 1992. Why Indians wear clothes: Managing identity across an ethnic boundary. *Ethnos* 57(1–2):51–60.

———. 1996. External enticement and non-westernization in the uses of the Ashéninka cushma. *Journal of Material Culture* 1(2):155–182.

Vedeld, Trond. 1992. Local institution-building and resource management in the West African Sahel. *Forum for Development Studies*, no. 1.

*Veja.* 1973. Interview with Cláudio Villas Boas, 14 February, pp. 28–33.

Verne, Jules. 1952. *Eight Hundred Leagues on the Amazon.* New York: Didier. (Original, 1883. Edited by Caxton. Chicago: Clarke.)

Vickers, W. T. 1981. The Jesuits and the SIL: External policies for Ecuador's Tucanoans through three centuries. In S. Hvalkof and P. Aaby, eds., *Is God an American? An An-*

*thropological Perspective on the Missionary Work of the Summer Institute of Linguistics*, pp. 51–61. Document 42. Copenhagen: International Work Group of Indigenous Affairs; London: Survival International.

———. 1983. The territorial dimensions of Siona-Secoya and Encabellado adaptation. In R. B. Hames and W. T. Vickers, eds., *Adaptive Responses of Native Amazonians*, pp. 451–478. New York: Academic Press.

Villavicencio, M. 1984. *Geografía de la Republica del Ecuador*. Quito: Corporación Editora Nacional.

Vivien, J. and J. J. Faure. 1985. *Arbres des Forêts Denses d'Afrique Centrale.* ACCT/Ministère des Relations Extérieures, Coopération et Développement.

Volkman, Toby Alice. 1998. *Crossing Borders: The Case for Area Studies*, pp. 28–29. New York: Ford Foundation Report, Winter.

Wade, R. 1997. Greening the bank. In J. Lewis, R. Kanbur, and R. Webb, eds., *The World Bank: Its First Half Century*, pp. 611–734. Washington, DC: Brookings.

Walter, S. and J.-C. R. Rakotonirina. 1995. *L'Exploitation de Prunus africanum à Madagascar.* PCDI Zahamena et la Direction des Eaux et Forêts, Antananarivo, Madagascar.

Wamulwange, G. L. 1993. Traditional Management System of Natural Resources: A Case for Western Province (Barotseland) Zambia. Presented at the Symposium for Indigenous Peoples, The Pueblo of Zuni, NM, July.

Warren, Carol and Kylie Elston. 1994. *Environmental Regulation in Indonesia.* Asia Paper 3, University of Australia Press and Asia Research Center on Social, Political, and Economic Change.

Warren, James. 1981. *The Sulu Zone: 1768–1898*. Singapore: Singapore University Press.

Watson-Verran, H. and D. Turnbull. 1994. Science and other indigenous knowledge systems. In S. Jassanoff et al., eds., *Handbook of Science and Technology Studies*, pp. 115–139. Sage, New Delhi: Thousand Oaks.

Watt, G. 1908. *Commercial Products of India.* New York: Dutton.

Watt, J. M. and M. M. Breyer-Brandwijk. 1962. *Medicinal and Poisonous Plants of Eastern and Southern Africa.* London: Livingstone.

Watts, J. 1994. *Developments Towards Participatory Forest Management on Mount Cameroon (The Limbe Botanic Garden and Rainforest Genetic Conservation Project 1988–1994).* London: Rural Development Forestry Network, Overseas Development Institute, Summer.

Watts, Michael J. 1983. *Silent Violence.* Berkeley: University of California Press.

———. 1993. Idioms of land and labor: Producing politics and rice in Senegambia. In T. J. Bassett and D. E. Crummey, eds., *Land in African Agrarian Systems*. Madison: University of Wisconsin Press.

Weckmann, Luis. 1996 (original, 1984). *La herencia medieval de México.* Mexico City: El Colegio de México/Fondo de Cultural Económica, p. 63.

Weinstock, Joseph A. 1983. Rattan: Ecological balance in a Borneo rainforest swidden. *Economic Botany* 37(1):58–68.

———. 1989. Shifting cultivation and agro-forestry in Indonesia, some notes and data. *Proceedings of the Joint Seminar on Watershed Research and Management*, Bogor and Balikpapan, Indonesia.

Welch, C. 1995. The Ogoni and self-determination. *Journal of Modern African Studies* 33(4):635–650.

Wells, M. 1994. Biodiversity conservation and local development aspirations: New priorities for the 1990s. In C. Perrings, K. G. Maler, C. Holling, et al., eds., *Biodiversity Conservation: Problems and Policies*. The Netherlands: Kluwer Academic Press.

Wells, M. and K. Brandon. 1992. People and Parks: Linking Protected Areas with Local Communities. Washington DC: World Bank.

West, Patrick C. and Steven R. Brechin, eds. 1991. *Resident Peoples and National Parks.* Tucson: The University of Arizona Press.

Western, David and R. Michael Wright, eds. 1994. *Natural Connections: Perspectives in Community-Based Conservation.* Washington, DC: Island Press.

Whelan, Tensie, ed. 1991. *Nature Tourism: Managing for the Environment.* Washington, DC: Island Press.

Whistler, W. A. 1984. Annotated list of Samoan plant names. *Economic Botany* 38:464–489.

Whitmore, T. C. 1972. *Thymelaeaceae. Tree Flora of Malaya*, vol. 2, 383–391. Kuala Lumpur, Malaysia: Longman.

Wicks, C. M., J. D. Dalton, and H. Macleod. 1986. Report on Preliminary Agricultural Survey of the Area Around the Korup National Park in Cameroon. London: Bioresources.

Widjono, Roedy Haryo. 1991. Undang-Undang Panji Selatan Kerajaan Kutai Kertanegara ing Martapura. *Gaharu* 02(1). Samarinda, East Kalimantan.

Wijayanto, N. 1993. Potensi Pohon Kebun Campuran Damar Mata Kucing di Desa Pahmungan, Krui, Lampung. Field report, ORSTOM/BIOTROP.

Williams, B. 1952. They still believe in "bush-medicine." *Pacific Discovery* 5:12–14.

Williams, J. 1837. *A Narrative of Missionary Enterprises in the South Sea Islands.* London: Snow.

Williams, R. 1976. *Keywords.* New York: Oxford University Press.

Williams, Raymond. 1980. Ideas of nature. In R. Williams, ed., *Problems in Materialism and Culture,* pp. 67–85. London: Verso.

Wolters, O. W. 1967. *Early Indonesian Commerce.* Ithaca, NY: Cornell University Press.

Wood, C. B. 1993. Managing the Forest Boundary: Case Study 1: Cameroon: Overseas Development.

Woods, L. A., J. M. Perry, and J. W. Steagall. 1992. International tourism and economic development: Belize ten years after independence. *SPEAReports* (Belize) 8:75–97.

Woods, M. and A. Warren. 1988. *Glasshouses—A History of Greenhouses, Orangeries and Conservatories.* New York: Rizzoli.

World Bank. 1994. *Indonesia, Environment and Development: Challenges for the Future.* Environment Unit, East Asia and Pacific Region. March.

World Congress on Environment and Development. 1987. *Our Common Future.* Oxford: Oxford University Press.

World Resources Institute (WRI). 1992. *Global Biodiversity Strategy: Guidelines for Action to Save, Study, and Use Earth's Biotic Wealth Sustainably and Equitably.* WRI, International Union for the Conservation of Nature and Natural Resources (IUCN) and UNEP with the Food and Agricultural Organizations of the United Nations (FAO) and the United Nations Educational, Scientific, and Cultural Organization (UNESCO).

Worster, D. 1990. The ecology of order and chaos. *Environmental History Review* 14(1/2):1–18.

Zamarenda, A. 1996. *Proyecto de Diagnostico para el Seguimento de Trabajo con los Siona-Secoya, Quichua-Shuar.* Project report. Fundación Solsticio/Confeniae-Onise-Oise, Unión Base, Puyo.

Zayed, S., B. Sorg, and E. Hecker. 1984. Structure activity relations of poly-functional diterpenes of the tigliane type. VI. Irritant and tumor-promoting activities of semisynthetic mono and diesters of 12-deosyphorbol. *Planta Medica* 34:65–59.

Zepernick, B. 1972. *Arzneipflanzen der Polynesier.* Berlin: Verlag von Deitrich Reimer.

Zerner, Charles. 1992. *Indigenous Forest-Dwelling Communities in Indonesia's Outer Islands: Livelihood, Rights, and Environmental Management Institutions in the Era of*

*Industrial Forest Exploitation.* A report commissioned by the World Bank, Forestry Sector Review.

———. 1994a. Telling stories about biological diversity. In S. Brush and D. Stabinsky, eds., *Valuing Local Knowledge: Indigenous People and Intellectual Property Rights*, pp. 68–101. Washington, DC: Island Press.

———. 1994b. Through a green lens: The construction of customary law and environment in Indonesia's Maluku Islands. *Law and Society Review* 28(9):1079–1122.

Zerner, Charles, ed. In press. *Culture and the Question of Rights: Forests, Coasts, and Seas in Southeast Asia.* Durham, NC: Duke University Press.

Zerner, Charles and Kelly Kennedy. 1996. Equity issues in bioprospecting. In M. Baumann, J. Bell, F. Koechlin, and M. Pimbert, eds., *The Life Industry: Biodiversity, People, and Profits,* pp. 96–109. Exeter: Intermediate Technology.

Zimmerer, Karl S. 1994. Human geography and the "New Ecology": The prospect and promise of integration. *Annals of the Association of American Geographers* 84(1):108–125.

———. 1998. *Nature's Geography: New Lessons of Conservation in Developing Countries.* Madison: University of Wisconsin Press.

Zisman, S. 1989. The Directory of Protected Areas and Sites of Nature Conservation Interest in Belize. Occasional Publication no. 10, Department of Geography, University of Edinburgh.

Zumaeta, P. 1913. *Las Cuestiones del Putumayo.* Segundo Memorial. Folleto no. 2. Barcelona: Imprenta Viuda de Luis Tasso.

# CONTRIBUTORS

**Jill M. Belsky** is an Associate Professor of Sociology at the University of Montana. Her teaching and research interests center on the political ecology of upland farming, non-timber forest products, and community conservation efforts in Indonesia, Belize, and the western United States. Her published works include the chapter "Non-timber Forest Products in Conservation and Community Development: Desmoncus sp. in Gales Point, Manatee, Belize" in *Timber, Tourists and Temples: Conservation and Development in the Maya Forest of Belize, Guatemala and Mexico* (1997).

**Paul Alan Cox** is Director of the National Tropical Botanical Garden, headquartered in Lawai, Kauai, where he continues his research on plant conservation and ethnobotany. He is the author of *Nafanua: Saving the Samoan Rain Forest* (1999) and coauthor of *Plants, People, and Culture: An Introduction to the Science of Ethnobotany* (1997).

**Anthony Balfour Cunningham** is Regional Coordinator for the WWF/UNESCO/ Kew People and Plants Initiative, an international program supporting ethnobotanical studies in developing countries at the interface between local communities and conservation areas. One of his most recent publications is the chapter entitled "Human Use of Plants" in *Vegetation of Southern Africa* (1997).

**Michelle Cunningham,** a Rangeland Ecologist, has worked in Namibia and South Africa. She recently reviewed the trade in *Prunus africana* bark in a report coauthored by A. B. Cunningham in a study for the German Nature Conservation Agency (1997).

**Hubert de Foresta** is a Forest Ecologist at the Research Institute for Development in France (IRD, previously ORSTOM). He has been based in Indonesia since 1990, and since 1994 he has been affiliated with the Policy Research Programme of the Southeast Asia branch of the International Center for Research in Agroforestry (ICRAF) in Sumatra, Indonesia. He coordinated a joint program of nongovernmental organizations and research institutes (Team Krui) focusing on damar agroforests in Krui (from 1996 to 1998) and supported by the Ford Foundation. He focuses mainly on policy issues linked to the designation of the State Forest Zone in Indonesia.

**Stephanie Fried** is a Senior Scientist in the International Program at the Environmental Defense Fund, a United States–based nonprofit environmental organization, and a Research Fellow at the East-West Center. Her work focuses on the environmental and social dimensions of economic development in Southeast Asia, including indigenous peoples and their rights to natural resources. Most recently she has produced independent analyses of environmental and social impacts of World Bank and International Monetary Fund programs in Indonesia.

**Søren Hvalkof** is a Senior Consultant at the Nordic Agency for Development and Ecology (NORDECO), a Copenhagen-based company specializing in natural resource management in developing countries, and an Adjunct Researcher at the Department of Anthropology, University of Massachusetts at Amherst, with a grant from the Danish Council for Development Research to write a political ecology of Gran Pajonal, central Peruvian Amazon. He is a board member and supervisor for the Solstice Foundation, which supports indigenous projects in the Ecuadorian Amazon region. His theoretical interests focus on political ecological perspectives of the Amazon basin region, where he has done extensive fieldwork with the Ashéninka Indians of Peru.

His published works include an anthology, *Is God an American? An Anthropological Perspective on the Missionary Work of the Summer Institute of Linguistics* (1981); a report, "Nature, Political Ecology, and Social Practice: Toward an Academic and Political Agenda" (1998); and *Liberation through Land Titling in the Peruvian Amazon* (1998).

**Timothy C. Jessup** is the Director of the Wallacea Bioregional Program at the World Wildlife Fund, Indonesia. He was previously Project Leader of the Kayan Mentarang Project, and Coordinator of the Conservation Science and Policy Program. His research interests and conservation work have focused on human ecology and forest history in Kalimantan, integrated conservation and development, and the conservation of landscapes and ecoregions. He is coeditor of a book on the economic damages of the 1997 forest fires and haze in Indonesia and neighboring countries.

**Kusworo** graduated from the Faculty of Agriculture in Lampung, Indonesia. He has been working on research and development for conservation and community development through local nongovernmental organizations in Lampung. He joined the Policy Research Programme at the Southeast Asia branch of the International Center for Research in Agroforestry (ICRAF) in 1996. He is working on Forestry Development and Policy in Lampung Province, Sumatra, Indonesia.

**Sarah A. Laird** is an Independent Consultant working in the field of forest and biodiversity conservation, with an emphasis on cultural and commercial aspects. She has authored or coauthored *Biodiversity Prospecting* (1993), *Commercial Use of Biodiversity* (1999), and *The Management of Forests for Timber and Non-Wood Forest Products in Central Africa* (1999). Her contribution to this book grew from a project undertaken for the World Wide Fund for Nature, and other research conducted between 1994 and 1997, in Cameroon.

**Patrice Levang** is an Agroeconomist at the Research Institute for Development in France (IRD, previously ORSTOM). Since 1986 he has worked on household economy and resource management in transmigration areas in Indonesia. He is affiliated with the Department of Transmigration in Jakarta, Indonesia, where he coordinates a program on rehabilitation of Imperata grasslands through agroforestry practices.

**Estherine Lisinge** is a Legal Analyst specializing in environmental law. She is currently Director of Policy at World Wide Fund for Nature (WWF) Cameroon Programme Office, overseeing implementation of international agreements, including the Convention on Biological Diversity. Before joining WWF in 1994, she was Tutorial Mistress in Law at the University of Yaounde II.

**Celia Lowe** is an Assistant Professor of Anthropology at the University of Washington, Seattle. She is working on an ethnography of conservation titled *Cultures of Nature: Identity, Representation, and Biodiversity Conservation in the Togean Islands of Sulawesi.* She also works with international and Indonesian conservation organizations to enhance possibilities for collaboration around shared desires for species abundance and diversity.

**Geneviève Michon** is an Ethnobotanist at the Research Institute for Development in France (IRD, previously ORSTOM). She has been based in Indonesia since 1989, and since 1994 she has been affiliated with the Southeast Asia branch of the International Center for Research in Agroforestry (ICRAF). She coordinates a research program on forest resource management by local communities in Indonesia and the Philippines, focusing on the transition from extraction to production systems for non-timber tropical forest products.

**Frank Momberg** is the Indochina Program Director for Flora-Fauna International. From 1990 to 1998, he worked in Indonesia, focusing on community-based conservation. He has worked with World Wildlife Fund Indonesia as mapping trainer for community resource rights throughout Indonesia, training nongovernmental organizations and indigenous communities, and as teamleader for Lorentz National Park Project in Irian Jaya and the Bioregional Planning Project for Kalimantan and Irian Jaya from 1992 to 1998.

**Roderick P. Neumann** is Associate Professor of Geography in the Department of International Relations at Florida International University. He has published *Imposing Wilderness: Struggles over Livelihood and Nature Preservation in Africa* (1998). His research interests focus on the interplay between landscape representation, ideas of nature, and environmental politics in Africa.

**Bronwyn Parry,** Research Fellow and Tutor in Human Geography at King's College, Cambridge, has undertaken extensive research into the collection, commercialization, and patenting of plant genetic materials. She is the author of *The Fate of the Collections: Revealing the Social and Spatial Dynamics of Genetic Resource Use* and the editor of a volume on the globalization of trade in genetic resources. She is a collaborator on a project examining the interrelationships between intellectual property rights and the protection of cultural property and other resources in Papua New Guinea, and she is also beginning research on the collection and commercial use of human genetic materials.

**Rajindra K. Puri** is a Visiting Fellow in the Program on Environment at the East-West Center in Honolulu, Hawai'i. Trained as an ecological anthropologist, he has conducted research in ethnobiology, historical ecology, and conservation social science in Kalimantan since 1990. He was previously a postdoctoral research fellow at the Center for International Forestry Research (CIFOR), and he has worked with World Wildlife Fund Indonesia, the Wildlife Conservation Society, and the United Nations Educational, Scientific, and Cultural Organization (UNESCO) People and Plants initiative on a variety of research and training projects. He has completed a field guide and vernacular dictionary of the ethnobiology of eighteen linguistic groups in the Bulungan Regency in East Kalimantan, to be jointly published by CIFOR and the East-West Center.

**Jesse C. Ribot** is a Research Affiliate at the Harvard Center for Population and Development Studies and a Senior Associate of the World Resources Institute. His recent works include *Decentralization and Participation in Sahelian Forestry: Legal Instruments of Central Political-Administrative Control* (1999), and *Theorizing Access: Forest Profits along Senegal's Charcoal Commodity Chain* (1998). He is also coeditor of *Climate and Social Vulnerability in Semi-arid Lands* (1996).

**Richard A. Schroeder** is Director of African Studies and Associate Professor of Geography at Rutgers University. He is author of *Shady Practices: Agroforestry and Gender Politics in The Gambia* (in press) and coeditor of *Producing Nature and Poverty in Africa* (in press).

**Candace Slater** is Chancellor's Professor of Spanish and Portuguese at the University of California, Berkeley. She is concluding a study of diverse images and ideas of the Amazon entitled *Entangled Edens*.

**Michael Watts** is Chancellor's Professor of Geography and Development Studies, and Director of the Institute of International Studies, at the University of California, Berkeley, where he has taught for twenty years. He has written extensively on questions of environmental change and political ecology in West Africa, and he is currently working on the oil communities and environmental mobilization in Nigeria. He is the author of *Globalizing Agro-Food* (1996) and *Reworking Modernity* (1990).

**Charles Zerner** is a Visiting Scholar at New York University's Institute for Law and Society, and Adjunct Professor at the Draper Program in Humanities and Social Thought at New York University and at the Center for Research and Conservation at Columbia University. He founded the Natural Resources and Rights Program (NRRP) of the Rainforest Alliance in 1993 and directed the NRRP until 1999. The NRRP worked to integrate concerns for justice, power, and culture into the policies and programs of international conservation and donor organizations, nongovernmental groups, and the scholarly community.

Zerner is the editor of *Culture and the Question of Rights: Forests, Coasts, and Seas in Southeast Asia* (Duke University Press, in press). He is editing, with Anna Lowenhaupt Tsing and J. Peter Brosius, *Representing Communities: Politics and Histories of Community-Based Resource Management,* and co-authoring, with Craig Thorburn, *Sea Change: The Role of Culture, Property, and Rights in Managing Indonesia's Marine Fisheries* (Cornell Program, in press).

# INDEX

*A. korupensis* collecting. *See Ancistrocladus korupensis* collecting
ABS issues. *See* Access/benefit-sharing (ABS) issues
Access/benefit-sharing (ABS) issues, 345, 348, 368–69n3. *See also Ancistrocladus korupensis* collecting; Benefit sharing; Biodiversity prospecting
Access rights: and *Ancistrocladus korupensis* collecting, 345, 347; and conservation, 118, 127; limitations of, 134, 135–36, 154; and Pesisir damar gardening, 187–90, 202nn38–40. *See also* Access/benefit-sharing (ABS) issues; Property rights
Accountability: and atrocities, 96–99, 112nn21, 23–24; and local governance, 134–35, 138–39, 153–55. *See also* Enfranchisement; Governance
Activism, transnational. *See* Transnational nongovernmental organizations
Actor-oriented interface analysis, 31
Adat law. *See* Customary law
Administrative structures, 8, 138–39, 155–56nn4, 5, 164–65. *See also* Enfranchisement; Governance
Agroforestry, 160, 166, 199n8. *See also* Bentian rattan agroforestry system; Pesisir damar gardening
Amazonia, 33–34, 67–79, 80n2; map, 68; population, 73, 77, 80nn1, 20, 81–82nn41, 42. *See also* Amazonia, images of; Upper Amazon
Amazonia, images of: current, 72–74; El Dorado, 74–79, 81nn30–32; historical, 69–72, 80nn7, 9, 14, 15
Analytical methods, 6–7
*Ancistrocladus korupensis* collecting, 345–60, 365–68; botanical information, 350, 369n6, 372n19; cultural context,

357–60, 371–72nn16–18, 20; intermediary roles, 354–55, 371n14; and law, 346–47, 355–57, 371n15; maps, 346; original collections, 350–51, 369n7; process benefits, 365; research, 351–52, 369n8, 370–71nn11, 13, 372n21; research agreements, 365–67; state role, 352–54, 360, 369–70n10
Antidevelopment thinking, 29–32, 50–51nn11, 12
Anti-Slavery International, 106, 114n40
Atrocities, 93–101, 112n22, 113nn27–31; accountability, 96–99, 112nn21, 23–24; as conscious strategy, 99–100, 114nn32, 33; current, 115–16n51; and dehumanization, 100, 114n34; and rejection of development, 18n17

Bark extraction. *See Prunus africana* bark extraction
Belizean ecotourism, 285–307, 307n1; alternative images of, 296–98, 307n7; community resentment, 301–3, 307–8n11; historical background, 290–93, 307nn4, 5; maps, 286, 287; origins of, 293–96; participation levels, 300–301; promotional images of, 286–89, 305–6; state role, 303–4, 308n12; unintended consequences, 298–300
Benefit sharing, 53–59, 63–64, 361–67, 373n24; agreements, 393–98, 401–2n11; and agroforestry, 199n8; bioregional approach, 362–63; and colonialism, 382; and commodity chains, 57–58; and elites, 56–57; and geographical distribution, 55; global fund approach, 363–65, 373nn26, 27; and indigeneity, 57; point of collection approach, 361–62, 373n25, 374; process benefits, 365; and provenance, 55–56,

**443**

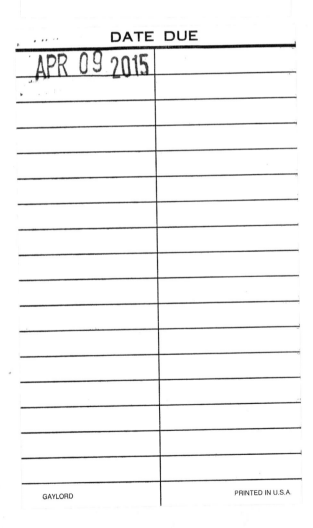